SCATTERING OF ELECTROMAGNETIC WAVES

WILEY SERIES IN REMOTE SENSING

Jin Au Kong, Editor

Asrar • THEORY AND APPLICATIONS OF OPTICAL REMOTE SENSING

Crane • ELECTROMAGNETIC WAVE PROPAGATION THROUGH RAIN

Curlander and McDonough • SYNTHETIC APERTURE RADAR: SYSTEMS AND SIGNAL PROCESSING

Elachi • INTRODUCTION TO THE PHYSICS AND TECHNIQUES OF REMOTE SENSING

Haykin, Lewis, Raney, and Rossiter • REMOTE SENSING OF SEA ICE AND ICEBERGS

Haykin and Steinhardt • ADAPTIVE RADAR DETECTION AND ESTIMATION

Hord • REMOTE SENSING: METHODS AND APPLICATIONS

Janssen • ATMOSPHERIC REMOTE SENSING BY MICROWAVE RADIOMETRY

Maffett • TOPICS FOR A STATISTICAL DESCRIPTION OF RADAR CROSS SECTIONS

Steinberg and Subbaram • MICROWAVE IMAGING TECHNIQUES

Szekielda • SATELLITE MONITORING OF THE EARTH

Tsang, Kong, and Shin • THEORY OF MICROWAVE REMOTE SENSING

Tsang, Kong, and Ding • SCATTERING OF ELECTROMAGNETIC WAVES: THEORIES AND APPLICATIONS

Tsang, Kong, Ding, and Ao • SCATTERING OF ELECTROMAGNETIC WAVES: NUMERICAL SIMULATIONS

Tsang and Kong • SCATTERING OF ELECTROMAGNETIC WAVES: ADVANCED TOPICS

SCATTERING OF ELECTROMAGNETIC WAVES

Advanced Topics

Leung Tsang

Jin Au Kong

A Wiley-Interscience Publication

JOHN WILEY & SONS, INC.

New York • Chichester • Weinheim • Brisbane • Singapore • Toronto

This text is printed on acid-free paper. ♾

For ordering and customer service, call 1-800-CALL-WILEY.

Library of Congress Cataloging in Publication Data

Tsang, Leung.
Scattering of electromagnetic waves: Advanced topics / L. Tsang, J. A. Kong.
p. cm. — (Wiley series in remote sensing)
Includes index.
ISBN 0-471-38801-7 (cloth: alk. paper)
1. Electromagnetic waves—Scattering. I. Kong, Jin Au, 1942– II. Title. III. Series.

QC665.S3 T72 2000
621.36'78—dc21 00-040864

10 9 8 7 6 5 4 3 2 1

To my family, Hannah, Clarisse, and Kaleb for their love.

— L. Tsang

To my family.

— J. A. Kong

CONTENTS

CHAPTER 8

Scattering of Electromagnetic Waves

Volume I: Theories and Applications (Tsang, Kong, and Ding)

Volume II: Numerical Simulations (Tsang, Kong, Ding, and Ao)

Volume III: Advanced Topics (Tsang and Kong)

PREFACE

Electromagnetic wave scattering is an active, interdisciplinary area of research with myriad practical applications in fields ranging from atomic physics to medical imaging to geoscience and remote sensing. In particular, the subject of wave scattering by random discrete scatterers and rough surfaces presents great theoretical challenges due to the large degrees of freedom in these systems and the need to include multiple scattering effects accurately. In the past three decades, considerable theoretical progress has been made in elucidating and understanding the scattering processes involved in such problems. Diagrammatic techniques and effective medium theories remain essential for analytical studies; however, rapid advances in computer technology have opened new doors for researchers with the full power of Monte Carlo simulations in the numerical analysis of random media scattering. Numerical simulations allow us to solve the Maxwell equations exactly without the limitations of analytical approximations, whose regimes of validity are often difficult to assess. Thus it is our aim to present in these three volumes a balanced picture of both theoretical and numerical methods that are commonly used for tackling electromagnetic wave scattering problems. While our book places an emphasis on remote sensing applications, the materials covered here should be useful for students and researchers from a variety of backgrounds as in, for example, composite materials, photonic devices, optical thin films, lasers, optical tomography, and X-ray lithography. Introductory chapters and sections are also added so that the materials can be readily understood by graduate students. We hope that our book would help stimulate new ideas and innovative approaches to electromagnetic wave scattering in the years to come.

The increasingly important role of numerical simulations in solving electromagnetic wave scattering problems has motivated us to host a companion web site that contains computer codes on topics relevant to the book. These computer codes are written in the MATLAB programming language and are available for download from our web site at `www.emwave.com`. They are provided to serve two main purposes. The first is to supply our readers a hands-on laboratory for performing numerical experiments, through which the concepts in the book can be more dynamically relayed. The second is to give new researchers a set of basic tools with which they could quickly build on projects of their own. The fluid nature of the web site would also allow us to regularly update the contents and keep pace with new research developments.

The present volume deals with multiple scattering and analytic wave theories. It also contains in-depth discussion of topics introduced in Volumes I and II. In Volume I, the small perturbation method (SPM) and Kirchhoff approach (KA) were introduced for wave scattering by one-dimensional random rough surfaces. In Chapters 1 and 2, these methods are extended to electromagnetic scattering by two-dimensional dielectric random rough surfaces. In recent years, there has been a great deal of theoretical progress aimed at bridging the parametric gap between the applicability of SPM and KA. The phase perturbation method bears much similarity to SPM with the important difference that the perturbation series is made in the exponent. It successfully reduces to SPM and KA in their respective regimes. On the other hand, some rough surfaces, e.g. ocean surfaces, are better characterized by small rapidly varying roughness superimposed on a smoothly undulating surface. The two-scale, or composite surface, model is used to compute emissivities from rough ocean surfaces.

Radiative transfer (RT) equation is usually derived heuristically by considering the scattering and attenuation of specific intensity in an elemental volume. However, at a more fundamental level, the fields satisfy the wave equation. Thus it is more rigorous to start with analytic wave theories and try to derive the RT equation from there. Before doing so, we consider in Chapter 3 a simple volume scattering model where scattering by a thick layer is replaced by a cascading of thin layers consisting of point scatterers. It is shown that the Foldy's approximation and the RT equation in difference form can be derived by this simple model. The cascading layer approach illustrates in a more intuitive way the role of multiple scattering in a thick layer and helps us understand when RT type equation is expected to be valid.

In Chapter 4, analytic wave theories are developed using diagrammatic expansion techniques. We derive Dyson's equation for the mean field and the Bethe-Salpeter equation for the field covariance. Practical computations usually require approximations be made to these equations. We show that the ladder approximation for the Bethe-Salpeter equation and nonlinear approximation for Dyson's equation are consistent with energy conservation. We also introduce the strong permittivity fluctuation theory where by properly extracting the singularity of the dyadic Green's function, the bilocally-approximated Dyson equation is made applicable for random medium with large permittivity fluctuations.

In Chapter 5, multiple scattering equations for random discrete scatterers are derived using the operator formalism. We discuss approximations of the multiple scattering equations through conditional averaging. This results in effective medium theories such as Foldy's approximation and the quasi-crystalline approximation (QCA). The method of coherent potential (CP) can be used in conjunction with these approximations to improve the results. QCA can also be conveniently formulated using the T-matrix multiple scattering approach introduced in Volume II. This is discussed in Chapter 6. We develop a QCA based dense medium radiative transfer (DMRT) theory and apply it to remote sensing problems.

In Chapter 7, the DMRT equation is derived from Dyson's equation with QCA-CP and the Bethe-Salpeter equation with the correlated ladder approximation. These approximations are shown to be consistent with energy conservation. Active and passive remote sensing applications are provided for illustrations. Finally, in Chapter 8, we address the interesting phenomenon of backscattering enhancement through multiple scattering theories for both isotropic and anisotropic scatterers.

Acknowledgments

We would like to acknowledge the collaboration with our colleagues and graduate students. In particular, we wish to thank Professor Chi Chan of City University of Hong Kong, Professor Joel T. Johnson of Ohio State University, Dr. Robert T. Shin of MIT Lincoln Laboratory, and Dr. Dale Winebrenner of University of Washington. The graduate students who completed their Ph.D. theses from the University of Washington on random media scattering include Boheng Wen (1989), Kung-Hau Ding (1989), Shu-Hsiang Lou (1991), Charles E. Mandt (1992), Richard D. West (1994), Zhengxiao Chen (1994), Lisa M. Zurk (1995), Kyung Pak (1996), Guifu Zhang (1998), and Qin Li (2000). Much of their dissertation works are included in this book. Financial supports from the Air Force Office of Scientific Research, Army Research Office, National Aeronautics and Space Administration, National Science Foundation, Office of Naval Research, and Schlumberger-Doll Research Center for research materials included in this book are gratefully acknowledged. We also want to acknowledge the current UW graduate students who have helped to develop the numerical codes used throughout this book. These include to Chi-Te Chen, Houfei Chen, Jianjun Guo, Chung-Chi Huang, and Lin Zhou. Special thanks are due to Kung-Hau Ding for careful proofreading of the manuscript and Bae-Ian Wu for production assistance. We would also like to thank Chi On Ao for his help in editing and typsetting the manuscript.

Leung Tsang
Seattle, Washington
Jin Au Kong
Cambridge, Massachusetts

February 2001

SCATTERING OF ELECTROMAGNETIC WAVES

Chapter 1

TWO-DIMENSIONAL RANDOM ROUGH SURFACE SCATTERING BASED ON SMALL PERTURBATION METHOD

In Volume I, we have studied the small perturbation method (SPM) for one-dimensional surface. The small perturbation method is valid for small rms height and small slope. Over the years, SPM has found extensive applications. In this chapter, we apply SPM to three-dimensional problem with two-dimensional random rough surfaces. The analyses are performed up to second order. Second order calculations are important to ensure energy conservation and also to calculate cross-polarization. We also give detailed derivation of the second order solution of dielectric surfaces. The formulation of SPM is based on extinction theorem.

The most common analytical treatments of scattering by random rough surfaces are the Kirchhoff approach (KA) and the small perturbation method (SPM). Improvements upon these two methods include the phase perturbation method [Winebrenner and Ishimaru, 1985a,b; Broschat et al. 1988], second order Kirchhoff method [Ishimaru and Chen, 1990, 1991]. Integral equation method (IEM) [Fung, 1994; Chen and Fung, 1995; Fung and Pan, 1987; Chen et al. 2000], Wiener method [Ito, 1985; Ogura and Takahashi, 1995], small slope approximation (SSA) [Broschat, 1993; Voronovich, 1994a,b; Thorsos and Broschat, 1995]. For the case of absorption and emissivity calculations, it was shown that the results of SSA and SPM are identical for the half-space case [Irisov, 1994, 1997]. Two-scale model is used to superimpose large scale roughness and small scale roughness [Plant, 1986; Yueh, 1997; Voronovich, 1996; Durden and Vesecky, 1985, 1990]. Other notable methods include the Feynman diagrammatic method [DeSanto, 1974; DeSanto and Shisha, 1974; DeSanto and Wombell, 1991], full wave methods [Collin, 1994; Bahar and Fitzwater, 1989] as well as the mean field theory [Sentenac and Greffet, 1998; Greffet and Nieto-Vesperinas, 1998].

In this chapter, we use the small perturbation method for electromagnetic wave scattering by two-dimensional random rough surfaces. In the next chapter, the Kirchhoff approach, the phase perturbation method, and the two-scale model are discussed.

1 Electromagnetic Wave Scattering by a Perfect Electric Conductor

Consider a plane electromagnetic wave

$$\overline{E}_i = \hat{e}_i \exp(ik_{ix}x + ik_{iy}y - ik_{iz}z) \tag{1.1.1}$$

incident upon a rough surface of perfect electric conductor. In (1.1.1), $k_{ix} = k\sin\theta_i\cos\phi_i$, $k_{iy} = k\sin\theta_i\sin\phi_i$, and $k_{iz} = k\cos\theta_i$. We also have $k_{i\rho} = k\sin\theta_i$. The rough surface is characterized by a random height function

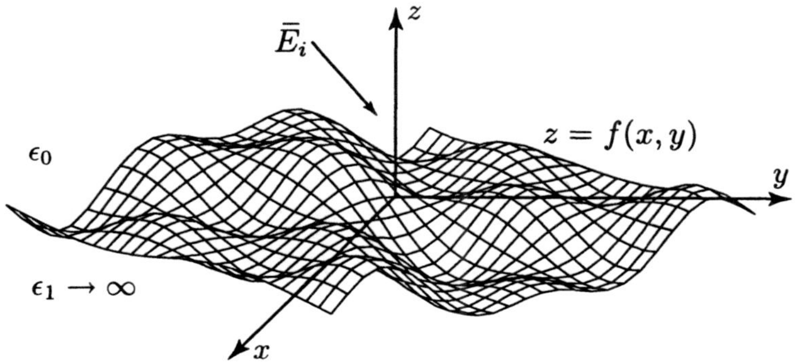

Figure 1.1.1 Electromagnetic scattering by a 2-D perfectly conducting rough surface.

$z = f(x,y)$, where $f(x,y)$ is a random function with zero mean $\langle f(x,y)\rangle = 0$ (Fig. 1.1.1). Let f_{min} and f_{max} be the minimum and maximum values, respectively, of the surface profile $f(x,y)$. From Huygens' principle, the extinction theorem, and setting the tangential electric field equal to zero, we have

$$\overline{E}_i(\overline{r}) + \int_S dS'\, ik\eta\overline{\overline{G}}(\overline{r},\overline{r}') \cdot [\hat{n} \times \overline{H}(\overline{r}')] = \begin{cases} \overline{E}(\overline{r}) & \text{for } z > f(\overline{r}_\perp) \quad (1.1.2a) \\ 0 & \text{for } z < f(\overline{r}_\perp) \quad (1.1.2b) \end{cases}$$

where $\overline{r}_\perp$ denotes vector $x\hat{x} + y\hat{y}$ in x-y plane.

We can define surface field unknowns. For $z' = f(x',y')$, the tangential magnetic field is defined as

$$dS'\eta\hat{n} \times \overline{H}(\overline{r}') = d\overline{r}'_\perp \overline{a}(\overline{r}'_\perp) \tag{1.1.3}$$

We make use of the integral representation of dyadic Green's function as given in equation (2.1.20) of Volume I. For $z < f_{min}$, we use the lower half of (1.1.2) to get

$$\begin{aligned} \overline{E}_i(\overline{r}) = \frac{1}{8\pi^2} \int d\overline{k}_\perp\, e^{i\overline{k}_\perp \cdot \overline{r}_\perp}\, e^{-ik_z z} \frac{k}{k_z} \int d\overline{r}'_\perp\, e^{-i\overline{k}_\perp \cdot \overline{r}'_\perp + ik_z f(\overline{r}'_\perp)} \\ \times [\hat{e}(-k_z)\hat{e}(-k_z) + \hat{h}(-k_z)\hat{h}(-k_z)] \cdot \overline{a}(\overline{r}'_\perp) \end{aligned} \tag{1.1.4}$$

Equation (1.1.4) is the extended boundary condition and can be used to solve for the surface field $\overline{a}(\overline{r}'_\perp)$. Since $\hat{n} \times \overline{H}$ is tangential to the surface, we also have the condition

$$\hat{n}(\overline{r}'_\perp) \cdot \overline{a}(\overline{r}'_\perp) = 0 \tag{1.1.5}$$

Using the definition of the normal vector $\hat{n}(\overline{r}'_\perp)$, we have

$$a_z(\overline{r}'_\perp) = \left(\hat{x}\, \frac{\partial f(\overline{r}'_\perp)}{\partial x'} + \hat{y}\, \frac{\partial f(\overline{r}'_\perp)}{\partial y'}\right) \cdot \overline{a}_\perp(\overline{r}'_\perp) \tag{1.1.6}$$

After the surface field is evaluated, the scattered field is, for $z > f_{max}$, using the plane wave representation of Green's function in equation (2.1.20) of Volume I.

$$\overline{E}_s(\overline{r}) = -\frac{1}{8\pi^2}\int d\overline{k}_\perp\, e^{i\overline{k}_\perp\cdot\overline{r}_\perp}\, e^{ik_z z}\frac{k}{k_z}\int d\overline{r}'_\perp\, e^{-i\overline{k}_\perp\cdot\overline{r}'_\perp}e^{-ik_z f(\overline{r}'_\perp)} \times\left[\hat{e}(k_z)\hat{e}(k_z)+\hat{h}(k_z)\hat{h}(k_z)\right]\cdot\overline{a}(\overline{r}'_\perp) \tag{1.1.7}$$

Equation (1.1.7) shows that the scattered wave consists of a spectrum of plane waves. The incident wave, on the other hand, as represented by (1.1.1), consists of a plane wave in a single direction. Equations (1.1.4) through (1.1.7) are exact. We shall solve (1.1.4) and (1.1.7) by the perturbation method to the second order. A higher-order solution can be calculated in a similar manner with more complicated algebra.

To solve for the surface fields, the perturbation method makes use of series expansions. Let

$$\overline{a}(\overline{r}'_\perp) = \sum_{m=0}^{\infty}\overline{a}^{(m)}(\overline{r}'_\perp) = \overline{a}^{(0)}(\overline{r}'_\perp)+\overline{a}^{(1)}(\overline{r}'_\perp)+\overline{a}^{(2)}(\overline{r}'_\perp)+\cdots \tag{1.1.8}$$

where $\overline{a}^{(m)}$ is the mth-order solution of $\overline{a}$. We also have

$$e^{\pm ik_z f(\overline{r}'_\perp)} = \sum_{m=0}^{\infty}\frac{[\pm ik_z f(\overline{r}'_\perp)]^m}{m!} \tag{1.1.9}$$

From (1.1.6) and (1.1.8), we have

$$a_z^{(0)}(\overline{r}'_\perp) = 0 \tag{1.1.10}$$

$$a_z^{(m)}(\overline{r}'_\perp) = \left(\hat{x}\,\frac{\partial f(\overline{r}'_\perp)}{\partial x'}+\hat{y}\,\frac{\partial f(\overline{r}'_\perp)}{\partial y'}\right)\cdot\overline{a}_\perp^{(m-1)}(\overline{r}'_\perp) \tag{1.1.11}$$

Equation (1.1.11) is a small slope approximation because the assumption of the smallness of the slope is used to balance the order of solutions. That is,

$$\frac{\partial f}{\partial x},\frac{\partial f}{\partial y} \ll 1 \tag{1.1.12}$$

As a result of the small slope approximation, we note that the $\hat{z}$ component of the surface field $\overline{a}$ is one order lower than the $\hat{x}$- and $\hat{y}$-components. We define the Fourier transform of the surface field by

$$\overline{A}(\overline{k}_\perp) = \frac{1}{(2\pi)^2}\int d\overline{r}'_\perp\,\overline{a}(\overline{r}'_\perp)\,e^{-i\overline{k}_\perp\cdot\overline{r}'_\perp} \tag{1.1.13}$$

Assuming small height, we obtain

$$|k_z f| \ll 1 \tag{1.1.14}$$

We note that both the extinction theorem of (1.1.4) and the scattered field of (1.1.7) are of similar form. On expansion to the second order,

$$\int d\overline{r}'_\perp\, \overline{a}(\overline{r}'_\perp)\, e^{-i\overline{k}_\perp\cdot\overline{r}'_\perp \pm ik_z f(\overline{r}'_\perp)}$$

$$= \int d\overline{r}'_\perp\, \overline{a}(\overline{r}'_\perp)\, e^{-i\overline{k}_\perp\cdot\overline{r}'_\perp}\left[1 \pm ik_z f(\overline{r}'_\perp) - \frac{k_z^2 f^2(\overline{r}'_\perp)}{2}\right]$$

$$= 4\pi^2\Big\{A(\overline{k}_\perp) \pm ik_z \int d\overline{k}'_\perp A(\overline{k}'_\perp)F(\overline{k}_\perp - \overline{k}'_\perp) - \frac{k_z^2}{2}\int d\overline{k}'_\perp A(\overline{k}'_\perp)F^{(2)}(\overline{k}_\perp - \overline{k}'_\perp)\Big\} \qquad (1.1.15)$$

where $F(\overline{k}_\perp)$ is the Fourier transform of $f(\overline{r}_\perp)$ and $F^{(2)}(\overline{k}_\perp)$ is the Fourier transform of $f^2(\overline{r}_\perp)$, that is,

$$F^{(2)}(\overline{k}_\perp) = \frac{1}{4\pi^2}\int_{-\infty}^{\infty} d\overline{r}_\perp\, e^{-i\overline{k}_\perp\cdot\overline{r}_\perp} f^2(\overline{r}_\perp) \qquad (1.1.16)$$

In (1.1.15), the + sign in the exponent corresponds to the extinction theorem and the − sign corresponds to the scattered field.

If we use (1.1.15), then (1.1.4) and (1.1.7) assume the following forms. For $z < f_{min}$ we have

$$\begin{aligned}\overline{E}_i(\overline{r}) &= \hat{e}_i \exp(ik_{ix}x + ik_{iy}y - ik_{iz}z)\\ &= \frac{1}{2}\int d\overline{k}_\perp\, e^{i\overline{k}_\perp\cdot\overline{r}_\perp - ik_z z}\frac{k}{k_z}[\hat{e}(-k_z)\hat{e}(-k_z) + \hat{h}(-k_z)\hat{h}(-k_z)]\\ &\quad\cdot\Big[\overline{A}(\overline{k}_\perp) + ik_z\int d\overline{k}'_\perp \overline{A}(\overline{k}'_\perp)F(\overline{k}_\perp - \overline{k}'_\perp)\\ &\quad - \frac{k_z^2}{2}\int d\overline{k}'_\perp \overline{A}(\overline{k}'_\perp)F^{(2)}(\overline{k}_\perp - \overline{k}'_\perp)\Big]\end{aligned} \qquad (1.1.17)$$

and for $z > f_{max}$ we have

$$\begin{aligned}\overline{E}_s(\overline{r}) &= -\frac{1}{2}\int d\overline{k}_\perp\, e^{i\overline{k}_\perp\cdot\overline{r}_\perp + ik_z z}\frac{k}{k_z}[\hat{e}(k_z)\hat{e}(k_z) + \hat{h}(k_z)\hat{h}(k_z)]\\ &\quad\cdot\Big[\overline{A}(\overline{k}_\perp) - ik_z\int d\overline{k}'_\perp \overline{A}(\overline{k}'_\perp)F(\overline{k}_\perp - \overline{k}'_\perp)\\ &\quad - \frac{k_z^2}{2}\int d\overline{k}'_\perp \overline{A}(\overline{k}'_\perp)F^{(2)}(\overline{k}_\perp - \overline{k}'_\perp)\Big]\end{aligned} \qquad (1.1.18)$$

We use (1.1.17) to solve for the unknown $\overline{A}(\overline{k}_\perp)$ in a perturbative manner. Then we use (1.1.10) to calculate the scattered fields. Taking the Fourier transform of (1.1.17) gives the simple relation

$$\hat{e}_i\delta(\overline{k}_\perp - \overline{k}_{i\perp}) = \frac{1}{2}\frac{k}{k_z}[\hat{e}(-k_z)\hat{e}(-k_z) + \hat{h}(-k_z)\hat{h}(-k_z)] \cdot \left[\overline{A}(\overline{k}_\perp) + ik_z \int d\overline{k}'_\perp \overline{A}(\overline{k}'_\perp)F(\overline{k}_\perp - \overline{k}'_\perp) - \frac{k_z^2}{2}\int d\overline{k}'_\perp \overline{A}(\overline{k}'_\perp)F^{(2)}(\overline{k}_\perp - \overline{k}'_\perp)\right] \tag{1.1.19}$$

Taking (1.1.11) in spectral domain gives

$$A_z^{(m)}(\overline{k}_\perp) = i\int d\overline{k}'_\perp F(\overline{k}_\perp - \overline{k}'_\perp)(\overline{k}_\perp - \overline{k}'_\perp)\cdot \overline{A}_\perp^{(m-1)}(\overline{k}'_\perp) \tag{1.1.20}$$

Equations (1.1.19) and (1.1.20) are the two equations which can be solved by balancing terms to the zeroth, first, and second orders. It is convenient to form an orthonormal vector system to calculate surface fields. Define

$$\overline{k}_\perp = k_x\hat{x} + k_y\hat{y} = k_\rho\cos\phi_k\hat{x} + k_\rho\sin\phi_k\hat{y} \tag{1.1.21a}$$

$$\hat{q}(\overline{k}_\perp) = \hat{e}(-k_z) = \frac{\hat{x}k_y - \hat{y}k_x}{k_\rho} = \hat{x}\sin\phi_k - \hat{y}\cos\phi_k \tag{1.1.21b}$$

$$\hat{p}(\overline{k}_\perp) = \hat{z}\times\hat{q}(\overline{k}_\perp) = \frac{\hat{x}k_x + \hat{y}k_y}{k_\rho} = \cos\phi_k\hat{x} + \sin\phi_k\hat{y} = \hat{k}_\perp \tag{1.1.21c}$$

so that

$$\hat{h}(\pm k_z) = \mp\frac{k_z}{k}\hat{p}(\overline{k}_\perp) + \frac{k_\rho}{k}\hat{z} = \mp\frac{k_z}{k}(\cos\phi_k\hat{x} + \sin\phi_k\hat{y}) + \frac{k_\rho}{k}\hat{z} \tag{1.1.22}$$

Thus we use two orthonormal systems, $(\hat{h}, \hat{e}, \hat{k})$ and $(\hat{z}, \hat{q}, \hat{p})$. The relation is $\hat{q} = \hat{e}$. Also $\hat{p}$ is the unit vector denoting the projection of $\hat{k}$ on the x-y plane. Hence $\hat{p}$ is the direction vector of the projection of $\hat{h}(-k_z)$ in x-y plane. We let

$$k_x = k_\rho\cos\phi_k \tag{1.1.22a}$$

$$k_y = k_\rho\sin\phi_k \tag{1.1.22b}$$

$$k'_x = k'_\rho\cos\phi'_k \tag{1.1.22c}$$

$$k'_y = k'_\rho\sin\phi'_k \tag{1.1.22d}$$

then

$$\hat{e}(\pm k_z)\cdot\hat{q}(\overline{k}'_\perp) = \frac{k_yk'_y + k_xk'_x}{k_\rho k'_\rho} = \cos(\phi_k - \phi'_k) \tag{1.1.22e}$$

$$\hat{e}(\pm k_z)\cdot\hat{p}(\overline{k}'_\perp) = \frac{k_yk'_x - k_xk'_y}{k_\rho k'_\rho} = \sin(\phi_k - \phi'_k) \tag{1.1.22f}$$

$$\hat{h}(\pm k_z) \cdot \hat{q}(\overline{k}'_\perp) = \mp \frac{k_z}{k k_\rho k'_\rho}(k_x k'_y - k_y k'_x) = \pm \frac{k_z}{k} \sin(\phi_k - \phi'_k) \tag{1.1.22g}$$

$$\hat{h}(\pm k_z) \cdot \hat{p}(\overline{k}'_\perp) = \mp \frac{k_z}{k k_\rho k'_\rho}(k_x k'_x + k_y k'_y) = \mp \frac{k_z}{k} \cos(\phi_k - \phi'_k) \tag{1.1.22h}$$

$$\overline{k}_\perp \cdot \hat{q}(\overline{k}'_\perp) = \frac{k_x k'_y - k_y k'_x}{k'_\rho} = -k_\rho \sin(\phi_k - \phi'_k) \tag{1.1.22i}$$

$$\overline{k}_\perp \cdot \hat{p}(\overline{k}'_\perp) = \frac{k_x k'_x + k_y k'_y}{k'_\rho} = k_\rho \cos(\phi_k - \phi'_k) \tag{1.1.22j}$$

1.1 Zeroth- and First-Order Solutions

Zeroth-Order Solution

Balancing (1.1.19) and (1.1.20) to the zeroth order, we obtain

$$A_z^{(0)}(\overline{k}_\perp) = 0 \tag{1.1.23}$$

$$\hat{e}_i \delta(\overline{k}_\perp - \overline{k}_{i\perp}) = \frac{1}{2}\frac{k}{k_z}[\hat{e}(-k_z)\hat{e}(-k_z) + \hat{h}(-k_z)\hat{h}(-k_z)] \cdot \overline{A}_\perp^{(0)}(\overline{k}_\perp) \tag{1.1.24}$$

Solution of (1.1.24) can easily be calculated in the $(\hat{q}, \hat{p}, \hat{z})$ system defined in (1.1.21) and (1.1.22). Let

$$\hat{q}_i = \hat{q}(\overline{k}_{i\perp}) \tag{1.1.25}$$

$$\hat{p}_i = \hat{p}(\overline{k}_{i\perp}) \tag{1.1.26}$$

be the $\hat{q}$ and $\hat{p}$ for the incident direction. Then the solution of (1.1.24) is

$$\overline{A}_\perp^{(0)}(\overline{k}_\perp) = (\hat{q}_i a_q^{(0)} + \hat{p}_i a_p^{(0)})\delta(\overline{k}_\perp - \overline{k}_{i\perp}) \tag{1.1.27}$$

where

$$a_q^{(0)} = 2\hat{e}(-k_{iz}) \cdot \hat{e}_i \frac{k_{iz}}{k} \tag{1.1.28a}$$

$$a_p^{(0)} = 2\hat{h}(-k_{iz}) \cdot \hat{e}_i \tag{1.1.28b}$$

The Dirac delta function in (1.1.27) indicates that the zeroth-order surface field consists of only a single spectral component that corresponds to specular reflection. Substituting (1.1.27) in (1.1.18) gives the zeroth-order solution of the scattered field as

$$\overline{E}_s^{(0)}(\overline{r}) = \left\{-\hat{e}(k_{iz})\Big(\hat{e}_i \cdot \hat{e}(-k_{iz})\Big) + \hat{h}(k_{iz})\Big(\hat{e}_i \cdot \hat{h}(-k_{iz})\Big)\right\} e^{i\overline{k}_{i\perp}\cdot\overline{r}_\perp + ik_{iz}z} \tag{1.1.29}$$

which is the response of a flat surface.

First-Order Solution

Balancing (1.1.19) to the first order gives

$$0 = \left\{\hat{e}(-k_z)\hat{e}(-k_z) + \hat{h}(-k_z)\hat{h}(-k_z)\right\} \cdot \left[\overline{A}^{(1)}(\overline{k}_\perp) + ik_z \int d\overline{k}'_\perp \overline{A}^{(0)}(\overline{k}'_\perp)F(\overline{k}_\perp - \overline{k}'_\perp)\right] \tag{1.1.30}$$

Note that (1.1.30) has only two independent components since it states that the projection of the vector in the square bracket onto the two polarization directions are equal to zero. The third component can be obtained by balancing (1.1.20) to the first order that gives

$$A_z^{(1)}(\overline{k}_\perp) = i \int d\overline{k}'_\perp F(\overline{k}_\perp - \overline{k}'_\perp)(\overline{k}_\perp - \overline{k}'_\perp) \cdot \overline{A}_\perp^{(0)}(\overline{k}'_\perp) \tag{1.1.31}$$

From (1.1.27) and (1.1.31), we have

$$\begin{aligned} A_z^{(1)}(\overline{k}_\perp) &= iF(\overline{k}_\perp - \overline{k}_{i\perp})(\overline{k}_\perp - \overline{k}_{i\perp}) \cdot \left(\hat{q}_i a_q^{(0)} + \hat{p}_i a_p^{(0)}\right) \\ &= iF(\overline{k}_\perp - \overline{k}_{i\perp})\Big\{-k_\rho \sin(\phi_k - \phi_i)a_q^{(0)} \\ &\quad + \left[k_\rho \cos(\phi_k - \phi_i) - k_{i\rho}\right] a_p^{(0)}\Big\} \end{aligned} \tag{1.1.32}$$

To solve (1.1.30), let

$$\overline{A}^{(1)}(\overline{k}_\perp) = A_q^{(1)}(\overline{k}_\perp)\hat{q}(\overline{k}_\perp) + A_p^{(1)}(\overline{k}_\perp)\hat{p}(\overline{k}_\perp) + A_z^{(1)}(\overline{k}_\perp)\hat{z} \tag{1.1.33}$$

Note that in here we use $\hat{q}(\overline{k}_\perp)$ and $\hat{p}(\overline{k}_\perp)$ as basis vectors. Previously [Shin, 1984; Tsang et al. 1985], the representation of $\overline{A}^{(1)}(\overline{k}_\perp)$ was made differently using $\hat{q}_i$ and $\hat{p}_i$ as basis vectors. The present set of basis vectors simplifies subsequent calculations. We find it more convenient to use $\hat{q}$ and $\hat{p}$ as defined by the scattered directions. Substituting (1.1.33) and (1.1.27) into (1.1.30) gives two equations in which the dot product of $\hat{e}(-k_z)$ with the square-bracketed terms in (1.1.30) gives zero and the dot product of $\hat{h}(-k_z)$ with the square-bracketed terms in (1.1.30) also gives zero.

From (1.1.33) and (1.1.30) the dot product of $\hat{e}(-k_z)$ with the square-bracketed terms gives

$$A_q^{(1)}(\overline{k}_\perp) = -ik_z\hat{e}(-k_z) \cdot \left[a_q^{(0)}\hat{q}_i + a_p^{(0)}\hat{p}_i\right]F(\overline{k}_\perp - \overline{k}_{i\perp}) \tag{1.1.34}$$

The dot product of $\hat{h}(-k_z)$ with the square-bracketed terms in (1.1.30) gives

$$\frac{k_z}{k}A_p^{(1)}(\overline{k}_\perp) = -\frac{k_\rho}{k}A_z^{(1)}(\overline{k}_\perp) - ik_z\hat{h}(-k_z) \cdot \left[a_q^{(0)}\hat{q}_i + a_p^{(0)}\hat{p}_i\right]F(\overline{k}_\perp - \overline{k}_{i\perp}) \tag{1.1.35}$$

Thus the two components $A_q^{(1)}$ and $A_p^{(1)}$ depend on the projection of the polarizations $\hat{e}(-k_z)$ and $\hat{h}(-k_z)$ on the incident polarizations as projected on the x-y plane. Substituting (1.1.35) into (1.1.34) and using (1.1.22) gives the explicit expressions for $A_q^{(1)}(\overline{k}_\perp)$ and $A_p^{(1)}(\overline{k}_\perp)$:

$$A_q^{(1)}(\overline{k}_\perp) = iF(\overline{k}_\perp - \overline{k}_{i\perp})\widetilde{A}_q^{(1)}(\overline{k}_\perp) \tag{1.1.36a}$$

$$A_p^{(1)}(\overline{k}_\perp) = iF(\overline{k}_\perp - \overline{k}_{i\perp})\widetilde{A}_p^{(1)}(\overline{k}_\perp) \tag{1.1.36b}$$

where

$$\widetilde{A}_q^{(1)}(\overline{k}_\perp) = -k_z\left[a_q^{(0)}\cos(\phi_k - \phi_i) + a_p^{(0)}\sin(\phi_k - \phi_i)\right] \tag{1.1.36c}$$

$$\widetilde{A}_p^{(1)}(\overline{k}_\perp) = \frac{k^2}{k_z}\left[a_q^{(0)}\sin(\phi_k - \phi_i) - a_p^{(0)}\cos(\phi_k - \phi_i)\right] + \frac{k_\rho k_{i\rho}}{k_z}a_p^{(0)} \tag{1.1.36d}$$

For TE excitation, $a_q^{(0)} = 2k_{iz}/k$, $a_p^{(0)} = 0$

$$\widetilde{A}_q^{(1)}(\overline{k}_\perp) = -\frac{2k_z k_{iz}}{k}\cos(\phi_k - \phi_i) \tag{1.1.36e}$$

$$\widetilde{A}_p^{(1)}(\overline{k}_\perp) = \frac{2kk_{iz}}{k_z}\sin(\phi_k - \phi_i) \tag{1.1.36f}$$

For TM excitation, $a_q^{(0)} = 0$, $a_p^{(0)} = 2$ we have

$$\widetilde{A}_q^{(1)}(\overline{k}_\perp) = -2k_z\sin(\phi_k - \phi_i) \tag{1.1.36g}$$

$$\widetilde{A}_p^{(1)}(\overline{k}_\perp) = \frac{2}{k_z}\left\{k_\rho k_{i\rho} - k^2\cos(\phi_k - \phi_i)\right\} \tag{1.1.36h}$$

Substituting into (1.1.18) gives the first-order scattered field as

$$\begin{aligned}
\overline{E}_s^{(1)}(\overline{r}) = &-\frac{1}{2}\int d\overline{k}_\perp\, e^{i\overline{k}_\perp\cdot\overline{r}_\perp + ik_z z}\frac{k}{k_z}\left[\hat{e}(k_z)\hat{e}(k_z) + \hat{h}(k_z)\hat{h}(k_z)\right] \\
&\cdot\left[A_q^{(1)}(\overline{k}_\perp)\hat{q}(\overline{k}_\perp) + A_p^{(1)}(\overline{k}_\perp)\hat{p}(\overline{k}_\perp) + \hat{z}A_z^{(1)}(\overline{k}_\perp)\right. \\
&\left. - ik_z F(\overline{k}_\perp - \overline{k}_{i\perp})\left(\hat{q}_i a_q^{(0)} + \hat{p}_i a_p^{(0)}\right)\right] \\
= &-\frac{1}{2}\int d\overline{k}_\perp\, e^{i\overline{k}_\perp\cdot\overline{r}_\perp + ik_z z}\frac{k}{k_z} \\
&\cdot\left\{\hat{e}(k_z)\left[A_q^{(1)}(\overline{k}_\perp) - ik_z F(\overline{k}_\perp - \overline{k}_{i\perp})\hat{e}(k_z)\cdot\left(\hat{q}_i a_q^{(0)} + \hat{p}_i a_p^{(0)}\right)\right]\right. \\
&+ \hat{h}(k_z)\left[-\frac{k_z}{k}A_p^{(1)}(\overline{k}_\perp) + \frac{k_\rho}{k}A_z^{(1)}(\overline{k}_\perp)\right. \\
&\left.\left. - ik_z F(\overline{k}_\perp - \overline{k}_{i\perp})\hat{h}(k_z)\cdot\left(\hat{q}_i a_q^{(0)} + \hat{p}_i a_p^{(0)}\right)\right]\right\}
\end{aligned} \tag{1.1.37a}$$

In view of (1.1.34), (1.1.35), and (1.1.22a)–(1.1.22d) and the fact that $\hat{h}(k_z) \cdot \hat{q}_i = -\hat{h}(-k_z) \cdot \hat{q}_i$ and $\hat{h}(k_z) \cdot \hat{p}_i = -\hat{h}(-k_z) \cdot \hat{p}_i$, the terms inside the two square brackets are the same and we get a factor of 2. Thus

$$\overline{E}_s^{(1)}(\overline{r}) = -\int d\overline{k}_\perp\, e^{i\overline{k}_\perp \cdot \overline{r}_\perp + ik_z z} \frac{k}{k_z} \left[\hat{e}(k_z)\widetilde{A}_q^{(1)}(\overline{k}_\perp) + \hat{h}(k_z)\left(-\frac{k_z}{k}\widetilde{A}_p^{(1)}(\overline{k}_\perp)\right)\right] \cdot iF(\overline{k}_\perp - \overline{k}_{i\perp}) \tag{1.1.37b}$$

The incident power per unit area is

$$\overline{S}_i \cdot \hat{z} = -\frac{\cos\theta_i}{2\eta} \tag{1.1.38}$$

The power per unit area associated with the first-order fields (which is also that of the incoherent wave) is,

$$\langle \overline{S}_s \cdot \hat{z} \rangle = \frac{1}{2}\mathrm{Re}\left\langle \overline{E}_s^{(1)} \times \overline{H}_s^{(1)*} \right\rangle \cdot \hat{z} \tag{1.1.39a}$$

Using (1.1.37b), we obtain

$$\overline{H}_s^{(1)}(\overline{r}) = \frac{1}{\eta}\int d\overline{k}'_\perp\, e^{i\overline{k}'_\perp \cdot \overline{r}_\perp + ik'_z z} \frac{k}{k'_z} \left[\hat{h}(k'_z)\widetilde{A}_q^{(1)}(\overline{k}'_\perp) + \hat{e}(k'_z)\frac{k'_z}{k}\widetilde{A}_p^{(1)}(\overline{k}'_\perp)\right] \cdot iF(\overline{k}'_\perp - \overline{k}_{i\perp}) \tag{1.1.39b}$$

Also

$$\langle F(\overline{k}_\perp - \overline{k}_{i\perp})F^*(\overline{k}'_\perp - \overline{k}'_{i\perp})\rangle = \delta(\overline{k}_\perp - \overline{k}'_\perp)W(\overline{k}_\perp - \overline{k}_{i\perp}) \tag{1.1.40}$$

and we have

$$\langle \overline{S}_s \cdot \hat{z} \rangle = \frac{1}{2\eta}\int_P d\overline{k}_\perp \frac{k}{k_z} W(\overline{k}_\perp - \overline{k}_{i\perp}) \left[|\widetilde{A}_q^{(1)}(\overline{k}_\perp)|^2 + \left|\widetilde{A}_p^{(1)}(\overline{k}_\perp)\frac{k_z}{k}\right|^2\right] \tag{1.1.41}$$

where P stands for propagating waves with $k_\rho \le k$. This is because the real part is taken to get power in (1.1.39), and only the propagating waves of the spectrum in (1.1.41) have nonzero real part and contribute to power. Casting (1.1.41) in terms of directions in angular variables (θ_s, ϕ_s), we have

$$k_x = k\sin\theta_s\cos\phi_s \tag{1.1.42a}$$

$$k_y = k\sin\theta_s\sin\phi_s \tag{1.1.42b}$$

$$k_z = k\cos\theta_s \tag{1.1.42c}$$

$$dk_x\, dk_y = k^2\sin\theta_s\cos\theta_s d\theta_s d\phi_s \tag{1.1.42d}$$

We can write (1.1.41) as

$$\langle \overline{S}_s \cdot \hat{z} \rangle = |\overline{S}_i \cdot \hat{z}| \int_0^{\pi/2} d\theta_s \sin\theta_s \int_0^{2\pi} d\phi_s \frac{\gamma(\hat{k}_s, \hat{k}_i)}{4\pi} \tag{1.1.43}$$

where

$$\gamma(\hat{k}_s, \hat{k}_i) = \frac{4\pi k^2}{\cos\theta_i} W(\overline{k}_\perp - \overline{k}_{i\perp}) \left[|\widetilde{A}_q^{(1)}(\overline{k}_\perp)|^2 + \left| \widetilde{A}_p^{(1)}(\overline{k}_\perp) \frac{k_z}{k} \right|^2 \right] \tag{1.1.44}$$

To get γ_{hh} and γ_{vh}, we let incident wave be TE with $\hat{e}_i = \hat{e}(-k_{iz})$ so that $a_q^{(0)} = 2k_{iz}/k$, $a_p^{(0)} = 0$. Then γ_{hh} corresponds to the $|\widetilde{A}_q^{(1)}|^2$ in (1.1.44) while γ_{vh} corresponds to the $|\widetilde{A}_p^{(1)} k_z/k|^2$ in (1.1.44). Here h stands for horizontal polarization (TE) and v stands for vertical polarization (TM).

$$\gamma_{hh} = 16\pi k^4 \, W(\overline{k}_\perp - \overline{k}_{i\perp}) \cos^2(\phi_s - \phi_i) \cos\theta_i \cos^2\theta_s \tag{1.1.45a}$$

$$\gamma_{vh} = 16\pi k^4 \, W(\overline{k}_\perp - \overline{k}_{i\perp}) \sin^2(\phi_s - \phi_i) \cos\theta_i \tag{1.1.45b}$$

To get γ_{vv} and γ_{hv}, we let $\hat{e}_i = \hat{h}(-k_{iz})$ so that $a_q^{(0)} = 0$ and $a_p^{(0)} = 2$. Then γ_{hv} corresponds to the $|\widetilde{A}_q^{(1)}|^2$ in (1.1.44) while γ_{vv} corresponds to the $|\widetilde{A}_p^{(1)} k_z/k|^2$ in (1.1.44). Thus

$$\gamma_{hv} = \frac{16\pi k^4}{\cos\theta_i} W(\overline{k}_\perp - \overline{k}_{i\perp}) \cos^2\theta_s \sin^2(\phi_s - \phi_i) \tag{1.1.45c}$$

$$\gamma_{vv} = \frac{16\pi k^4}{\cos\theta_i} W(\overline{k}_\perp - \overline{k}_{i\perp}) [\sin\theta_s \sin\theta_i - \cos(\phi_s - \phi_i)]^2 \tag{1.1.45d}$$

In the backscattering direction ($\theta_s = \theta_i$ and $\phi_s = \pi + \phi_i$) one has $\overline{k}_\perp = -\overline{k}_{i\perp}$ and

$$\sigma_{\alpha\beta} = \gamma_{\alpha\beta} \cos\theta_i \tag{1.1.46}$$

so that

$$\sigma_{vv} = 16\pi k^4 \, W(-2\overline{k}_{i\perp})(1 + \sin^2\theta_i)^2 \tag{1.1.47a}$$

$$\sigma_{vh} = 0 \tag{1.1.47b}$$

$$\sigma_{hv} = 0 \tag{1.1.47c}$$

$$\sigma_{hh} = 16\pi k^4 \, W(-2\overline{k}_{i\perp}) \cos^4\theta_i \tag{1.1.47d}$$

It is noteworthy that in the backscattering direction, there is no depolarization for a linearly polarized incident wave. Also, σ_{vv} is larger than σ_{hh}. If θ_i is close to grazing, so that $\theta_i \to 90°$, then σ_{vv} is much larger than σ_{hh}.

1.2 Second-Order Solutions

For the second-order solution, let

$$\overline{A}^{(2)}(\overline{k}_\perp) = A_q^{(2)}(\overline{k}_\perp)\hat{q}(\overline{k}_\perp) + A_p^{(2)}(\overline{k}_\perp)\hat{p}(\overline{k}_\perp) + A_z^{(2)}(\overline{k}_\perp)\hat{z} \tag{1.1.48}$$

Balancing (1.1.19) to the second order, we get two equations (one for $\hat{e}(-k_z)$ component, the other for $\hat{h}(-k_z)$ component):

$$A_q^{(2)}(\overline{k}_\perp) + ik_z \int d\overline{k}'_\perp \Big[\left(\hat{e}(-k_z) \cdot \hat{q}(\overline{k}'_\perp) \right) A_q^{(1)}(\overline{k}'_\perp) + \left(\hat{e}(-k_z) \cdot \hat{p}(\overline{k}'_\perp) \right) A_p^{(1)}(\overline{k}'_\perp) \Big] F(\overline{k}_\perp - \overline{k}'_\perp) - \left[\frac{k_z^2}{2} \left(\hat{e}(-k_z) \cdot \hat{q}_i \right) a_q^{(0)} + \frac{k_z^2}{2} \left(\hat{e}(-k_z) \cdot \hat{p}_i \right) a_p^{(0)} \right] F^{(2)}(\overline{k}_\perp - \overline{k}_{i\perp}) = 0 \tag{1.1.49a}$$

$$\frac{k_z}{k} A_p^{(2)}(\overline{k}_\perp) + \frac{k_\rho}{k} A_z^{(2)}(\overline{k}_\perp) + ik_z \int d\overline{k}'_\perp \Big[\left(\hat{h}(-k_z) \cdot \hat{q}(\overline{k}'_\perp) \right) A_q^{(1)}(\overline{k}'_\perp) + \left(\hat{h}(-k_z) \cdot \hat{p}(\overline{k}'_\perp) \right) A_p^{(1)}(\overline{k}'_\perp) + \frac{k_\rho}{k} A_z^{(1)}(\overline{k}'_\perp) \Big] F(\overline{k}_\perp - \overline{k}'_\perp) - \left[\frac{k_z^2}{2} \left(\hat{h}(-k_z) \cdot \hat{q}_i \right) a_q^{(0)} + \frac{k_z^2}{2} \left(\hat{h}(-k_z) \cdot \hat{p}_i \right) a_p^{(0)} \right] F^{(2)}(\overline{k}_\perp - \overline{k}_{i\perp}) = 0 \tag{1.1.49b}$$

The third equation for the three components is from (1.1.20)

$$A_z^{(2)}(\overline{k}_\perp) = i \int d\overline{k}'_\perp F(\overline{k}_\perp - \overline{k}'_\perp)(\overline{k}_\perp - \overline{k}'_\perp) \cdot \left\{ \hat{q}(\overline{k}'_\perp) A_q^{(1)}(\overline{k}'_\perp) + \hat{p}(\overline{k}'_\perp) A_p^{(1)}(\overline{k}'_\perp) \right\} \tag{1.1.50}$$

We can solve the three equations (1.1.49*a*), (1.1.49*b*), and (1.1.50) for the three unknowns $A_z^{(2)}$, $A_p^{(2)}$, and $A_q^{(2)}$. However, it is easier to use (1.1.18) and make use of the similarity of the integrand of (1.1.18) to (1.1.49*a*) and (1.1.49*b*). From (1.1.18), the second-order scattered field is

$$\overline{E}_s^{(2)} = -\frac{1}{2} \int d\overline{k}_\perp \, e^{i\overline{k}_\perp \cdot \overline{r}_\perp + ik_z z} \frac{k}{k_z} \Bigg\{ \hat{e}(k_z) \Bigg(A_q^{(2)}(\overline{k}_\perp) - ik_z \int d\overline{k}'_\perp \left[\hat{e}(k_z) \cdot \hat{q}(\overline{k}'_\perp) A_q^{(1)}(\overline{k}'_\perp) + \hat{e}(k_z) \cdot \hat{p}(\overline{k}'_\perp) A_p^{(1)}(\overline{k}'_\perp) \right] F(\overline{k}_\perp - \overline{k}'_\perp) - \left[\frac{k_z^2}{2} \left(\hat{e}(k_z) \cdot \hat{q}_i \right) a_q^{(0)} + \frac{k_z^2}{2} \left(\hat{e}(k_z) \cdot \hat{p}_i \right) a_p^{(0)} \right] F^{(2)}(\overline{k}_\perp - \overline{k}_{i\perp}) \Bigg)$$

$$+ \hat{h}(k_z) \Bigg(-A_p^{(2)}(\overline{k}_\perp) \frac{k_z}{k} + A_z^{(2)}(\overline{k}_\perp) \frac{k_\rho}{k} - ik_z \int d\overline{k}'_\perp \Big[\hat{h}(k_z) \cdot \hat{q}(\overline{k}'_\perp) A_q^{(1)}(\overline{k}'_\perp) + \hat{h}(k_z) \cdot \hat{p}(\overline{k}'_\perp) A_p^{(1)}(\overline{k}'_\perp) + \frac{k_\rho}{k} A_z^{(1)}(\overline{k}'_\perp) \Big] F(\overline{k}_\perp - \overline{k}'_\perp)$$

$$-\left[\frac{k_z^2}{2}\Big(\hat{h}(k_z)\cdot\hat{q}_i\Big)a_q^{(0)}+\frac{k_z^2}{2}\Big(\hat{h}(k_z)\cdot\hat{p}_i\Big)a_p^{(0)}\right]F^{(2)}(\overline{k}_\perp-\overline{k}_{i\perp})\Big)\Bigg\} \tag{1.1.51}$$

We can use (1.1.49a) and (1.1.49b) directly in (1.1.51). Note that $\hat{e}(-k_z)=\hat{e}(k_z)$, $\hat{h}(-k_z)\cdot\hat{q}_i=-\hat{h}(k_z)\cdot\hat{q}_i$, and $\hat{h}(-k_z)\cdot\hat{p}_i=-\hat{h}(k_z)\cdot\hat{p}_i$. A factor of 2 is obtained. Equation (1.1.51) then simplifies to

$$\begin{aligned}\overline{E}_s^{(2)}(\overline{r}) = &-\int d\overline{k}_\perp\, e^{i\overline{k}\cdot\overline{r}}\frac{k}{k_z}\Bigg\{\hat{e}(k_z)\Big(-ik_z\int d\overline{k}'_\perp\Big[\Big(\hat{e}(k_z)\cdot\hat{q}(\overline{k}'_\perp)\Big)A_q^{(1)}(\overline{k}'_\perp)\\ &+\Big(\hat{e}(k_z)\cdot\hat{p}(\overline{k}'_\perp)\Big)A_p^{(1)}(\overline{k}'_\perp)\Big]F(\overline{k}_\perp-\overline{k}'_\perp)\Big)\\ &+\hat{h}(k_z)\Big(\frac{k_\rho}{k}A_z^{(2)}(\overline{k}_\perp)-ik_z\int d\overline{k}'_\perp\Big[\Big(\hat{h}(k_z)\cdot\hat{q}(\overline{k}'_\perp)\Big)A_q^{(1)}(\overline{k}'_\perp)\\ &+\Big(\hat{h}(k_z)\cdot\hat{p}(\overline{k}'_\perp)\Big)A_p^{(1)}(\overline{k}'_\perp)\Big]F(\overline{k}_\perp-\overline{k}'_\perp)\Big)\Bigg\}\end{aligned} \tag{1.1.52}$$

Thus substituting (1.1.36) and (1.1.50) into (1.1.52) gives

$$\begin{aligned}\overline{E}_s^{(2)}(\overline{r}) = &\int d\overline{k}_\perp\, e^{i\overline{k}_\perp\cdot\overline{r}_\perp+ik_zz}\\ &\Big[\hat{e}(k_z)\int d\overline{k}'_\perp M_e(\overline{k}_\perp,\overline{k}'_\perp)F(\overline{k}_\perp-\overline{k}'_\perp)F(\overline{k}'_\perp-\overline{k}_{i\perp})\\ &+\hat{h}(k_z)\int d\overline{k}'_\perp M_h(\overline{k}_\perp,\overline{k}'_\perp)F(\overline{k}_\perp-\overline{k}'_\perp)F(\overline{k}'_\perp-\overline{k}_{i\perp})\Big]\end{aligned} \tag{1.1.53a}$$

where the coupling coefficients are

$$M_e(\overline{k}_\perp,\overline{k}'_\perp) = -k\left[\cos(\phi_k-\phi'_k)\widetilde{A}_q^{(1)}(\overline{k}'_\perp)+\sin(\phi_k-\phi'_k)\widetilde{A}_p^{(1)}(\overline{k}'_\perp)\right] \tag{1.1.53b}$$

$$\begin{aligned}M_h(\overline{k}_\perp,\overline{k}'_\perp) = &-\frac{k^2}{k_z}\sin(\phi_k-\phi'_k)\widetilde{A}_q^{(1)}(\overline{k}'_\perp)\\ &+\frac{k^2}{k_z}\cos(\phi_k-\phi'_k)\widetilde{A}_p^{(1)}(\overline{k}'_\perp)-\frac{k_\rho k'_\rho}{k_z}\widetilde{A}_p^{(1)}(\overline{k}'_\perp)\end{aligned} \tag{1.1.53c}$$

For a horizontally polarized incident wave we have $\hat{e}_i=\hat{e}(-k_{iz})$, and from (1.1.36$e$) and (1.1.36$f$) we obtain

$$\begin{aligned}M_e(\overline{k}_\perp,\overline{k}'_\perp) &= M_{ee}(\overline{k}_\perp,\overline{k}'_\perp)\\ &= -2k_{iz}\Big\{-k'_z\cos(\phi_k-\phi'_k)\cos(\phi'_k-\phi_i)\\ &\quad+\frac{k^2}{k'_z}\sin(\phi_k-\phi'_k)\sin(\phi'_k-\phi_i)\Big\}\end{aligned} \tag{1.1.53d}$$

$$\begin{aligned}M_h(\overline{k}_\perp,\overline{k}'_\perp) &= M_{he}(\overline{k}_\perp,\overline{k}'_\perp)\\ &= \frac{2k_{iz}k}{k_z}\Big\{k'_z\sin(\phi_k-\phi'_k)\cos(\phi'_k-\phi_i)\\ &\quad+\frac{k^2}{k'_z}\cos(\phi_k-\phi'_k)\sin(\phi'_k-\phi_i)-\frac{k_\rho k'_\rho}{k'_z}\sin(\phi'_k-\phi_i)\Big\}\end{aligned} \tag{1.1.53e}$$

For a vertically polarized incident wave we have $\hat{e}_i = \hat{h}(-k_{iz})$, and from (1.1.36$g$) and (1.1.36$h$) we obtain

$$\begin{aligned}M_e(\overline{k}_\perp,\overline{k}'_\perp) &= M_{eh}(\overline{k}_\perp,\overline{k}'_\perp)\\ &= -2k\Big\{-k'_z\cos(\phi_k-\phi'_k)\sin(\phi'_k-\phi_i)\\ &\quad+\frac{\sin(\phi_k-\phi'_k)}{k'_z}\left[k'_\rho k_{i\rho}-k^2\cos(\phi'_k-\phi_i)\right]\Big\}\end{aligned} \tag{1.1.53f}$$

$$\begin{aligned}M_h(\overline{k}_\perp,\overline{k}'_\perp) &= M_{hh}(\overline{k}_\perp,\overline{k}'_\perp)\\ &= \frac{2}{k_z}\left\{k^2k'_z\sin(\phi_k-\phi'_k)\sin(\phi'_k-\phi_i)+\left[k^2\cos(\phi_k-\phi'_k)-k_\rho k'_\rho\right]\right.\\ &\quad\left.\cdot\left[\frac{k'_\rho k_{i\rho}-k^2\cos(\phi'_k-\phi_i)}{k'_z}\right]\right\}\end{aligned} \tag{1.1.53g}$$

First we calculate the second order coherent wave $\langle\overline{E}_s^{(2)}(\overline{r})\rangle$. Since

$$\langle F(\overline{k}_\perp-\overline{k}'_\perp)F(\overline{k}'_\perp-\overline{k}_{i\perp})\rangle = W(\overline{k}_{i\perp}-\overline{k}'_\perp)\delta(\overline{k}_\perp-\overline{k}_{i\perp}) \tag{1.1.54}$$

the second-order coherent wave is in the specular direction.

$$\begin{aligned}\langle\overline{E}_s^{(2)}(\overline{r})\rangle = e^{i\overline{k}_{i\perp}\cdot\overline{r}_\perp+ik_{iz}\cdot z}\Big[&\hat{e}(k_{iz})\int d\overline{k}'_\perp W(\overline{k}_{i\perp}-\overline{k}'_\perp)M_e(\overline{k}_{i\perp},\overline{k}'_\perp)\\ &+\hat{h}(k_{iz})\int d\overline{k}'_\perp W(\overline{k}_{i\perp}-\overline{k}'_\perp)M_h(\overline{k}_{i\perp},\overline{k}'_\perp)\Big]\end{aligned} \tag{1.1.54a}$$

For horizontal incident polarization, from (1.1.53d) and (1.1.53e) we obtain

$$M_e(\overline{k}_{i\perp},\overline{k}'_\perp) = M_{ee}(\overline{k}_{i\perp},\overline{k}'_\perp) = 2k_{iz}\left[k'_z\cos^2(\phi'_k-\phi_i)+\frac{k^2}{k'_z}\sin^2(\phi'_k-\phi_i)\right] \tag{1.1.55a}$$

$$M_h(\overline{k}_{i\perp},\overline{k}'_\perp) = M_{he}(\overline{k}_{i\perp},\overline{k}'_\perp) = \frac{2kk'_\rho}{k'_z}\sin(\phi'_k-\phi_i)\left[k'_\rho\cos(\phi'_k-\phi_i)-k_{i\rho}\right] \tag{1.1.55b}$$

For a vertically polarized incident wave, from (1.1.53f) and (1.1.53g) we obtain

$$M_e(\overline{k}_{i\perp},\overline{k}'_\perp) = M_{eh}(\overline{k}_{i\perp},\overline{k}'_\perp)$$
$$= -\frac{2kk'_\rho}{k'_z}\sin(\phi'_k-\phi_i)\left[k'_\rho\cos(\phi'_k-\phi_i)-k_{i\rho}\right] \tag{1.1.56a}$$
$$M_h(\overline{k}_{i\perp},\overline{k}'_\perp) = M_{hh}(\overline{k}_{i\perp},\overline{k}'_\perp)$$
$$= -\frac{2}{k_{iz}}\left\{k^2k'_z\sin^2(\phi'_k-\phi_i)+\frac{1}{k'_z}\left[k'_\rho k_{i\rho}-k^2\cos(\phi'_k-\phi_i)\right]^2\right\} \tag{1.1.56b}$$

If the spectral density is isotropic, that is,

$$W(\overline{k}_{i\perp}-\overline{k}'_\perp) = W(|\overline{k}_{i\perp}-\overline{k}'_\perp|) \tag{1.1.57}$$

then the cross-polarized component is equal to zero so that there is no depolarization in the coherent reflected wave. To see that, we can set $\phi_i = 0$. Then W depends on $\cos\phi'_k$ only, and the rest of the integrand is proportional to $\sin\phi'_k$. Thus the integration of $d\phi'_k$ gives zero. To check energy conservation, we include coherent and incoherent reflectivity. The incoherent bistatic intensity is as given in (1.1.45). For coherent reflectivity, we have

$$\langle\overline{S}_s\cdot\hat{z}\rangle_{\text{coherent}}$$
$$= \frac{1}{2}\text{Re}\left(\langle\overline{E}_s^{(0)}\rangle\times\langle\overline{H}_s^{(0)*}\rangle+\langle\overline{E}_s^{(2)}\rangle\times\langle\overline{H}_s^{(0)*}\rangle+\langle\overline{E}_s^{(0)}\rangle\times\langle\overline{H}_s^{(2)*}\rangle\right)\cdot\hat{z}$$
$$= \frac{1}{2}\text{Re}\left(\langle\overline{E}_s^{(0)}\rangle\times\langle\overline{H}_s^{(0)*}\rangle+2\langle\overline{E}_s^{(2)}\rangle\times\langle\overline{H}_s^{(0)*}\rangle\right)\cdot\hat{z} \tag{1.1.58}$$

First we consider the TE incidence. Then from (1.1.29), letting $\hat{e}_i = \hat{e}(-k_{iz})$, we have

$$\overline{E}_s^{(0)}(\overline{r}) = -\hat{e}(k_{iz})e^{i\overline{k}_{i\perp}\cdot\overline{r}_\perp+ik_{iz}z} \tag{1.1.59}$$

The zeroth-order magnetic field is then

$$\overline{H}_s^{(0)}(\overline{r}) = \frac{1}{\eta}\hat{h}(k_{iz})e^{i\overline{k}_{i\perp}\cdot\overline{r}_\perp+ik_{iz}z} \tag{1.1.60}$$

Thus

$$\frac{1}{2}\text{Re}\langle\overline{E}_s^{(0)}\rangle\times\langle\overline{H}_s^{(0)*}\rangle\cdot\hat{z} = \frac{1}{2\eta}\cos\theta_i \tag{1.1.61}$$

Using (1.1.60) and (1.1.55), we have

$$\text{Re}\left\{\hat{z}\cdot\langle\overline{E}_s^{(2)}\rangle\times\overline{H}_s^{(0)*}\right\} = \text{Re}\left\{-\frac{1}{\eta}\cos\theta_i\int d\overline{k}'_\perp W(\overline{k}_{i\perp}-\overline{k}'_\perp)M_{ee}(\overline{k}_{i\perp},\overline{k}'_\perp)\right\}$$
$$= -\frac{2k\cos^2\theta_i}{\eta}\int_{|\overline{k}'_\perp|<k} d\overline{k}'_\perp W(\overline{k}_{i\perp}-\overline{k}'_\perp)$$
$$\cdot\left[k'_z\cos^2(\phi'_k-\phi_i)+\frac{k^2}{k'_z}\sin^2(\phi'_k-\phi_i)\right] \tag{1.1.62}$$

Note that since only the real part contributes to power, only the propagating waves contribute in the plane wave spectrum. Note that (1.1.62) is a negative contribution making the coherent reflectivity lower than that of a flat surface. Putting (1.1.61) and (1.1.62) into (1.1.58), after making the transformation of $d\overline{k}_\perp$ to angular integrals, the coherent reflectivity is

$$\frac{\langle \overline{S}_s \cdot \hat{z} \rangle_{\text{coherent}}}{\left(\frac{\cos\theta_i}{2\eta}\right)} = 1 - 4k_{iz}k^3 \int_0^{\pi/2} d\theta_s \sin\theta_s \cos\theta_s \int_0^{2\pi} d\phi_s W(\overline{k}_{i\perp} - \overline{k}_\perp)$$
$$\cdot \left[\cos\theta_s \cos^2(\phi_s - \phi_i) + \frac{\sin^2(\phi_s - \phi_i)}{\cos\theta_s}\right] \tag{1.1.63}$$

On the other hand, from (1.1.45a) and (1.1.45b), the incoherent field reflectivity is

$$\frac{1}{4\pi}\int_0^{\pi/2} d\theta_s \sin\theta_s \int_0^{2\pi} d\phi_s (\gamma_{hh} + \gamma_{vh})$$
$$= 4k^4 \cos\theta_i \int_0^{\pi/2} d\theta_s \sin\theta_s \int_0^{2\pi} d\phi_s W(\overline{k}_\perp - \overline{k}_{i\perp})$$
$$\cdot \left[\cos^2(\phi_s - \phi_i)\cos^2\theta_s + \sin^2(\phi_s - \phi_i)\right] \tag{1.1.64}$$

Thus combining (1.1.63) and (1.1.64) indicates that energy conservation is exactly obeyed. The incoherent scattering intensity due to first order fields exactly balance the intensity of the second order coherent field to obey energy conservation. To calculate the second-order intensity, we need to calculate

$$\hat{z} \cdot \overline{E}_s^{(2)}(\overline{r}) \times \frac{\nabla \times \overline{E}_s^{(2)*}(\overline{r})}{-i\omega\mu}$$
$$= \frac{1}{\eta}\hat{z} \cdot \int d\overline{k}_{1\perp} e^{i\overline{k}_{1\perp}\cdot\overline{r}_\perp + ik_{1z}z}$$
$$\left[\hat{e}(k_{1z}) \int d\overline{k}'_\perp M_e(\overline{k}_{1\perp}, \overline{k}'_\perp) + \hat{h}(k_{1z}) \int d\overline{k}'_\perp M_h(\overline{k}_{1\perp}, \overline{k}'_\perp)\right]$$
$$\times \int d\overline{k}_{2\perp} e^{-i\overline{k}_{2\perp}\cdot\overline{r}_\perp - ik_{2z}^* z}$$
$$\left[-\hat{h}(k_{2z}) \int d\overline{k}''_\perp M_e^*(\overline{k}_{2\perp}, \overline{k}''_\perp) + \hat{e}(_{2z}) \int d\overline{k}''_\perp M_h^*(\overline{k}_{2\perp}, \overline{k}''_\perp)\right]$$
$$F(\overline{k}_{1\perp} - \overline{k}'_\perp) F(\overline{k}'_\perp - \overline{k}_{i\perp}) F(\overline{k}''_\perp - \overline{k}_{2\perp}) F(\overline{k}_{i\perp} - \overline{k}''_\perp) \tag{1.1.65a}$$

This consists of four integrals over $d\overline{k}_{1\perp} d\overline{k}'_\perp d\overline{k}_{2\perp} d\overline{k}''_\perp$. That requires taking the average of the product of F's. From

$$\begin{aligned}&\langle F(\overline{k}_{1\perp}-\overline{k}'_\perp)F(\overline{k}'_\perp-\overline{k}_{i\perp})F(\overline{k}''_\perp-\overline{k}_{2\perp})F(\overline{k}_{i\perp}-\overline{k}''_\perp)\rangle=\\&\quad\delta(\overline{k}_{1\perp}-\overline{k}_{i\perp})W(\overline{k}_{1\perp}-\overline{k}'_\perp)\delta(\overline{k}_{2\perp}-\overline{k}_{i\perp})W(\overline{k}_{2\perp}-\overline{k}''_\perp)\\&\quad+\delta(\overline{k}_{1\perp}-\overline{k}_{2\perp})\delta(\overline{k}'_\perp-\overline{k}''_\perp)W(\overline{k}_{1\perp}-\overline{k}'_\perp)W(\overline{k}'_\perp-\overline{k}_{i\perp})\\&\quad+\delta(\overline{k}_{1\perp}-\overline{k}_{2\perp})\delta(\overline{k}''_\perp+\overline{k}'_\perp-\overline{k}_{i\perp}-\overline{k}_{1\perp})W(\overline{k}_{1\perp}-\overline{k}'_\perp)W(\overline{k}'_\perp-\overline{k}_{i\perp})\end{aligned} \tag{1.1.65b}$$

The first term of (1.1.65b) is that of the coherent wave. The second and third terms are proportional to $\delta(\overline{k}_{1\perp}-\overline{k}_{2\perp})$. We can subtract out the coherent intensity to form the incoherent intensity. That means we only include the second and third terms of (1.1.65b) with $\overline{k}_{1\perp}=\overline{k}_{2\perp}$. We have incoherent field

$$\overline{\mathcal{E}}_s^{(2)}=\overline{E}_s^{(2)}-\langle\overline{E}_s^{(2)}\rangle \tag{1.1.66}$$

The power per unit area associated with the second-order incoherent field is

$$\begin{aligned}&\frac{1}{2}\operatorname{Re}\left(\hat{z}\cdot\overline{\mathcal{E}}_s^{(2)}(\overline{r})\times\frac{\nabla\times\overline{\mathcal{E}}_s^{(2)*}(\overline{r})}{-i\omega\mu}\right)\\&=\frac{1}{2\eta}\operatorname{Re}\left\{\int_p d\overline{k}_\perp\hat{z}\cdot\hat{k}\int d\overline{k}'_\perp W(\overline{k}_\perp-\overline{k}'_\perp)W(\overline{k}'_\perp-\overline{k}_{i\perp})\right.\\&\quad\times\Big[|M_e(\overline{k}_\perp,\overline{k}'_\perp)|^2+M_e(\overline{k}_\perp,\overline{k}'_\perp)M_e^*(\overline{k}_\perp,-\overline{k}'_\perp+\overline{k}_{i\perp}+\overline{k}_\perp)\\&\quad\left.+|M_h(\overline{k}_\perp,\overline{k}'_\perp)|^2+M_h(\overline{k}_\perp,\overline{k}'_\perp)M_h^*(\overline{k}_\perp,-\overline{k}'_\perp+\overline{k}_{i\perp}+\overline{k}_\perp)\Big]\right\}\end{aligned} \tag{1.1.67}$$

where $\int_p d\overline{k}_\perp$ denotes integration over propagating waves only. In (1.1.67) the coherent intensity has been subtracted out to form the incoherent intensity. The first two terms in the square bracket in (1.1.67) correspond to horizontally polarized scattered waves and the third and fourth terms correspond to vertically polarized waves. The first and third terms in the square bracket of (1.1.67) correspond to the second term of (1.1.65) while the second and fourth terms in (1.1.67) correspond to the third term of (1.1.65).

Converting to angular integration as done in (1.1.42)–(1.1.44), we can calculate the bistatic scattering coefficients. Thus the horizontally polarized scattered wave bistatic coefficient is

$$\begin{aligned}\gamma_{h\beta}^{(2)}(\hat{k}_s,\hat{k}_i)&=\frac{4\pi k^2\cos^2\theta_s}{\cos\theta_i}\operatorname{Re}\left\{\int d\overline{k}'_\perp W(\overline{k}_\perp-\overline{k}'_\perp)W(\overline{k}'_\perp-\overline{k}_{i\perp})\right.\\&\quad\left.\cdot\Big[|M_e(\overline{k}_\perp,\overline{k}'_\perp)|^2+M_e(\overline{k}_\perp,\overline{k}'_\perp)M_e^*(\overline{k}_\perp,-\overline{k}'_\perp+\overline{k}_{i\perp}+\overline{k}_\perp)\Big]\right\}\end{aligned} \tag{1.1.68}$$

where $\beta=v,h$.

Similarly, the vertically polarized scattered wave bistatic coefficient is

$$\gamma_{v\beta}^{(2)}(\hat{k}_s, \hat{k}_i) = \frac{4\pi k^2 \cos^2\theta_s}{\cos\theta_i} \operatorname{Re}\left\{ \int d\overline{k}'_\perp W(\overline{k}_\perp - \overline{k}'_\perp) W(\overline{k}'_\perp - \overline{k}_{i\perp}) \cdot \left[|M_h(\overline{k}_\perp, \overline{k}'_\perp)|^2 + M_h(\overline{k}_\perp, \overline{k}'_\perp) M_h^*(\overline{k}_\perp, -\overline{k}'_\perp + \overline{k}_{i\perp} + \overline{k}_\perp) \right] \right\} \quad (1.1.69)$$

For horizontal incident polarization, $M_e = M_{ee}$ and $M_h = M_{he}$. For vertical incident polarization, $M_e = M_{eh}$ and $M_h = M_{hh}$.

2 Electromagnetic Wave Scattering by a Dielectric Rough Surface

In this section we extend the method of Section 1 to the case of a dielectric rough surface. Let ϵ_1 denote the permittivity of the lower half space (Fig. 1.2.1). The geometrical configuration of the rough surface is as described in Section 1. We shall derive the zeroth-order, the first-order, and the second-order bistatic scattering amplitudes. Consider an incident plane wave

$$\overline{E}_i = \hat{e}_i \exp(i\overline{k}_{i\perp} \cdot \overline{r}_\perp - ik_{iz}z) \quad (1.2.1)$$

From Huygens' principle and the extinction theorem for the field $\overline{E}(\overline{r})$ in region 0 and the transmitted field $\overline{E}_1(\overline{r})$ in the dielectric medium, we obtain

$$\overline{E}_i(\overline{r}) + \int_{S'} dS' \left\{ i\omega\mu_o \overline{\overline{G}}(\overline{r}, \overline{r}') \cdot [\hat{n} \times \overline{H}(\overline{r}')] + \nabla \times \overline{\overline{G}}(\overline{r}, \overline{r}') \cdot [\hat{n} \times \overline{E}(\overline{r}')] \right\}$$

$$= \begin{cases} \overline{E}(\overline{r}) & z > f(\overline{r}_\perp) \quad (1.2.2a) \\ 0 & z < f(\overline{r}_\perp) \quad (1.2.2b) \end{cases}$$

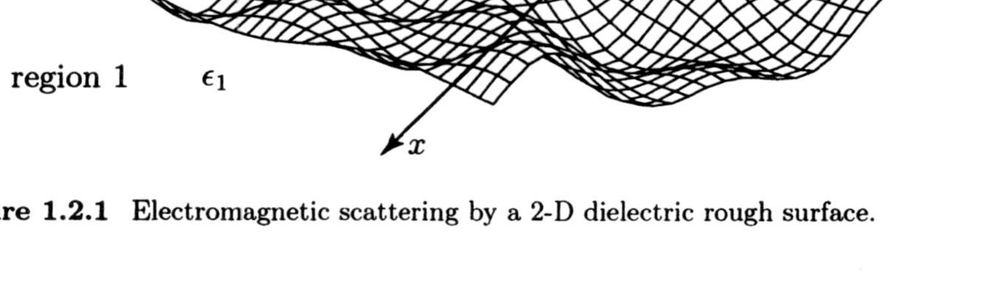

Figure 1.2.1 Electromagnetic scattering by a 2-D dielectric rough surface.

$$\int_{S'} dS' \left\{ i\omega\mu_1 \overline{\overline{G}}_1(\overline{r},\overline{r}') \cdot [\hat{n}_d \times \overline{H}_1(\overline{r}')] + \nabla \times \overline{\overline{G}}_1(\overline{r},\overline{r}') \cdot [\hat{n}_d \times \overline{E}_1(\overline{r}')] \right\}$$

$$= \begin{cases} 0 & z > f(\overline{r}_\perp) \quad (1.2.3a) \\ \overline{E}_1(\overline{r}) & z < f(\overline{r}_\perp) \quad (1.2.3b) \end{cases}$$

where S' denotes the rough surface in which the surface integration is to be carried out, $\hat{n}$ and $\hat{n}_d$ are the unit vectors normal to the rough surface and pointing into the free space and the dielectric medium, respectively, and $\overline{\overline{G}}(\overline{r},\overline{r}')$ and $\overline{\overline{G}}_1(\overline{r},\overline{r}')$ are the respective dyadic Green's functions for free space and the homogeneous dielectric of region 1.

Since tangential electric and magnetic fields are continuous, we can define surface field unknowns as follows

$$dS'\eta\,\hat{n} \times \overline{H}(\overline{r}') = d\overline{r}'_\perp\,\overline{a}(\overline{r}'_\perp) = dS'\eta\,\hat{n} \times \overline{H}_1(\overline{r}') \quad (1.2.4a)$$

$$dS'\,\hat{n} \times \overline{E}(\overline{r}') = d\overline{r}'_\perp\,\overline{b}(\overline{r}'_\perp) = dS'\,\hat{n} \times \overline{E}_1(\overline{r}') \quad (1.2.4b)$$

Next we make use of the integral representation of dyadic Green's function as given in Chapter 2 of Volume I. For $z < f(\overline{r}'_\perp)$ we have

$$\overline{\overline{G}}(\overline{r},\overline{r}') = \frac{i}{8\pi^2} \int d\overline{k}_\perp e^{i\overline{k}_\perp\cdot\overline{r}_\perp} e^{-ik_z z} e^{-i\overline{k}_\perp\cdot\overline{r}'_\perp} e^{ik_z f(\overline{r}'_\perp)} \cdot \frac{1}{k_z} \left[\hat{e}(-k_z)\hat{e}(-k_z) + \hat{h}(-k_z)\hat{h}(-k_z) \right] \quad (1.2.5a)$$

$$\nabla \times \overline{\overline{G}}(\overline{r},\overline{r}') = -\frac{1}{8\pi^2} \int d\overline{k}_\perp e^{i\overline{k}_\perp\cdot\overline{r}_\perp} e^{-ik_z z} e^{-i\overline{k}_\perp\cdot\overline{r}'_\perp} e^{ik_z f(\overline{r}'_\perp)} \cdot \frac{k}{k_z} \left[-\hat{h}(-k_z)\hat{e}(-k_z) + \hat{e}(-k_z)\hat{h}(-k_z) \right] \quad (1.2.5b)$$

On the other hand, for $z > f(\overline{r}'_\perp)$ we obtain

$$\overline{\overline{G}}_1(\overline{r},\overline{r}') = \frac{i}{8\pi^2} \int d\overline{k}_\perp e^{i\overline{k}_\perp\cdot\overline{r}_\perp} e^{ik_{1z} z} e^{-i\overline{k}_\perp\cdot\overline{r}'_\perp} e^{-ik_{1z} f(\overline{r}'_\perp)} \cdot \frac{1}{k_{1z}} \left[\hat{e}(k_{1z})\hat{e}(k_{1z}) + \hat{h}_1(k_{1z})\hat{h}_1(k_{1z}) \right] \quad (1.2.6a)$$

$$\nabla \times \overline{\overline{G}}_1(\overline{r},\overline{r}') = -\frac{1}{8\pi^2} \int d\overline{k}_\perp e^{i\overline{k}_\perp\cdot\overline{r}_\perp} e^{ik_{1z} z} e^{-i\overline{k}_\perp\cdot\overline{r}'_\perp} e^{-ik_{1z} f(\overline{r}'_\perp)} \cdot \frac{k_1}{k_{1z}} \left[-\hat{h}_1(k_{1z})\hat{e}_1(k_{1z}) + \hat{e}_1(k_{1z})\hat{h}_1(k_{1z}) \right] \quad (1.2.6b)$$

Substitute these into (1.2.2) and (1.2.3). Let f_{min} and f_{max} be, respectively, the minimum and maximum values of the surface profile $f(\overline{r}'_\perp)$. Evaluating (1.2.2b) for $z < f_{min}$ and (1.2.3a) for $z > f_{max}$, we obtain the equations of the two extinction theorems.

$$\overline{E}_i(\overline{r}) = \frac{1}{8\pi^2}\int d\overline{k}_\perp\, e^{i\overline{k}_\perp\cdot\overline{r}_\perp}\, e^{-ik_z z}\frac{k}{k_z}\int d\overline{r}'_\perp\, e^{-i\overline{k}_\perp\cdot\overline{r}'_\perp}\, e^{ik_z f(\overline{r}'_\perp)}$$
$$\times\left\{\left[\hat{e}(-k_z)\hat{e}(-k_z)+\hat{h}(-k_z)\hat{h}(-k_z)\right]\cdot\overline{a}(\overline{r}'_\perp)\right.$$
$$\left.+\left[-\hat{h}(-k_z)\hat{e}(-k_z)+\hat{e}(-k_z)\hat{h}(-k_z)\right]\cdot\overline{b}(\overline{r}'_\perp)\right\} \qquad (1.2.7a)$$

$$0 = \frac{1}{8\pi^2}\int d\overline{k}_\perp\, e^{i\overline{k}_\perp\cdot\overline{r}_\perp}\, e^{ik_{1z} z}\frac{k_1}{k_{1z}}\int d\overline{r}'_\perp\, e^{-i\overline{k}_\perp\cdot\overline{r}'_\perp}\, e^{-ik_{1z} f(\overline{r}'_\perp)}$$
$$\times\left\{\frac{k}{k_1}\left[\hat{e}_1(k_{1z})\hat{e}_1(k_{1z})+\hat{h}_1(k_{1z})\hat{h}_1(k_{1z})\right]\cdot\overline{a}(\overline{r}'_\perp)\right.$$
$$\left.+\left[-\hat{h}_1(k_{1z})\hat{e}_1(k_{1z})+\hat{e}_1(k_{1z})\hat{h}_1(k_{1z})\right]\cdot\overline{b}(\overline{r}'_\perp)\right\} \qquad (1.2.7b)$$

The above equations are the extended boundary conditions and can be used to solve for the surface fields along with the following equations, which are results of (1.2.4*a*) and (1.2.4*b*),

$$\hat{n}(\overline{r}'_\perp)\cdot\overline{a}(\overline{r}'_\perp) = 0 \qquad (1.2.8a)$$
$$\hat{n}(\overline{r}'_\perp)\cdot\overline{b}(\overline{r}'_\perp) = 0 \qquad (1.2.8b)$$

Since

$$\hat{n} = \frac{\left(-\dfrac{\partial f}{\partial x}\hat{x}-\dfrac{\partial f}{\partial y}\hat{y}+\hat{z}\right)}{\sqrt{1+\left(\dfrac{\partial f}{\partial x}\right)^2+\left(\dfrac{\partial f}{\partial y}\right)^2}} \qquad (1.2.9)$$

from (1.2.8*a*) and (1.2.8*b*) we have

$$a_z(\overline{r}'_\perp) = \left(\hat{x}\,\frac{\partial f(\overline{r}'_\perp)}{\partial x'}+\hat{y}\,\frac{\partial f(\overline{r}'_\perp)}{\partial y'}\right)\cdot\overline{a}_\perp(\overline{r}'_\perp) \qquad (1.2.10)$$
$$b_z(\overline{r}'_\perp) = \left(\hat{x}\,\frac{\partial f(\overline{r}'_\perp)}{\partial x'}+\hat{y}\,\frac{\partial f(\overline{r}'_\perp)}{\partial y'}\right)\cdot\overline{b}_\perp(\overline{r}'_\perp) \qquad (1.2.11)$$

with a_z and b_z as the z-components of $\overline{a}$ and $\overline{b}$, respectively.

Once the surface fields are obtained, the scattered field in region 0 and the transmitted field in medium 1 can be derived by using (1.2.2*a*) and (1.2.3*b*). The scattered fields and transmitted fields are calculated by evaluating (1.2.2*a*) and (1.2.3*b*) for $z > f_{max}$ and $z < f_{min}$, respectively. For

$z > f(\overline{r}'_\perp)$

$$\overline{\overline{G}}(\overline{r},\overline{r}') = \frac{i}{8\pi^2}\int d\overline{k}_\perp e^{i\overline{k}_\perp\cdot\overline{r}_\perp} e^{ik_z z} e^{-i\overline{k}_\perp\cdot\overline{r}'_\perp} e^{-ik_{1z}f(\overline{r}'_\perp)} \cdot \frac{1}{k_z}\left[\hat{e}(k_z)\hat{e}(k_z) + \hat{h}(k_z)\hat{h}(k_z)\right] \tag{1.2.12a}$$

$$\nabla\times\overline{\overline{G}}(\overline{r},\overline{r}') = -\frac{1}{8\pi^2}\int d\overline{k}_\perp e^{i\overline{k}_\perp\cdot\overline{r}_\perp} e^{ik_z z} e^{-i\overline{k}_\perp\cdot\overline{r}'_\perp} e^{-ik_z f(\overline{r}'_\perp)} \cdot \frac{k}{k_z}\left[-\hat{h}(k_z)\hat{e}(k_z) + \hat{e}(k_z)\hat{h}(k_z)\right] \tag{1.2.12b}$$

$$\begin{aligned}\overline{E}_s(\overline{r}) = &-\frac{1}{8\pi^2}\int d\overline{k}_\perp\, e^{i\overline{k}_\perp\cdot\overline{r}_\perp}\, e^{ik_z z}\frac{k}{k_z}\int d\overline{r}'_\perp\, e^{-i\overline{k}_\perp\cdot\overline{r}'_\perp} e^{-ik_z f(\overline{r}'_\perp)}\\ &\times\left\{\left[\hat{e}(k_z)\hat{e}(k_z) + \hat{h}(k_z)\hat{h}(k_z)\right]\cdot\overline{a}(\overline{r}'_\perp)\right.\\ &\left.+\left[-\hat{h}(k_z)\hat{e}(k_z) + \hat{e}(k_z)\hat{h}(k_z)\right]\cdot\overline{b}(\overline{r}'_\perp)\right\}\end{aligned} \tag{1.2.13a}$$

We note that the expression for the scattered field is similar to that of the extinction theorem. The property can be exploited later on in simplifying the analysis. Similarly for transmitted fields, we have

$$\begin{aligned}\overline{E}_t(\overline{r}) = &\frac{1}{8\pi^2}\int d\overline{k}_\perp\, e^{i\overline{k}_\perp\cdot\overline{r}_\perp}\, e^{-ik_{1z}z}\frac{k_1}{k_{1z}}\int d\overline{r}'_\perp\, e^{-i\overline{k}_\perp\cdot\overline{r}'_\perp} e^{ik_{1z} f(\overline{r}'_\perp)}\\ &\times\left\{\frac{k}{k_1}\left[\hat{e}_1(-k_{1z})\hat{e}_1(-k_{1z}) + \hat{h}_1(-k_{1z})\hat{h}_1(-k_{1z})\right]\cdot\overline{a}(\overline{r}'_\perp)\right.\\ &\left.+\left[-\hat{h}_1(-k_{1z})\hat{e}_1(-k_{1z}) + \hat{e}_1(-k_{1z})\hat{h}_1(-k_{1z})\right]\cdot\overline{b}(\overline{r}'_\perp)\right\}\end{aligned} \tag{1.2.13b}$$

The objective is to solve for the surface fields using (1.2.7) and (1.2.8) and then to solve for the scattered and transmitted fields using (1.2.13).

Equations (1.2.7) and (1.2.10) through (1.2.13) are exact. To solve for the surface fields, the perturbation method makes use of series expansions. Let

$$\overline{a}(\overline{r}'_\perp) = \sum_{m=0}^{\infty}\overline{a}^{(m)}(\overline{r}'_\perp) \tag{1.2.14a}$$

$$\overline{b}(\overline{r}'_\perp) = \sum_{m=0}^{\infty}\overline{b}^{(m)}(\overline{r}'_\perp) \tag{1.2.14b}$$

where superscript (m) denotes mth order solution. We also have

$$e^{\pm ik_z f(\overline{r}'_\perp)} = \sum_{m=0}^{\infty} \frac{[\pm ik_z f(\overline{r}'_\perp)]^m}{m!} \tag{1.2.15a}$$

$$e^{\pm ik_{1z} f(\overline{r}'_\perp)} = \sum_{m=0}^{\infty} \frac{[\pm ik_{1z} f(\overline{r}'_\perp)]^m}{m!} \tag{1.2.15b}$$

In SPM, $f(\overline{r}_\perp)$ and its derivatives are regarded as small parameters. Thus not only the rms heights are small but also the slopes have to be small. The expansions of (1.2.14) and (1.2.15) are substituted into (1.2.7) to obtain the set of equations for the different-order solutions. From (1.2.10), (1.2.11) and (1.2.14) we obtain

$$a_z^{(0)}(\overline{r}'_\perp) = b_z^{(0)}(\overline{r}'_\perp) = 0 \tag{1.2.16}$$

$$a_z^{(m)}(\overline{r}'_\perp) = \left(\hat{x}\,\frac{\partial f(\overline{r}'_\perp)}{\partial x'} + \hat{y}\,\frac{\partial f(\overline{r}'_\perp)}{\partial y'}\right) \cdot \overline{a}_\perp^{(m-1)}(\overline{r}'_\perp) \tag{1.2.17a}$$

$$b_z^{(m)}(\overline{r}'_\perp) = \left(\hat{x}\,\frac{\partial f(\overline{r}'_\perp)}{\partial x'} + \hat{y}\,\frac{\partial f(\overline{r}'_\perp)}{\partial y'}\right) \cdot \overline{b}_\perp^{(m-1)}(\overline{r}'_\perp) \tag{1.2.17b}$$

The assumptions are

$$k_z f(\overline{r}'_\perp),\ k_{1z} f(\overline{r}'_\perp),\ \frac{\partial f}{\partial x'},\ \frac{\partial f}{\partial y'} \ll 1 \tag{1.2.18}$$

Note that the $\hat{z}$ components of the surface fields are of a lower order than the horizontal components because of the small slope approximation. Substituting (1.2.14) and (1.2.15) into (1.2.7) and (1.2.10) through (1.2.11) and equating the same-order terms, we can calculate the surface fields to the zeroth-order, the first-order, and so on. Then, from (1.2.13), the scattered fields can be obtained to different orders. In the following, we solve for the surface fields and scattered fields up to the second order. The zeroth-order solutions are just the reflected and transmitted fields of a flat surface. The first-order solution gives the lowest-order incoherent scattered intensities. However, the first-order solution does not give the depolarization effect in the backscattering direction. The second-order solution gives the lowest-order correction to the coherent reflection and transmission coefficients. Also, the depolarization of the backscattered power is manifested. As shown in previous section, the second order solution is needed to ensure energy conservation. This is particularly important for the calculation of emissivity.

We define Fourier transforms of surface fields, or surface fields in the spectral domain:

$$\overline{A}(\overline{k}_\perp) = \frac{1}{(2\pi)^2} \int d\overline{r}'_\perp \overline{a}(\overline{r}'_\perp) e^{-i\overline{k}_\perp \cdot \overline{r}'_\perp} \tag{1.2.19a}$$

$$\overline{B}(\overline{k}_\perp) = \frac{1}{(2\pi)^2} \int d\overline{r}'_\perp \overline{b}(\overline{r}'_\perp) e^{-i\overline{k}_\perp \cdot \overline{r}'_\perp} \tag{1.2.19b}$$

In spectral domain, (1.2.17) gives

$$A_z(\overline{k}_\perp) = i \int_{-\infty}^{\infty} d\overline{k}'_\perp (\overline{k}_\perp - \overline{k}'_\perp) \cdot \overline{A}_\perp(\overline{k}'_\perp) F(\overline{k}_\perp - \overline{k}'_\perp) \tag{1.2.19c}$$

$$B_z(\overline{k}_\perp) = i \int_{-\infty}^{\infty} d\overline{k}'_\perp (\overline{k}_\perp - \overline{k}'_\perp) \cdot \overline{B}_\perp(\overline{k}'_\perp) F(\overline{k}_\perp - \overline{k}'_\perp) \tag{1.2.19d}$$

Using the Fourier transform of the surface fields and expanding

$$e^{\pm ik_z f(\overline{r}'_\perp)} = 1 \pm ik_z f(\overline{r}'_\perp) - \frac{k_z^2}{2} f^2(\overline{r}'_\perp)$$

in (1.2.7*a*) and (1.2.7*b*), we have, up to the second order,

$$\begin{aligned}\overline{E}_i(\overline{r}) = \frac{1}{2} \int d\overline{k}_\perp e^{i\overline{k}_\perp \cdot \overline{r}_\perp + ik_z z} \frac{k}{k_z} \Big\{ \Big[\hat{e}(-k_z)\hat{e}(-k_z) + \hat{h}(-k_z)\hat{h}(-k_z)\Big] \\ \cdot \Big[\overline{A}(\overline{k}_\perp) + ik_z \int d\overline{k}'_\perp \overline{A}(\overline{k}'_\perp) F(\overline{k}_\perp - \overline{k}'_\perp) - \frac{k_z^2}{2} \int d\overline{k}'_\perp \overline{A}(\overline{k}'_\perp) F^{(2)}(\overline{k}_\perp - \overline{k}'_\perp)\Big] \\ + \Big[-\hat{h}(-k_z)\hat{e}(-k_z) + \hat{e}(-k_z)\hat{h}(-k_z)\Big]\Big[\overline{B}(\overline{k}_\perp) + ik_z \int d\overline{k}'_\perp \overline{B}(\overline{k}'_\perp) F(\overline{k}_\perp - \overline{k}'_\perp) \\ - \frac{k_z^2}{2} \int d\overline{k}'_\perp \overline{B}(\overline{k}'_\perp) F^{(2)}(\overline{k}_\perp - \overline{k}'_\perp)\Big]\Big\}\end{aligned} \tag{1.2.19e}$$

The extinction theorem of (1.2.7*b*) is

$$\begin{aligned}0 = \frac{1}{2} \int d\overline{k}_\perp e^{i\overline{k}_\perp \cdot \overline{r}_\perp + ik_{1z} z} \frac{k_1}{k_{1z}} \Big\{ \frac{k}{k_1} \Big[\hat{e}_1(k_{1z})\hat{e}_1(k_{1z}) + \hat{h}_1(k_{1z})\hat{h}_1(k_{1z})\Big] \\ \cdot \Big[\overline{A}(\overline{k}_\perp) - ik_{1z} \int d\overline{k}'_\perp \overline{A}(\overline{k}'_\perp) F(\overline{k}_\perp - \overline{k}'_\perp) - \frac{k_{1z}^2}{2} \int d\overline{k}'_\perp \overline{A}(\overline{k}'_\perp) F^{(2)}(\overline{k}_\perp - \overline{k}'_\perp)\Big] \\ + \Big[-\hat{h}_1(k_{1z})\hat{e}_1(k_{1z}) + \hat{e}_1(k_{1z})\hat{h}_1(k_{1z})\Big] \Big[\overline{B}(\overline{k}_\perp) - ik_{1z} \int d\overline{k}'_\perp \overline{B}(\overline{k}'_\perp) F(\overline{k}_\perp - \overline{k}'_\perp) \\ - \frac{k_{1z}^2}{2} \int d\overline{k}'_\perp \overline{B}(\overline{k}'_\perp) F^{(2)}(\overline{k}_\perp - \overline{k}'_\perp)\Big]\Big\}\end{aligned} \tag{1.2.19f}$$

Similarly, for the scattered field we have

$$\overline{E}_s(\overline{r}) = -\frac{1}{2}\int d\overline{k}_\perp e^{i\overline{k}_\perp\cdot\overline{r}_\perp} e^{ik_z z}\frac{k}{k_z}\Bigg\{\left[\hat{e}(k_z)\hat{e}(k_z)+\hat{h}(k_z)\hat{h}(k_z)\right]$$

$$\cdot\left[\overline{A}(\overline{k}_\perp)-ik_z\int d\overline{k}'_\perp\overline{A}(\overline{k}'_\perp)F(\overline{k}_\perp-\overline{k}'_\perp)-\frac{k_z^2}{2}\int d\overline{k}'_\perp\overline{A}(\overline{k}'_\perp)F^{(2)}(\overline{k}_\perp-\overline{k}'_\perp)\right]$$

$$+\left[-\hat{h}(k_z)\hat{e}(k_z)+\hat{e}(k_z)\hat{h}(k_z)\right]\left[\overline{B}(\overline{k}_\perp)-ik_z\int d\overline{k}'_\perp\overline{B}(\overline{k}'_\perp)F(\overline{k}_\perp-\overline{k}'_\perp)\right.$$

$$\left.-\frac{k_z^2}{2}\int d\overline{k}'_\perp\overline{B}(\overline{k}'_\perp)F^{(2)}(\overline{k}_\perp-\overline{k}'_\perp)\right]\Bigg\} \tag{1.2.19g}$$

To solve the surface fields, it is convenient to define $\hat{q}$ and $\hat{p}$, which are the unit vectors that are parallel to the projections of $\hat{e}$ and $\hat{h}$ on the horizontal plane.

$$\hat{q}(\overline{k}_\perp) = \hat{e}(-k_z) = \frac{\hat{x}k_y - \hat{y}k_x}{k_\rho} \tag{1.2.20a}$$

is the same as $\hat{e}$, and

$$\hat{p}(\overline{k}_\perp) = \hat{z}\times\hat{q}(\overline{k}_\perp) = \frac{\hat{x}k_x + \hat{y}k_y}{k_\rho} = \hat{k}_\perp \tag{1.2.20b}$$

is the projection of $\hat{h}(-k_z)$ on the x-y plane and is also the unit vector in the projection of $\overline{k}$ onto x-y plane. We use two orthonormal systems $(\hat{h}, \hat{e}, \hat{k})$ and $(\hat{z}, \hat{q}, \hat{p})$. Let $\overline{k}_\perp = k_\rho\cos\phi_k\hat{x} + k_\rho\sin\phi_k\hat{y}$ and $\overline{k}'_\perp = k'_\rho\cos\phi'_k\hat{x} + k'_\rho\sin\phi'_k\hat{y}$. Then

$$\hat{h}(\pm k_z) = \mp\frac{k_z}{k}\hat{p}(\overline{k}_\perp) + \frac{k_\rho}{k}\hat{z} \tag{1.2.21a}$$

$$\hat{h}_1(\pm k_{1z}) = \mp\frac{k_{1z}}{k_1}\hat{p}(\overline{k}_\perp) + \frac{k_\rho}{k_1}\hat{z} \tag{1.2.21b}$$

$$\hat{e}(\pm k_z)\cdot\hat{q}(\overline{k}'_\perp) = \frac{k_y k'_y + k_x k'_x}{k_\rho k'_\rho} = \cos(\phi_k - \phi'_k) \tag{1.2.22a}$$

$$\hat{e}(\pm k_z)\cdot\hat{p}(\overline{k}'_\perp) = \frac{k_y k'_x - k_x k'_y}{k_\rho k'_\rho} = \sin(\phi_k - \phi'_k) \tag{1.2.22b}$$

$$\hat{h}(\pm k_z)\cdot\hat{q}(\overline{k}'_\perp) = \mp\frac{k_z}{k k_\rho k'_\rho}(k_x k'_y - k_y k'_x) = \pm\frac{k_z}{k}\sin(\phi_k - \phi'_k) \tag{1.2.22c}$$

$$\hat{h}(\pm k_z)\cdot\hat{p}(\overline{k}'_\perp) = \mp\frac{k_z}{k k_\rho k'_\rho}(k_x k'_x + k_y k'_y) = \mp\frac{k_z}{k}\cos(\phi_k - \phi'_k) \tag{1.2.22d}$$

$$\hat{h}_1(\pm k_{1z})\cdot\hat{q}(\overline{k}'_\perp) = \mp\frac{k_{1z}}{k_1 k_\rho k'_\rho}(k_x k'_y - k_y k'_x) = \pm\frac{k_{1z}}{k_1}\sin(\phi_k - \phi'_k) \tag{1.2.22e}$$

$$\hat{h}_1(\pm k_{1z}) \cdot \hat{p}(\overline{k}'_\perp) = \mp \frac{k_{1z}}{k_1 k_\rho k'_\rho}(k_x k'_x + k_y k'_y) = \mp \frac{k_{1z}}{k_1} \cos(\phi_k - \phi'_k) \quad (1.2.22f)$$

$$\overline{k}_\perp \cdot \hat{q}(\overline{k}'_\perp) = \frac{k_x k'_y - k_y k'_x}{k'_\rho} = -k_\rho \sin(\phi_k - \phi'_k) \quad (1.2.22g)$$

$$\overline{k}_\perp \cdot \hat{p}(\overline{k}'_\perp) = \frac{k_x k'_x + k_y k'_y}{k'_\rho} = k_\rho \cos(\phi_k - \phi'_k) \quad (1.2.22h)$$

$$(\overline{k}_\perp - \overline{k}'_\perp) \cdot \hat{q}(\overline{k}'_\perp) = -k_\rho \sin(\phi_k - \phi'_k) \quad (1.2.22i)$$

$$(\overline{k}_\perp - \overline{k}'_\perp) \cdot \hat{p}(\overline{k}'_\perp) = k_\rho \cos(\phi_k - \phi'_k) - k'_\rho \quad (1.2.22j)$$

We decompose surface fields in the spectral domains into $\hat{q}$-, $\hat{p}$- and $\hat{z}$-components

$$\overline{A}(\overline{k}_\perp) = A_q(\overline{k}_\perp)\hat{q}(\overline{k}_\perp) + A_p(\overline{k}_\perp)\hat{p}(\overline{k}_\perp) + A_z(\overline{k}_\perp)\hat{z} \quad (1.2.23a)$$

$$\overline{B}(\overline{k}_\perp) = B_q(\overline{k}_\perp)\hat{q}(\overline{k}_\perp) + B_p(\overline{k}_\perp)\hat{p}(\overline{k}_\perp) + B_z(\overline{k}_\perp)\hat{z} \quad (1.2.23b)$$

The $\hat{z}$-components of surface fields, in terms of the horizontal components, using (1.2.19a), are

$$A_z^{(m)}(\overline{k}_\perp) = i \int d\overline{k}'_\perp F(\overline{k}_\perp - \overline{k}'_\perp)(\overline{k}_\perp - \overline{k}'_\perp) \cdot \overline{A}_\perp^{(m-1)}(\overline{k}'_\perp) \quad (1.2.24a)$$

$$B_z^{(m)}(\overline{k}_\perp) = i \int d\overline{k}'_\perp F(\overline{k}_\perp - \overline{k}'_\perp)(\overline{k}_\perp - \overline{k}'_\perp) \cdot \overline{B}_\perp^{(m-1)}(\overline{k}'_\perp) \quad (1.2.24b)$$

This means that the z-components of a given order are calculated readily from the horizontal components of a lower order. We next use the following results:

$$\begin{aligned} &\Big[\hat{e}(\pm k_z)\hat{e}(\pm k_z) + \hat{h}(\pm k_z)\hat{h}(\pm k_z)\Big] \cdot \overline{A}(\overline{k}_\perp) \\ &= \left\{\hat{e}(\pm k_z)A_q(\overline{k}_\perp) + \hat{h}(\pm k_z)\Big(\frac{k_\rho}{k}A_z(\overline{k}_\perp) \mp \frac{k_z}{k}A_p(\overline{k}_\perp)\Big)\right\} \end{aligned} \quad (1.2.24c)$$

$$\begin{aligned} &\Big[-\hat{h}(\pm k_z)\hat{e}(\pm k_z) + \hat{e}(\pm k_z)\hat{h}(\pm k_z)\Big] \cdot \overline{B}(\overline{k}_\perp) \\ &= \left\{-\hat{h}(\pm k_z)B_q(\overline{k}_\perp) + \hat{e}(\pm k_z)\Big(\frac{k_\rho}{k}B_z(\overline{k}_\perp) \mp \frac{k_z}{k}B_p(\overline{k}_\perp)\Big)\right\} \end{aligned} \quad (1.2.24d)$$

$$\begin{aligned} &\Big[\hat{e}_1(k_{1z})\hat{e}_1(k_{1z}) + \hat{h}_1(k_{1z})\hat{h}_1(k_{1z})\Big] \cdot \overline{A}(\overline{k}_\perp) \\ &= \hat{e}_1(k_{1z})A_q(\overline{k}_\perp) + \hat{h}_1(k_{1z})\Big[-\frac{k_{1z}}{k_1}A_p(\overline{k}_\perp) + \frac{k_\rho}{k_1}A_z(\overline{k}_\perp)\Big] \end{aligned} \quad (1.2.24e)$$

$$\begin{aligned} &\Big[-\hat{h}_1(k_{1z})\hat{e}_1(k_{1z}) + \hat{e}_1(k_{1z})\hat{h}_1(k_{1z})\Big] \cdot \overline{B}(\overline{k}_\perp) \\ &= -\hat{h}_1(k_{1z})B_q(\overline{k}_\perp) + \hat{e}_1(k_{1z})\Big[-\frac{k_{1z}}{k_1}B_p(\overline{k}_\perp) + \frac{k_\rho}{k_1}B_z(\overline{k}_\perp)\Big] \end{aligned} \quad (1.2.24f)$$

The $\exp(ik_z f)$ and the $\exp(-ik_{iz} f)$ terms, respectively, in (1.2.7a) and (1.2.7b) are expanded to the second order in f. For example, (1.2.7a) becomes

$$\int_{-\infty}^{\infty} d\overline{k}_\perp e^{i\overline{k}_\perp\cdot\overline{r}_\perp - ik_z z}\hat{e}_i\delta(\overline{k}_\perp - \overline{k}_{i\perp})$$
$$= \frac{1}{8\pi^2}\int d\overline{k}_\perp e^{i\overline{k}_\perp\cdot\overline{r}_\perp - ik_z z}\frac{k}{k_z}\int d\overline{r}'_\perp e^{-i\overline{k}_\perp\cdot\overline{r}'_\perp}\left(1 + ik_z f(\overline{r}'_\perp) - \frac{k_z^2}{2}f^2(\overline{r}'_\perp)\right)$$
$$\times\left\{\left[\hat{e}(-k_z)\hat{e}(-k_z) + \hat{h}(-k_z)\hat{h}(-k_z)\right]\cdot\int d\overline{k}'_\perp\overline{A}(\overline{k}'_\perp)e^{+i\overline{k}'_\perp\cdot\overline{r}'_\perp}\right.$$
$$\left. + \left[-\hat{h}(-k_z)\hat{e}(-k_z) + \hat{e}(-k_z)\hat{h}(-k_z)\right]\cdot\int d\overline{k}'_\perp\overline{B}(\overline{k}'_\perp)e^{+i\overline{k}'_\perp\cdot\overline{r}'_\perp}\right\} \quad (1.2.25a)$$

The Fourier transforms of the height function and the products of height functions are defined as

$$\frac{1}{4\pi^2}\int d\overline{r}_\perp e^{-i\overline{k}_\perp\cdot\overline{r}_\perp}f(\overline{r}_\perp) = F(\overline{k}_\perp) \quad (1.2.25b)$$
$$\frac{1}{4\pi^2}\int d\overline{r}_\perp e^{-i\overline{k}_\perp\cdot\overline{r}_\perp}f^2(\overline{r}_\perp) = F^{(2)}(\overline{k}_\perp) \quad (1.2.25c)$$
$$F^{(2)}(\overline{k}_\perp) = \int_{-\infty}^{\infty} d\overline{k}'_\perp F(\overline{k}'_\perp)F(\overline{k}_\perp - \overline{k}'_\perp) \quad (1.2.25d)$$

Using (1.2.25b–d) in (1.2.25a) gives

$$\hat{e}_i\delta(\overline{k}_\perp - \overline{k}_{i\perp}) = \frac{1}{2}\frac{k}{k_z}[\hat{e}(-k_z)\hat{e}(-k_z) + \hat{h}(-k_z)\hat{h}(-k_z)]$$
$$\cdot\left[\overline{A}(\overline{k}_\perp) + ik_z\int d\overline{k}'_\perp\overline{A}(\overline{k}'_\perp)F(\overline{k}_\perp - \overline{k}'_\perp)\right.$$
$$\left. - \frac{k_z^2}{2}\int d\overline{k}'_\perp\overline{A}(\overline{k}'_\perp)F^{(2)}(\overline{k}_\perp - \overline{k}'_\perp)\right]$$
$$+ \frac{1}{2}\frac{k}{k_z}[-\hat{h}(-k_z)\hat{e}(-k_z) + \hat{e}(-k_z)\hat{h}(-k_z)]$$
$$\cdot\left[\overline{B}(\overline{k}_\perp) + ik_z\int d\overline{k}'_\perp\overline{B}(\overline{k}'_\perp)F(\overline{k}_\perp - \overline{k}'_\perp)\right.$$
$$\left. - \frac{k_z^2}{2}\int d\overline{k}'_\perp\overline{B}(\overline{k}'_\perp)F^{(2)}(\overline{k}_\perp - \overline{k}'_\perp)\right] \quad (1.2.25e)$$

Similarly, (1.2.7b) becomes

$$
\begin{aligned}
0 = &\frac{1}{2}\frac{k}{k_{1z}}[\hat{e}_1(k_{1z})\hat{e}_1(k_{1z}) + \hat{h}_1(k_{1z})\hat{h}_1(k_{1z})] \\
&\cdot \left[\overline{A}(\overline{k}_\perp) - ik_{1z}\int d\overline{k}'_\perp \overline{A}(\overline{k}'_\perp)F(\overline{k}_\perp - \overline{k}'_\perp)\right. \\
&\left. - \frac{k_{1z}^2}{2}\int d\overline{k}'_\perp \overline{A}(\overline{k}'_\perp)F^{(2)}(\overline{k}_\perp - \overline{k}'_\perp)\right] \\
&+ \frac{1}{2}\frac{k_1}{k_{1z}}[-\hat{h}_1(k_{1z})\hat{e}_1(k_{1z}) + \hat{e}_1(k_{1z})\hat{h}_1(k_{1z})] \\
&\cdot \left[\overline{B}(\overline{k}_\perp) - ik_{1z}\int d\overline{k}'_\perp \overline{B}(\overline{k}'_\perp)F(\overline{k}_\perp - \overline{k}'_\perp)\right. \\
&\left. - \frac{k_{1z}^2}{2}\int d\overline{k}'_\perp \overline{B}(\overline{k}'_\perp)F^{(2)}(\overline{k}_\perp - \overline{k}'_\perp)\right] \qquad (1.2.25f)
\end{aligned}
$$

In the following, we will solve (1.2.25e) and (1.2.25f) up to the second order.

2.1 Zeroth- and First-Order Solutions

Zeroth-Order Solution

From (1.2.24), let

$$A_z^{(0)} = B_z^{(0)} = 0 \qquad (1.2.26a)$$

$$\overline{A}_\perp^{(0)}(\overline{k}_\perp) = (\hat{q}_i a_q^{(0)} + \hat{p}_i a_p^{(0)})\delta(\overline{k}_\perp - \overline{k}_{i\perp}) \qquad (1.2.26b)$$

$$\overline{B}_\perp^{(0)}(\overline{k}_\perp) = (\hat{q}_i b_q^{(0)} + \hat{p}_i b_p^{(0)})\delta(\overline{k}_\perp - \overline{k}_{i\perp}) \qquad (1.2.26c)$$

Balance (1.2.25e) and (1.2.25f) to the zeroth order

$$
\begin{aligned}
&\hat{e}_i\delta(\overline{k}_\perp - \overline{k}_{i\perp}) \\
&\quad = \frac{k}{2k_z}\left[\hat{e}(-k_z)\hat{e}(-k_z) + \hat{h}(-k_z)\hat{h}(-k_z)\right] \cdot \overline{A}_\perp^{(0)}(\overline{k}_\perp) \\
&\quad + \frac{k}{2k_z}\left[-\hat{h}(-k_z)\hat{e}(-k_z) + \hat{e}(-k_z)\hat{h}(-k_z)\right] \cdot \overline{B}_\perp^{(0)}(\overline{k}_\perp) \qquad (1.2.26d)
\end{aligned}
$$

$$
\begin{aligned}
0 = &\frac{k}{2k_{1z}}\left[\hat{e}_1(k_{1z})\hat{e}_1(k_{1z}) + \hat{h}_1(k_{1z})\hat{h}_1(k_{1z})\right] \cdot \overline{A}_\perp^{(0)}(\overline{k}_\perp) \\
&+ \frac{k_1}{2k_{1z}}\left[-\hat{h}_1(k_{1z})\hat{e}_1(k_{1z}) + \hat{e}_1(k_{1z})\hat{h}_1(k_{1z})\right] \cdot \overline{B}_\perp^{(0)}(\overline{k}_\perp) \qquad (1.2.26e)
\end{aligned}
$$

Substitute (1.2.26b) and (1.2.26c) into (1.2.26d) and (1.2.26e) and make use of (1.2.22a)–(1.2.22f); on taking a dot product of (1.2.26d) with $\hat{e}(-k_{iz})$ and

$\hat{h}(-k_{iz})$, we have

$$\hat{e}(-k_{iz}) \cdot \hat{e}_i = \frac{1}{2} \frac{k}{k_{iz}} a_q^{(0)} + \frac{1}{2} b_p^{(0)} \tag{1.2.27a}$$

$$\hat{h}(-k_{iz}) \cdot \hat{e}_i = \frac{1}{2} a_p^{(0)} - \frac{1}{2} \frac{k}{k_{iz}} b_q^{(0)} \tag{1.2.27b}$$

On taking a dot product of (1.2.26e) with $\hat{e}_1(k_{1zi})$ and $\hat{h}_1(k_{1zi})$ gives

$$0 = \frac{1}{2} \frac{k}{k_{1zi}} a_q^{(0)} - \frac{1}{2} b_p^{(0)} \tag{1.2.27c}$$

$$0 = -\frac{1}{2} \frac{k}{k_1} a_p^{(0)} - \frac{1}{2} \frac{k_1}{k_{1zi}} b_q^{(0)} \tag{1.2.27d}$$

Solving (1.2.27a) and (1.2.27c) gives

$$\overline{a}_q^{(0)}(\overline{k}_{i\perp}) = [\hat{e}(-k_{iz}) \cdot \hat{e}_i] \frac{k_{iz}}{k} (1 - R_{ho}) \tag{1.2.28a}$$

$$\overline{b}_p^{(0)}(\overline{k}_{i\perp}) = [\hat{e}(-k_{iz}) \cdot \hat{e}_i] (1 + R_{ho}) \tag{1.2.28b}$$

where R_{ho} is the Fresnel reflection coefficient for the TE waves:

$$R_{ho} = \frac{k_{iz} - k_{1zi}}{k_{iz} + k_{1zi}} \tag{1.2.29}$$

Solving (1.2.27b) and (1.2.27d) gives

$$\overline{a}_p^{(0)}(\overline{k}_{i\perp}) = [\hat{h}(-k_{iz}) \cdot \hat{e}_i](1 + R_{vo}) \tag{1.2.30a}$$

$$\overline{b}_q^{(0)}(\overline{k}_{i\perp}) = -[\hat{h}(-k_{iz}) \cdot \hat{e}_i] \frac{k_{iz}}{k} (1 - R_{vo}) \tag{1.2.30b}$$

where R_{vo} is the Fresnel reflection coefficient for the TM waves:

$$R_{vo} = \frac{k_1^2 k_{iz} - k^2 k_{1zi}}{k_1^2 k_{iz} + k^2 k_{1zi}} \tag{1.2.31}$$

The zeroth-order surface fields are

$$\overline{a}^{(0)}(\overline{r}_\perp) = (\hat{q}_i a_q^{(0)} + \hat{p}_i a_p^{(0)}) e^{i\overline{k}_{i\perp} \cdot \overline{r}_\perp} \tag{1.2.32a}$$

$$\overline{b}^{(0)}(\overline{r}_\perp) = (\hat{q}_i b_q^{(0)} + \hat{p}_i b_p^{(0)}) e^{i\overline{k}_{i\perp} \cdot \overline{r}_\perp} \tag{1.2.32b}$$

Substitution into (1.2.13a) gives the zeroth order scattered field

$$\begin{aligned} \overline{E}_s^{(0)}(\overline{r}) = & -\frac{1}{8\pi^2} \int d\overline{k}_\perp e^{i\overline{k}_\perp \cdot \overline{r}_\perp + ik_z z} \frac{k}{k_z} \int d\overline{r}'_\perp e^{-i\overline{k}_\perp \cdot \overline{r}'_\perp} \\ & \cdot \left\{ \left[\hat{e}(k_z)\hat{e}(k_z) + \hat{h}(k_z)\hat{h}(k_z) \right] \cdot \left(\hat{q}_i a_q^{(0)} + \hat{p}_i a_p^{(0)} \right) e^{i\overline{k}_{i\perp} \cdot \overline{r}'_\perp} \right. \\ & \left. + \left[-\hat{h}(k_z)\hat{e}(k_z) + \hat{e}(k_z)\hat{h}(k_z) \right] \cdot \left(\hat{q}_i b_q^{(0)} + \hat{p}_i b_p^{(0)} \right) e^{i\overline{k}_{i\perp} \cdot \overline{r}'_\perp} \right\} \end{aligned} \tag{1.2.33}$$

Integrating over $d\overline{r}'_\perp$ gives $\delta(\overline{k}_\perp - \overline{k}_{i\perp})$. Simplifying we get

$$\overline{E}_s^{(0)}(\overline{r}) = -\frac{1}{2}\frac{k}{k_{iz}} e^{i\overline{k}_{i\perp}\cdot\overline{r}_\perp + ik_{iz}z} \cdot \left\{ \hat{e}(k_{iz})\left(a_q^{(0)} - \frac{k_{iz}}{k} b_p^{(0)} \right) - \hat{h}(k_{iz})\left(a_p^{(0)}\frac{k_{iz}}{k} + b_q^{(0)} \right) \right\} \quad (1.2.34)$$

Substitutions of (1.2.28*a*) and (1.2.28*b*) gives

$$\overline{E}_s^{(0)} = \left\{ R_{ho}[\hat{e}(-k_{iz})\cdot\hat{e}_i]\hat{e}(k_{iz}) + R_{vo}[\hat{h}(-k_{iz})\cdot\hat{e}_i]\hat{h}(k_{iz}) \right\} e^{i\overline{k}_{i\perp}\cdot\overline{r}_\perp + ik_{iz}z} \quad (1.2.35)$$

Similarly, for the transmitted field we have

$$\overline{E}_t^{(0)} = \left\{ (1+R_{ho})[\hat{e}(-k_{iz})\cdot\hat{e}_i]\hat{e}_1(-k_{1zi}) + \frac{k}{k_1}(1+R_{vo})[\hat{h}(-k_{iz})\cdot\hat{e}_i]\hat{h}_1(-k_{1zi}) \right\} e^{i\overline{k}_{i\perp}\cdot\overline{r}_\perp - ik_{1zi}z} \quad (1.2.36)$$

The zeroth order solution of (1.2.35) and (1.2.36) are just the reflected and transmitted fields for a flat surface.

First-Order Solution

We first calculate the first-order solution of the z-component of the surface fields $\overline{A}$ and $\overline{B}$ using (1.2.24*a*) and (1.2.26*b*). Let

$$A_z^{(1)}(\overline{k}_\perp) = iF(\overline{k}_\perp - \overline{k}_{i\perp})\widetilde{A}_z(\overline{k}_\perp) \quad (1.2.37a)$$

$$B_z^{(1)}(\overline{k}_\perp) = iF(\overline{k}_\perp - \overline{k}_{i\perp})\widetilde{B}_z(\overline{k}_\perp) \quad (1.2.37b)$$

Then

$$\begin{aligned}\widetilde{A}_z(\overline{k}_\perp) &= (\overline{k}_\perp - \overline{k}_{i\perp})\cdot\hat{q}_i a_q^{(0)} + (\overline{k}_\perp - \overline{k}_{i\perp})\cdot\hat{p}_i a_p^{(0)} \\ &= -k_\rho \sin(\phi_k - \phi_i) a_q^{(0)} + \Big(k_\rho\cos(\phi_k - \phi_i) - k_{\rho i}\Big) a_p^{(0)}\end{aligned} \quad (1.2.37c)$$

$$\begin{aligned}\widetilde{B}_z(\overline{k}_\perp) &= (\overline{k}_\perp - \overline{k}_{i\perp})\cdot\hat{q}_i b_q^{(0)} + (\overline{k}_\perp - \overline{k}_{i\perp})\cdot\hat{p}_i b_p^{(0)}(\overline{k}_{i\perp}) \\ &= -k_\rho \sin(\phi_k - \phi_i) b_q^{(0)} + \Big(k_\rho\cos(\phi_k - \phi_i) - k_{\rho i}\Big) b_p^{(0)}\end{aligned} \quad (1.2.37d)$$

Balancing (1.2.25*e*) to the first order, we have an equation for $\overline{A}_\perp^{(1)}(\overline{k}_\perp)$ and $\overline{B}_\perp^{(1)}(\overline{k}_\perp)$. We further let

$$\overline{A}_\perp^{(1)}(\overline{k}_\perp) = iF(\overline{k}_\perp - \overline{k}_{i\perp})\overline{\widetilde{A}}_\perp(\overline{k}_\perp) \quad (1.2.38a)$$

$$\overline{B}_\perp^{(1)}(\overline{k}_\perp) = iF(\overline{k}_\perp - \overline{k}_{i\perp})\overline{\widetilde{B}}_\perp(\overline{k}_\perp) \quad (1.2.38b)$$

$$\overline{\widetilde{A}}_\perp(\overline{k}_\perp) = \widetilde{A}_q(\overline{k}_\perp)\hat{q}(\overline{k}_\perp) + \widetilde{A}_p(\overline{k}_\perp)\hat{p}(\overline{k}_\perp) \quad (1.2.38c)$$

$$\overline{\widetilde{B}}_\perp(\overline{k}_\perp) = \widetilde{B}_q(\overline{k}_\perp)\hat{q}(\overline{k}_\perp) + \widetilde{B}_p(\overline{k}_\perp)\hat{p}(\overline{k}_\perp) \quad (1.2.38d)$$

The first-order equation of (1.2.25e) on using (1.2.38) is

$$
\begin{aligned}
& -\frac{1}{2}\frac{k}{k_z}\Big[\hat{e}(-k_z)\hat{e}(-k_z)+\hat{h}(-k_z)\hat{h}(-k_z)\Big] iF(\overline{k}_\perp-\overline{k}_{i\perp}) \\
& \cdot\Big[\widetilde{A}_q(\overline{k}_\perp)\hat{q}(\overline{k}_\perp)+\widetilde{A}_p(\overline{k}_\perp)\hat{p}(\overline{k}_\perp)\Big] \\
& -\frac{1}{2}\frac{k}{k_z}\Big[-\hat{h}(-k_z)\hat{e}(-k_z)+\hat{e}(-k_z)\hat{h}(-k_z)\Big] iF(\overline{k}_\perp-\overline{k}_{i\perp}) \\
& \cdot\Big[\widetilde{B}_q(\overline{k}_\perp)\hat{q}(\overline{k}_\perp)+\widetilde{B}_p(\overline{k}_\perp)\hat{p}(\overline{k}_\perp)\Big] \\
= & \frac{1}{2}\frac{k_\rho}{k_z}A_z^{(1)}(\overline{k}_\perp)\hat{h}(-k_z)+\frac{1}{2}\frac{k_\rho}{k_z}B_z^{(1)}(\overline{k}_\perp)\hat{e}(-k_z) \\
& +\frac{1}{2}\frac{k}{k_z}\Big[\hat{e}(-k_z)\hat{e}(-k_z)+\hat{h}(-k_z)\hat{h}(-k_z)\Big] \\
& \cdot ik_zF(\overline{k}_\perp-\overline{k}_{i\perp})\left(\hat{q}_i a_q^{(0)}+\hat{p}_i a_p^{(0)}\right) \\
& +\frac{1}{2}\frac{k}{k_z}\Big[-\hat{h}(-k_z)\hat{e}(-k_z)+\hat{e}(-k_z)\hat{h}(-k_z)\Big] \\
& \cdot ik_zF(\overline{k}_\perp-\overline{k}_{i\perp})\left(\hat{q}_i b_q^{(0)}+\hat{p}_i b_p^{(0)}\right)
\end{aligned}
\tag{1.2.39}
$$

Simplifying (1.2.39), we have

$$
\begin{aligned}
& -\frac{k}{k_z}\left[\hat{e}(-k_z)\widetilde{A}_q(\overline{k}_\perp)+\hat{h}(-k_z)\widetilde{A}_p(\overline{k}_\perp)\frac{k_z}{k}\right] \\
& -\frac{k}{k_z}\left[-\hat{h}(-k_z)\widetilde{B}_q(\overline{k}_\perp)+\hat{e}(-k_z)\widetilde{B}_p(\overline{k}_\perp)\frac{k_z}{k}\right] \\
= & \frac{k_\rho}{k_z}\widetilde{A}_z(\overline{k}_\perp)\hat{h}(-k_z)+\frac{k_\rho}{k_z}\widetilde{B}_z(\overline{k}_\perp)\hat{e}(-k_z) \\
& +k\Big[\hat{e}(-k_z)\hat{e}(-k_z)+\hat{h}(-k_z)\hat{h}(-k_z)\Big]\cdot\left(\hat{q}_i a_q^{(0)}+\hat{p}_i a_p^{(0)}\right) \\
& +k\Big[-\hat{h}(-k_z)\hat{e}(-k_z)+\hat{e}(-k_z)\hat{h}(-k_z)\Big]\cdot\left(\hat{q}_i b_q^{(0)}+\hat{p}_i b_p^{(0)}\right)
\end{aligned}
\tag{1.2.40}
$$

Matching the $\hat{e}(-k_z)$ and $\hat{h}(-k_z)$ components for both sides of (1.2.40) gives:

$$
\begin{aligned}
& -\frac{k}{k_z}\widetilde{A}_q(\overline{k}_\perp)-\widetilde{B}_p(\overline{k}_\perp) \\
& \quad =\frac{k_\rho}{k_z}\widetilde{B}_z(\overline{k}_\perp)+k\left[\left(\hat{e}(-k_z)\cdot\hat{q}_i\right)a_q^{(0)}+\left(\hat{e}(-k_z)\cdot\hat{p}_i\right)a_p^{(0)}\right] \\
& \quad\quad +k\left[\left(\hat{h}(-k_z)\cdot\hat{q}_i\right)b_q^{(0)}+\left(\hat{h}(-k_z)\cdot\hat{p}_i\right)b_p^{(0)}\right]
\end{aligned}
\tag{1.2.41a}
$$

$$-\widetilde{A}_p(\overline{k}_\perp)+\frac{k}{k_z}\widetilde{B}_q(\overline{k}_\perp)$$
$$=\frac{k_\rho}{k_z}\widetilde{A}_z(\overline{k}_\perp)+k\left[\left(\hat{h}(-k_z)\cdot\hat{q}_i\right)a_q^{(0)}+\left(\hat{h}(-k_z)\cdot\hat{p}_i\right)a_p^{(0)}\right]$$
$$+k\left[-\left(\hat{e}(-k_z)\cdot\hat{q}_i\right)b_q^{(0)}-\left(\hat{e}(-k_z)\cdot\hat{p}_i\right)b_p^{(0)}\right] \qquad (1.2.41b)$$

Similarly balancing (1.2.25f) to the first order, we have

$$-\frac{k}{k_{1z}}\widetilde{A}_q(\overline{k}_\perp)+\widetilde{B}_p(\overline{k}_\perp)$$
$$=\frac{k_\rho}{k_{1z}}\widetilde{B}_z(\overline{k}_\perp)-k\left[\left(\hat{e}_1(k_{1z})\cdot\hat{q}_i\right)a_q^{(0)}+\left(\hat{e}_1(k_{1z})\cdot\hat{p}_i\right)a_p^{(0)}\right]$$
$$-k_1\left[\left(\hat{h}_1(k_{1z})\cdot\hat{q}_i\right)b_q^{(0)}+\left(\hat{h}_1(k_{1z})\cdot\hat{p}_i\right)b_p^{(0)}\right] \qquad (1.2.41c)$$

$$\frac{k}{k_1}\widetilde{A}_p(\overline{k}_\perp)+\frac{k_1}{k_{1z}}\widetilde{B}_q(\overline{k}_\perp)$$
$$=\frac{kk_\rho}{k_1k_{1z}}\widetilde{A}_z(\overline{k}_\perp)-k\left[\left(\hat{h}_1(k_{1z})\cdot\hat{q}_i\right)a_q^{(0)}+\left(\hat{h}_1(k_{1z})\cdot\hat{p}_i\right)a_p^{(0)}\right]$$
$$+k_1\left[\left(\hat{e}_1(k_{1z})\cdot\hat{q}_i\right)b_q^{(0)}+\left(\hat{e}_1(k_{1z})\cdot\hat{p}_i\right)b_p^{(0)}\right] \qquad (1.2.41d)$$

Equations (1.2.41a)–(1.2.41d) provide four equations for the four unknowns $\widetilde{A}_q(\overline{k}_\perp)$, $\widetilde{B}_p(\overline{k}_\perp)$, $\widetilde{A}_p(\overline{k}_\perp)$, and $\widetilde{B}_q(\overline{k}_\perp)$. Also in (1.2.41$a$–$d$), we have

$$\hat{e}(-k_z)\cdot\hat{q}_i=\hat{e}_1(k_{1z})\cdot\hat{q}_i \qquad (1.2.42a)$$
$$\hat{e}(-k_z)\cdot\hat{p}_i=\hat{e}_1(k_{1z})\cdot\hat{p}_i \qquad (1.2.42b)$$
$$\hat{h}_1(k_{1z})\cdot\hat{q}_i=-\frac{k_{1z}}{k_1}\frac{k}{k_z}\hat{h}(-k_z)\cdot\hat{q}_i \qquad (1.2.42c)$$
$$\hat{h}_1(k_{1z})\cdot\hat{p}_i=-\frac{k_{1z}}{k_1}\frac{k}{k_z}\hat{h}(-k_z)\cdot\hat{p}_i \qquad (1.2.42d)$$

Adding (1.2.41a) and (1.2.41c), we found that the $a_q^{(0)}$, $a_p^{(0)}$ terms cancel each other. Thus,

$$-k\left(\frac{1}{k_z}+\frac{1}{k_{1z}}\right)\widetilde{A}_q(\overline{k}_\perp)$$
$$=k_\rho\left(\frac{1}{k_z}+\frac{1}{k_{1z}}\right)\widetilde{B}_z(\overline{k}_\perp)+b_q^{(0)}\left[k\left(h(-k_z)\cdot\hat{q}_i\right)-k_1\left(\hat{h}_1(k_{1z})\cdot\hat{q}_i\right)\right]$$
$$+b_p^{(0)}\left[k\left(h(-k_z)\cdot\hat{p}_i\right)-k_1\left(\hat{h}_1(k_{1z})\cdot\hat{p}_i\right)\right] \qquad (1.2.43)$$

Using (1.2.42e) and (1.2.42f) in (1.2.43), we get

$$\widetilde{A}_q(\overline{k}_\perp)=-\frac{k_\rho}{k}\widetilde{B}_z(\overline{k}_\perp)-k_{1z}b_q^{(0)}\left(\hat{h}(-k_z)\cdot\hat{q}_i\right)-k_{1z}b_p^{(0)}\left(\hat{h}(-k_z)\cdot\hat{p}_i\right) \quad (1.2.44)$$

On the other hand, multiply (1.2.41a) by $1/k_{1z}$ and (1.2.41c) by $1/k_z$ and subtract, we get

$$\widetilde{B}_p(\overline{k}_\perp) = -k\left[\left(\hat{e}(-k_z)\cdot\hat{q}_i\right)a_q^{(0)} + \left(\hat{e}(-k_z)\cdot\hat{p}_i\right)a_p^{(0)}\right] + \frac{k}{k_z(k_z+k_{1z})}(k_1^2-k^2)\left[\left(\hat{h}(-k_z)\cdot\hat{q}_i\right)b_q^{(0)} + \left(\hat{h}(-k_z)\cdot\hat{p}_i\right)b_p^{(0)}\right] \quad (1.2.45)$$

Similarly, by using (1.2.41b) and (1.2.41d) we obtain

$$\widetilde{A}_p(\overline{k}_\perp) = -\frac{k_\rho(k_1^2-k^2)}{k_1^2k_z+k^2k_{1z}}\widetilde{A}_z(\overline{k}_\perp) + \frac{kk_1^2(k_{1z}+k_z)}{k_1^2k_z+k^2k_{1z}} \cdot\left[\left(\hat{e}(-k_z)\cdot\hat{q}_i\right)b_q^{(0)} + \left(\hat{e}(-k_z)\cdot\hat{p}_i\right)b_p^{(0)}\right] - \frac{k}{k_z}\frac{(k_1^2k_z^2-k^2k_{1z}^2)}{k_1^2k_z+k^2k_{1z}} \cdot\left[\left(\hat{h}(-k_z)\cdot\hat{q}_i\right)a_q^{(0)} + \left(\hat{h}(-k_z)\cdot\hat{p}_i\right)a_p^{(0)}\right] \quad (1.2.46)$$

$$\widetilde{B}_q(\overline{k}_\perp) = \frac{1}{k_1^2k_z+k^2k_{1z}}kk_\rho\,(k_{1z}+k_z)\,\widetilde{A}_z(\overline{k}_\perp) + \frac{k_zk_{1z}}{k_1^2k_z+k^2k_{1z}}(k_1^2-k^2) \cdot\left[\left(\hat{e}(-k_z)\cdot\hat{q}_i\right)b_q^{(0)} + \left(\hat{e}(-k_z)\cdot\hat{p}_i\right)b_p^{(0)}\right] + \frac{k_zk_{1z}}{k_1^2k_z+k^2k_{1z}} \cdot k^2\left(1+\frac{k_{1z}}{k_z}\right)\left[\left(\hat{h}(-k_z)\cdot\hat{q}_i\right)a_q^{(0)} + \left(\hat{h}(-k_z)\cdot\hat{p}_i\right)a_p^{(0)}\right] \quad (1.2.47)$$

Next we examine the two cases of TE and TM incident polarizations. First let excitation be TE wave of $\hat{e}_i = \hat{e}(-k_{iz})$. Then $a_q^{(0)} = \frac{k_{iz}}{k}(1-R_{ho})$, $b_p^{(0)} = (1+R_{ho})$, $a_p^{(0)} = 0$, $b_q^{(0)} = 0$. Thus, from (1.2.45) we obtain

$$\begin{aligned}\widetilde{B}_p^{(e)}(\overline{k}_\perp) &= -k\left(\hat{e}(-k_z)\cdot\hat{q}_i\right)\frac{k_{iz}}{k}(1-R_{ho}) + \frac{k(k_1^2-k^2)}{k_z(k_z+k_{1z})}\left(\hat{h}(-k_z)\cdot\hat{p}_i\right)(1+R_{ho}) \\ &= \cos(\phi_k-\phi_i)\left[-k_{iz}(1-R_{ho}) + \frac{k_1^2-k^2}{k_z+k_{1z}}(1+R_{ho})\right] \end{aligned} \quad (1.2.48a)$$

Similarly, from (1.2.44) we have

$$\begin{aligned}\widetilde{A}_q^{(e)}(\overline{k}_\perp) &= -\frac{k_\rho}{k}(\overline{k}_\perp-\overline{k}_{i\perp})\cdot\hat{p}_i(1+R_{ho}) - k_{1z}\left(\hat{h}(-k_z)\cdot\hat{p}_i\right)(1+R_{ho}) \\ &= \frac{(1+R_{ho})}{k}\left\{-(k_\rho^2+k_{1z}k_z)\cos(\phi_k-\phi_i) + k_\rho k_{\rho i}\right\} \end{aligned} \quad (1.2.48b)$$

From (1.2.46) we obtain

$$\begin{aligned}\widetilde{A}_p^e(\overline{k}_\perp) = &-\frac{k_\rho(k_1^2-k^2)}{k_1^2k_z+k^2k_{1z}}\left[\Big((\overline{k}_\perp-\overline{k}_{i\perp})\cdot\hat{q}_i\Big)\frac{k_{iz}}{k}(1-R_{ho})\right]\\ &+\frac{kk_1^2(k_{1z}+k_z)}{k_1^2k_z+k^2k_{1z}}\Big(\hat{e}(-k_z)\cdot\hat{p}_i\Big)(1+R_{ho})\\ &-\frac{k}{k_z}\frac{(k_1^2k_z^2-k^2k_{1z}^2)}{k_1^2k_z+k^2k_{1z}}\Big(\hat{h}(-k_z)\cdot\hat{q}_i\Big)\frac{k_{iz}}{k}(1-R_{ho})\\ = &\frac{kk_1^2(k_{1z}+k_z)}{k_1^2k_z+k^2k_{1z}}\sin(\phi_k-\phi_i)(1+R_{ho})\end{aligned} \tag{1.2.48c}$$

From (1.2.47) we have

$$\begin{aligned}\widetilde{B}_q^e(\overline{k}_\perp) = &\frac{kk_\rho}{k_1^2k_z+k^2k_{1z}}(k_{1z}+k_z)\Big((\overline{k}_\perp-\overline{k}_{i\perp})\cdot\hat{q}_i\Big)\frac{k_{iz}}{k}(1-R_{ho})\\ &+\frac{k_zk_{1z}(k_1^2-k^2)}{k_1^2k_z+k^2k_{1z}}\Big(\hat{e}(-k_z)\cdot\hat{p}_i\Big)(1+R_{ho})\\ &+\frac{k_{1z}k^2}{k_1^2k_z+k^2k_{1z}}(k_z+k_{1z})\Big(\hat{h}(-k_z)\cdot\hat{q}_i\Big)\frac{k_{iz}}{k}(1-R_{ho})\\ = &\frac{\sin(\phi_k-\phi_i)}{k_1^2k_z+k^2k_{1z}}\{-(k_z+k_{1z})(1-R_{ho})k_{iz}\left[k_zk_{1z}+k_\rho^2\right]\\ &+k_zk_{1z}(k_1^2-k^2)(1+R_{ho})\}\\ = &\sin(\phi_k-\phi_i)\left\{-k_{iz}(1-R_{ho})+\frac{k_zk_{1z}(k_1^2-k^2)}{k_1^2k_z+k^2k_{1z}}(1+R_{ho})\right\}\end{aligned} \tag{1.2.48d}$$

In (1.2.48*d*) we have made use of the following relation

$$k_1^2k_z+k^2k_{1z}=(k_z+k_{1z})(k_\rho^2+k_zk_{1z})$$

Next we let incident polarization be that of a TM wave with $\hat{e}_i=\hat{h}(-k_{iz})$, $a_q^{(0)}=0$, $b_p^{(0)}=0$, $a_p^{(0)}=(1+R_{vo})$, and $b_q^{(0)}=-\frac{k_{iz}}{k}(1-R_{vo})$. Then from (1.2.44)–(1.2.47) we have

$$\begin{aligned}\widetilde{B}_p^{(h)} = &-k\Big(\hat{e}(-k_z)\cdot\hat{p}_i\Big)(1+R_{vo})\\ &+\frac{k(k_1^2-k^2)}{k_z(k_z+k_{1z})}\Big(\hat{h}(-k_z)\cdot\hat{q}_i\Big)\left(-\frac{k_{iz}}{k}\right)(1-R_{vo})\\ = &\sin(\phi_k-\phi_i)\left\{-k(1+R_{vo})+\frac{(k_1^2-k^2)}{k_{1z}+k_z}\frac{k_{iz}}{k}(1-R_{vo})\right\}\end{aligned} \tag{1.2.48e}$$

$$\begin{aligned}
\widetilde{A}_q^{(h)}(\overline{k}_\perp) = & -\frac{k_\rho}{k}\Big((\overline{k}_\perp - \overline{k}_{i\perp})\cdot\hat{q}_i\Big)\left(-\frac{k_{iz}}{k}\right)(1-R_{vo}) \\
& - k_{1z}\left(\hat{h}(-k_z)\cdot\hat{q}_i\right)\left(-\frac{k_{iz}}{k}\right)(1-R_{vo}) \\
= & -\sin(\phi_k-\phi_i)(1-R_{vo})\frac{k_{iz}}{k^2}(k_\rho^2 + k_z k_{1z}) \qquad (1.2.48f)
\end{aligned}$$

$$\begin{aligned}
\widetilde{A}_p^{(h)}(\overline{k}_\perp) = & -\frac{k_\rho(k_1^2-k^2)}{k_1^2 k_z + k^2 k_{1z}}\Big((\overline{k}_\perp - \overline{k}_{i\perp})\cdot\hat{p}_i\Big)(1+R_{vo}) \\
& + \frac{kk_1^2(k_{1z}+k_z)}{k_1^2 k_z + k^2 k_{1z}}\left(\hat{e}(-k_z)\cdot\hat{q}_i\right)\left(-\frac{k_{iz}}{k}(1-R_{vo})\right) \\
& - \frac{k}{k_z}\frac{(k_1^2 k_z^2 - k^2 k_{1z}^2)}{k_1^2 k_z + k^2 k_{1z}}\left(\hat{h}(-k_z)\cdot\hat{p}_i\right)(1+R_{vo}) \\
= & \frac{1}{k_1^2 k_z + k^2 k_{1z}}\Big\{(1+R_{vo})k_\rho k_{\rho i}(k_1^2-k^2) \\
& + kk_1^2(k_{1z}+k_z)\cos(\phi_k-\phi_i)\left(-\frac{k_{iz}}{k}\right)(1-R_{vo})\Big\} \qquad (1.2.48g)
\end{aligned}$$

$$\begin{aligned}
\widetilde{B}_q^{(h)}(\overline{k}_\perp) = & \frac{kk_\rho(k_{1z}+k_z)}{k_1^2 k_z + k^2 k_{1z}}(\overline{k}_\perp - \overline{k}_{i\perp})\cdot\hat{p}_i(1+R_{vo}) \\
& + \frac{k_z k_{1z}(k_1^2-k^2)}{k_1^2 k_z + k^2 k_{1z}}\left(\hat{e}(-k_z)\cdot\hat{q}_i\right)\left(-\frac{k_{iz}}{k}\right)(1-R_{vo}) \\
& + \frac{k_{1z}k^2(k_z+k_{1z})}{k_1^2 k_z + k^2 k_{1z}}\left(\hat{h}(-k_z)\cdot\hat{p}_i\right)(1+R_{vo}) \\
= & k(1+R_{vo})\cos(\phi_k-\phi_i) - \frac{k_z k_{1z}(k_1^2-k^2)}{k(k_1^2 k_z + k^2 k_{1z})}k_{iz} \\
& \cdot(1-R_{vo})\cos(\phi_k-\phi_i) - \frac{kk_\rho(k_{1z}+k_z)}{k_1^2 k_z + k^2 k_{1z}}k_{\rho i}(1+R_{vo}) \qquad (1.2.48h)
\end{aligned}$$

Solution in (1.2.19g), the first-order scattered field is

$$\begin{aligned}
\overline{E}_s^{(1)}(\overline{r}) = & -\frac{1}{2}\int d\overline{k}_\perp e^{i\overline{k}_\perp\cdot\overline{r}_\perp + ik_z z} iF(\overline{k}_\perp - \overline{k}_{i\perp})\Bigg\{\hat{e}(k_z)\Bigg[\widetilde{A}_q(\overline{k}_\perp)\frac{k}{k_z} \\
& - k\left((\hat{e}(k_z)\cdot\hat{q}_i)\, a_q^{(0)} + \hat{e}(k_z)\cdot\hat{p}_i a_p^{(0)}\right) + \frac{k_\rho}{k_z}\widetilde{B}_z(\overline{k}_\perp) \\
& - \widetilde{B}_p(\overline{k}_\perp) - k\left(\left(\hat{h}(k_z)\cdot\hat{q}_i\right) b_q^{(0)} + \left(\hat{h}(k_z)\cdot\hat{p}_i\right) b_p^{(0)}\right)\Bigg]
\end{aligned}$$

$$+ \hat{h}(k_z)\left[\frac{k_\rho}{k_z}\widetilde{A}_z(\overline{k}_\perp) - \widetilde{A}_p(\overline{k}_\perp) - k\left(\left(\hat{h}(k_z)\cdot\hat{q}_i\right)a_q^{(0)} + \left(\hat{h}(k_z)\cdot\hat{p}_i\right)a_p^{(0)}\right)\right.$$

$$\left.\left. - \widetilde{B}_q(\overline{k}_\perp)\frac{k}{k_z} + k\left((\hat{e}(k_z)\cdot\hat{q}_i)\, b_q^{(0)} + (\hat{e}(k_z)\cdot\hat{p}_i)\, b_p^{(0)}\right)\right]\right\} \tag{1.2.49}$$

Before we substitute in the expansion, the results can be simplified if we compare with (1.2.41*a*) and (1.2.41*b*). We can simplify more readily if we notice the similarity between the extinction equation and the scattered field equation. Note that $\hat{h}(k_z)\cdot\hat{q}_i = -\hat{h}(-k_z)\cdot\hat{q}_i$; $\hat{h}(k_z)\cdot\hat{p}_i = -\hat{h}(-k_z)\cdot\hat{p}_i$; $\hat{e}(k_z) = \hat{e}(-k_z)$. Hence we will get a factor of 2 in (1.2.49), and the first-order scattered field is

$$\overline{E}_s^{(1)}(\overline{r}) = \int d\overline{k}_\perp e^{i\overline{k}_\perp\cdot\overline{r}_\perp + ik_z z} iF(\overline{k}_\perp - \overline{k}_{i\perp})$$

$$\left\{\hat{e}(k_z)\left(\widetilde{B}_p(\overline{k}_\perp) + k\left((\hat{e}(k_z)\cdot\hat{q}_i)\, a_q^{(0)} + (\hat{e}(k_z)\cdot\hat{p}_i)\, a_p^{(0)}\right)\right)\right.$$

$$\left. + \hat{h}(k_z)\left(\widetilde{A}_p(\overline{k}_\perp) - k\left((\hat{e}(k_z)\cdot\hat{q}_i)\, b_q^{(0)} + (\hat{e}(k_z)\cdot\hat{p}_i)\, b_p^{(0)}\right)\right)\right\} \tag{1.2.50}$$

For incident field of TE polarization $\hat{e}_i = \hat{e}(-k_{iz})$, we let

$$f_{ee}^{(1)}(\overline{k}_\perp, \overline{k}_{i\perp}) = \widetilde{B}_p^e(\overline{k}_\perp) + k\left(\hat{e}(k_z)\cdot\hat{q}_i\right)a_q^{(0)} \tag{1.2.51}$$

Using (1.2.48*a*) and (1.2.28*a*) in (1.2.51), we have

$$f_{ee}^{(1)}(\overline{k}_\perp, \overline{k}_{i\perp}) = \frac{(k_1^2 - k^2)}{k_z + k_{1z}}\left(\frac{2k_{iz}}{k_{iz} + k_{1zi}}\right)\cos(\phi_k - \phi_i) \tag{1.2.52a}$$

Similarly,

$$f_{he}^{(1)}(\overline{k}_\perp, \overline{k}_{i\perp}) = \widetilde{A}_p^e(\overline{k}_\perp) - k\left(\hat{e}(\hat{k}_z)\cdot\hat{p}_i\right)b_p^{(0)}$$

$$= (k_1^2 - k^2)\frac{k_{1z}k}{k_1^2 k_z + k^2 k_{1z}}\frac{2k_{iz}}{k_{iz} + k_{1zi}}\sin(\phi_k - \phi_i) \tag{1.2.52b}$$

For an incident field of TM polarization $\hat{e}_i = \hat{h}(-k_{iz})$, we have

$$f_{eh}^{(1)}(\overline{k}_\perp, \overline{k}_{i\perp}) = \widetilde{B}_p^{(h)}(\overline{k}_\perp) + k\left(\hat{e}(k_z)\cdot\hat{p}_i\right)a_p^{(0)}$$

$$= \frac{(k_1^2 - k^2)}{(k_z + k_{1z})}k_{iz}\frac{2kk_{1zi}}{k^2 k_{1zi} + k_1^2 k_{iz}}\sin(\phi_k - \phi_i) \tag{1.2.52c}$$

$$f_{hh}^{(1)}(\overline{k}_\perp, \overline{k}_{i\perp}) = \widetilde{A}_p^{(h)}(\overline{k}_\perp) - k\left(\hat{e}(k_z)\cdot\hat{q}_i\right)b_q^{(0)}$$

$$= \frac{k_1^2 - k^2}{k_1^2 k_z + k^2 k_{1z}}\frac{2k^2 k_{iz}}{k_1^2 k_{iz} + k^2 k_{1zi}}\left\{-k_{1z}k_{1zi}\cos(\phi_k - \phi_i) + k_\rho k_{\rho i}\frac{k_1^2}{k^2}\right\} \tag{1.2.52d}$$

The simplifications of (1.2.52c) and (1.2.52d) are due to cancellation of some of the terms. Using (1.2.52a, b, c, d) in (1.2.50), we have

$$\overline{E}_s^{(1)}(\overline{r}) = \int d\overline{k}_\perp e^{i\overline{k}_\perp \cdot \overline{r}_\perp + ik_z z} iF(\overline{k}_\perp - \overline{k}_{i\perp}) \Big\{ \hat{e}(k_z) \Big[f_{ee}^{(1)}(\overline{k}_\perp, \overline{k}_{i\perp}) \Big(\hat{e}(-k_{iz}) \cdot \hat{e}_i \Big) + f_{eh}^{(1)}(\overline{k}_\perp, \overline{k}_{i\perp}) \Big(\hat{h}(-k_{iz}) \cdot \hat{e}_i \Big) \Big] + \hat{h}(k_z) \Big[f_{he}^{(1)}(\overline{k}_\perp, \overline{k}_{i\perp}) \Big(\hat{e}(-k_{iz}) \cdot \hat{e}_i \Big) + f_{hh}^{(1)}(\overline{k}_\perp, \overline{k}_{i\perp}) \Big(\hat{h}(-k_{iz}) \cdot \hat{e}_i \Big) \Big] \Big\} \tag{1.2.53}$$

We next find the power scattered using definitions as in Section 1.

$$\overline{S}_i \cdot \hat{z} = -\frac{\cos\theta_i}{2\eta} \tag{1.2.54a}$$

$$\langle \overline{S}_s \cdot \hat{z} \rangle = \frac{1}{2} \text{Re} \, \langle \overline{E}_s^{(1)} \times \overline{H}_s^{(1)*} \rangle \cdot \hat{z} \tag{1.2.54b}$$

Thus

$$\begin{aligned} \langle \overline{S}_s \cdot \hat{z} \rangle &= \frac{1}{2\eta} \text{Re} \int_{|\overline{k}_\perp| \le k} d\overline{k}_\perp \frac{k_z}{k} W(\overline{k}_\perp - \overline{k}_{i\perp}) \\ &\quad \Big\{ \Big| f_{ee}^{(1)}(\overline{k}_\perp, \overline{k}_{i\perp}) \Big(\hat{e}(-k_{iz}) \cdot \hat{e}_i \Big) + f_{eh}^{(1)}(\overline{k}_\perp, \overline{k}_{i\perp}) \Big(\hat{h}(-k_{iz}) \cdot \hat{e}_i \Big) \Big|^2 \\ &\quad + \Big| f_{he}^{(1)}(\overline{k}_\perp, \overline{k}_{i\perp}) \Big(\hat{e}(-k_{iz}) \cdot \hat{e}_i \Big) + f_{hh}^{(1)}(\overline{k}_\perp, \overline{k}_{i\perp}) \Big(\hat{h}(-k_{iz}) \cdot \hat{e}_i \Big) \Big|^2 \Big\} \\ &= \frac{\cos\theta_i}{2\eta} \int_0^{\pi/2} d\theta_s \sin\theta_s \int_0^{2\pi} d\phi_s \frac{\gamma(\hat{k}_s, \hat{k}_i)}{4\pi} \end{aligned} \tag{1.2.55}$$

where $|\overline{k}_\perp| \le k$ stands for propagating waves only. The bistatic scattering coefficients are

$$\begin{aligned} \gamma(\hat{k}_s, \hat{k}_i) &= \frac{4\pi k^2 \cos\theta_s}{\cos\theta_i} W(\overline{k}_\perp - \overline{k}_{i\perp}) \\ &\quad \cdot \Big\{ \Big| f_{ee}^{(1)}(\overline{k}_\perp, \overline{k}_{i\perp}) (\hat{e}(-k_{iz}) \cdot \hat{e}_i) + f_{eh}^{(1)}(\overline{k}_\perp, \overline{k}_{i\perp}) \Big(\hat{h}(-k_z) \cdot \hat{e}_i \Big) \Big|^2 \Big\} \\ &\qquad \text{for scattered wave in } \hat{e}(k_z) \text{ polarization} \end{aligned} \tag{1.2.56a}$$

$$\begin{aligned} &= \frac{4\pi k^2 \cos\theta_s}{\cos\theta_i} W(\overline{k}_\perp - \overline{k}_{i\perp}) \\ &\quad \cdot \Big\{ \Big| f_{he}^{(1)}(\overline{k}_\perp, \overline{k}_{i\perp}) (\hat{e}(-k_{iz}) \cdot \hat{e}_i) + f_{hh}^{(1)}(\overline{k}_\perp, \overline{k}_{i\perp}) \Big(\hat{h}(-k_z) \cdot \hat{e}_i \Big) \Big|^2 \Big\} \\ &\qquad \text{for scattered wave in } \hat{h}(k_z) \text{ polarization} \end{aligned} \tag{1.2.56b}$$

2.2 Second-Order Solutions

Next, we derive the second-order solutions. Balancing (1.2.19e) to the second order, we have

$$
\begin{aligned}
0 = &\left[\hat{e}(-k_z)A_q^{(2)}(\overline{k}_\perp) + \hat{h}(-k_z)\left(\frac{k_\rho}{k}A_z^{(2)}(\overline{k}_\perp) + \frac{k_z}{k}A_p^{(2)}(\overline{k}_\perp)\right)\right] \\
&+ ik_z\left[\hat{e}(-k_z)\hat{e}(-k_z) + \hat{h}(-k_z)\hat{h}(-k_z)\right]\cdot\int d\overline{k}'_\perp \overline{A}^{(1)}(\overline{k}'_\perp)F(\overline{k}_\perp - \overline{k}'_\perp) \\
&- \frac{k_z^2}{2}\left[\hat{e}(-k_z)\hat{e}(-k_z) + \hat{h}(-k_z)\hat{h}(-k_z)\right]\cdot\int d\overline{k}'_\perp \overline{A}^{(0)}(\overline{k}'_\perp)F^{(2)}(\overline{k}_\perp - \overline{k}'_\perp) \\
&+ \left[-\hat{h}(-k_z)B_q^{(2)}(\overline{k}_\perp) + \hat{e}(-k_z)\left(\frac{k_\rho}{k}B_z^{(2)}(\overline{k}_\perp) + \frac{k_z}{k}B_p^{(2)}(\overline{k}_\perp)\right)\right] \\
&+ ik_z\left[-\hat{h}(-k_z)\hat{e}(-k_z) + \hat{e}(-k_z)\hat{h}(-k_z)\right]\cdot\int d\overline{k}'_\perp \overline{B}^{(1)}(\overline{k}'_\perp)F(\overline{k}_\perp - \overline{k}'_\perp) \\
&- \frac{k_z^2}{2}\left[-\hat{h}(-k_z)\hat{e}(-k_z) + \hat{e}(-k_z)\hat{h}(-k_z)\right]\cdot\int d\overline{k}'_\perp \overline{B}^{(0)}(\overline{k}'_\perp)F^{(2)}(\overline{k}_\perp - \overline{k}'_\perp)
\end{aligned}
\tag{1.2.57}
$$

Balancing coefficients of $\hat{e}(-k_z)$, we obtain

$$
\begin{aligned}
-A_q^{(2)}(\overline{k}_\perp) - &\frac{k_z}{k}B_p^{(2)}(\overline{k}_\perp) \\
= &\frac{k_\rho}{k}B_z^{(2)}(\overline{k}_\perp) + ik_z\int d\overline{k}'_\perp \left(\hat{e}(-k_z)\cdot\overline{A}_\perp^{(1)}(\overline{k}'_\perp)\right)F(\overline{k}_\perp - \overline{k}'_\perp) \\
&- \frac{k_z^2}{2}\int d\overline{k}'_\perp \left(\hat{e}(-k_z)\cdot\overline{A}^{(0)}(\overline{k}'_\perp)\right)F^{(2)}(\overline{k}_\perp - \overline{k}'_\perp) \\
&+ ik_z\int d\overline{k}'_\perp \left(\hat{h}(-k_z)\cdot\overline{B}_\perp^{(1)}(\overline{k}'_\perp)\right)F(\overline{k}_\perp - \overline{k}'_\perp) \\
&+ ik_z\int d\overline{k}'_\perp \left(\hat{h}(-k_z)\cdot\hat{z}\overline{B}_z^{(1)}(\overline{k}'_\perp)\right)F(\overline{k}_\perp - \overline{k}'_\perp) \\
&- \frac{k_z^2}{2}\int d\overline{k}'_\perp \left(\hat{h}(-k_z)\cdot\overline{B}_\perp^{(0)}(\overline{k}'_\perp)\right)F^{(2)}(\overline{k}_\perp - \overline{k}'_\perp)
\end{aligned}
\tag{1.2.58a}
$$

Balancing coefficients of $\hat{h}(-k_z)$, we have

$$
\begin{aligned}
-\frac{k_z}{k}A_p^{(2)}(\overline{k}_\perp) + &B_q^{(2)}(\overline{k}_\perp) \\
= &\frac{k_\rho}{k}A_z^{(2)}(\overline{k}_\perp) + ik_z\int d\overline{k}'_\perp \hat{h}(-k_z)\cdot\overline{A}_\perp^{(1)}(\overline{k}'_\perp)F(\overline{k}_\perp - \overline{k}'_\perp) \\
&+ ik_z\int d\overline{k}'_\perp \hat{h}(-k_z)\cdot\hat{z}A_z^{(1)}(\overline{k}'_\perp)F(\overline{k}_\perp - \overline{k}'_\perp)
\end{aligned}
$$

$$
\begin{aligned}
&- \frac{k_z^2}{2} \int d\overline{k}'_\perp \hat{h}(-k_z) \cdot \overline{A}_\perp^{(0)}(\overline{k}'_\perp) F^{(2)}(\overline{k}_\perp - \overline{k}'_\perp) \\
&- ik_z \int d\overline{k}'_\perp \hat{e}(-k_z) \cdot \overline{B}_\perp^{(1)}(\overline{k}'_\perp) F(\overline{k}_\perp - \overline{k}'_\perp) \\
&+ \frac{k_z^2}{2} \int d\overline{k}'_\perp \hat{e}(-k_z) \cdot \overline{B}_\perp^{(0)}(\overline{k}'_\perp) F^{(2)}(\overline{k}_\perp - \overline{k}'_\perp) \qquad (1.2.58b)
\end{aligned}
$$

Balancing (1.2.19f) to the second order and using (1.2.24e) and (1.2.24b), we obtain

$$
\begin{aligned}
0 = &\frac{k}{k_1} \left\{ \hat{e}_1(k_{1z}) A_q^{(2)}(\overline{k}_\perp) + \hat{h}_1(k_{1z}) \left(-\frac{k_{1z}}{k_1} A_p^{(2)}(\overline{k}_\perp) + \frac{k_\rho}{k_1} A_z^{(2)}(\overline{k}_\perp) \right) \right\} \\
&- \frac{k}{k_1} ik_{1z} \int d\overline{k}'_\perp \left(\hat{e}_1(k_{1z})\hat{e}_1(k_{1z}) + \hat{h}_1(k_{1z})\hat{h}_1(k_{1z}) \right) \cdot \overline{A}^{(1)}(\overline{k}'_\perp) F(\overline{k}_\perp - \overline{k}'_\perp) \\
&- \frac{k}{k_1} \frac{k_{1z}^2}{2} \int d\overline{k}'_\perp \left(\hat{e}_1(k_{1z})\hat{e}_1(k_{1z}) + \hat{h}_1(k_{1z})\hat{h}_1(k_{1z}) \right) \cdot \overline{A}^{(0)}(\overline{k}'_\perp) F^{(2)}(\overline{k}_\perp - \overline{k}'_\perp) \\
&- \hat{h}_1(k_{1z}) \overline{B}_q^{(2)}(\overline{k}_\perp) + \hat{e}_1(k_{1z}) \left(-\frac{k_{1z}}{k_1} B_p^{(2)}(\overline{k}_\perp) + \frac{k_\rho}{k_1} B_z^{(2)}(\overline{k}_\perp) \right) \\
&- ik_{1z} \int d\overline{k}'_\perp \left(-\hat{h}_1(k_{1z})\hat{e}_1(k_{1z}) + \hat{e}_1(k_{1z})\hat{h}_1(k_{1z}) \right) \cdot \overline{B}^{(1)}(\overline{k}'_\perp) F(\overline{k}_\perp - \overline{k}'_\perp) \\
&- \frac{k_{1z}^2}{2} \int d\overline{k}'_\perp \left(-\hat{h}_1(k_{1z})\hat{e}_1(k_{1z}) + \hat{e}_1(k_{1z})\hat{h}_1(k_{1z}) \right) \overline{B}^{(0)}(\overline{k}'_\perp) F^{(2)}(\overline{k}_\perp - \overline{k}'_\perp)
\end{aligned}
\qquad (1.2.59)
$$

Balancing coefficients of $\hat{e}_1(k_{1z})$ yields

$$
\begin{aligned}
&-A_q^{(2)}(\overline{k}_\perp) \frac{k}{k_1} + \frac{k_{1z}}{k_1} B_p^{(2)}(\overline{k}_\perp) \\
&= \frac{k_\rho}{k_1} B_z^{(2)}(\overline{k}_\perp) - ik_{1z} \frac{k}{k_1} \int d\overline{k}'_\perp \left(\hat{e}_1(k_{1z}) \cdot \overline{A}_\perp^{(1)}(\overline{k}'_\perp) \right) F(\overline{k}_\perp - \overline{k}'_\perp) \\
&\quad - \frac{k_{1z}^2}{2} \frac{k}{k_1} \int d\overline{k}'_\perp \left(\hat{e}_1(k_{1z}) \cdot \overline{A}^{(0)}(\overline{k}'_\perp) \right) F^{(2)}(\overline{k}_\perp - \overline{k}'_\perp) \\
&\quad - ik_{1z} \int d\overline{k}'_\perp \left(\hat{h}_1(k_{1z}) \cdot \overline{B}_\perp^{(1)}(\overline{k}'_\perp) + \hat{h}_1(k_{1z}) \cdot \hat{z} B_z^{(1)}(\overline{k}'_\perp) \right) F(\overline{k}_\perp - \overline{k}'_\perp) \\
&\quad - \frac{k_{1z}^2}{2} \int d\overline{k}'_\perp \left(\hat{h}_1(k_{1z}) \cdot \overline{B}^{(0)}(\overline{k}'_\perp) \right) F^{(2)}(\overline{k}_\perp - \overline{k}'_\perp) \qquad (1.2.60a)
\end{aligned}
$$

Balancing coefficient of $\hat{h}_1(k_{1z})$ gives

$$\frac{k_{1z}k}{k_1^2}A_p^{(2)}(\overline{k}_\perp)+B_q^{(2)}(\overline{k}_\perp)$$
$$=\frac{k_\rho k}{k_1^2}A_z^{(2)}(\overline{k}_\perp)-ik_{1z}\frac{k}{k_1}$$
$$\cdot\int d\overline{k}'_\perp\left(\hat{h}_1(k_{1z})\cdot\overline{A}_\perp^{(1)}(\overline{k}'_\perp)+\hat{h}_1(k_{1z})\cdot\hat{z}A_z^{(1)}(\overline{k}'_\perp)\right)F(\overline{k}_\perp-\overline{k}'_\perp)$$
$$-\frac{k_{1z}^2}{2}\frac{k}{k_1}\int d\overline{k}'_\perp\left(\hat{h}_1(k_{1z})\cdot\overline{A}^{(0)}(\overline{k}'_\perp)\right)F^{(2)}(\overline{k}_\perp-\overline{k}'_\perp)$$
$$+ik_{1z}\int d\overline{k}'_\perp\left(\hat{e}_1(k_{1z})\cdot\overline{B}^{(1)}(\overline{k}'_\perp)\right)F(\overline{k}_\perp-\overline{k}'_\perp)$$
$$+\frac{k_{1z}^2}{2}\int d\overline{k}'_\perp\left(\hat{e}_1(k_{1z})\cdot\overline{B}^{(0)}(\overline{k}'_\perp)\right)F^{(2)}(\overline{k}_\perp-\overline{k}'_\perp)\qquad(1.2.60b)$$

For the scattered wave, the second-order scattered field is, from (1.2.19g) and (1.2.24c) and (1.2.24d),

$$\overline{E}_s(\overline{r})=-\frac{1}{2}\int d\overline{k}_\perp e^{i\overline{k}_\perp\cdot\overline{r}_\perp}e^{ik_zz}\frac{k}{k_z}\cdot\left\{\hat{e}(k_z)A_q^{(2)}(\overline{k}_\perp)\right.$$
$$+\hat{h}(k_z)\left[\frac{k_\rho}{k}A_z^{(2)}(\overline{k}_\perp)-\frac{k_z}{k}A_p^{(2)}(\overline{k}_\perp)\right]$$
$$-\hat{e}(k_z)ik_z\int d\overline{k}'_\perp\left(\hat{e}(k_z)\cdot\overline{A}^{(1)}(\overline{k}'_\perp)\right)F(\overline{k}_\perp-\overline{k}'_\perp)$$
$$-\hat{h}(k_z)ik_z\int d\overline{k}'_\perp\left(\hat{h}(k_z)\cdot\overline{A}^{(1)}(\overline{k}'_\perp)\right)F(\overline{k}_\perp-\overline{k}'_\perp)$$
$$-\hat{e}(k_z)\frac{k_z^2}{2}\int d\overline{k}'_\perp\left(\hat{e}(k_z)\cdot\overline{A}^{(0)}(\overline{k}'_\perp)\right)F^{(2)}(\overline{k}_\perp-\overline{k}'_\perp)$$
$$-\hat{h}(k_z)\frac{k_z^2}{2}\int d\overline{k}'_\perp\left(\hat{h}(k_z)\cdot\overline{A}^{(0)}(\overline{k}'_\perp)\right)F^{(2)}(\overline{k}_\perp-\overline{k}'_\perp)$$
$$-\hat{h}(k_z)B_q^{(2)}(\overline{k}_\perp)+\hat{e}(k_z)\left(\frac{k_\rho}{k}B_z^{(2)}(\overline{k}_\perp)-\frac{k_z}{k}B_p^{(2)}(\overline{k}_\perp)\right)$$
$$+\hat{h}(k_z)ik_z\int d\overline{k}'_\perp\left(\hat{e}(k_z)\cdot\overline{B}^{(1)}(\overline{k}'_\perp)\right)F(\overline{k}_\perp-\overline{k}'_\perp)$$
$$-\hat{e}(k_z)ik_z\int d\overline{k}'_\perp\left(\hat{h}(k_z)\cdot\overline{B}^{(1)}(\overline{k}'_\perp)\right)F(\overline{k}_\perp-\overline{k}'_\perp)$$
$$+\hat{h}(k_z)\frac{k_z^2}{2}\int d\overline{k}'_\perp\left(\hat{e}(k_z)\cdot\overline{B}^{(0)}(\overline{k}'_\perp)\right)F^{(2)}(\overline{k}_\perp-\overline{k}'_\perp)$$
$$\left.-\hat{e}(k_z)\frac{k_z^2}{2}\int d\overline{k}'_\perp\left(\hat{h}(k_z)\cdot\overline{B}^{(0)}(\overline{k}'_\perp)\right)F^{(2)}(\overline{k}_\perp-\overline{k}'_\perp)\right\}\qquad(1.2.61)$$

Next, we put (1.2.61) in a form that resembles the extinction equation of (1.2.57).

$$\begin{aligned}
\overline{E}_s(\overline{r}) = & -\frac{1}{2}\int d\overline{k}_\perp e^{i\overline{k}_\perp\cdot\overline{r}_\perp+ik_z z}\frac{k}{k_z}\cdot\Bigg\{\hat{e}(k_z)\Bigg\{A_q^{(2)}(\overline{k}_\perp)+\frac{k_\rho}{k}B_z^{(2)}(\overline{k}_\perp) \\
& -\frac{k_z}{k}B_p^{(2)}(\overline{k}_\perp)-ik_z\int d\overline{k}'_\perp\left(\hat{e}(-k_z)\cdot\overline{A}^{(1)}(\overline{k}'_\perp)\right)F(\overline{k}_\perp-\overline{k}'_\perp) \\
& -\frac{k_z^2}{2}\int d\overline{k}'_\perp\left(\hat{e}(-k_z)\cdot\overline{A}^{(0)}(\overline{k}'_\perp)\right)F^{(2)}(\overline{k}_\perp-\overline{k}'_\perp) \\
& -ik_z\int d\overline{k}'_\perp\left[-\hat{h}(-k_z)\cdot\overline{B}_\perp^{(1)}(\overline{k}'_\perp)+\hat{h}(-k_z)\cdot\hat{z}B_z^{(1)}(\overline{k}'_\perp)\right]F(\overline{k}_\perp-\overline{k}'_\perp) \\
& -\frac{k_z^2}{2}\int d\overline{k}'_\perp\left[-\hat{h}(-k_z)\cdot\overline{B}_\perp^{(0)}(\overline{k}'_\perp)+F^{(2)}(\overline{k}_\perp-\overline{k}'_\perp)\right]\Bigg\} \\
& +\hat{h}(k_z)\Bigg\{\frac{k_\rho}{k}A_z^{(2)}(\overline{k}_\perp)-\frac{k_z}{k}A_p^{(2)}(\overline{k}_\perp)-B_q^{(2)}(\overline{k}_\perp) \\
& -ik_z\int d\overline{k}'_\perp\left[-\hat{h}(-k_z)\cdot\overline{A}_\perp^{(1)}(\overline{k}'_\perp)+\hat{h}(-k_z)\cdot\hat{z}A_z^{(1)}(\overline{k}'_\perp)\right]F(\overline{k}_\perp-\overline{k}'_\perp) \\
& -\frac{k_z^2}{2}\int d\overline{k}'_\perp\left[-\hat{h}(-k_z)\cdot\overline{A}_\perp^{(0)}(\overline{k}'_\perp)\right]F^{(2)}(\overline{k}_\perp-\overline{k}'_\perp) \\
& +ik_z\int d\overline{k}'_\perp\left[\hat{e}(-k_z)\cdot\overline{B}^{(1)}(\overline{k}'_\perp)\right]F(\overline{k}_\perp-\overline{k}'_\perp) \\
& +\frac{k_z^2}{2}\int d\overline{k}'_\perp\left[\hat{e}(-k_z)\cdot\overline{B}_\perp^{(0)}(\overline{k}'_\perp)\right]F^{(2)}(\overline{k}_\perp-\overline{k}'_\perp)\Bigg\}\Bigg\}
\end{aligned} \tag{1.2.62}$$

Using (1.2.58*a*) and (1.2.58*b*) in (1.2.62), we get a factor of 2. Thus

$$\begin{aligned}
\overline{E}_s(\overline{r}) = & -\int d\overline{k}_\perp e^{i\overline{k}_\perp\cdot\overline{r}_\perp+ik_z z}\frac{k}{k_z}\Bigg\{\hat{e}(k_z)\Bigg[A_q^{(2)}(\overline{k}_\perp)+\frac{k_\rho}{k}B_z^{(2)}(\overline{k}_\perp) \\
& -\frac{k_z^2}{2}\int d\overline{k}'_\perp\left(\hat{e}(-k_z)\cdot\overline{A}_\perp^{(0)}(\overline{k}'_\perp)\right)F^{(2)}(\overline{k}_\perp-\overline{k}'_\perp) \\
& +ik_z\int d\overline{k}'_\perp\hat{h}(-k_z)\cdot\overline{B}_\perp^{(1)}(\overline{k}'_\perp)F(\overline{k}_\perp-\overline{k}'_\perp)\Bigg] \\
& +\hat{h}(k_z)\Bigg[\frac{k_\rho}{k}A_z^{(2)}(\overline{k}_\perp)-B_q^{(2)}(\overline{k}_\perp) \\
& +ik_z\int d\overline{k}'_\perp\hat{h}(-k_z)\cdot\overline{A}_\perp^{(1)}(\overline{k}'_\perp)F(\overline{k}_\perp-\overline{k}'_\perp) \\
& +\frac{k_z^2}{2}\int d\overline{k}'_\perp\hat{e}(-k_z)\cdot\overline{B}_\perp^{(0)}(\overline{k}'_\perp)F^{(2)}(\overline{k}_\perp-\overline{k}'_\perp)\Bigg]\Bigg\}
\end{aligned} \tag{1.2.63}$$

We next eliminate $B_p^{(2)}$ from (1.2.58a) and (1.2.60a) :

$$-A_q^{(2)}(\overline{k}_\perp)\frac{1}{k_1}(k_z+k_{1z}) = \frac{k_\rho}{kk_1}(k_z+k_{1z})B_z^{(2)}(\overline{k}_\perp)$$
$$-\frac{k_{1z}k_z}{2k_1}(k_{1z}+k_z)\int d\overline{k}'_\perp\left(\hat{e}(-k_z)\cdot\overline{A}_\perp^{(0)}(\overline{k}'_\perp)\right)F^{(2)}(\overline{k}_\perp-\overline{k}'_\perp)$$
$$+\frac{ik_{1z}}{k_1}(k_z+k_{1z})\int d\overline{k}'_\perp\left(\hat{h}(-k_z)\cdot\overline{B}_\perp^{(1)}(\overline{k}'_\perp)\right)F(\overline{k}_\perp-\overline{k}'_\perp)$$
$$+\frac{k_{1z}}{2k_1}(k_{1z}^2-k_z^2)\int d\overline{k}'_\perp\left(\hat{h}(-k_z)\cdot\overline{B}_\perp^{(0)}(\overline{k}'_\perp)\right)F^{(2)}(\overline{k}_\perp-\overline{k}'_\perp) \quad (1.2.64)$$

Thus

$$A_q^{(2)}(\overline{k}_\perp) = -\frac{k_\rho}{k}B_z^{(2)}(\overline{k}_\perp)+\frac{k_{1z}k_z}{2}\int d\overline{k}'_\perp\left(\hat{e}(-k_z)\cdot\overline{A}_\perp^{(0)}(\overline{k}'_\perp)\right)F^{(2)}(\overline{k}_\perp-\overline{k}'_\perp)$$
$$-ik_{1z}\int d\overline{k}'_\perp\left(\hat{h}(-k_z)\cdot\overline{B}_\perp^{(1)}(\overline{k}'_\perp)\right)F(\overline{k}_\perp-\overline{k}'_\perp)$$
$$-\frac{k_{1z}}{2}(k_{1z}-k_z)\int d\overline{k}'_\perp\left(\hat{h}(-k_z)\cdot\overline{B}_\perp^{(0)}(\overline{k}'_\perp)\right)F^{(2)}(\overline{k}_\perp-\overline{k}'_\perp) \quad (1.2.65)$$

Similarly, using (1.2.58b) and (1.2.60b)

$$B_q^{(2)}(\overline{k}_\perp) = k_\rho k A_z^{(2)}(\overline{k}_\perp)\frac{k_{1z}+k_z}{k_1^2k_z+k^2k_{1z}}$$
$$+ik_{1z}k^2\frac{k_z+k_{1z}}{k_1^2k_z+k^2k_{1z}}\int d\overline{k}'_\perp\left(\hat{h}(-k_z)\cdot\overline{A}_\perp^{(1)}(\overline{k}'_\perp)\right)F(\overline{k}_\perp-\overline{k}'_\perp)$$
$$+\frac{k^2k_{1z}}{2}\frac{(k_1^2-k^2)}{k_1^2k_z+k^2k_{1z}}\int d\overline{k}'_\perp\left(\hat{h}(-k_z)\cdot\overline{A}_\perp^{(0)}(\overline{k}'_\perp)\right)F^{(2)}(\overline{k}_\perp-\overline{k}'_\perp)$$
$$+ik_{1z}k_z\frac{(k_1^2-k^2)}{k_1^2k_z+k^2k_{1z}}\int d\overline{k}'_\perp\left(\hat{e}(-k_z)\cdot\overline{B}_\perp^{(1)}(\overline{k}'_\perp)\right)F(\overline{k}_\perp-\overline{k}'_\perp)$$
$$+\frac{k_zk_{1z}}{2}\frac{(k_1^2k_{1z}+k_zk^2)}{k_1^2k_z+k^2k_{1z}}\int d\overline{k}'_\perp\left(\hat{e}(-k_z)\cdot\overline{B}_\perp^{(0)}(\overline{k}'_\perp)\right)F^{(2)}(\overline{k}_\perp-\overline{k}'_\perp) \quad (1.2.66)$$

Thus, substituting (1.2.65) and (1.2.66) into (1.2.63), we have

$$\overline{E}_s^{(2)}(\overline{r}) = -\int d\overline{k}_\perp e^{i\overline{k}_\perp\cdot\overline{r}_\perp+ik_zz}\frac{k}{k_z}$$
$$\cdot\left\{\hat{e}(k_z)\left[\frac{k_z}{2}(k_{1z}-k_z)\int d\overline{k}'_\perp\left(\hat{e}(-k_z)\cdot\overline{A}_\perp^{(0)}(\overline{k}'_\perp)\right)F^{(2)}(\overline{k}_\perp-\overline{k}'_\perp)\right.\right.$$
$$-i(k_{1z}-k_z)\int d\overline{k}'_\perp\left(\hat{h}(-k_z)\cdot\overline{B}_\perp^{(1)}(\overline{k}'_\perp)\right)F(\overline{k}_\perp-\overline{k}'_\perp)$$
$$\left.-\frac{k_{1z}}{2}(k_{1z}-k_z)\int d\overline{k}'_\perp\left(\hat{h}(-k_z)\cdot\overline{B}_\perp^{(0)}(\overline{k}'_\perp)\right)F^{(2)}(\overline{k}_\perp-\overline{k}'_\perp)\right]$$

$$
\begin{aligned}
&+ \hat{h}(k_z)\left[\frac{k_\rho}{k} A_z^{(2)}(\overline{k}_\perp)\frac{k_z(k_1^2-k^2)}{k_1^2k_z+k^2k_{1z}}\right. \\
&+ \frac{i}{k_1^2k_z+k^2k_{1z}} k_\rho^2(k^2-k_1^2)\int d\overline{k}'_\perp \left(\hat{h}(-k_z)\cdot\overline{A}_\perp^{(1)}(\overline{k}'_\perp)\right) F(\overline{k}_\perp-\overline{k}'_\perp) \\
&+ \frac{k_zk_1^2(k^2-k_1^2)}{2(k_1^2k_z+k^2k_{1z})}\int d\overline{k}'_\perp \left(\hat{e}(-k_z)\cdot\overline{B}_\perp^{(0)}(\overline{k}'_\perp)\right) F^{(2)}(\overline{k}_\perp-\overline{k}'_\perp) \\
&- \frac{k^2k_{1z}}{2}\frac{(k_1^2-k^2)}{k_1^2k_z+k^2k_{1z}}\int d\overline{k}'_\perp \left(\hat{h}(-k_z)\cdot\overline{A}_\perp^{(0)}(\overline{k}'_\perp)\right) F^{(2)}(\overline{k}_\perp-\overline{k}'_\perp) \\
&\left.\left. - ik_{1z}k_z\frac{(k_1^2-k^2)}{k_1^2k_z+k^2k_{1z}}\int d\overline{k}'_\perp \left(\hat{e}(-k_z)\cdot\overline{B}_\perp^{(1)}(\overline{k}'_\perp)\right) F(\overline{k}_\perp-\overline{k}'_\perp)\right]\right\} \qquad (1.2.67)
\end{aligned}
$$

Also note that

$$
\begin{aligned}
A_z^{(2)}(\overline{k}_\perp) &= \int d\overline{k}'_\perp F(\overline{k}_\perp-\overline{k}'_\perp) i(\overline{k}_\perp-\overline{k}'_\perp)\cdot\overline{A}_\perp^{(1)}(\overline{k}'_\perp) \\
&= -\int d\overline{k}'_\perp F(\overline{k}_\perp-\overline{k}'_\perp)F(\overline{k}'_\perp-\overline{k}_{i\perp}) \\
&\quad \cdot\left[(\overline{k}_\perp-\overline{k}'_\perp)\cdot\left(\widetilde{A}_q(\overline{k}'_\perp)\hat{q}(\overline{k}'_\perp)+\widetilde{A}_p(\overline{k}'_\perp)\hat{p}(\overline{k}'_\perp)\right)\right] \qquad (1.2.68)
\end{aligned}
$$

Substituting the zeroth- and the first-order solutions in (1.2.67) and using (1.2.68), we have

$$
\begin{aligned}
&\overline{E}_s^{(2)}(\overline{r}) = -\int d\overline{k}_\perp e^{i\overline{k}_\perp\cdot\overline{r}_\perp+ik_zz}\frac{k}{k_z} \\
&\cdot\left\{\hat{e}(k_z)\left\{\frac{k_z}{2}(k_{1z}-k_z)\left[\left(\hat{e}(-k_z)\cdot\hat{q}_i\right)a_q^{(0)}+\left(\hat{e}(-k_z)\cdot\hat{p}_i\right)a_p^{(0)}\right]F^{(2)}(\overline{k}_\perp-\overline{k}_{i\perp})\right.\right. \\
&+ (k_{1z}-k_z)\int d\overline{k}'_\perp\left[\left(\hat{h}(-k_z)\cdot\hat{q}(\overline{k}'_\perp)\right)\widetilde{B}_q(\overline{k}'_\perp)+\left(\hat{h}(-k_z)\cdot\hat{p}(\overline{k}'_\perp)\right)\widetilde{B}_p(\overline{k}'_\perp)\right] \\
&\qquad \cdot F(\overline{k}'_\perp-\overline{k}_{i\perp})F(\overline{k}_\perp-\overline{k}'_\perp) \\
&\left. - \frac{k_{1z}}{2}(k_{1z}-k_z)\left[\left(\hat{h}(-k_z)\cdot\hat{q}_i\right)b_q^{(0)}+\left(\hat{h}(-k_z)\cdot\hat{p}_i\right)b_p^{(0)}\right]F^{(2)}(\overline{k}_\perp-\overline{k}_{i\perp})\right\} \\
&+ \hat{h}(k_z)\left\{-\frac{k_\rho}{k}\frac{k_z(k_1^2-k^2)}{(k_1^2k_z+k^2k_{1z})}\int_{-\infty}^{\infty} d\overline{k}'_\perp(\overline{k}_\perp-\overline{k}'_\perp)\cdot F(\overline{k}_\perp-\overline{k}'_\perp)\right. \\
&\qquad \cdot F(\overline{k}'_\perp-\overline{k}_{i\perp})\left(\widetilde{A}_q(\overline{k}'_\perp)\hat{q}(\overline{k}'_\perp)+\widetilde{A}_p(\overline{k}'_\perp)\hat{p}(\overline{k}'_\perp)\right) \\
&- \frac{k_\rho^2(k^2-k_1^2)}{k_1^2k_z+k^2k_{1z}}\int d\overline{k}'_\perp F(\overline{k}_\perp-\overline{k}'_\perp) \\
&\cdot\left[\left(\hat{h}(-k_z)\cdot\hat{q}(\overline{k}'_\perp)\right)\widetilde{A}_q(\overline{k}'_\perp)+\left(\hat{h}(-k_z)\cdot\hat{p}(\overline{k}'_\perp)\right)\widetilde{A}_p(\overline{k}'_\perp)\right]F(\overline{k}'_\perp-\overline{k}_{i\perp})
\end{aligned}
$$

$$+\frac{k_z k_1^2(k^2-k_1^2)}{2(k_1^2 k_z+k^2 k_{1z})}\left[\left(\hat{e}(-k_z)\cdot b_q^{(0)}\right)\hat{q}_i+\left(\hat{e}(-k_z)\cdot\hat{p}_i\right)b_p^{(0)}\right]F^{(2)}(\overline{k}_\perp-\overline{k}_{i\perp})$$
$$-\frac{k^2 k_{1z}}{2}\frac{(k_1^2-k^2)}{k_1^2 k_z+k^2 k_{1z}}\left[\left(\hat{h}(-k_z)\cdot a_q^{(0)}\right)\hat{q}_i+\left(\hat{h}(-k_z)\cdot\hat{p}_i\right)a_p^{(0)}\right]F^{(2)}(\overline{k}_\perp-\overline{k}_{i\perp})$$
$$+k_z k_{1z}\frac{(k_1^2-k^2)}{k_1^2 k_z+k^2 k_{1z}}\int d\overline{k}'_\perp F(\overline{k}_\perp-\overline{k}'_\perp)\left[\left(\hat{e}(-k_z)\cdot\hat{q}(\overline{k}'_\perp)\right)\widetilde{B}_q(\overline{k}'_\perp)\right.$$
$$\left.\left.\left.+\left(\hat{e}(-k_z)\cdot\hat{p}(\overline{k}'_\perp)\right)\widetilde{B}_p(\overline{k}'_\perp)\right]F(\overline{k}'_\perp-\overline{k}_{i\perp})\right\}\right\} \tag{1.2.69}$$

Let

$$\overline{E}_s^{(2)}(\overline{r})=\int d\overline{k}_\perp e^{i\overline{k}_\perp\cdot\overline{r}_\perp+ik_z z}$$
$$\cdot\left\{\hat{e}(k_z)\left[f_{ee}^{(2)}(\overline{k}_\perp,\overline{k}_{i\perp})\left(\hat{e}(-k_{iz})\cdot\hat{e}_i\right)+f_{eh}^{(2)}(\overline{k}_\perp,\overline{k}_{i\perp})\left(\hat{h}(-k_{iz})\cdot\hat{e}_i\right)\right]\right.$$
$$\left.+\hat{h}(k_z)\left[f_{he}^{(2)}(\overline{k}_\perp,\overline{k}_{i\perp})\left(\hat{e}(-k_{iz})\cdot\hat{e}_i\right)+f_{hh}^{(2)}(\overline{k}_\perp,\overline{k}_{i\perp})\left(\hat{h}(-k_{iz})\cdot\hat{e}_i\right)\right]\right\} \tag{1.2.70}$$

Then comparing (1.2.69) and (1.2.70), we obtain,

$$f_{ee}^{(2)}(\overline{k}_\perp,\overline{k}_{i\perp})=-\frac{k}{k_z}(k_{1z}-k_z)\left\{\frac{k_z}{2}F^{(2)}(\overline{k}_\perp-\overline{k}_{i\perp})\left(\hat{e}(-k_z)\cdot\hat{q}_i\right)\frac{k_{iz}}{k}(1-R_{ho})\right.$$
$$+\int d\overline{k}'_\perp\left[\left(\hat{h}(-k_z)\cdot\hat{q}(\overline{k}'_\perp)\right)\widetilde{B}_q^e(\overline{k}'_\perp)+\left(\hat{h}(-k_z)\cdot\hat{p}(\overline{k}'_\perp)\right)\widetilde{B}_p^e(\overline{k}'_\perp)\right]$$
$$\cdot F(\overline{k}'_\perp-\overline{k}_{i\perp})F(\overline{k}_\perp-\overline{k}'_\perp)$$
$$\left.-\frac{k_{1z}}{2}F^{(2)}(\overline{k}_\perp-\overline{k}_{i\perp})\left(\hat{h}(-k_z)\cdot\hat{p}_i\right)(1+R_{ho})\right\} \tag{1.2.71a}$$

and

$$f_{eh}^{(2)}(\overline{k}_\perp,\overline{k}_{i\perp})=-\frac{k}{k_z}(k_{1z}-k_z)\left\{\frac{k_z}{2}F^{(2)}(\overline{k}_\perp-\overline{k}_{i\perp})\left(\hat{e}(-k_z)\cdot\hat{p}_i\right)(1+R_{vo})\right.$$
$$+\int d\overline{k}'_\perp\left[\left(\hat{h}(-k_z)\cdot\hat{q}(\overline{k}'_\perp)\right)\widetilde{B}_q^h(\overline{k}'_\perp)+\left(\hat{h}(-k_z)\cdot\hat{p}(\overline{k}'_\perp)\right)\widetilde{B}_p^h(\overline{k}'_\perp)\right]$$
$$\cdot F(\overline{k}'_\perp-\overline{k}_{i\perp})F(\overline{k}_\perp-\overline{k}'_\perp)$$
$$\left.-\frac{k_{1z}}{2}F^{(2)}(\overline{k}_\perp-\overline{k}_{i\perp})\left(\hat{h}(-k_z)\cdot\hat{q}_i\right)\left(-\frac{k_{iz}}{k}\right)(1-R_{vo})\right\} \tag{1.2.71b}$$

$$
\begin{aligned}
f_{he}^{(2)}(\overline{k}_\perp,\overline{k}_{i\perp}) = &-\frac{k}{k_z}\frac{(k_1^2-k^2)}{(k_1^2 k_z + k^2 k_{1z})} \\
&\left\{-\frac{k_\rho}{k}k_z\int_{-\infty}^{\infty} d\overline{k}'_\perp F(\overline{k}_\perp-\overline{k}'_\perp)F(\overline{k}'_\perp-\overline{k}_{i\perp})(\overline{k}_\perp-\overline{k}'_\perp)\right. \\
&\cdot\left(\hat{q}(\overline{k}'_\perp)\widetilde{A}_q^e(\overline{k}'_\perp)+\hat{p}(\overline{k}'_\perp)\widetilde{A}_p^e(\overline{k}'_\perp)\right) \\
&+k_\rho^2\int_{-\infty}^{\infty} d\overline{k}'_\perp F(\overline{k}_\perp-\overline{k}'_\perp)F(\overline{k}'_\perp-\overline{k}_{i\perp}) \\
&\cdot\left[\left(\hat{h}(-k_z)\cdot\hat{q}(\overline{k}'_\perp)\right)\widetilde{A}_q^e(\overline{k}'_\perp)+\left(\hat{h}(-k_z)\cdot\hat{p}(\overline{k}'_\perp)\right)\widetilde{A}_p^e(\overline{k}'_\perp)\right] \\
&-\frac{k_z k_1^2}{2}F^{(2)}(\overline{k}_\perp-\overline{k}_{i\perp})\left(\hat{e}(-k_z)\cdot\hat{p}_i\right)(1+R_{ho}) \\
&-\frac{k^2 k_{1z}}{2}F^{(2)}(\overline{k}_\perp-\overline{k}_{i\perp})\left(\hat{h}(-k_z)\cdot\hat{q}_i\right)\frac{k_{iz}}{k}(1-R_{ho}) \\
&+k_z k_{1z}\int d\overline{k}'_\perp F(\overline{k}_\perp-\overline{k}'_\perp)F(\overline{k}'_\perp-\overline{k}_{i\perp}) \\
&\left.\cdot\left[\left(\hat{e}(-k_z)\cdot\hat{q}(\overline{k}'_\perp)\right)\widetilde{B}_q^e(\overline{k}'_\perp)+\left(\hat{e}(-k_z)\cdot\hat{p}(\overline{k}'_\perp)\right)\widetilde{B}_p^e(\overline{k}'_\perp)\right]\right\}
\end{aligned}
\tag{1.2.71c}
$$

$$
\begin{aligned}
f_{hh}^{(2)}(\overline{k}_\perp,\overline{k}_{i\perp}) = &-\frac{k}{k_z}\frac{(k_1^2-k^2)}{(k_1^2 k_z + k^2 k_{1z})} \\
&\left\{-\frac{k_\rho}{k}k_z\int_{-\infty}^{\infty} d\overline{k}'_\perp F(\overline{k}_\perp-\overline{k}'_\perp)F(\overline{k}'_\perp-\overline{k}_{i\perp})(\overline{k}_\perp-\overline{k}'_\perp)\right. \\
&\cdot\left(\hat{q}(\overline{k}'_\perp)\widetilde{A}_q^h(\overline{k}'_\perp)+\hat{p}(\overline{k}'_\perp)\widetilde{A}_p^h(\overline{k}'_\perp)\right) \\
&+k_\rho^2\int_{-\infty}^{\infty} d\overline{k}'_\perp F(\overline{k}_\perp-\overline{k}'_\perp)F(\overline{k}'_\perp-\overline{k}_{i\perp}) \\
&\cdot\left[\left(\hat{h}(-k_z)\cdot\hat{q}(\overline{k}'_\perp)\right)\widetilde{A}_q^h(\overline{k}'_\perp)+\left(\hat{h}(-k_z)\cdot\hat{p}(\overline{k}'_\perp)\right)\widetilde{A}_p^h(\overline{k}'_\perp)\right] \\
&-\frac{k_z k_1^2}{2}F^{(2)}(\overline{k}_\perp-\overline{k}_{i\perp})\left(\hat{e}(-k_z)\cdot\hat{q}_i\right)\left(-\frac{k_{iz}}{k}\right)(1-R_{vo}) \\
&-\frac{k^2 k_{1z}}{2}F^{(2)}(\overline{k}_\perp-\overline{k}_{i\perp})\left(\hat{h}(-k_z)\cdot\hat{p}_i\right)(1+R_{vo}) \\
&+k_z k_{1z}\int d\overline{k}'_\perp F(\overline{k}_\perp-\overline{k}'_\perp)F(\overline{k}'_\perp-\overline{k}_{i\perp}) \\
&\left.\cdot\left[\left(\hat{e}(-k_z)\cdot\hat{q}(\overline{k}'_\perp)\right)\widetilde{B}_q^h(\overline{k}'_\perp)+\left(\hat{e}(-k_z)\cdot\hat{p}(\overline{k}'_\perp)\right)\widetilde{B}_p^h(\overline{k}'_\perp)\right]\right\}
\end{aligned}
\tag{1.2.71d}
$$

Note that in the above there are wavevectors, $\overline{k}_\perp$, $\overline{k}'_\perp$, and $\overline{k}_{i\perp}$. The wavevector $\overline{k}_\perp$ denotes scattered direction. The wavevector $\overline{k}_{i\perp}$ denotes incident direction. Since this is the second-order scattering, there is an intermediate direction $\overline{k}'_\perp$, of which an integration is required in (1.2.71). Substituting (1.2.48) into (1.2.71a) and also using $k_{1z} - k_z = (k_1^2 - k^2)/(k_{1z} + k_z)$, we obtain

$$f_{ee}^{(2)}(\overline{k}_\perp, \overline{k}_{i\perp}) = -\left(\frac{k_1^2 - k^2}{k_{1z} + k_z}\right)\frac{k_{iz}}{k_{1zi} + k_{iz}}$$
$$\Bigg\{\cos(\phi_k - \phi_i)F^{(2)}(\overline{k}_\perp - \overline{k}_{i\perp})(k_{1zi} - k_{1z}) + \int d\overline{k}'_\perp F(\overline{k}_\perp - \overline{k}'_\perp)F(\overline{k}'_\perp - \overline{k}_{i\perp})$$
$$\left[-2\sin(\phi_k - \phi'_k)\sin(\phi'_k - \phi_i)\left(-k_{1zi} + \frac{k'_z k'_{1z}(k_1^2 - k^2)}{k_1^2 k'_z + k^2 k'_{1z}}\right)\right.$$
$$\left.\left. + 2\cos(\phi_k - \phi'_k)\cos(\phi'_k - \phi_i)\left(-k_{1zi} + \frac{(k_1^2 - k^2)}{k'_{1z} + k'_z}\right)\right]\right\} \tag{1.2.72}$$

The above can be further simplified by noting that

$$\int d\overline{k}'_\perp F(\overline{k}_\perp - \overline{k}'_\perp)F(\overline{k}'_\perp - \overline{k}_{i\perp})$$
$$\cdot \left[\sin(\phi_k - \phi'_k)\sin(\phi'_k - \phi_i) - \cos(\phi_k - \phi'_k)\cos(\phi'_k - \phi_i)\right] 2k_{1zi}$$
$$= -2k_{1zi}\cos(\phi_k - \phi_i)\int d\overline{k}'_\perp F(\overline{k}_\perp - \overline{k}'_\perp)F(\overline{k}'_\perp - \overline{k}_{i\perp})$$
$$= -2k_{1zi}\cos(\phi_k - \phi_i)F^{(2)}(\overline{k}_\perp - \overline{k}_{i\perp}) \tag{1.2.73}$$

since

$$F^{(2)}(\overline{k}_\perp - \overline{k}_{i\perp}) = \int_{-\infty}^{\infty} d\overline{k}'_\perp F(\overline{k}_\perp - \overline{k}'_\perp)F(\overline{k}'_\perp - \overline{k}_{i\perp}) \tag{1.2.74}$$

Putting (1.2.74) into (1.2.73), we can further simplify and get the expression of $f_{ee}^{(2)}(\overline{k}_\perp, \overline{k}_{i\perp})$ in a more symmetrical form of

$$f_{ee}^{(2)}(\overline{k}_\perp, \overline{k}_{i\perp}) = -\frac{k_{iz}}{k_{iz} + k_{1zi}}\left(\frac{k_1^2 - k^2}{k_{1z} + k_z}\right)\Bigg\{-\cos(\phi_k - \phi_i)F^{(2)}(\overline{k}_\perp - \overline{k}_{i\perp})$$
$$\cdot (k_{1zi} + k_{1z}) + 2(k_1^2 - k^2)\int d\overline{k}'_\perp F(\overline{k}_\perp - \overline{k}'_\perp)F(\overline{k}'_\perp - \overline{k}_{i\perp})\Big[-\sin(\phi_k - \phi'_k)$$
$$\left.\left. \cdot \sin(\phi'_k - \phi_i)\frac{k'_z k'_{1z}}{k_1^2 k'_z + k^2 k'_{1z}} + \cos(\phi_k - \phi'_k)\cos(\phi'_k - \phi_i)\frac{1}{k'_{1z} + k'_z}\right]\right\} \tag{1.2.75}$$

We can write

$$f^{(2)}_{\alpha\beta}(\overline{k}_\perp, \overline{k}_{i\perp}) = \int_{-\infty}^{\infty} d\overline{k}'_\perp \left[\tilde{\tilde{f}}_{\alpha\beta a}(\overline{k}_\perp, \overline{k}_{i\perp}) + \tilde{\tilde{f}}_{\alpha\beta b}(\overline{k}_\perp, \overline{k}'_\perp, \overline{k}_{i\perp})\right]$$
$$\cdot F(\overline{k}_\perp - \overline{k}'_\perp)F(\overline{k}'_\perp - \overline{k}_{i\perp})$$
$$= \tilde{\tilde{f}}_{\alpha\beta a}(\overline{k}_\perp, \overline{k}_{i\perp})F^{(2)}(\overline{k}_\perp - \overline{k}_{i\perp})$$
$$+ \int_{-\infty}^{\infty} d\overline{k}'_\perp \tilde{\tilde{f}}_{\alpha\beta b}(\overline{k}_\perp, \overline{k}'_\perp, \overline{k}_{i\perp})F(\overline{k}_\perp - \overline{k}'_\perp)F(\overline{k}'_\perp - \overline{k}_{i\perp}) \quad (1.2.75a)$$

where $\alpha, \beta = v, h$. Then

$$\tilde{\tilde{f}}^{(2)}_{eea}(\overline{k}_\perp, \overline{k}_{i\perp}) = \frac{k_{iz}}{k_{iz} + k_{1zi}} \left(\frac{k_1^2 - k^2}{k_{1z} + k_z}\right) \left\{\cos(\phi_k - \phi_i)(k_{1zi} + k_{1z})\right\} \quad (1.2.75b)$$

$$\tilde{\tilde{f}}^{(2)}_{eeb}(\overline{k}_\perp, \overline{k}'_\perp, \overline{k}_{i\perp}) = -\frac{2k_{iz}}{k_{iz} + k_{1zi}} \left(\frac{k_1^2 - k^2}{k_{1z} + k_z}\right)(k_1^2 - k^2)$$
$$\cdot \left[-\sin(\phi_k - \phi'_k)\sin(\phi'_k - \phi_i)\frac{k'_z k'_{1z}}{k_1^2 k'_z + k^2 k'_{1z}}\right.$$
$$\left. + \cos(\phi_k - \phi'_k)\cos(\phi'_k - \phi_i)\frac{1}{k'_{1z} + k'_z}\right] \quad (1.2.75c)$$

$$\tilde{\tilde{f}}^{(2)}_{eha}(\overline{k}_\perp, \overline{k}_{i\perp}) = \frac{k_1^2 - k^2}{(k_{1z} + k_z)} k k_{iz} \frac{1}{(k_1^2 k_{iz} + k^2 k_{1zi})}$$
$$\cdot \left\{\sin(\phi_k - \phi_i)(k_1^2 + k_{1z}k_{1zi})\right\} \quad (1.2.75d)$$

$$\tilde{\tilde{f}}^{(2)}_{ehb}(\overline{k}_\perp, \overline{k}'_\perp, \overline{k}_{i\perp}) = -\frac{2k_1^2 - k^2}{(k_{1z} + k_z)} k k_{iz} \frac{1}{(k_1^2 k_{iz} + k^2 k_{1zi})} \left[\sin(\phi_k - \phi'_k)\frac{k_1^2 k'_\rho k_{\rho i}}{k'^2_\rho + k'_z k'_{1z}}\right.$$
$$+ \frac{(k_1^2 - k^2)k_{1zi}}{k_1^2 k'_z + k^2 k'_{1z}}\Big(\sin(\phi_k - \phi'_k)\cos(\phi'_k - \phi_i)k'_z k'_{1z}$$
$$\left. + \cos(\phi_k - \phi'_k)\sin(\phi'_k - \phi_i)(k'^2_\rho + k'_{1z}k'_z)\Big)\right] \quad (1.2.75e)$$

Also

$$f^{(2)}_{he}(\overline{k}_\perp, \overline{k}_{i\perp}) = -\frac{k}{k_z}\frac{(k_1^2 - k^2)}{(k_1^2 k_z + k^2 k_{1z})}\left\{-\sin(\phi_k - \phi_i)F^{(2)}(\overline{k}_\perp - \overline{k}_{i\perp})\right.$$
$$\cdot \frac{k_{iz}k_z}{k_{iz} + k_{1zi}}(k_1^2 - k_{1z}k_{1zi}) + k_z \int_{-\infty}^{\infty} d\overline{k}'_\perp F(\overline{k}_\perp - \overline{k}'_\perp)F(\overline{k}'_\perp - \overline{k}_{i\perp})$$
$$\left.\cdot \left[k_{1z}\cos(\phi_k - \phi'_k)\widetilde{B}^e_q(\overline{k}'_\perp) + k_{1z}\sin(\phi_k - \phi'_k)\widetilde{B}^e_p(\overline{k}'_\perp) + \frac{k_\rho k'_\rho}{k}\widetilde{A}^e_p(\overline{k}'_\perp)\right]\right\}$$

$$f_{hh}^{(2)}(\overline{k}_\perp, \overline{k}_{i\perp}) = -\frac{k}{k_z}\frac{(k_1^2 - k^2)}{(k_1^2 k_z + k^2 k_{1z})}\Bigg\{\cos(\phi_k - \phi_i)F^{(2)}(\overline{k}_\perp - \overline{k}_{i\perp})$$
$$\cdot \frac{k_1^2 k k_z k_{iz}(k_{1zi} - k_{1z})}{(k^2 k_{1zi} + k_1^2 k_{iz})} + k_z \int d\overline{k}'_\perp F(\overline{k}_\perp - \overline{k}'_\perp)F(\overline{k}'_\perp - \overline{k}_{i\perp})$$
$$\cdot \left[\frac{k_\rho k'_\rho}{k}\widetilde{A}_p^h(\overline{k}'_\perp) + \cos(\phi_k - \phi'_k)\widetilde{B}_q^h(\overline{k}'_\perp)k_{1z} + k_{1z}\sin(\phi_k - \phi'_k)\widetilde{B}_p^h(\overline{k}'_\perp)\right]\Bigg\}$$

Then

$$\tilde{\tilde{f}}_{hea}^{(2)}(\overline{k}_\perp, \overline{k}_{i\perp}) = \frac{k}{k_z}\frac{(k_1^2 - k^2)}{(k_1^2 k_z + k^2 k_{1z})}\frac{k_z k_{iz}}{(k_{iz} + k_{1zi})}\sin(\phi_k - \phi_i)(k_1^2 + k_{1z}k_{1zi}) \tag{1.2.75f}$$

$$\tilde{\tilde{f}}_{heb}^{(2)}(\overline{k}_\perp, \overline{k}'_\perp, \overline{k}_{i\perp}) = -\frac{2k}{k_z}\frac{(k_1^2 - k^2)}{(k_1^2 k_z + k^2 k_{1z})}\frac{k_z k_{iz}}{k_{iz} + k_{1zi}}$$
$$\Bigg[\sin(\phi'_k - \phi_i)\frac{k_1^2 k_\rho k'_\rho}{k'^2_\rho + k'_{1z}k'_z}$$
$$+ \frac{k_{1z}(k_1^2 - k^2)}{k_1^2 k'_z + k^2 k'_{1z}}\Big(k'_z k'_{1z}\cos(\phi_k - \phi'_k)\sin(\phi'_k - \phi_i)$$
$$+ (k'^2_\rho + k'_z k'_{1z})\sin(\phi_k - \phi'_k)\cos(\phi'_k - \phi_i)\Big)\Bigg] \tag{1.2.75g}$$

$$\tilde{\tilde{f}}_{hha}^{(2)}(\overline{k}_\perp, \overline{k}_{i\perp}) = -\frac{k}{k_z}\frac{(k_1^2 - k^2)}{(k_1^2 k_z + k^2 k_{1z})}\frac{1}{k_1^2 k_{iz} + k^2 k_{1zi}}$$
$$\cdot \cos(\phi_k - \phi_i)(k_{1zi} + k_{1z})k_1^2 k k_z k_{iz} \tag{1.2.75h}$$

$$\tilde{\tilde{f}}_{hhb}^{(2)}(\overline{k}_\perp, \overline{k}'_\perp, \overline{k}_{i\perp}) = -\frac{2k}{k_z}\frac{(k_1^2 - k^2)}{(k_1^2 k_z + k^2 k_{1z})}\frac{1}{k_1^2 k_{iz} + k^2 k_{1zi}}k_z k_{iz}$$
$$\Bigg[k k_{1z} k_{1zi}\frac{(k_1^2 - k^2)}{(k'_{1z} + k'_z)}\sin(\phi_k - \phi'_k)\sin(\phi'_k - \phi_i)$$
$$+ \frac{1}{k_1^2 k'_z + k^2 k'_{1z}}\Big(\frac{k_1^2}{k}(k_1^2 - k^2)k_\rho k_{\rho i}k'^2_\rho$$
$$- k_\rho k'_\rho k k_1^2 (k'_{1z} + k'_z)k_{1zi}\cos(\phi'_k - \phi_i)$$
$$- k_{\rho i}k'_\rho k k_1^2 (k'_{1z} + k'_z)k_{1z}\cos(\phi_k - \phi'_k)$$
$$- k k'_z k'_{1z}(k_1^2 - k^2)k_{1z}k_{1zi}\cos(\phi'_k - \phi_i)\cos(\phi_k - \phi'_k)\Big)\Bigg] \tag{1.2.75i}$$

The expressions of $f_{ee}^{(2)}$, $f_{he}^{(2)}$, $f_{eh}^{(2)}$, and $f_{hh}^{(2)}$ exhibit symmetry of reciprocity on interchanging $\overline{k}_\perp$ and $\overline{k}_{i\perp}$ with the third wave vector $\overline{k}'_\perp$ as a dummy variable. Numerical results of $f_{ee}^{(2)}$, $f_{he}^{(2)}$, $f_{eh}^{(2)}$, and $f_{hh}^{(2)}$ can be calculated by substituting $(1.2.75b, c, de, f, g, h, i)$ in $(1.2.75a)$.

3 Coherent Reflection, Emissivities, and Bistatic Scattering Coefficients of Random Dielectric Surfaces

In this section we will calculate the emissivities and bistatic scattering coefficients based on the second-order SPM that are derived in the previous sections. We will take statistical averages of the field solutions.

3.1 Coherent Reflection

In taking statistical averages, we make use of the following relations

$$F(\overline{k}_\perp) = \frac{1}{4\pi^2}\int d\overline{r}_\perp f(\overline{r}_\perp) e^{-i\overline{k}_\perp\cdot\overline{r}_\perp} \tag{1.3.1}$$

$$F^*(\overline{k}_\perp) = \frac{1}{4\pi^2}\int d\overline{r}_\perp f(\overline{r}_\perp) e^{i\overline{k}_\perp\cdot\overline{r}_\perp} = F(-\overline{k}_\perp) \tag{1.3.2}$$

$$\langle F(\overline{k}_{\perp 1}) F^*(\overline{k}_{\perp 2})\rangle = \delta(\overline{k}_{\perp 1} - \overline{k}_{\perp 2}) W(\overline{k}_{\perp 1}) \tag{1.3.3}$$

$$\langle F(\overline{k}_{\perp 1}) F(\overline{k}_{\perp 2})\rangle = \delta(\overline{k}_{\perp 1} + \overline{k}_{\perp 2}) W(\overline{k}_{\perp 1}) \tag{1.3.4}$$

where $f(\overline{r}_\perp)$ is the rough surface profile. Taking averages, we obtain

$$\begin{aligned}\langle F^{(2)}(\overline{k}_\perp - \overline{k}_{i\perp})\rangle &= \int_{-\infty}^{\infty} d\overline{k}'_\perp \langle F(\overline{k}'_\perp) F(\overline{k}_\perp - \overline{k}_{i\perp} - \overline{k}'_\perp)\rangle \\ &= \delta(\overline{k}_\perp - \overline{k}_{i\perp}) \int_{-\infty}^{\infty} d\overline{k}'_\perp W(\overline{k}'_\perp) \\ &= \delta(\overline{k}_\perp - \overline{k}_{i\perp}) \int_{-\infty}^{\infty} d\overline{k}'_\perp W(\overline{k}'_\perp - \overline{k}_{i\perp})\end{aligned} \tag{1.3.5}$$

Also, we have

$$\langle F(\overline{k}'_\perp - \overline{k}_{i\perp}) F(\overline{k}_\perp - \overline{k}'_\perp)\rangle = W(\overline{k}'_\perp - \overline{k}_{i\perp})\delta(\overline{k}_\perp - \overline{k}_{i\perp}) \tag{1.3.6}$$

Note that when $\overline{k}_\perp = \overline{k}_{i\perp}$, $\sin(\phi_k - \phi_i) = 0$ and $\cos(\phi_k - \phi_i) = 1$. Then $\langle f^{(2)}\rangle$ is proportional to Dirac delta function, where $f^{(2)}$ denotes the second-order scattering amplitude. Let

$$\langle f_{ab}^{(2)}(\overline{k}_\perp, \overline{k}_{i\perp})\rangle = \tilde{f}_{ab}^{(2)}(\overline{k}_{i\perp})\delta(\overline{k}_\perp - \overline{k}_{i\perp}) \tag{1.3.7}$$

where $a, b = e, h$. From (1.2.75), since $\overline{k}_\perp = \overline{k}_{i\perp}$ in (1.3.7) for the coherent wave, we have

$$\tilde{f}_{ee}^{(2)}(\overline{k}_{i\perp}) = -\frac{2k_{iz}(k_1^2-k^2)}{(k_{iz}+k_{1zi})^2}\left\{-k_{1zi}\int_{-\infty}^{\infty} d\overline{k}_\perp W(\overline{k}_\perp-\overline{k}_{i\perp}) \right.$$
$$+ (k_1^2-k^2)\int_{-\infty}^{\infty} d\overline{k}_\perp W(\overline{k}_\perp-\overline{k}_{i\perp})$$
$$\left.\cdot\left[\sin^2(\phi_k-\phi_i)\frac{k_zk_{1z}}{k_1^2k_z+k^2k_{1z}}+\cos^2(\phi_k-\phi_i)\frac{1}{k_{1z}+k_z}\right]\right\} \quad (1.3.8a)$$

$$\tilde{f}_{eh}^{(2)}(\overline{k}_{i\perp}) = -2\frac{k(k_1^2-k^2)k_{iz}}{k_{1zi}+k_{iz}}\frac{1}{k_1^2k_{iz}+k^2k_{1zi}}\int_{-\infty}^{\infty} d\overline{k}_\perp W(\overline{k}_\perp-\overline{k}_{i\perp})$$
$$\cdot\left[-\sin(\phi_k-\phi_i)\frac{k_1^2k_\rho k_{\rho i}}{k_\rho^2+k_zk_{1z}}+\frac{(k_1^2-k^2)k_{1zi}}{k_1^2k_z+k^2k_{1z}}\right.$$
$$\left.\cdot\sin(\phi_k-\phi_i)\cos(\phi_k-\phi_i)k_\rho^2\right] \quad (1.3.8b)$$

$$\tilde{f}_{he}^{(2)}(\overline{k}_{i\perp}) = -2\frac{kk_{iz}(k_1^2-k^2)}{(k_{1zi}+k_{iz})}\frac{1}{(k_1^2k_{iz}+k^2k_{1zi})}\int_{-\infty}^{\infty} d\overline{k}_\perp W(\overline{k}_\perp-\overline{k}_{i\perp})$$
$$\cdot\left[\sin(\phi_k-\phi_i)\frac{k_1^2k_{\rho i}k_\rho}{k_\rho^2+k_{1z}k_z}-\frac{(k_1^2-k^2)k_{1zi}k_\rho^2}{k_1^2k_z+k^2k_{1z}}\sin(\phi_k-\phi_i)\cos(\phi_k-\phi_i)\right]$$
$$= -\tilde{f}_{eh}^{(2)}(\overline{k}_{i\perp}) \quad (1.3.8c)$$

$$\tilde{f}_{hh}^{(2)}(\overline{k}_{i\perp}) = -\frac{2k(k_1^2-k^2)}{k_{iz}(k_1^2k_{iz}+k^2k_{1zi})^2}\left\{k_{1zi}\int_{-\infty}^{\infty} d\overline{k}_\perp W(\overline{k}_\perp-\overline{k}_{i\perp})k_1^2kk_{iz}^2\right.$$
$$+ k_{iz}^2\int_{-\infty}^{\infty} d\overline{k}_\perp W(\overline{k}_\perp-\overline{k}_{i\perp})\left[-kk_{1zi}^2\frac{(k_1^2-k^2)}{k_{1z}+k_z}\sin^2(\phi_k-\phi_i)\right.$$
$$+\frac{1}{k_1^2k_z+k^2k_{1z}}\left(\frac{k_1^2}{k}(k_1^2-k^2)k_{\rho i}^2k_\rho^2\right.$$
$$-2k_{\rho i}k_\rho kk_1^2(k_{1z}+k_z)k_{1zi}\cos(\phi_k-\phi_i)$$
$$\left.\left.\left.-kk_zk_{1z}(k_1^2-k^2)k_{1zi}^2\cos^2(\phi_k-\phi_i)\right)\right]\right\} \quad (1.3.8d)$$

From (1.2.70) to (1.3.7), the second order coherent reflected wave is

$$\langle\overline{E}_s^{(2)}(\overline{r})\rangle = e^{i\overline{k}_{i\perp}\cdot\overline{r}_\perp+ik_{iz}z}\left[\hat{e}(k_{iz})\tilde{f}_{ee}^{(2)}(\overline{k}_{i\perp})+\hat{h}(k_{iz})\tilde{f}_{he}^{(2)}(\overline{k}_{i\perp})\right] \quad (1.3.9a)$$

if incident wave is $\hat{e}(-k_{iz})$ polarized and

$$\langle\overline{E}_s^{(2)}(\overline{r})\rangle = e^{i\overline{k}_{i\perp}\cdot\overline{r}_\perp+ik_{iz}z}\left[\hat{e}(k_{iz})\tilde{f}_{eh}^{(2)}(\overline{k}_{i\perp})+\hat{h}(k_{iz})\tilde{f}_{hh}^{(2)}(\overline{k}_{i\perp})\right] \quad (1.3.9b)$$

if incident wave is $\hat{h}(-k_{iz})$ polarized. Thus in general, there can be cross-polarizations in the coherent reflection depending on the statistical symmetry of the rough surface.

Coherent Reflection for Isotropic Surfaces

For the special case of isotropic surfaces, the spectral density has no preferred direction.

$$W(\overline{k}_\perp) = W(|\overline{k}_\perp|) = W(k_\rho) \tag{1.3.10}$$

Then in (1.3.8)–(1.3.10) we have

$$W(\overline{k}_\perp - \overline{k}_{i\perp}) = W\left(k_\rho^2 + k_{\rho i}^2 - 2k_\rho k_{\rho i}\cos(\phi_k - \phi_i)\right) \tag{1.3.11}$$

The integrands in (1.3.8) and (1.3.10) depend on ϕ_k only through $\cos(\phi_k - \phi_i)$ and $\sin(\phi_k - \phi_i)$. We use the property that

$$\int_0^{2\pi} d\phi_k \sin(\phi_k - \phi_i) f\left(\cos(\phi_k - \phi_i)\right) = 0 \tag{1.3.12}$$

for any function f of $\cos(\phi_k - \phi_i)$. We have

$$\tilde{f}_{he}^{(2)}(\overline{k}_{i\perp}) = \tilde{f}_{eh}^{(2)}(\overline{k}_{i\perp}) = 0 \tag{1.3.13}$$

Thus the coherent reflection has no cross-polarization for isotropic surfaces.

For the special case of Gaussian correlation function, we have

$$W(\overline{k}_\perp) = \frac{h^2 l^2}{4\pi}\exp\left(-\frac{k_\rho^2 l^2}{4}\right) \tag{1.3.14}$$

$$W(\overline{k}_\perp - \overline{k}_{i\perp}) = \frac{h^2 l^2}{4\pi}\exp\left(-\frac{(k_\rho^2 + k_{\rho i}^2)l^2}{4} + x\cos(\phi_k - \phi_i)\right) \tag{1.3.15}$$

where

$$x = \frac{k_\rho k_{\rho i} l^2}{2} \tag{1.3.16}$$

We next use the following integrals

$$\frac{1}{2\pi}\int_0^{2\pi} e^{x\cos\theta}\cos n\theta d\theta = I_n(x) \tag{1.3.17}$$

$$\frac{1}{2\pi}\int_0^{2\pi} e^{x\cos\theta}\sin^2\theta d\theta = \frac{I_1(x)}{x} \tag{1.3.18}$$

$$\frac{1}{2\pi}\int_0^{2\pi} e^{x\cos\theta}\cos^2\theta d\theta = I_0(x) - \frac{I_1(x)}{x} \tag{1.3.19}$$

where I_n are the n-th order modified Bessel functions. Then

$$\tilde{f}_{eh}^{(2)}(\overline{k}_i) = 0 \tag{1.3.20}$$

$$\tilde{f}_{he}^{(2)}(\overline{k}_i) = 0 \tag{1.3.21}$$

$$\tilde{f}_{ee}^{(2)}(\overline{k}_i) = -\frac{k_{iz}(k_1^2 - k^2)}{(k_{iz} + k_{1zi})^2} h^2 l^2 \int_0^\infty dk_\rho k_\rho \exp\left(-\frac{(k_\rho^2 + k_{\rho i}^2)l^2}{4}\right) \cdot \left\{ I_0(x)\left[-k_{1zi} + \frac{(k_1^2 - k^2)}{k_{1z} + k_z}\right] - (k_1^2 - k^2)\frac{I_1(x)}{x}\frac{k_\rho^2}{k_{1z}k^2 + k_1^2 k_z}\right\} \tag{1.3.22}$$

$$\tilde{f}_{hh}^{(2)}(\overline{k}_{i\perp}) = -\frac{k_{iz}(k_1^2 - k^2)}{(k^2 k_{1zi} + k_1^2 k_{iz})^2} k_1^2 k^2 h^2 l^2 \int_0^\infty dk_\rho k_\rho \exp\left(-\frac{(k_\rho^2 + k_{\rho i}^2)l^2}{4}\right) \cdot \left\{ k_{1zi}\left(I_0(x) - \frac{2k_\rho k_{\rho i}}{(k_\rho^2 + k_{1z}k_z)} I_1(x)\right) + \frac{I_0(x)(k_1^2 - k^2)}{k_1^2 k_z + k^2 k_{1z}} \cdot \left[\frac{k_{\rho i}^2 k_\rho^2}{k^2} - \frac{k_z k_{1z} k_{1zi}^2}{k_1^2}\right] - \frac{I_1(x)}{x}\frac{(k_1^2 - k^2)}{k_1^2} k_{1zi}^2 \frac{k_\rho^2}{k_1^2 k_z + k^2 k_{1z}}\right\} \tag{1.3.23}$$

3.2 Emissivities of Four Stokes Parameters

We will calculate the emissivities of the four Stokes parameters by using SPM to the second-order. This includes the first-order scattering amplitude squared and the second-order coherent reflectivity. For formulas of emissivity of four stokes parameters in terms of bistatic scattering coefficients refer to Chapter 3, Section 5.4 of Volume I.

First consider an incident wave that is $\hat{e}(-k_{iz})$ polarized—that is, horizontally polarized. Then, to the required order, we obtain

$$\overline{E}_s(\overline{r}) = \overline{E}_s^{(0)}(\overline{r}) + \overline{E}_s^{(1)}(\overline{r}) + \overline{E}_s^{(2)}(\overline{r}) \tag{1.3.24}$$

$$= R_{ho}\hat{e}(k_{iz})e^{i\overline{k}_{i\perp}\cdot\overline{r}_\perp + ik_{iz}z} + \int d\overline{k}_\perp e^{i\overline{k}_\perp\cdot\overline{r}_\perp + ik_z z} iF(\overline{k}_\perp - \overline{k}_{i\perp}) \cdot \left[\hat{e}(k_z)f_{ee}^{(1)}(\overline{k}_\perp, \overline{k}_{i\perp}) + \hat{h}(k_z)f_{he}^{(1)}(\overline{k}_\perp, \overline{k}_{i\perp})\right] + \int d\overline{k}_\perp e^{i\overline{k}_\perp\cdot\overline{r}_\perp + ik_z z}\left[\hat{e}(k_z)f_{ee}^{(2)}(\overline{k}_\perp, \overline{k}_{i\perp}) + \hat{h}(k_z)f_{he}^{(2)}(\overline{k}_\perp, \overline{k}_{i\perp})\right] \tag{1.3.25}$$

$$\overline{H}_s(\overline{r}) = -\frac{R_{ho}}{\eta}\hat{h}(k_{iz})e^{i\overline{k}_{i\perp}\cdot\overline{r}_\perp + ik_{iz}z} + \frac{1}{\eta}\int d\overline{k}_\perp e^{i\overline{k}_\perp\cdot\overline{r}_\perp + ik_z z} iF(\overline{k}_\perp - \overline{k}_{i\perp})$$
$$\cdot\left[-\hat{h}(k_z)f_{ee}^{(1)}(\overline{k}_\perp, \overline{k}_{i\perp}) + \hat{e}(k_z)f_{he}^{(1)}(\overline{k}_\perp, \overline{k}_{i\perp})\right]$$
$$+\frac{1}{\eta}\int d\overline{k}_\perp e^{i\overline{k}_\perp\cdot\overline{r}_\perp + ik_z z}\left[-\hat{h}(k_z)f_{ee}^{(2)}(\overline{k}_\perp, \overline{k}_{i\perp}) + \hat{e}(k_z)f_{he}^{(2)}(\overline{k}_\perp, \overline{k}_{i\perp})\right] \tag{1.3.26}$$

The incident and scattered power per unit area as received by the surface, respectively, are

$$\overline{S}_i\cdot\hat{z} = -\frac{\cos\theta_i}{2\eta} \tag{1.3.27}$$

$$\begin{aligned}
\langle\overline{S}_s\cdot\hat{z}\rangle &= \frac{1}{2}\,\mathrm{Re}\langle\overline{E}_s\times\overline{H}_s^*\rangle\cdot\hat{z}\\
&= \frac{1}{2}\,\mathrm{Re}\,\overline{E}_s^{(0)}\times\overline{H}_s^{(0)*}\cdot\hat{z} + \frac{1}{2}\,\mathrm{Re}\,\overline{E}_s^{(0)}\times\langle\overline{H}_s^{(2)*}\rangle\cdot\hat{z}\\
&\quad+\frac{1}{2}\,\mathrm{Re}\langle\overline{E}_s^{(2)}\rangle\times\overline{H}_s^{(0)*}\cdot\hat{z} + \frac{1}{2}\,\mathrm{Re}\,\langle\overline{E}_s^{(1)}\times\overline{H}_s^{(1)*}\rangle\cdot\hat{z}\\
&= \frac{\cos\theta_i}{2\eta}\left\{|R_{ho}|^2 + 2\,\mathrm{Re}\left(R_{ho}\tilde{f}_{ee}^{(2)*}(\overline{k}_{i\perp})\right)\right\}\\
&\quad+\frac{1}{2\eta}\int_{k_\rho\le k} d\overline{k}_\perp W(\overline{k}_\perp - \overline{k}_{i\perp})\cos\theta_s\\
&\quad\cdot\left[|f_{ee}^{(1)}(\overline{k}_\perp, \overline{k}_{i\perp})|^2 + |f_{he}^{(1)}(\overline{k}_\perp, \overline{k}_{i\perp})|^2\right]
\end{aligned} \tag{1.3.28}$$

We have used the substitutions $k_x = k_\rho\cos\phi_k$, $k_y = k_\rho\sin\phi_k$, $k_\rho = k\sin\theta_k$, and $k_z = k\cos\theta_k$. Thus the emissivity of horizontally polarized wave is, in the direction of $(\theta_i, \pi + \phi_i)$,

$$\begin{aligned}
e_h(\theta_i, \pi+\phi_i) &= 1 - |R_{ho}|^2 - 2\,\mathrm{Re}\left(R_{ho}\tilde{f}_{ee}^{(2)*}(\overline{k}_{i\perp})\right)\\
&\quad-\frac{1}{\cos\theta_i}\int_0^{\pi/2} d\theta_k\sin\theta_k\int_0^{2\pi} d\phi_k k^2\cos^2\theta_k W(\overline{k}_\perp - \overline{k}_{i\perp})\\
&\quad\left[|f_{ee}^{(1)}(\overline{k}_\perp, \overline{k}_{i\perp})|^2 + |f_{he}^{(1)}(\overline{k}_\perp, \overline{k}_{i\perp})|^2\right]
\end{aligned} \tag{1.3.29}$$

For the case of vertically polarized incident wave we have

$$\overline{E}_s(\overline{r}) = R_{vo}\hat{h}(k_{iz}) + \int d\overline{k}_\perp e^{i\overline{k}_\perp\cdot\overline{r}_\perp + ik_z z} iF(\overline{k}_\perp - \overline{k}_{i\perp})$$
$$\cdot\left[\hat{e}(k_z)f_{eh}^{(1)}(\overline{k}_\perp, \overline{k}_{i\perp}) + \hat{h}(k_z)f_{hh}^{(1)}(\overline{k}_\perp, \overline{k}_{i\perp})\right] + \int d\overline{k}_\perp e^{i\overline{k}_\perp\cdot\overline{r}_\perp + ik_z z}$$
$$\cdot\left[\hat{e}(k_z)f_{eh}^{(2)}(\overline{k}_\perp, \overline{k}_{i\perp}) + \hat{h}(k_z)f_{hh}^{(2)}(\overline{k}_\perp, \overline{k}_{i\perp})\right] \tag{1.3.30}$$

The magnetic field can be calculated similarly. Thus, the emissivity of a vertically polarized wave is

$$e_v(\theta_i, \pi + \phi_i) = 1 - |R_{vo}|^2 - 2\ \mathrm{Re}\left(R_{vo}\tilde{f}_{hh}^{(2)*}(\overline{k}_{i\perp})\right)$$
$$- \frac{1}{\cos\theta_i}\int_0^{\pi/2} d\theta_k \sin\theta_k \int_0^{2\pi} d\phi_k k^2 \cos^2\theta_k W(\overline{k}_\perp - \overline{k}_{i\perp})$$
$$\left[|f_{eh}^{(1)}(\overline{k}_\perp, \overline{k}_{i\perp})|^2 + |f_{hh}^{(1)}(\overline{k}_\perp, \overline{k}_{i\perp})|^2\right] \quad (1.3.31)$$

We clarify the notations for polarization and observation directions. For rough surface scattering using dyadic Green's function representation, we have been using $\hat{e}(k_z)$ and $\hat{h}(k_z)$. However, for volume scattering and also for emissivity calculations we have used $\hat{h}(\theta, \phi)$ and $\hat{v}(\theta, \phi)$. The relations between the two sets of notations are as follows

$$\hat{k}(\theta, \phi) = \hat{k} = \sin\theta\cos\phi\hat{x} + \sin\theta\sin\phi\hat{y} + \cos\theta\hat{z} \quad (1.3.32)$$
$$\hat{v}(\theta, \phi) = \hat{v} = \cos\theta\cos\phi\hat{x} + \cos\theta\sin\phi\hat{y} - \sin\theta\hat{z} \quad (1.3.33)$$
$$\hat{h}(\theta, \phi) = \hat{h} = -\sin\phi\hat{x} + \cos\phi\hat{y} \quad (1.3.34)$$
$$\hat{e}(k_z) = \hat{x}\sin\phi - \hat{y}\cos\phi = -\hat{h}(\theta, \phi) \quad (1.3.35)$$
$$\hat{h}(k_z) = -\cos\theta\cos\phi\hat{x} - \cos\theta\sin\phi\hat{y} + \sin\theta\hat{z} = -\hat{v}(\theta, \phi) \quad (1.3.36)$$
$$\hat{e}(-k_z) = -\hat{h}(\pi - \theta, \phi) \quad (1.3.37)$$
$$\hat{h}(-k_z) = \cos\theta\cos\phi\hat{x} + \cos\theta\sin\phi\hat{y} + \sin\theta\hat{z} = -\hat{v}(\pi - \theta, \phi) \quad (1.3.38)$$
$$\hat{v}(\pi - \theta, \phi) = -\cos\theta\cos\phi\hat{x} - \cos\theta\sin\phi\hat{y} - \sin\theta\hat{z} \quad (1.3.39)$$

For the present case, since the incident wave direction is

$$\hat{k}_i = \sin\theta_i\cos\phi_i\hat{x} + \sin\theta_i\sin\phi_i\hat{y} - \cos\theta_i\hat{z} = \hat{s}_{ob} = \hat{k}(\pi - \theta_i, \phi_i) \quad (1.3.40)$$

the observed emissivity is in the direction

$$\hat{s}_o = \hat{k}(\theta_i, \pi + \phi_i) \quad (1.3.41)$$

We note that

$$4\pi\gamma_{hh}(\theta_s, \phi_s; \pi - \theta_i, \phi_i) = \left[|R_{ho}|^2 + 2\mathrm{Re}\left(R_{ho}\tilde{f}_{ee}^{(2)*}(\overline{k}_{i\perp})\right)\right]$$
$$\cdot\, \delta(\cos\theta_s - \cos\theta_i)\delta(\phi_s - \phi_i)$$
$$+ \frac{k^2\cos^2\theta_s}{\cos\theta_i} W(\overline{k}_\perp - \overline{k}_{i\perp}) \left|f_{ee}^{(1)}(\overline{k}_\perp, \overline{k}_{i\perp})\right|^2 \quad (1.3.42a)$$

$$4\pi\gamma_{vh}(\theta_s, \phi_s; \pi - \theta_i, \phi_i) = \frac{k^2\cos^2\theta_s}{\cos\theta_i} W(\overline{k}_\perp - \overline{k}_{i\perp}) \left|f_{he}^{(1)}(\overline{k}_\perp, \overline{k}_{i\perp})\right|^2 \quad (1.3.42b)$$

$$4\pi\gamma_{vv}(\theta_s,\phi_s;\pi-\theta_i,\phi_i)=\left[|R_{vo}|^2+2\mathrm{Re}\left(R_{vo}\tilde{f}_{hh}^{(2)*}(\overline{k}_{i\perp})\right)\right] \cdot\delta(\cos\theta_s-\cos\theta_i)\delta(\phi_s-\phi_i) + \frac{k^2\cos^2\theta_s}{\cos\theta_i}W(\overline{k}_\perp-\overline{k}_{i\perp})\left|f_{hh}^{(1)}(\overline{k}_\perp,\overline{k}_{i\perp})\right|^2 \quad (1.3.42c)$$

$$4\pi\gamma_{hv}(\theta_s,\phi_s;\pi-\theta_i,\phi_i)=\frac{k^2\cos^2\theta_s}{\cos\theta_i}W(\overline{k}_\perp-\overline{k}_{i\perp})\left|f_{eh}^{(1)}(\overline{k}_\perp,\overline{k}_{i\perp})\right|^2 \quad (1.3.42d)$$

with $(\theta_s,\phi_s)=(\theta_k,\phi_k)$ indicating the angular variables of the scattered wave direction. For a P-polarized incident wave (i.e., 45°) we have

$$\hat{p}=\frac{1}{\sqrt{2}}\left\{\hat{v}(\hat{s}_{ob})+\hat{h}(\hat{s}_{ob})\right\} \quad (1.3.43)$$

$$=\frac{1}{\sqrt{2}}\left\{\hat{v}(\pi-\theta_i,\phi_i)+\hat{h}(\pi-\theta_i,\phi_i)\right\}=\frac{1}{\sqrt{2}}\left\{-\hat{h}(-k_{iz})-\hat{e}(-k_{iz})\right\} \quad (1.3.44)$$

Because of the linearity of Maxwell's equations we have

$$\begin{aligned}\overline{E}_s(\overline{r})=\frac{1}{\sqrt{2}}\Bigg\{&\left(-R_{ho}\hat{e}(k_{iz})-R_{vo}\hat{h}(k_{iz})\right)e^{i\overline{k}_{i\perp}\cdot\overline{r}_\perp+ik_{iz}z}\\&+\int d\overline{k}_\perp e^{i\overline{k}_\perp\cdot\overline{r}_\perp+ik_zz}iF(\overline{k}_\perp-\overline{k}_{i\perp})\\&\cdot\Big[\hat{e}(k_z)\left(-f_{ee}^{(1)}(\overline{k}_\perp,\overline{k}_{i\perp})-f_{eh}^{(1)}(\overline{k}_\perp,\overline{k}_{i\perp})\right)\\&+\hat{h}(k_z)\left(-f_{he}^{(1)}(\overline{k}_\perp,\overline{k}_{i\perp})-f_{hh}^{(1)}(\overline{k}_\perp,\overline{k}_{i\perp})\right)\Big]\\&+\int d\overline{k}_\perp e^{i\overline{k}_\perp\cdot\overline{r}_\perp+ik_zz}\Big[\hat{e}(k_z)\left(-f_{ee}^{(2)}(\overline{k}_\perp,\overline{k}_{i\perp})-f_{eh}^{(2)}(\overline{k}_\perp,\overline{k}_{i\perp})\right)\\&+\hat{h}(k_z)\left(-f_{he}^{(2)}(\overline{k}_\perp,\overline{k}_{i\perp})-f_{hh}^{(2)}(\overline{k}_\perp,\overline{k}_{i\perp})\right)\Big]\end{aligned} \quad (1.3.45)$$

The scattered power per unit area is

$$\begin{aligned}\langle\overline{S}_s\cdot\hat{z}\rangle=\frac{1}{4\eta}\Bigg\{&\cos\theta_i\left(|R_{ho}|^2+|R_{vo}|^2\right)\\&+\mathrm{Re}\Big(R_{ho}\left(\tilde{f}_{ee}^{(2)*}(\overline{k}_{i\perp})+\tilde{f}_{eh}^{(2)*}(\overline{k}_{i\perp})\right)\Big)\cos\theta_i\\&+\mathrm{Re}\Big(R_{ho}^*\left(\tilde{f}_{ee}^{(2)}(\overline{k}_{i\perp})+\tilde{f}_{eh}^{(2)}(\overline{k}_{i\perp})\right)\cos\theta_i\Big)\\&+\mathrm{Re}\Big(R_{vo}\left(\tilde{f}_{he}^{(2)*}(\overline{k}_{i\perp})+\tilde{f}_{hh}^{(2)*}(\overline{k}_{i\perp})\right)\cos\theta_i\Big)\\&+\mathrm{Re}\Big(R_{vo}^*\left(\tilde{f}_{he}^{(2)*}(\overline{k}_{i\perp})+\tilde{f}_{hh}^{(2)*}(\overline{k}_{i\perp})\right)\cos\theta_i\Big)\end{aligned}$$

$$+\,\mathrm{Re}\int d\overline{k}_\perp \cos\theta_s W(\overline{k}_\perp - \overline{k}_{i\perp})\left[\left|f^{(1)}_{ee}(\overline{k}_\perp,\overline{k}_{i\perp}) + f^{(1)}_{eh}(\overline{k}_\perp,\overline{k}_{i\perp})\right|^2 + \left|f^{(1)}_{he}(\overline{k}_\perp,\overline{k}_{i\perp}) + f^{(1)}_{hh}(\overline{k}_\perp,\overline{k}_{i\perp})\right|^2\right]\Bigg\} \tag{1.3.46}$$

Thus

$$4\pi\gamma_{hP}(\theta_s,\phi_s;\pi-\theta_i,\phi_i) = \frac{1}{2}\Bigg\{\Big[|R_{ho}|^2 + 2\mathrm{Re}\Big(R_{ho}\left(\tilde{f}^{(2)*}_{ee}(\overline{k}_{i\perp}) + \tilde{f}^{(2)*}_{eh}(\overline{k}_{i\perp})\right)\Big)\Big] \cdot \delta(\cos\theta_s - \cos\theta_i)\delta(\phi_s-\phi_i) + \frac{k^2\cos^2\theta_s}{\cos\theta_i} W(\overline{k}_\perp - \overline{k}_{i\perp})\left|f^{(1)}_{ee}(\overline{k}_\perp,\overline{k}_{i\perp}) + f^{(1)}_{eh}(\overline{k}_\perp,\overline{k}_{i\perp})\right|^2\Bigg\} \tag{1.3.47}$$

$$4\pi\gamma_{vP}(\theta_s,\phi_s;\pi-\theta_i,\phi_i) = \frac{1}{2}\Bigg\{\Big[|R_{vo}|^2 + 2\mathrm{Re}\Big(R_{vo}\left(\tilde{f}^{(2)*}_{he}(\overline{k}_{i\perp}) + \tilde{f}^{(2)*}_{hh}(\overline{k}_{i\perp})\right)\Big)\Big] \cdot \delta(\cos\theta_s - \cos\theta_i)\delta(\phi_s-\phi_i) + \frac{k^2\cos^2\theta_s}{\cos\theta_i} W(\overline{k}_\perp - \overline{k}_{i\perp})\left|f^{(1)}_{he}(\overline{k}_\perp,\overline{k}_{i\perp}) + f^{(1)}_{hh}(\overline{k}_\perp,\overline{k}_{i\perp})\right|^2\Bigg\} \tag{1.3.48}$$

Hence the emissivity of the third Stokes parameter is, from (3.5.67) of Volume I and the bistatic scattering coefficients in this section,

$$\begin{aligned}
&e_3(\theta_i,\pi+\phi_i) \\
&= \frac{1}{4\pi}\sum_{\beta=v,h}\int_0^{2\pi} d\phi_s \int_0^{\pi/2} d\theta_s \sin\theta_s\Big[2\gamma_{\beta P}(\hat{s},\hat{s}_{ob}) - \gamma_{\beta v}(\hat{s},\hat{s}_{ob}) - \gamma_{\beta h}(\hat{s},\hat{s}_{ob})\Big] \\
&= \int_0^{2\pi} d\phi_s \int_0^{\pi/2} d\theta_s \sin\theta_s \cdot \\
&\quad\Big[2\mathrm{Re}\Big(R_{ho}\tilde{f}^{(2)*}_{eh}(\overline{k}_{\perp i})\Big)\delta(\cos\theta_s - \cos\theta_i)\delta(\phi_s-\phi_i) \\
&\quad + \frac{k^2\cos^2\theta_s}{\cos\theta_i}W(\overline{k}_\perp - \overline{k}_{i\perp})2\mathrm{Re}\Big(f^{(1)}_{ee}(\overline{k}_\perp,\overline{k}_{i\perp})f^{(1)*}_{eh}(\overline{k}_\perp,\overline{k}_{i\perp})\Big) \\
&\quad + 2\mathrm{Re}\Big(R_{vo}\tilde{f}^{(2)*}_{he}(\overline{k}_{i\perp})\Big)\delta(\cos\theta_s - \cos\theta_i)\delta(\phi_s-\phi_i) \\
&\quad + \frac{k^2\cos^2\theta_s}{\cos\theta_i}W(\overline{k}_\perp - \overline{k}_{i\perp})2\mathrm{Re}\Big(f^{(1)}_{hh}(\overline{k}_\perp,\overline{k}_{i\perp})f^{(1)*}_{he}(\overline{k}_\perp,\overline{k}_{i\perp})\Big)\Big]
\end{aligned} \tag{1.3.49}$$

Simplifying, we get

$$\begin{aligned}
e_3(\theta_i, \pi+\phi_i) = {} & 2\mathrm{Re}\Big(R_{ho}\tilde{f}_{eh}^{(2)*}(\overline{k}_{i\perp})\Big) \\
& + \int_0^{2\pi} d\phi_s \int_0^{\pi/2} d\theta_s \sin\theta_s \frac{k^2\cos^2\theta_s}{\cos\theta_i} W(\overline{k}_\perp - \overline{k}_{i\perp}) \\
& \cdot 2\mathrm{Re}\Big(f_{ee}^{(1)}(\overline{k}_\perp, \overline{k}_{i\perp}) f_{eh}^{(1)*}(\overline{k}_\perp, \overline{k}_{i\perp})\Big) + 2\mathrm{Re}\Big(R_{vo}\tilde{f}_{he}^{(2)*}(\overline{k}_{i\perp})\Big) \\
& + \int_0^{2\pi} d\phi_s \int_0^{\pi/2} d\theta_s \sin\theta_s \frac{k^2\cos\theta_s}{\cos\theta_i} W(\overline{k}_\perp - \overline{k}_{i\perp}) \\
& \cdot 2\mathrm{Re}\Big(f_{hh}^{(1)}(\overline{k}_\perp, \overline{k}_{i\perp}) f_{he}^{(1)*}(\overline{k}_\perp, \overline{k}_{i\perp})\Big)
\end{aligned} \tag{1.3.50}$$

For an R-polarized circularly polarized incident wave, we obtain

$$\begin{aligned}
\hat{R} &= \frac{1}{\sqrt{2}}\Big\{\hat{v}(\hat{s}_{ob}) + i\hat{h}(\hat{s}_{ob})\Big\} = \frac{1}{\sqrt{2}}\Big\{\hat{v}(\pi-\theta_i,\phi_i) + i\hat{h}(\pi-\theta_i,\phi_i)\Big\} \\
&= \frac{1}{\sqrt{2}}\Big\{-\hat{h}(-k_{iz}) - i\hat{e}(-k_{iz})\Big\}
\end{aligned} \tag{1.3.51}$$

Because of the linearity of Maxwell's equations, we have

$$\begin{aligned}
\overline{E}_s(\overline{r}) = {} & \frac{1}{\sqrt{2}}\Big\{\Big(-iR_{ho}\hat{e}(k_{iz}) - R_{vo}\hat{h}(k_{iz})\Big)e^{i\overline{k}_{i\perp}\cdot\overline{r}_\perp + ik_{iz}z} \\
& + \int d\overline{k}_\perp e^{i\overline{k}_\perp\cdot\overline{r}_\perp + ik_z z} iF(\overline{k}_\perp - \overline{k}_{i\perp})\Big[\hat{e}(k_z)\Big(-if_{ee}^{(1)}(\overline{k}_\perp, \overline{k}_{i\perp}) - f_{eh}^{(1)}(\overline{k}_\perp, \overline{k}_{i\perp})\Big) \\
& + \hat{h}(k_z)\Big(-if_{he}^{(1)}(\overline{k}_\perp, \overline{k}_{i\perp}) - f_{hh}^{(1)}(\overline{k}_\perp, \overline{k}_{i\perp})\Big)\Big] \\
& + \int d\overline{k}_\perp e^{i\overline{k}_\perp\cdot\overline{r}_\perp + ik_z z}\Big[\hat{e}(k_z)\Big(-if_{ee}^{(2)}(\overline{k}_\perp, \overline{k}_{i\perp}) - f_{eh}^{(2)}(\overline{k}_\perp, \overline{k}_{i\perp})\Big) \\
& + \hat{h}(k_z)\Big(-if_{he}^{(2)}(\overline{k}_\perp, \overline{k}_{i\perp}) - f_{hh}^{(2)}(\overline{k}_\perp, \overline{k}_{i\perp})\Big)\Big]\Big\}
\end{aligned} \tag{1.3.52}$$

The magnetic field can be calculated similarly. The scattered power per unit area is

$$\begin{aligned}
\langle\overline{S}_s\cdot\hat{z}\rangle = {} & \frac{1}{4\eta}\Big\{\cos\theta_i\left(|R_{ho}|^2 + |R_{vo}|^2\right) \\
& + \mathrm{Re}\Big(R_{ho}\left(\tilde{f}_{ee}^{(2)*}(\overline{k}_{i\perp}) + i\tilde{f}_{eh}^{(2)*}(\overline{k}_{i\perp})\right)\cos\theta_i\Big) \\
& + \mathrm{Re}\Big(R_{ho}^*\left(\tilde{f}_{ee}^{(2)}(\overline{k}_{i\perp}) - i\tilde{f}_{eh}^{(2)}(\overline{k}_{i\perp})\right)\cos\theta_i\Big) \\
& + \mathrm{Re}\Big(R_{vo}\left(-i\tilde{f}_{he}^{(2)*}(\overline{k}_{i\perp}) + \tilde{f}_{hh}^{(2)*}(\overline{k}_{i\perp})\right)\cos\theta_i\Big)
\end{aligned}$$

$$+\operatorname{Re}\Big(R_{vo}^*\left(i\tilde{f}_{he}^{(2)}(\overline{k}_{i\perp})+\tilde{f}_{hh}^{(2)}(\overline{k}_{i\perp})\right)\cos\theta_i\Big)$$
$$+\operatorname{Re}\int d\overline{k}_\perp\cos\theta_s W(\overline{k}_\perp-\overline{k}_{i\perp})\Big(\left|f_{ee}^{(1)}(\overline{k}_\perp,\overline{k}_{i\perp})-if_{eh}^{(1)}(\overline{k}_\perp,\overline{k}_{i\perp})\right|^2$$
$$+\left|f_{he}^{(1)}(\overline{k}_\perp,\overline{k}_{i\perp})-if_{hh}^{(1)}(\overline{k}_\perp,\overline{k}_{i\perp})\right|^2\Big)\Big\} \tag{1.3.53}$$

The bistatic scattering coefficients are

$$4\pi\gamma_{hR}(\theta_s,\phi_s;\pi-\theta_i,\phi_i)$$
$$=\frac{1}{2}\Big\{\Big[|R_{ho}|^2+2\operatorname{Re}\Big(R_{ho}\left(\tilde{f}_{ee}^{(2)*}(\overline{k}_{i\perp})+i\tilde{f}_{eh}^{(2)*}(\overline{k}_{i\perp})\right)\Big)\Big]$$
$$\cdot\delta(\cos\theta_s-\cos\theta_i)\delta(\phi_s-\phi_i)$$
$$+\frac{k^2\cos^2\theta_s}{\cos\theta_i}W(\overline{k}_\perp-\overline{k}_{i\perp})\left|f_{ee}^{(1)}(\overline{k}_\perp,\overline{k}_{i\perp})-if_{eh}^{(1)}(\overline{k}_\perp,\overline{k}_{i\perp})\right|^2\Big\} \tag{1.3.54}$$

$$4\pi\gamma_{vR}(\theta_s,\phi_s;\pi-\theta_i,\phi_i)$$
$$=\frac{1}{2}\Big\{\Big[|R_{vo}|^2+2\operatorname{Re}\Big(R_{vo}\left(-i\tilde{f}_{he}^{(2)*}(\overline{k}_{i\perp})+\tilde{f}_{hh}^{(2)*}(\overline{k}_{i\perp})\right)\Big)\Big]$$
$$\cdot\delta(\cos\theta_s-\cos\theta_i)\delta(\phi_s-\phi_i)$$
$$+\frac{k^2\cos^2\theta_s}{\cos\theta_i}W(\overline{k}_\perp-\overline{k}_{i\perp})\left|f_{he}^{(1)}(\overline{k}_\perp,\overline{k}_{i\perp})-if_{hh}^{(1)}(\overline{k}_\perp,\overline{k}_{i\perp})\right|^2\Big\} \tag{1.3.55}$$

Hence the emissivity of the fourth Stokes parameter is, from (3.5.68) of Volume I and the bistatic scattering coefficients in this section,

$$e_4(\theta_i,\pi+\phi_i)$$
$$=\frac{1}{4\pi}\sum_{\beta=v,h}\int_0^{2\pi}d\phi_s\int_0^{\pi/2}d\theta_s\sin\theta_s\Big[2\gamma_{\beta R}(\hat{s},\hat{s}_{ob})-\gamma_{\beta h}(\hat{s},\hat{s}_{ob})-\gamma_{\beta v}(\hat{s},\hat{s}_{ob})\Big]$$
$$=\int_0^{2\pi}d\phi_s\int_0^{\pi/2}d\theta_s\sin\theta_s\cdot$$
$$\Big[-2\operatorname{Im}\Big(R_{ho}\tilde{f}_{eh}^{(2)*}(\overline{k}_{\perp i})\Big)\delta(\cos\theta_s-\cos\theta_i)\delta(\phi_s-\phi_i)$$
$$-\frac{k^2\cos^2\theta_s}{\cos\theta_i}W(\overline{k}_\perp-\overline{k}_{i\perp})2\operatorname{Im}\Big(f_{ee}^{(1)}(\overline{k}_\perp,\overline{k}_{i\perp})f_{eh}^{(1)*}(\overline{k}_\perp,\overline{k}_{i\perp})\Big)$$
$$+2\operatorname{Im}\Big(R_{vo}\tilde{f}_{he}^{(2)*}(\overline{k}_{i\perp})\Big)\delta(\cos\theta_s-\cos\theta_i)\delta(\phi_s-\phi_i)$$
$$+\frac{k^2\cos^2\theta_s}{\cos\theta_i}W(\overline{k}_\perp-\overline{k}_{i\perp})2\operatorname{Im}\Big(f_{hh}^{(1)}(\overline{k}_\perp,\overline{k}_{i\perp})f_{he}^{(1)*}(\overline{k}_\perp,\overline{k}_{i\perp})\Big)\Big] \tag{1.3.56}$$

Simplifying, we get

$$\begin{aligned} e_4(\theta_i, \pi + \phi_i) = & -2\mathrm{Im}\Big(R_{ho}\tilde{f}^{(2)*}_{eh}(\overline{k}_{i\perp})\Big) + 2\mathrm{Im}\Big(R_{vo}\tilde{f}^{(2)*}_{he}(\overline{k}_{i\perp})\Big) \\ & + \int_0^{2\pi} d\phi_s \int_0^{\pi/2} d\theta_s \sin\theta_s \frac{k^2\cos^2\theta_s}{\cos\theta_i} W(\overline{k}_\perp - \overline{k}_{i\perp}) \\ & \cdot \Big[-2\mathrm{Im}\Big(f^{(1)}_{ee}(\overline{k}_\perp, \overline{k}_{i\perp}) f^{(1)*}_{eh}(\overline{k}_\perp, \overline{k}_{i\perp})\Big) \\ & + 2\mathrm{Im}\Big(f^{(1)}_{hh}(\overline{k}_\perp, \overline{k}_{i\perp}) f^{(1)*}_{he}(\overline{k}_\perp, \overline{k}_{i\perp})\Big)\Big] \end{aligned} \tag{1.3.57}$$

3.3 Bistatic Scattering Coefficients

We have presented the bistatic scattering coefficients by listing $\langle |E_s^{(0)}|^2\rangle$, $E_s^{(0)}\langle E_s^{(2)*}\rangle$, and $\langle E_s^{(1)} E_s^{(1)*}\rangle$. To go to a higher order presents a problem, because even assuming that the odd moments of F to be zero, we need to have $\langle E_s^{(2)} E_s^{(2)*}\rangle$ and $\langle E_s^{(1)} E_s^{(3)}\rangle$. However, $E_s^{(3)}$ is complicated. Thus in the following, we shall only include $\langle E_s^{(2)} E_s^{(2)*}\rangle$.

We shall label $\gamma^{(2)}$ as given by

$$\langle \overline{S}_s \cdot \hat{z}\rangle = \frac{1}{2}\mathrm{Re}\Big[\langle \overline{E}_s^{(2)} \times \overline{H}_s^{(2)*}\rangle \cdot \hat{z} - \langle \overline{E}_s^{(2)}\rangle \times \langle \overline{H}_s^{(2)*}\rangle \cdot \hat{z}\Big] \tag{1.3.58}$$

which represents incoherent scattering.

From (1.2.75a) we have

$$\begin{aligned} f^{(2)}_{\alpha\beta}(\overline{k}_\perp, \overline{k}_{i\perp}) = & \int_{-\infty}^{\infty} d\overline{k}'_\perp \left[\tilde{\tilde{f}}_{\alpha\beta}(\overline{k}_\perp, \overline{k}'_\perp, \overline{k}_{i\perp})\right] \\ & \cdot F(\overline{k}_\perp - \overline{k}'_\perp) F(\overline{k}'_\perp - \overline{k}_{i\perp}) \end{aligned} \tag{1.3.59}$$

$$\tilde{\tilde{f}}_{\alpha\beta}(\overline{k}_\perp, \overline{k}'_\perp, \overline{k}_{i\perp}) = \tilde{\tilde{f}}_{\alpha\beta a}(\overline{k}_\perp, \overline{k}_{i\perp}) + \tilde{\tilde{f}}_{\alpha\beta b}(\overline{k}_\perp, \overline{k}'_\perp, \overline{k}_{i\perp}) \tag{1.3.60}$$

where $\tilde{\tilde{f}}_{\alpha\beta a}$'s and $\tilde{\tilde{f}}_{\alpha\beta b}$'s are listed in (1.2.75). Then, for example,

$$\begin{aligned} & \langle f^{(2)}_{ee}(\overline{k}_{\perp 1}, \overline{k}_{i\perp}) f^{(2)*}_{ee}(\overline{k}_{\perp 2}, \overline{k}_{i\perp})\rangle - \langle f^{(2)}_{ee}(\overline{k}_{\perp 1}, \overline{k}_{i\perp})\rangle\langle f^{(2)*}_{ee}(\overline{k}_{\perp 2}, \overline{k}_{i\perp})\rangle \\ & = \int_{-\infty}^{\infty} d\overline{k}'_\perp \int_{-\infty}^{\infty} d\overline{k}''_\perp\, \tilde{\tilde{f}}_{ee}(\overline{k}_{\perp 1}, \overline{k}'_\perp, \overline{k}_{i\perp}) \tilde{\tilde{f}}^*_{ee}(\overline{k}_{\perp 2}, \overline{k}''_\perp, \overline{k}_{i\perp}) \\ & \Big\{\langle F(\overline{k}_{\perp 1} - \overline{k}'_\perp) F(\overline{k}'_\perp - \overline{k}_\perp) F^*(\overline{k}_{\perp 2} - \overline{k}''_\perp) F^*(\overline{k}''_\perp - \overline{k}_{i\perp})\rangle \\ & - \langle F(\overline{k}_{\perp 1} - \overline{k}'_\perp) F(\overline{k}'_\perp - \overline{k}_\perp)\rangle\langle F^*(\overline{k}_{\perp 2} - \overline{k}''_\perp) F^*(\overline{k}''_\perp - \overline{k}_{i\perp})\rangle\Big\} \end{aligned} \tag{1.3.61}$$

$$= \int_{-\infty}^{\infty} d\overline{k}'_\perp \int_{-\infty}^{\infty} d\overline{k}''_\perp \bar{\bar{f}}_{ee}(\overline{k}_{\perp 1}, \overline{k}'_\perp, \overline{k}_{i\perp}) \bar{\bar{f}}^*_{ee}(\overline{k}_{\perp 2}, \overline{k}''_\perp, \overline{k}_{i\perp})$$
$$\Big\{ \delta(\overline{k}_{1\perp} - \overline{k}_{2\perp}) \delta(\overline{k}'_\perp - \overline{k}''_\perp) W(\overline{k}_{1\perp} - \overline{k}'_\perp) W(\overline{k}'_\perp - \overline{k}_{i\perp})$$
$$+ \delta(\overline{k}_{1\perp} - \overline{k}_{2\perp}) \delta(\overline{k}''_\perp + \overline{k}'_\perp - \overline{k}_{i\perp} - \overline{k}_{1\perp}) W(\overline{k}_{1\perp} - \overline{k}'_\perp) W(\overline{k}'_\perp - \overline{k}_{i\perp}) \Big\}$$
$$= \delta(\overline{k}_{1\perp} - \overline{k}_{2\perp}) \int_{-\infty}^{\infty} d\overline{k}'_\perp \Big[\bar{\bar{f}}_{ee}(\overline{k}_{\perp 1}, \overline{k}'_\perp, \overline{k}_{i\perp}) \bar{\bar{f}}^*_{ee}(\overline{k}_{\perp 1}, \overline{k}'_\perp, \overline{k}_{i\perp})$$
$$+ \bar{\bar{f}}_{ee}(\overline{k}_{\perp 1}, \overline{k}'_\perp, \overline{k}_{i\perp}) \bar{\bar{f}}^*_{ee}(\overline{k}_{\perp 1}, \overline{k}_{1\perp} + \overline{k}_{i\perp} - \overline{k}'_\perp, \overline{k}_{i\perp}) \Big]$$
$$\cdot W(\overline{k}_{\perp 1} - \overline{k}'_\perp) W(\overline{k}'_\perp - \overline{k}_{i\perp}) \tag{1.3.62}$$

Thus if we define incoherent field as

$$\overline{\mathcal{E}}^{(2)}_s = \overline{E}^{(2)}_s - \langle \overline{E}^{(2)}_s \rangle \tag{1.3.63}$$

and calculate Poynting's vector for an incoherent wave as

$$\frac{1}{2} \text{Re}\Big(\hat{z} \cdot \overline{\mathcal{E}}^{(2)}_s \times \frac{\nabla \times \mathcal{E}^{(2)*}_s(\overline{r})}{-i\omega\mu} \Big)$$

then the product is similar to that of the previous section. For horizontal polarization incidence $\hat{e}_i = \hat{e}(-k_{iz})$ we have

$$\gamma_{hh}(\theta_s, \phi_s; \pi - \phi_i, \phi_i)$$
$$= \frac{4\pi k^2 \cos^2\theta_s}{\cos\theta_i} \Big\{ \int_{-\infty}^{\infty} d\overline{k}'_\perp W(\overline{k}_\perp - \overline{k}'_\perp) W(\overline{k}'_\perp - \overline{k}_{i\perp}) \bar{\bar{f}}_{ee}(\overline{k}_\perp, \overline{k}'_\perp, \overline{k}_{i\perp})$$
$$\Big[\bar{\bar{f}}^*_{ee}(\overline{k}_\perp, \overline{k}'_\perp, \overline{k}_{i\perp}) + \bar{\bar{f}}^*_{ee}(\overline{k}_\perp, \overline{k}_\perp - \overline{k}'_\perp + \overline{k}_{i\perp}, \overline{k}_{i\perp}) \Big] \Big\} \tag{1.3.64a}$$

$$\gamma_{vh}(\theta_s, \phi_s; \pi - \phi_i, \phi_i)$$
$$= \frac{4\pi k^2 \cos^2\theta_s}{\cos\theta_i} \Big\{ \int_{-\infty}^{\infty} d\overline{k}'_\perp W(\overline{k}_\perp - \overline{k}'_\perp) W(\overline{k}'_\perp - \overline{k}_{i\perp}) \bar{\bar{f}}_{he}(\overline{k}_\perp, \overline{k}'_\perp, \overline{k}_{i\perp})$$
$$\Big[\bar{\bar{f}}^*_{he}(\overline{k}_\perp, \overline{k}'_\perp, \overline{k}_{i\perp}) + \bar{\bar{f}}^*_{he}(\overline{k}_\perp, \overline{k}_\perp - \overline{k}'_\perp + \overline{k}_{i\perp}, \overline{k}_{i\perp}) \Big] \Big\} \tag{1.3.64b}$$

Thus for vertical polarization incidence $\hat{e}_i = \hat{h}(-k_{iz})$

$$\gamma_{hv}(\theta_s, \phi_s; \pi - \phi_i, \phi_i)$$
$$= \frac{4\pi k^2 \cos^2\theta_s}{\cos\theta_i} \Big\{ \int_{-\infty}^{\infty} d\overline{k}'_\perp W(\overline{k}_\perp - \overline{k}'_\perp) W(\overline{k}'_\perp - \overline{k}_{i\perp}) \bar{\bar{f}}_{eh}(\overline{k}_\perp, \overline{k}'_\perp, \overline{k}_{i\perp})$$
$$\cdot \Big[\bar{\bar{f}}^*_{eh}(\overline{k}_\perp, \overline{k}'_\perp, \overline{k}_{i\perp}) + \bar{\bar{f}}^*_{eh}(\overline{k}_\perp, \overline{k}_\perp - \overline{k}'_\perp + \overline{k}_{i\perp}, \overline{k}_{i\perp}) \Big] \Big\} \tag{1.3.64c}$$

$$\gamma_{vv}(\theta_s,\phi_s;\pi-\phi_i,\phi_i)$$
$$= \frac{4\pi k^2\cos^2\theta_s}{\cos\theta_i}\left\{\int_{-\infty}^{\infty} d\overline{k}'_\perp W(\overline{k}_\perp-\overline{k}'_\perp)W(\overline{k}'_\perp-\overline{k}_{i\perp})\tilde{\tilde{f}}_{hh}(\overline{k}_\perp,\overline{k}'_\perp,\overline{k}_{i\perp})\right.$$
$$\left.\cdot\left[\tilde{\tilde{f}}^*_{hh}(\overline{k}_\perp,\overline{k}'_\perp,\overline{k}_{i\perp})+\tilde{\tilde{f}}^*_{hh}(\overline{k}_\perp,\overline{k}_\perp-\overline{k}'_\perp+\overline{k}_{i\perp},\overline{k}_{i\perp})\right]\right\} \tag{1.3.64d}$$

To compute numerical results from (1.3.64*a*–*d*), Eqs. (1.3.60) and (1.2.75) need to be used.

REFERENCES AND ADDITIONAL READINGS

Bahar, E. and M. A. Fitzwater (1989), Depolarization and backscatter enhancement in light scattering from random rough surfaces: Comparison of full-wave theory with experiment, *J. Opt. Soc. Am. A*, **6**, 33–43.

Bass, F. G. and I. M. Fuks (1979), *Wave Scattering from Statistically Rough Surfaces*, translated by C. B. Vesecky and J. F. Vesecky, Pergamon Press, Oxford, UK.

Baylard, C., J.-J. Greffet, and A. A. Maradudin (1993), Coherent reflection factor of a random rough surface: applications, *J. Opt. Soc. Am. A*, **10**, 2637–2647.

Beckmann, P. and A. Spizzichino (1963), *The Scattering of Electromagnetic Waves from Rough Surfaces*, Pergamon Press, Oxford.

Broschat, S. L. (1993), The small slope approximation reflection coefficient for scattering from a "Pierson-Moskowitz" sea surface, *IEEE Trans. Geosci. Remote Sens.*, **31(5)**, 1112–1114.

Broschat, S. L., L. Tsang, A. Ishimaru, and E. I. Thorsos (1988), A numerical comparison of the phase perturbation technique with the classical field perturbation and Kirchhoff approximations for random rough surface scattering problems, *J. Electromag. Waves and Appl.*, **2(1)**, 85–102.

Brown, G. S. (1978), Backscattering from a Gaussian distributed perfectly conducting rough surface, *IEEE Trans. Antennas Propagat.*, **26**, 472–482.

Brown, G. S. (1985), Simplification in the stochastic Fourier transform approach to random rough surface scattering, *IEEE Trans. Antennas Propagat.*, **33**, 48–55.

Chen, K. S. and A. K. Fung (1995), A comparison of backscattering models for rough surfaces, *IEEE Trans. Geosci. Remote Sens.*, **33(1)**, 195–199.

Chen, K. S., T. D. Wu, M. K. Tsay, and A. K. Fung (2000), A note on the multiple scattering of the IEM model, *IEEE Trans. Geosci. Remote Sens.*, **38(1)**, 249–256.

Collin, R. E. (1994), Full wave theories for rough surface scattering: An updated assessment, *Radio Sci.*, **29**, 1237–1254.

De Roo, R. D. and F. T. Ulaby (1994), Bistatic specular scattering from rough dielectric surfaces, *IEEE Trans. Antennas Propagat.*, **42(2)**, 220–231.

DeSanto, J. A. (1974), Green's function for electromagnetic scattering from a random rough surfaces, *J. Math. Phys.*, **15**, 283–288.

DeSanto, J. A. and O. Shisha (1974), Numerical solution of a singular integral equation in random rough surface scattering theory, *J. of Comp. Phys.*, **15**, 286–292.

DeSanto, J. A. and R. J. Wombell (1991), Rough surface scattering, *Waves in Random Media*, **1**, 41–56.

Durden, S. L. and J. F. Vesecky (1985), A physical radar cross-section model for a wind driven sea with swell, *IEEE J. Ocean. Eng.*, **10**, 445–451.

Durden, S. L. and J. F. Vesecky (1990), A numerical study of the separation wavenumber in the two scale scattering approximation, *IEEE Trans. Geosci. Remote Sens.*, **28**, 271–272.

Fung, A. K. (1994), *Microwave Scattering and Emission Models and Their Applications*, Artech House, Norwood, Massachusetts.

Fung, A. K. and G. W. Pan (1987), A scatter model for perfectly conducting random surface: I. Model development. II. Range of validity, *Int. J. Remote Sens.*, **8(11)**, 1579–1605.

Greffet, J. J. and M. Nieto-Vesperinas (1998), Field theory for generalized bidirectional reflectivity: derivation of Helmholtz's reciprocity principle and Kirchhoff's law, *J. Opt. Soc. Am. A*, **15(10)**, 2735–2744.

Irisov, V. G. (1994), Small slope expansion for electromagnetic-wave diffraction on a rough surface, *Waves in Random Media*, **4(4)**, 441–452.

Irisov, V. G. (1997), Small-slope expansion for thermal and reflected radiation from a rough surface, *Waves in Random Media*, **7(1)**, 1–10.

Ishimaru, A. (1978), *Wave Propagation and Scattering in Random Media*, **1** and **2**, Academic Press, New York.

Ishimaru, A. and J. S. Chen (1990), Scattering from very rough surfaces based on the modified second order Kirchhoff approximation with angular and propagation shadowing, *J. Acous. Soc. Am. A*, **88**, 1877–1883.

Ishimaru, A. and J. S. Chen (1991), Scattering from very rough metallic and dielectric surfaces: a theory based on the modified Kirchhoff approximation, *Waves in Random Media*, **1(1)**, 21–34.

Ito, S. (1985), Analysis of scalar wave scattering from slightly rough random surface: A multiple scattering theory, *Radio Sci.*, **20**, 1–12.

Jackson, D. R., D. P. Winebrenner, and A. Ishimaru (1988), Comparison of perturbation theories for rough-surface scattering, *J. Acous. Soc. Am.*, **83(3)**, 961–969.

Johnson, J. T. (1999), Third order small perturbation method for scattering from dielectric rough surfaces, *J. Opt. Soc. Am. A*, **16**, 2720–2736.

Li, Z. and A. K. Fung (1991), A reformulation of the surface field integral equation, *J. Electromag. Waves and Appl.*, **5**, 195–203.

Nieto-Vesperinas, M. (1991), *Scattering and Diffraction in Physical Optics*, Wiley-Interscience, New York.

Ogilvy, J. A. (1991), *Theory of Wave Scattering from Random Rough Surfaces*, Adam Hilger, Bristol.

Ogura, H. and N. Takahashi (1995), Wave scattering from a random rough surface: reciprocal theorem and backscattering enhancement, *Waves in Random Media*, **5(2)**, 223–242.

Plant, W. J. (1986), A two-scale model of short wind generated waves and scatterometry, *J. Geophys. Res.*, **91(C9)**, 10735–10749.

Rice, S. O. (1951), Reflection of electromagnetic waves from slightly rough surfaces, *Comm. Pure Appl. Math.*, **4**, 351–378.

Rice, S. O. (1963), Reflection of EM waves by slightly rough surfaces, *The Theory of Electromagnetic Waves*, M. Kline, Ed., Interscience, New York.

Rodriguez, E. and Y. Kim (1992), A unified perturbation expansion for surface scattering, *Radio Sci.*, **27**, 79–93.

Sentenac, A. and J.-J. Greffet (1998), Mean-field theory of light scattering by one-dimensional rough surfaces, *J. Opt. Soc. Am. A*, **15**, 528–532.

Shin, R. T. (1984), Theoretical models for microwave remote sensing of earth terrain, Ph.D. thesis, Department of Electrical Engineering and Computer Science, Massachusetts Institute of Technology.

Soto-Crespo, J. M., M. Nieto-Vesperinas, and A. Friberg (1990), Scattering from slightly rough random surfaces: a detailed study on the validity of the small perturbation method, *J. Opt. Soc. Am. A*, **7**, 1185–1201.

Thorsos, E. I. and S. L. Broschat (1995), An investigation of the small slope approximation for scattering from rough surfaces. Part I. Theory, *J. Acous. Soc. Am.*, **97(4)**, 2082–2093.

Thorsos, E. I. and D. Winebrenner (1991), An examination of the full-wave method for rough surface scattering in the case of small roughness, *J. Geophys. Res.*, **19(C9)**, 17107–17121.

Tsang, L., J. A. Kong, and R. T. Shin (1985), *Theory of Microwave Remote Sensing*, Wiley-Interscience, New York.
Ulaby, F. T., R. K. Moore, and A. K. Fung (1981), *Microwave Remote Sensing: Active and Passive*, **1** and **2**, Addison-Wesley, Reading, MA.
Ulaby, F. T., R. K. Moore, and A. K. Fung (1986), *Microwave Remote Sensing: Active and Passive*, **3**, Artech House, Norwood, Massachusetts.
Valenzuela, G. R. (1967), Depolarization of EM waves by slightly rough surfaces, *IEEE Trans. Antennas Propagat.*, **15**, 552–557.
Valenzuela, G. R. (1968), Scattering of electromagnetic waves from a tilted slightly rough surface, *Radio Sci.*, **3**, 1057–1066.
Voronovich, A. G. (1994a), Small-slope approximation for electromagnetic wave scattering at a rough interface of two dielectric half-spaces, *Waves in Random Media*, **4(3)**, 337–367.
Voronovich, A. G. (1994b), *Wave Scattering from Rough Surfaces*, Springer-Verlag, Berlin.
Voronovich, A. G. (1996), A two-scale model from the point of view of the small-slope approximation, *Waves in Random Media*, **6(1)**, 73–83.
Winebrenner, D. P. and A. Ishimaru (1985a), Application of the phase-perturbation technique to randomly rough surface, *J. Opt. Soc. Am. A*, **2**, 2285–2294.
Winebrenner, D. P. and A. Ishimaru (1985b), Investigation of a surface field perturbation technique for scattering from rough surface, *Radio Sci.*, **20**, 161–170.
Yueh, S. H. (1997), Modeling of wind direction signal in polarimetric sea surface brightness temperatures, *IEEE Trans. Geosci. Remote Sens.*, **35(6)**, 1400–1418.

Chapter 2

KIRCHHOFF APPROACH AND RELATED METHODS FOR ROUGH SURFACE SCATTERING

In this chapter, we discuss Kirchhoff approach for two-dimensional surfaces. We studied both perfectly conducting surfaces and dielectric surfaces and also including higher order slope corrections. We also illustrate the phase perturbation method [Winebrenner and Ishimaru, 1985a,b] for a one-dimensional Dirichlet problem. In Section 3, the composite surface model is used to calculate emissivity of a two-scale random rough surface. The composite surface model has been extensively used in ocean scattering and emission problems.

1 Kirchhoff Approach

The Kirchhoff approach applies the tangent plane approximation to calculate the surface fields. This is valid if the radius of curvature of the surface is larger than a wavelength. In Volume I, we studied the case of one-dimensional surface. In this section we apply the Kirchhoff method to the vector electromagnetic case for two-dimensional random rough surface.

1.1 Perfectly Conducting Rough Surface

Consider an incident electromagnetic wave

$$\overline{E}_i = \hat{e}_i \exp(i\overline{k}_i \cdot \overline{r}) \tag{2.1.1}$$

($\overline{k}_i = k_{ix}\hat{x} + k_{iy}\hat{y} - k_{iz}\hat{z}$) impinging upon a perfectly conducting random rough surface of $z = f(x, y)$. From Huygens' principle and the boundary condition that $\hat{n} \times \overline{E} = 0$, we have

$$\overline{E}_s(\overline{r}') = \int_S dS\, i\omega\mu\overline{\overline{G}}(\overline{r}', \overline{r}) \cdot \hat{n} \times \overline{H}(\overline{r}) \tag{2.1.2}$$

The Kirchhoff approach does not use the extinction theorem equation and uses the Huygen's principle directly. This can be done by approximating the surface fields. We let

$$\overline{N} = -\frac{\partial f}{\partial x}\hat{x} - \frac{\partial f}{\partial y}\hat{y} + \hat{z} \tag{2.1.3}$$

The Kirchhoff approximation is

$$\hat{n} \times \overline{H}(\overline{r}) = 2\hat{n} \times \overline{H}_{inc}(\overline{r}) \tag{2.1.4}$$

Substitution into (2.1.2) gives

$$\begin{aligned}\overline{E}_s(\overline{r}') &= 2\int dx\, dy\, i\omega\mu\overline{\overline{G}}(\overline{r}', \overline{r}) \cdot \overline{N} \times \overline{H}_{inc}(\overline{r}) \\ &= 2i\int dx\, dy \overline{\overline{G}}(\overline{r}', \overline{r}) \cdot \overline{N} \times (\overline{k}_i \times \hat{e}_i)e^{i\overline{k}_i \cdot \overline{r}}\end{aligned} \tag{2.1.5}$$

Using dyadic Green's function of Eq. (2.1.20) in Volume I, with the spectral representation, we have

$$\overline{E}_s(\overline{r}') = \int_{-\infty}^{\infty} d\overline{k}_\perp \int_{-\infty}^{\infty} d\overline{r}_\perp \, e^{i\overline{k}_\perp\cdot\overline{r}'_\perp + ik_z z'} \Big(-\frac{1}{4\pi^2}\Big)\frac{k}{k_z}\Big[\hat{e}(k_z)\hat{e}(k_z) + \hat{h}(k_z)\hat{h}(k_z)\Big]$$
$$\cdot\Big(-\frac{\partial f}{\partial x}\hat{x} - \frac{\partial f}{\partial y}\hat{y} + \hat{z}\Big) \times (\hat{k}_i \times \hat{e}_i)\, e^{i\overline{k}_{d\perp}\cdot\overline{r}_\perp + ik_{dz} f(\overline{r}_\perp)} \tag{2.1.6}$$

In (2.1.6), $\overline{k}_d$ is the difference wave vector between the incident wave vector and the scattered wave vector. Let the scattered wave vector be $\overline{k}$. Then

$$\overline{k}_d = \overline{k}_i - \overline{k} \tag{2.1.7}$$
$$\overline{k}_{d\perp} = (k_{ix} - k_x)\hat{x} + (k_{iy} - k_y)\hat{y} \tag{2.1.8}$$
$$k_{dz} = -(k_{iz} + k_z) \tag{2.1.9}$$

Note that it is additive inside the parenthesis in (2.1.9) because the incident wave is downward-going while the scattered wave is upward-going. We use integration by parts

$$-\int dx \frac{\partial f}{\partial x} e^{ik_{dx}x + ik_{dz} f(x,y)} = \frac{k_{dx}}{k_{dz}} \int dx e^{ik_{dx}x + ik_{dz} f(x,y)}$$

Thus

$$\overline{E}_s(\overline{r}) = \int_{-\infty}^{\infty} d\overline{k}_\perp \, e^{i\overline{k}_\perp\cdot\overline{r}_\perp + ik_z z}\overline{E}_s(\overline{k}_\perp) \tag{2.1.10}$$

where

$$\overline{E}_s(\overline{k}_\perp) = \overline{E}_{sI}(\overline{k}_\perp) I(\overline{k}_\perp) \tag{2.1.11a}$$

$$\overline{E}_{sI}(\overline{k}_\perp) = \Big(-\frac{1}{4\pi^2}\Big)\frac{k}{k_z}\Big[\hat{e}(k_z)\hat{e}(k_z) + \hat{h}(k_z)\hat{h}(k_z)\Big] \cdot \frac{\overline{k}_d}{k_{dz}} \times (\hat{k}_i \times \hat{e}_i) \tag{2.1.11b}$$

and

$$I(\overline{k}_\perp) = \int_{-\infty}^{\infty} d\overline{r}_\perp \, e^{i\overline{k}_{d\perp}\cdot\overline{r}_\perp + ik_{dz} f(\overline{r}_\perp)} \tag{2.1.12}$$

is the diffraction integral. Let h be the rms height of the surface so that $\langle \exp(i\nu f)\rangle = \exp\Big(-\frac{h^2\nu^2}{2}\Big)$. Taking an average of the diffraction integral I gives

$$\langle I\rangle = \int d\overline{r}_\perp \, e^{i\overline{k}_{d\perp}\cdot\overline{r}_\perp}\langle e^{ik_{dz} f(\overline{r}_\perp)}\rangle = \int d\overline{r}_\perp \, e^{i\overline{k}_{d\perp}\cdot\overline{r}_\perp} \exp\Big(-\frac{h^2 k_{dz}^2}{2}\Big)$$
$$= 4\pi^2 \exp(-2h^2 k_{iz}^2)\delta(\overline{k}_{d\perp}) \tag{2.1.13}$$

Thus, on putting (2.1.13) into (2.1.10), the coherent scattered field is

$$\begin{aligned}\langle \overline{E}_s(\overline{k}_\perp)\rangle = & -\delta(\overline{k}_\perp - \overline{k}_{i\perp})\exp(-2h^2k_{iz}^2) \\ & \frac{k}{k_{iz}}\Big(\hat{e}(k_{iz})\hat{e}(k_{iz}) + \hat{h}(k_{iz})\hat{h}(k_{iz})\Big)\cdot\Big(\hat{z}\times(\hat{k}_i\times\hat{e}_i)\Big) \end{aligned} \tag{2.1.14}$$

$$\begin{aligned}\langle \overline{E}_s(\overline{r})\rangle = & -e^{i\overline{k}_{i\perp}\cdot\overline{r}_\perp + ik_{iz}z}\exp(-2h^2k_{iz}^2) \\ & \frac{k}{k_{iz}}\Big(\hat{e}(k_{iz})\hat{e}(k_{iz}) + \hat{h}(k_{iz})\hat{h}(k_{iz})\Big)\cdot\Big(\hat{z}\times(\hat{k}_i\times\hat{e}_i)\Big) \end{aligned} \tag{2.1.15}$$

For TE incidence $\hat{e}_i = \hat{e}(-k_{iz})$

$$\hat{z}\times(\hat{k}_i\times\hat{e}_i) = \frac{k_{iz}}{k}\hat{e}(k_{iz})$$

Hence

$$\langle \overline{E}_s(\overline{r})\rangle = -\hat{e}(k_{iz})e^{i\overline{k}_{i\perp}\cdot\overline{r}_\perp + ik_{iz}z}\exp(-2h^2k_{iz}^2) \tag{2.1.16}$$

For TM incidence $\hat{e}_i = \hat{h}(-k_{iz})$, $\hat{k}_i\times\hat{e}_i = \hat{e}(-k_{iz})$. Thus $-\hat{z}\times(\hat{k}_i\times\hat{e}_i) = -\frac{1}{k_{i\perp}}(\hat{x}k_{ix} + \hat{y}k_{iy})$, $-\hat{h}(k_{iz})\cdot\hat{z}\times(\hat{k}_i\times\hat{e}_i) = \frac{k_{iz}}{k}$. Hence

$$\langle \overline{E}_s(\overline{r})\rangle = \hat{h}(k_{iz})e^{i\overline{k}_{i\perp}\cdot\overline{r}_\perp + ik_{iz}z}\exp(-2h^2k_{iz}^2) \tag{2.1.17}$$

The scattered power per unit area is

$$\langle \overline{S}_s\cdot\hat{z}\rangle = \frac{1}{2}\mathrm{Re}\langle \overline{E}_s\times\overline{H}_s^*\rangle\cdot\hat{z} \tag{2.1.18}$$

The magnetic field is

$$\overline{H}_s(\overline{r}) = \int_{-\infty}^{\infty} d\overline{k}_\perp\, e^{i\overline{k}_\perp\cdot\overline{r}_\perp + ik_zz}\frac{1}{\omega\mu}\overline{k}\times\overline{E}_s(\overline{k}_\perp) = \int_{-\infty}^{\infty} d\overline{k}_\perp\, e^{i\overline{k}_\perp\cdot\overline{r}_\perp + ik_zz}\overline{H}_s(\overline{k}_\perp) \tag{2.1.19}$$

where

$$\overline{H}_s(\overline{k}_\perp) = \overline{H}_{sI}(\overline{k}_\perp)I(\overline{k}_\perp) \tag{2.1.20a}$$

$$\begin{aligned}\overline{H}_{sI}(\overline{k}_\perp) = & \frac{1}{\eta}\Big(-\frac{1}{4\pi^2}\Big)\frac{k}{k_z}\Big[-\hat{h}(k_z)\hat{e}(k_z) + \hat{e}(k_z)\hat{h}(k_z)\Big] \\ & \cdot\frac{\overline{k}_d}{k_{dz}}\times(\hat{k}_i\times\hat{e}_i)\end{aligned} \tag{2.1.20b}$$

$$\begin{aligned}\langle \overline{S}_s\cdot\hat{z}\rangle = & \frac{1}{2}\mathrm{Re}\int_{-\infty}^{\infty} d\overline{k}_{1\perp}\int_{-\infty}^{\infty} d\overline{k}_{2\perp}\, e^{i(\overline{k}_{1\perp}-\overline{k}_{2\perp})\cdot\overline{r}_\perp}e^{i(k_{1z}-k_{2z}^*)z} \\ & \cdot\Big[(\overline{E}_{sI}(\overline{k}_{1\perp})\times\overline{H}_{sI}^*(\overline{k}_{2\perp}))\cdot\hat{z}\Big] \\ & \cdot\Big(\langle I(\overline{k}_{1\perp})I^*(\overline{k}_{2\perp})\rangle - \langle I(\overline{k}_{1\perp})\rangle\langle I^*(\overline{k}_{2\perp})\rangle\Big)\end{aligned} \tag{2.1.21}$$

Taking averages, we have

$$\begin{aligned}&\langle I(\overline{k}_{1\perp})I^*(\overline{k}_{2\perp})\rangle - \langle I(\overline{k}_{1\perp})\rangle\langle I^*(\overline{k}_{2\perp})\rangle\\&= \int_{-\infty}^{\infty} d\overline{r}_{1\perp} \int_{-\infty}^{\infty} d\overline{r}_{2\perp}\, e^{i\overline{k}_{d\perp 1}\cdot\overline{r}_{\perp 1}}\, e^{-i\overline{k}_{d\perp 2}\cdot\overline{r}_{\perp 2}}\\&\quad\cdot\left\{\langle e^{ik_{dz1}f(\overline{r}_{\perp 1})}\, e^{-ik^*_{dz2}f(\overline{r}_{\perp 2})}\rangle - \langle e^{ik_{dz1}f(\overline{r}_{\perp 1})}\rangle\langle e^{-ik^*_{dz2}f(\overline{r}_{\perp 2})}\rangle\right\}\\&= \int_{-\infty}^{\infty} d\overline{r}_{1\perp} \int_{-\infty}^{\infty} d\overline{r}_{2\perp}\, e^{i\overline{k}_{d\perp 1}\cdot\overline{r}_{\perp 1} - i\overline{k}_{d\perp 2}\cdot\overline{r}_{\perp 2}} M(\overline{k}_{1\perp}, \overline{k}_{2\perp}, \overline{r}_{1\perp} - \overline{r}_{2\perp})\end{aligned} \tag{2.1.22}$$

where

$$\begin{aligned}&M(\overline{k}_{1\perp}, \overline{k}_{2\perp}, \overline{r}_{1\perp} - \overline{r}_{2\perp})\\&\quad= \left\{\langle e^{ik_{dz1}f(\overline{r}_{\perp 1}) - ik^*_{dz2}f(\overline{r}_{\perp 2})}\rangle - \langle e^{ik_{dz1}f(\overline{r}_{\perp 1})}\rangle\langle e^{-ik^*_{dz2}f(\overline{r}_{\perp 2})}\rangle\right\}\end{aligned} \tag{2.1.23}$$

depends only on the difference coordinate $\overline{r}_{1\perp} - \overline{r}_{2\perp}$. Using difference and average coordinates

$$\overline{r}_{1\perp} - \overline{r}_{2\perp} = \overline{r}_{\perp} \tag{2.1.24}$$

$$\frac{\overline{r}_{1\perp} + \overline{r}_{2\perp}}{2} = \overline{r}_{A\perp} \tag{2.1.25}$$

gives

$$\begin{aligned}&\langle I(\overline{k}_{1\perp})I^*(\overline{k}_{2\perp})\rangle - \langle I(\overline{k}_{1\perp})\rangle\langle I^*(\overline{k}_{2\perp})\rangle\\&= \int_{-\infty}^{\infty} d\overline{r}_{A\perp} \int_{-\infty}^{\infty} d\overline{r}_{\perp}\, M(\overline{k}_{1\perp}, \overline{k}_{2\perp}, \overline{r}_{\perp}) e^{i(\overline{k}_{d\perp 1} - \overline{k}_{d\perp 2})\cdot\overline{r}_{A\perp}}\, e^{i(\overline{k}_{d\perp 1} + \overline{k}_{d\perp 2})\cdot\overline{r}_{\perp}/2}\\&= 4\pi^2\delta(\overline{k}_{d\perp 1} - \overline{k}_{d\perp 2}) \int_{-\infty}^{\infty} d\overline{r}_{\perp}\, M(\overline{k}_{1\perp}, \overline{k}_{2\perp}, \overline{r}_{\perp})\, e^{i\overline{k}_{d\perp 1}\cdot\overline{r}_{\perp}}\end{aligned} \tag{2.1.26}$$

Substituting (2.1.26) into (2.1.21), the Poynting's vector becomes

$$\begin{aligned}\langle\overline{S}_s\cdot\hat{z}\rangle &= 2\pi^2\,\mathrm{Re}\int_{-\infty}^{\infty} d\overline{k}_{\perp}\left((\overline{E}_{sI}(\overline{k}_{\perp})\times\overline{H}_{sI}(\overline{k}_{\perp}))\cdot\hat{z}\right)\\&\quad\cdot\int_{-\infty}^{\infty} d\overline{r}_{\perp}\, M(\overline{k}_{\perp}, \overline{k}_{\perp}, \overline{r}_{\perp})\, e^{i\overline{k}_{d\perp}\cdot\overline{r}_{\perp}}\\&= \frac{1}{2\eta}\mathrm{Re}\int_{-\infty}^{\infty} d\overline{k}_{\perp}\, \frac{k}{k_z}\frac{1}{4\pi^2}\left\{\left(\hat{e}(k_z)\cdot\frac{\overline{k}_d\times(\hat{k}_i\times\hat{e}_i)}{k_{dz}}\right)^2\right.\\&\quad\left.+\left(\hat{h}(k_z)\cdot\frac{\overline{k}_d\times(\hat{k}_i\times\hat{e}_i)}{k_{dz}}\right)^2\right\}\\&\quad\cdot\int_{-\infty}^{\infty} d\overline{r}_{\perp}\, M(\overline{k}_{\perp}, \overline{k}_{\perp}, \overline{r}_{\perp})\, e^{i\overline{k}_{d\perp}\cdot\overline{r}_{\perp}}\end{aligned} \tag{2.1.27}$$

where

$$M(\overline{k}_\perp, \overline{k}_\perp, \overline{r}_\perp) = \left\{ \langle e^{ik_{dz}f(\overline{r}_{1\perp}) - ik_{dz}f(\overline{r}_{1\perp} - \overline{r}_\perp)} \rangle - \exp(-h^2 k_{dz}^2) \right\}$$
$$= \exp(-h^2 k_{dz}^2) \left\{ \exp(k_{dz}^2 h^2 C(\overline{r}_\perp)) - 1 \right\} \tag{2.1.28}$$

and $C(\overline{r}_\perp)$ is the covariance function. Then, we make a Taylor series expansion

$$\int_{-\infty}^{\infty} d\overline{r}_\perp \, M(\overline{k}, \overline{k}, \overline{r}_\perp) \, e^{i\overline{k}_{d\perp} \cdot \overline{r}_\perp}$$
$$= \exp(-h^2 k_{dz}^2) \sum_{m=1}^{\infty} \int_{-\infty}^{\infty} d\overline{r}_\perp \, e^{i\overline{k}_{d\perp} \cdot \overline{r}_\perp} \frac{h^{2m} k_{dz}^{2m}}{m!} C^m(\overline{r}_\perp)$$
$$= \exp(-h^2 k_{dz}^2) \sum_{m=1}^{\infty} 4\pi^2 \frac{k_{dz}^{2m}}{m!} W^{(m)}(\overline{k}_{d\perp}) \tag{2.1.29}$$

where

$$W^{(m)}(\overline{k}_{d\perp}) = \frac{1}{4\pi^2} \int_{-\infty}^{\infty} d\overline{r}_\perp \, e^{-i\overline{k}_\perp \cdot \overline{r}_\perp} \, h^{2m} C^m(\overline{r}_\perp) \tag{2.1.30}$$

Hence, substitution into (2.1.27) gives

$$\langle \overline{S}_s \cdot \hat{z} \rangle = \frac{1}{2\eta} \mathrm{Re} \int_{-\infty}^{\infty} d\overline{k}_\perp \frac{k}{k_z} \left\{ \left(\hat{e}(k_z) \cdot \frac{\overline{k}_d \times (\hat{k}_i \times \hat{e}_i)}{k_{dz}} \right)^2 \right.$$
$$\left. + \left(\hat{h}(k_z) \cdot \frac{\overline{k}_d \times (\hat{k}_i \times \hat{e}_i)}{k_{dz}} \right)^2 \right\}$$
$$\cdot \exp(-h^2 k_{dz}^2) \sum_{m=1}^{\infty} \frac{k_{dz}^{2m}}{m!} W^{(m)}(\overline{k}_{d\perp}) \tag{2.1.31}$$

Transforming to angular variables $k_x = k \sin\theta_s \cos\phi_s$, $k_y = k \sin\theta_s \sin\phi_s$, $k_z = k\cos\theta_s$, and $dk_x \, dk_y = k^2 \sin\theta_s \cos\theta_s d\theta_s d\phi_s$.

$$\langle \overline{S}_s \cdot \hat{z} \rangle = \frac{1}{2\eta} \mathrm{Re} \int_0^{\pi/2} d\theta_s \sin\theta_s \int_0^{2\pi} d\phi_s k^2 \left\{ \left(\hat{e}(k_z) \cdot \frac{\overline{k}_d \times (\hat{k}_i \times \hat{e}_i)}{k_{dz}} \right)^2 \right.$$
$$\left. + \left(\hat{h}(k_z) \cdot \frac{\overline{k}_d \times (\hat{k}_i \times \hat{e}_i)}{k_{dz}} \right)^2 \right\} \exp(-h^2 k_{dz}^2) \sum_{m=1}^{\infty} \frac{k_{dz}^{2m}}{m!} W^{(m)}(\overline{k}_{d\perp}) \tag{2.1.32}$$

Since

$$\overline{S}_i \cdot \hat{z} = \frac{\cos\theta_i}{2\eta} \tag{2.1.33}$$

and letting γ be the bistatic scattering coefficient, we obtain

$$\langle \overline{S}_s \cdot \hat{z} \rangle = \int_0^{\pi/2} d\theta_s \sin\theta_s \int_0^{2\pi} d\phi_s (\overline{S}_i \cdot \hat{z}) \frac{\gamma(\theta_s, \phi_s)}{4\pi} \tag{2.1.34}$$

where

$$\gamma(\theta_s,\phi_s;\theta_i,\phi_i) = 4\pi k^2 \left\{ \left(\hat{e}(k_z) \cdot \frac{\overline{k}_d \times (\hat{k}_i \times \hat{e}_i)}{k_{dz}} \right)^2 + \left(\hat{h}(k_z) \cdot \frac{\overline{k}_d \times (\hat{k}_i \times \hat{e}_i)}{k_{dz}} \right)^2 \right\} \cdot \exp(-h^2 k_{dz}^2) \sum_{m=1}^{\infty} \frac{k_{dz}^{2m}}{m!} W^{(m)}(\overline{k}_{d\perp}) \tag{2.1.35}$$

(a) For TE incidence $\hat{e}_i = \hat{e}(-k_{iz})$

$$\hat{e}(k_z) \cdot \overline{k}_d \times (\hat{k}_i \times \hat{e}_i) = k \Big\{ -\hat{e}(k_z) \cdot \hat{e}(k_{iz}) + \hat{h}(k_z) \cdot \hat{h}(-k_{iz}) \Big\} \tag{2.1.36a}$$

$$\hat{e}(k_z) \cdot \hat{e}(k_{iz}) = \cos(\phi_s - \phi_i) \tag{2.1.36b}$$

$$\hat{h}(k_z) \cdot \hat{h}(-k_{iz}) = -\cos\theta_s \cos\theta_i \cos(\phi_s - \phi_i) + \sin\theta_s \sin\theta_i \tag{2.1.36c}$$

$$\hat{h}(k_z) \cdot \overline{k}_d \times (\hat{k}_i \times \hat{e}_i) = k \Big\{ -\hat{h}(k_z) \cdot \hat{e}(-k_{iz}) - \hat{e}(k_z) \cdot \hat{h}(-k_{iz}) \Big\} \tag{2.1.37a}$$

$$\hat{h}(k_z) \cdot \hat{e}(-k_{iz}) = -\cos\theta_s \sin(\phi_i - \phi_s) \tag{2.1.37b}$$

$$\hat{e}(k_z) \cdot \hat{h}(-k_{iz}) = -\cos\theta_i \sin(\phi_i - \phi_s) \tag{2.1.37c}$$

Then, the bistatic scattering coefficients are

$$\gamma_{hh}(\theta_s, \phi_s; \theta_i, \phi_i) = \frac{4\pi k^2}{\cos\theta_i} \frac{1}{(\cos\theta_i + \cos\theta_s)^2} \Big\{ \sin\theta_s \sin\theta_i - (\cos\theta_s \cos\theta_i + 1) \cos(\phi_s - \phi_i) \Big\}^2 \times \exp(-h^2 k_{dz}^2) \sum_{m=1}^{\infty} \frac{k_{dz}^{2m}}{m!} W^{(m)}(\overline{k}_{d\perp}) \tag{2.1.38a}$$

$$\gamma_{vh}(\theta_s, \phi_s; \theta_i, \phi_i) = \frac{4\pi k^2}{\cos\theta_i} \sin^2(\phi_s - \phi_i) \exp(-h^2 k_{dz}^2) \sum_{m=1}^{\infty} \frac{k_{dz}^{2m}}{m!} W^{(m)}(\overline{k}_{d\perp}) \tag{2.1.38b}$$

(b) For TM incidence $\hat{e}_i = \hat{h}(-k_{iz})$ we have

$$\hat{e}(k_z) \cdot \overline{k}_d \times (\hat{k}_i \times \hat{e}_i) = k(\cos\theta_i + \cos\theta_s) \sin(\phi_i - \phi_s)$$

$$\begin{aligned} \hat{h}(k_z) \cdot \overline{k}_d \times (\hat{k}_i \times \hat{e}_i) &= k \Big(-\hat{h}(k_z) \cdot \hat{h}(-k_{iz}) + \hat{e}(k_z) \cdot \hat{e}(-k_{iz}) \Big) \\ &= k \Big\{ -\sin\theta_s \sin\theta_i + \cos\theta_s \cos\theta_i \cos(\phi_s - \phi_i) + \cos(\phi_s - \phi_i) \Big\} \end{aligned}$$

Hence the bistatic coefficients are

$$\gamma_{hv}(\theta_s,\phi_s;\theta_i,\phi_i) = \frac{4\pi k^2}{\cos\theta_i}\sin^2(\phi_i-\phi_s)\exp(-h^2k_{dz}^2)\sum_{m=1}^{\infty}\frac{k_{dz}^{2m}}{m!}W^{(m)}(\overline{k}_{d\perp}) \quad (2.1.38c)$$

$$\gamma_{vv}(\theta_s,\phi_s;\theta_i,\phi_i) = \frac{4\pi k^2}{\cos\theta_i}\frac{1}{(\cos\theta_i+\cos\theta_s)^2}\left\{\sin\theta_s\sin\theta_i-\cos(\phi_s-\phi_i)(1+\cos\theta_i\cos\theta_s)\right\}^2 \times\exp(-h^2k_{dz}^2)\sum_{m=1}^{\infty}\frac{k_{dz}^{2m}}{m!}W^{(m)}(\overline{k}_{d\perp}) \quad (2.1.38d)$$

One interesting observation of the results in (2.1.38) is the lack of distinct polarization dependence. This is different from the small perturbation method, which shows strong polarization dependence. For the case of $C(\rho) = e^{-\rho^2/l^2}$, by making use of the integral identity we obtain

$$\int_0^{\infty} d\rho\rho J_o(k_\rho\rho)e^{-m\rho^2/l^2} = \frac{l^2}{2m}\exp\left[-\frac{k_\rho^2 l^2}{4m}\right] \quad (2.1.39)$$

we have

$$W^{(m)}(\overline{k}_\perp) = \frac{l^2h^{2m}}{4\pi m}\exp\left(-\frac{|\overline{k}_\perp|^2l^2}{4m}\right) \quad (2.1.40)$$

1.2 Dielectric Rough Surfaces

Consider a plane wave incident upon a random rough surface overlying a dielectric half-space (Fig. 2.1.1) The electric field of the incident wave is given by

$$\overline{E}_i = \hat{e}_i E_o e^{i\overline{k}_i\cdot\overline{r}} \quad (2.1.41)$$

where $\overline{k}_i$ denotes the incident wave vector and $\hat{e}_i$ is the polarization of the electric field vector. The rough surface is characterized by a random height distribution $z = f(\overline{r}_\perp)$ with zero mean, $\langle f(\overline{r}_\perp)\rangle = 0$. From Huygen's principle we have

$$\overline{E}_s(\overline{r}) = \int_{S'} dS'\left\{i\omega\mu_o\overline{\overline{G}}(\overline{r},\overline{r}')\cdot[\hat{n}\times\overline{H}(\overline{r}')] + \nabla\times\overline{\overline{G}}(\overline{r},\overline{r}')\cdot[\hat{n}\times\overline{E}(\overline{r}')]\right\} \quad (2.1.42a)$$

$$\overline{E}_t(\overline{r}) = \int_{S'} dS'\left\{i\omega\mu_o\overline{\overline{G}}_1(\overline{r},\overline{r}')\cdot[\hat{n}_d\times\overline{H}(\overline{r}')] + \nabla\times\overline{\overline{G}}_1(\overline{r},\overline{r}')\cdot[\hat{n}_d\times\overline{E}(\overline{r}')]\right\} \quad (2.1.42b)$$

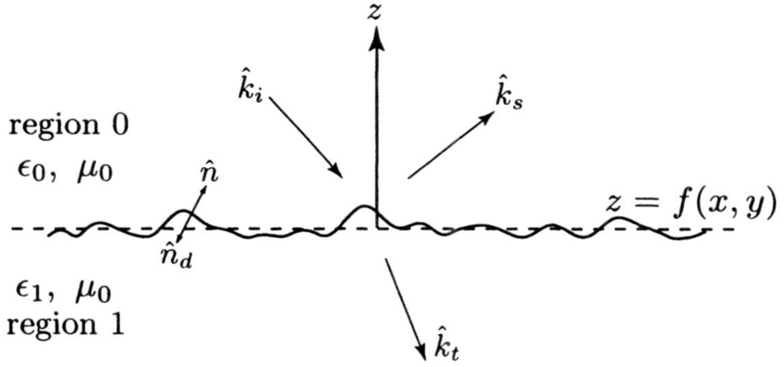

Figure 2.1.1 Geometrical configuration of the problem.

where S' denotes the rough surface on which the surface integration is to be carried out, and $\hat{n}$ and $\hat{n}_d$ are the unit vectors normal to the rough surface and pointing into the reflected and transmitted regions, respectively (Fig. 2.1.1). The dyadic Green's functions for homogeneous space of the regions 0 and 1, $\overline{\overline{G}}(\overline{r}, \overline{r}')$ and $\overline{\overline{G}}_1(\overline{r}, \overline{r}')$, are

$$\overline{\overline{G}}(\overline{r}, \overline{r}') = \left[\overline{\overline{I}} + \frac{\nabla\nabla}{k^2}\right] \frac{e^{ik|\overline{r}-\overline{r}'|}}{4\pi|\overline{r}-\overline{r}'|} \tag{2.1.43a}$$

and

$$\overline{\overline{G}}_1(\overline{r}, \overline{r}') = \left[\overline{\overline{I}} + \frac{\nabla\nabla}{k_1^2}\right] \frac{e^{ik_1|\overline{r}-\overline{r}'|}}{4\pi|\overline{r}-\overline{r}'|} \tag{2.1.43b}$$

where $k = \omega\sqrt{\mu_o \epsilon_o}$ and $k_1 = \omega\sqrt{\mu_o \epsilon_1}$. If the observation point is in the far-field region, then the dyadic Green's functions simplify to

$$\overline{\overline{G}}(\overline{r}, \overline{r}') \simeq (\overline{\overline{I}} - \hat{k}_s \hat{k}_s) \frac{e^{ikr}}{4\pi r} e^{-i\overline{k}_s \cdot \overline{r}'} \tag{2.1.44}$$

$$\overline{\overline{G}}_1(\overline{r}, \overline{r}') \simeq (\overline{\overline{I}} - \hat{k}_t \hat{k}_t) \frac{e^{ik_1 r}}{4\pi r} e^{-i\overline{k}_t \cdot \overline{r}'} \tag{2.1.45}$$

where $\hat{k}_s$ and $\hat{k}_t$ denote the scattered and transmitted directions in region 0 and region 1, respectively.

Substituting (2.1.44) and (2.1.45) into the diffraction integral (2.1.42), we obtain respectively the scattered field and transmitted field in directions $\hat{k}_s$ and $\hat{k}_t$,

$$\overline{E}_s(\overline{r}) = \frac{ik\, e^{ikr}}{4\pi r}(\overline{\overline{I}} - \hat{k}_s\hat{k}_s) \cdot \int_{S'} dS' \left\{\hat{k}_s \times [\hat{n} \times \overline{E}(\overline{r}')] + \eta[\hat{n} \times \overline{H}(\overline{r}')]\right\} e^{-i\overline{k}_s\cdot\overline{r}'} \tag{2.1.46a}$$

$$\overline{E}_t(\overline{r}) = \frac{ik_1\, e^{ik_1r}}{4\pi r}(\overline{\overline{I}} - \hat{k}_t\hat{k}_t) \cdot \int_{S'} dS' \left\{\hat{k}_t \times [\hat{n}_d \times \overline{E}(\overline{r}')] + \eta_1[\hat{n}_d \times \overline{H}(\overline{r}')]\right\} e^{-i\overline{k}_t\cdot\overline{r}'} \tag{2.1.46b}$$

where η and η_1 are the wave impedances in the regions 0 and 1, respectively.

Tangent Plane Approximation

In the Kirchhoff approach, approximate expressions for the surface fields are obtained using the tangent plane approximation. The fields at any point of the surface are approximated by the fields that would be present on the tangent plane at that point. Thus, the tangent plane approximation requires a large radius of curvature relative to the incident wavelength at every point on the surface.

Let an orthonormal system $(\hat{p}_i, \hat{q}_i, \hat{k}_i)$ be defined at the point $\overline{r}'$, with

$$\hat{q}_i = \frac{\hat{k}_i \times \hat{n}}{|\hat{k}_i \times \hat{n}|} \tag{2.1.47}$$

$$\hat{p}_i = \hat{q}_i \times \hat{k}_i \tag{2.1.48}$$

The unit vectors $\hat{q}_i$ and $\hat{p}_i$ are the local perpendicular and parallel polarization vectors at the point $\overline{r}'$. In applying the tangent plane approximation, we solve the boundary value problem for the TE and TM polarization of a wave incident onto an infinite planar interface with the tangent plane as the interface. To get the surface fields the incident field is decomposed into locally perpendicular and parallel polarization fields.

The perpendicular component of the incident field is

$$(\hat{e}_i \cdot \hat{q}_i)\hat{q}_i\, E_o\, e^{i\overline{k}_i\cdot\overline{r}'}$$

so that the local reflected wave is

$$(\hat{e}_i \cdot \hat{q}_i)\hat{q}_i\, E_o R_h\, e^{i\overline{k}_i\cdot\overline{r}'}$$

where R_h is the local Fresnel reflection coefficient for perpendicular polarization.

$$R_h = \frac{k\cos\theta_{li} - \sqrt{k_1^2 - k^2\sin^2\theta_{li}}}{k\cos\theta_{li} + \sqrt{k_1^2 - k^2\sin^2\theta_{li}}} \tag{2.1.49a}$$

with θ_{li} as the local angle of incidence at the point $\overline{r}'$. The magnetic fields associated with the incident wave is

$$\frac{1}{\eta}\hat{k}_i \times (\hat{e}_i \cdot \hat{q}_i)\hat{q}_i\, E_o\, e^{i\overline{k}_i\cdot\overline{r}'}$$

To get the magnetic field for the reflected wave, we need the local reflected direction using the tangent plane as the interface

$$\hat{k}_r = \hat{k}_i - 2\hat{n}(\hat{n}\cdot\hat{k}_i)$$

Thus the magnetic field associated with the reflected field is

$$\frac{1}{\eta}\hat{k}_r \times (\hat{e}_i \cdot \hat{q}_i)\hat{q}_i\, E_o R_h\, e^{i\overline{k}_i\cdot\overline{r}'}$$

Hence, the tangential electric field of this perpendicular component at the point $\overline{r}'$ is

$$\hat{n}\times\overline{E} = (\hat{n}\times\hat{q}_i)(\hat{e}_i\cdot\hat{q}_i)(1+R_h)\, E_o\, e^{i\overline{k}_i\cdot\overline{r}'}$$

and the associated magnetic field is

$$\begin{aligned}\hat{n}\times\overline{H} &= \frac{1}{\eta}(\hat{e}_i\cdot\hat{q}_i)\hat{n}\times\Big[(\hat{k}_i\times\hat{q}_i) + R_h(\hat{k}_r\times\hat{q}_i)\Big]\, E_o\, e^{i\overline{k}_i\cdot\overline{r}'}\\ &= -(1-R_h)(\hat{n}\cdot\hat{k}_i)\frac{(\hat{e}_i\cdot\hat{q}_i)}{\eta}\hat{q}_i\, E_o\, e^{i\overline{k}_i\cdot\overline{r}'}\end{aligned}$$

We have made use of the relations $\hat{n}\cdot\hat{q}_i = 0$ and $\hat{n}\cdot\hat{k}_r = -\hat{n}\cdot\hat{k}_i$. The calculations can be repeated for the local parallel polarized component with local reflection coefficient

$$R_v = \frac{\epsilon_1 k\cos\theta_{li} - \epsilon_o\sqrt{k_1^2 - k^2\sin^2\theta_{li}}}{\epsilon_1 k\cos\theta_{li} + \epsilon_o\sqrt{k_1^2 - k^2\sin^2\theta_{li}}} \tag{2.1.49b}$$

Summing the local parallel and perpendicular polarized components, we obtain

$$\begin{aligned}&\hat{n}\times\overline{E}(\overline{r}')\\ &= E_o\Big\{(\hat{e}_i\cdot\hat{q}_i)(\hat{n}\times\hat{q}_i)(1+R_h) + (\hat{e}_i\cdot\hat{p}_i)(\hat{n}\cdot\hat{k}_i)\hat{q}_i(1-R_v)\Big\}e^{i\overline{k}_i\cdot\overline{r}'}\end{aligned} \tag{2.1.50a}$$

$$\begin{aligned}&\hat{n}\times\overline{H}(\overline{r}')\\ &= \frac{E_o}{\eta}\Big\{-(\hat{e}_i\cdot\hat{q}_i)(\hat{n}\cdot\hat{k}_i)\hat{q}_i(1-R_h) + (\hat{e}_i\cdot\hat{p}_i)(\hat{n}\times\hat{q}_i)(1+R_v)\Big\}e^{i\overline{k}_i\cdot\overline{r}'}\end{aligned} \tag{2.1.50b}$$

The local angle of incidence can be calculated from the formula

$$\cos\theta_{li} = -\hat{n}\cdot\hat{k}_i \tag{2.1.51}$$

The normal vector at the point $\overline{r}'$ is given by

$$\hat{n}(\overline{r}') = \frac{-\alpha\hat{x} - \beta\hat{y} + \hat{z}}{\sqrt{1 + \alpha^2 + \beta^2}} \tag{2.1.52}$$

where α and β are the local slopes in the $\hat{x}$- and $\hat{y}$-directions,

$$\alpha = \frac{\partial f(x', y')}{\partial x'} \tag{2.1.53a}$$

$$\beta = \frac{\partial f(x', y')}{\partial y'} \tag{2.1.53b}$$

Substituting (2.1.50) into (2.1.46), we obtain, after some algebraic manipulations,

$$\overline{E}_s(\overline{r}) = \frac{ik\, e^{ikr}}{4\pi r} E_o(\overline{\overline{I}} - \hat{k}_s\hat{k}_s) \cdot \int_{A_o} d\overline{r}'_\perp \overline{F}(\alpha, \beta)\, e^{i(\overline{k}_i - \overline{k}_s)\cdot\overline{r}'} \tag{2.1.54a}$$

Similarly, for the transmitted field we obtain

$$\overline{E}_t(\overline{r}) = -\frac{ik_1\, e^{ik_1 r}}{4\pi r} E_o(\overline{\overline{I}} - \hat{k}_t\hat{k}_t) \cdot \int_{A_o} d\overline{r}'_\perp \overline{N}(\alpha, \beta)\, e^{i(\overline{k}_i - \overline{k}_t)\cdot\overline{r}'} \tag{2.1.54b}$$

In (2.1.54) we have

$$\begin{aligned}\overline{F}(\alpha, \beta) = (1 + \alpha^2 + \beta^2)^{1/2} \Big\{ &-(\hat{e}_i \cdot \hat{q}_i)(\hat{n} \cdot \hat{k}_i)\hat{q}_i(1 - R_h) \\ &+ (\hat{e}_i \cdot \hat{p}_i)(\hat{n} \times \hat{q}_i)(1 + R_v) + (\hat{e}_i \cdot \hat{q}_i)(\hat{k}_s \times (\hat{n} \times \hat{q}_i))(1 + R_h) \\ &+ (\hat{e}_i \cdot \hat{p}_i)(\hat{n} \cdot \hat{k}_i)(\hat{k}_s \times \hat{q}_i)(1 - R_v) \Big\}\end{aligned} \tag{2.1.55a}$$

$$\begin{aligned}\overline{N}(\alpha, \beta) = (1 + \alpha^2 + \beta^2)^{1/2} \Big\{ &-\frac{\eta_1}{\eta}(\hat{e}_i \cdot \hat{q}_i)(\hat{n} \cdot \hat{k}_i)\hat{q}_i(1 - R_h) \\ &+ \frac{\eta_1}{\eta}(\hat{e}_i \cdot \hat{p}_i)(\hat{n} \times \hat{q}_i)(1 + R_v) + (\hat{e}_i \cdot \hat{q}_i)(\hat{k}_t \times (\hat{n} \times \hat{q}_i))(1 + R_h) \\ &+ (\hat{e}_i \cdot \hat{p}_i)(\hat{n} \cdot \hat{k}_i)(\hat{k}_t \times \hat{q}_i)(1 - R_v) \Big\}\end{aligned} \tag{2.1.55b}$$

The orthonormal systems for the incident, scattered, and transmitted fields are given respectively by ($\hat{v}_i$, $\hat{h}_i$, $\hat{k}_i$), ($\hat{v}_s$, $\hat{h}_s$, $\hat{k}_s$), and ($\hat{v}_t$, $\hat{h}_t$, $\hat{k}_t$) with

$$\hat{k}_i = \hat{x}\sin\theta_i\cos\phi_i + \hat{y}\sin\theta_i\sin\phi_i - \hat{z}\cos\theta_i \tag{2.1.56a}$$

$$\hat{h}_i = -\hat{x}\sin\phi_i + \hat{y}\cos\phi_i \tag{2.1.56b}$$

$$\hat{v}_i = -\hat{x}\cos\theta_i\cos\phi_i - \hat{y}\cos\theta_i\sin\phi_i - \hat{z}\sin\theta_i \tag{2.1.56c}$$

$$\hat{k}_s = \hat{x}\sin\theta_s\cos\phi_s + \hat{y}\sin\theta_s\sin\phi_s + \hat{z}\cos\theta_s \tag{2.1.57a}$$

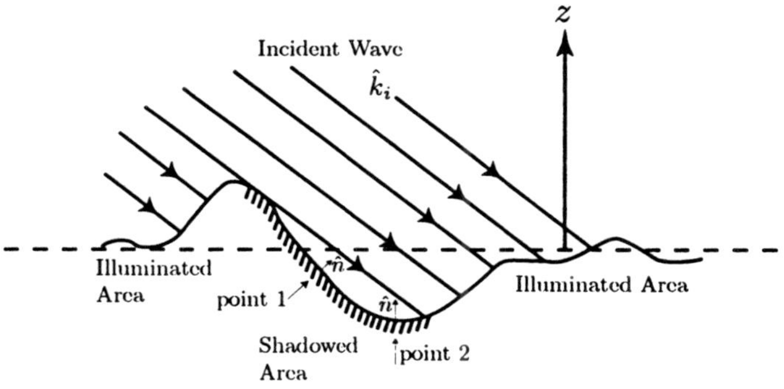

Figure 2.1.2 The shadowed area on the rough surface.

$$\hat{h}_s = -\hat{x}\sin\phi_s + \hat{y}\cos\phi_s \tag{2.1.57b}$$
$$\hat{v}_s = \hat{x}\cos\theta_s\cos\phi_s + \hat{y}\cos\theta_s\sin\phi_s - \hat{z}\sin\theta_s \tag{2.1.57c}$$

$$\hat{k}_t = \hat{x}\sin\theta_t\cos\phi_t + \hat{y}\sin\theta_t\sin\phi_t - \hat{z}\cos\theta_t \tag{2.1.58a}$$
$$\hat{h}_t = -\hat{x}\sin\phi_t + \hat{y}\cos\phi_t \tag{2.1.58b}$$
$$\hat{v}_t = -\hat{x}\cos\theta_t\cos\phi_t - \hat{y}\cos\theta_t\sin\phi_t - \hat{z}\sin\theta_t \tag{2.1.58c}$$

We note that except for the phase factors, the expressions in the integrands of the diffraction integral (2.1.54) are not explicit functions of $\overline{r}'$. They are explicit functions of the slopes α and β, which are functions of $\overline{r}'$. The tangent plane-approximated diffraction integrals, as expressed in (2.1.54), do not take into account the effects of shadowing and multiple scattering.

When the angle of incidence is not normal to the x-y plane, some points on the rough surface will not be illuminated directly because of shadowing (Fig. 2.1.2). For some points (point 1, Fig. 2.1.2), the local angle of incidence θ_{li} is not defined since

$$\cos\theta_{li} = -\hat{n}\cdot\hat{k}_i < 0 \tag{2.1.59}$$

All the points on the rough surface with such local slopes will not be illuminated directly. Some other points (point 2, Fig. 2.1.2) are not directly illuminated, even though the local angle of incidence is well defined, because of the height of rough surface at that point relative to the heights of the surrounding points.

For the dielectric random rough surface, various approximations have been applied to the Kirchhoff-approximated diffraction integrals. The integrands that depend on the local surface slopes can be expanded about zero slopes and can then be integrated by parts with the edge effect discarded.

In the high-frequency limit, the geometrical optics solution can be obtained from (2.1.54) with the stationary phase method.

In the calculation of the scattered fields, the expression for the diffraction integral (2.1.54a) contains the total field (incident and scattered) on the surface. An alternative is to use the scattered field. The scattered field evaluated from (2.1.54a) with the exact $\overline{F}(\alpha,\beta)$ is the same whether one uses the total field or the scattered field on the surface. However, when the integrand $\overline{F}(\alpha,\beta)$ is approximated, the results using total or scattered surface fields can give different results. This is also true when shadowing is present.

Coherent and Incoherent Scattering Coefficients

The scattered intensities from a random rough surface can in general be decomposed into coherent and incoherent components. Coherent components only contribute in the specular reflected or transmitted directions, whereas incoherent components contribute in all directions. In the limiting case of a flat surface, the scattered intensity consists of only the specularly reflected and transmitted coherent intensities. In the other limiting case of a very rough surface, the coherent components almost vanish and only incoherent components remain.

A commonly used approximation is to expand the integrands $\overline{F}(\alpha,\beta)$ and $\overline{N}(\alpha,\beta)$ about zero slopes and to keep only the first few terms. In later sections, we will include higher-order terms. In the present section we approximate $\overline{F}$ and $\overline{N}$ by zero slopes. We have

$$\overline{F}(\alpha,\beta) \cong \overline{F}(0,0) \tag{2.1.60a}$$

$$\overline{N}(\alpha,\beta) \cong \overline{N}(0,0) \tag{2.1.60b}$$

where $\overline{F}(0,0)$ is $\overline{F}(\alpha,\beta)$ evaluated at $\alpha=\beta=0$ and $\overline{N}(0,0)$ is $\overline{N}(\alpha,\beta)$ evaluated at $\alpha=\beta=0$. At $\alpha=\beta=0$, $\hat{q}_i=-\hat{h}_i$, $\hat{p}_i=-\hat{v}_i$, and $\hat{n}=\hat{z}$. Then

$$\begin{aligned}&\overline{F}(0,0)\\ &= -(\hat{e}_i\cdot\hat{h}_i)(\hat{z}\cdot\hat{k}_i)\hat{h}_i(1-R_{ho})) + (\hat{e}_i\cdot\hat{v}_i)(\hat{z}\times\hat{h}_i)(1+R_{vo}) + (\hat{e}_i\cdot\hat{h}_i)\\ &\quad\cdot\left(\hat{k}_s\times(\hat{z}\times\hat{h}_i)\right)(1+R_{ho}) + (\hat{e}_i\cdot\hat{v}_i)(\hat{z}\cdot\hat{k}_i)(\hat{k}_s\times\hat{h}_i)(1-R_{vo})\end{aligned} \tag{2.1.60c}$$

The R_{vo} and R_{ho} of the above equations are, respectively, the Fresnel reflection coefficients of a flat surface for vertically and horizontally polarized incident waves and are equal to the expressions in (2.1.49a) and (2.1.49b) with θ_{li} replaced by θ_i. Thus, from (2.1.54) we have

$$\overline{E}_s = \frac{ik\,e^{ikr}}{4\pi r}E_o(\overline{\overline{I}}-\hat{k}_s\hat{k}_s)\cdot\overline{F}(0,0)I \tag{2.1.61a}$$

$$\overline{E}_t = -\frac{ik_1 e^{ik_1 r}}{4\pi r} E_o(\overline{\overline{I}} - \hat{k}_t\hat{k}_t) \cdot \overline{N}(0,0) I_t \tag{2.1.61b}$$

where the integrals I and I_t are given by

$$I = \int_{A_o} e^{i(\overline{k}_i - \overline{k}_s)\cdot\overline{r}'} d\overline{r}'_\perp \tag{2.1.62a}$$

$$I_t = \int_{A_o} e^{i(\overline{k}_i - \overline{k}_t)\cdot\overline{r}'} d\overline{r}'_\perp \tag{2.1.62b}$$

The scattered and transmitted fields are next separated into a mean field and a fluctuating part of the field

$$\overline{E}_s(\overline{r}) = \overline{E}_{sm}(\overline{r}) + \overline{\mathcal{E}}_s(\overline{r}) \tag{2.1.63}$$

$$\overline{E}_t(\overline{r}) = \overline{E}_{tm}(\overline{r}) + \overline{\mathcal{E}}_t(\overline{r}) \tag{2.1.64}$$

with

$$\langle \overline{\mathcal{E}}_s(\overline{r}) \rangle = \langle \overline{\mathcal{E}}_t(\overline{r}) \rangle = 0 \tag{2.1.65}$$

and $\overline{E}_{sm}$ and $\overline{E}_{tm}$ denote the mean scattered and transmitted fields, respectively. The total scattered intensity is a sum of coherent and incoherent intensities

$$\langle |\overline{E}_s(\overline{r})|^2 \rangle = |E_{sm}|^2 + \langle |\overline{\mathcal{E}}_s(\overline{r})|^2 \rangle \tag{2.1.66a}$$

$$\langle |\overline{E}_t(\overline{r})|^2 \rangle = |E_{tm}|^2 + \langle |\overline{\mathcal{E}}_t(\overline{r})|^2 \rangle \tag{2.1.66b}$$

In view of (2.1.61) and (2.1.62), we have

$$|\overline{E}_{sm}(\overline{r})|^2 = \frac{k^2 |E_o|^2}{16\pi^2 r^2} \left\{ \left| \hat{v}_s \cdot \overline{F}(0,0) \right|^2 + \left| \hat{h}_s \cdot \overline{F}(0,0) \right|^2 \right\} |\langle I \rangle|^2 \tag{2.1.67a}$$

$$\langle |\overline{\mathcal{E}}_s(\overline{r})|^2 \rangle = \frac{k^2 |E_o|^2}{16\pi^2 r^2} \left\{ \left| \hat{v}_s \cdot \overline{F}(0,0) \right|^2 + \left| \hat{h}_s \cdot \overline{F}(0,0) \right|^2 \right\} D_I \tag{2.1.67b}$$

$$|\overline{E}_{tm}(\overline{r})|^2 = \frac{k_1^2 |E_o|^2}{16\pi^2 r^2} \left\{ \left| \hat{v}_t \cdot \overline{N}(0,0) \right|^2 + \left| \hat{h}_t \cdot \overline{N}(0,0) \right|^2 \right\} |\langle I_t \rangle|^2 \tag{2.1.68a}$$

$$\langle |\overline{\mathcal{E}}_t(\overline{r})|^2 \rangle = \frac{k_1^2 |E_o|^2}{16\pi^2 r^2} \left\{ \left| \hat{v}_t \cdot \overline{N}(0,0) \right|^2 + \left| \hat{h}_t \cdot \overline{N}(0,0) \right|^2 \right\} D_{I_t} \tag{2.1.68b}$$

where

$$D_I = \langle |I|^2 \rangle - |\langle I \rangle|^2 \tag{2.1.69a}$$

$$D_{I_t} = \langle |I_t|^2 \rangle - |\langle I_t \rangle|^2 \tag{2.1.69b}$$

To derive analytic expression of scattering coefficients, we assume that the rough surface is a stationary Gaussian process.

$$p(f(\overline{r}_\perp)) = \frac{1}{\sqrt{2\pi} h} e^{-f^2/2h^2} \tag{2.1.70}$$

where h is the standard deviation of the surface height. For two points on the surface, $\overline{r}_{\perp 1}$ and $\overline{r}_{\perp 2}$, we have

$$\langle f(\overline{r}_{\perp 1}) f(\overline{r}_{\perp 2}) \rangle = h^2 C(\rho) \tag{2.1.71}$$

where $\rho = \sqrt{(x_1 - x_2)^2 + (y_1 - y_2)^2}$. The characteristic functions are

$$\langle e^{i\nu f(\overline{r}_\perp)} \rangle = \int_{-\infty}^{\infty} df\, p(f) e^{i\nu f} = \exp\left[-\frac{1}{2} h^2 \nu^2\right] \tag{2.1.72}$$

$$\langle e^{i\nu (f(\overline{r}_{\perp 1}) - f(\overline{r}_{\perp 2}))} \rangle = \exp\left[-h^2 \nu^2 \Big(1 - C(\rho)\Big)\right] \tag{2.1.73}$$

The expressions for $|\langle I \rangle|^2$, D_I, $|\langle I_t \rangle|^2$, and D_{I_t} can be derived in terms of the statistical moments of the height distribution.

The integral I is

$$I = \int_{A_o} e^{i\overline{k}_{d\perp} \cdot \overline{r}'_\perp} e^{ik_{dz} f(\overline{r}'_\perp)} d\overline{r}'_\perp \tag{2.1.74}$$

where

$$\overline{k}_d = \overline{k}_i - \overline{k}_s = k_{dx}\hat{x} + k_{dy}\hat{y} + k_{dz}\hat{z} \tag{2.1.75}$$

The ensemble average of I is given by

$$\langle I \rangle = \int_{A_o} e^{i\overline{k}_{d\perp} \cdot \overline{r}'_\perp} \langle e^{ik_{dz} f(\overline{r}'_\perp)} \rangle d\overline{r}'_\perp \tag{2.1.76}$$

In view of (2.1.72), we obtain

$$\langle I \rangle = 4L_x L_y \exp\left[-\frac{1}{2} k_{dz}^2 h^2\right] \operatorname{sinc}(k_{dx} L_x) \operatorname{sinc}(k_{dy} L_y) \tag{2.1.77}$$

where $\operatorname{sinc} x = \sin x / x$ and $2L_x$ and $2L_y$ are the lengths of rough surface illuminated in the $\hat{x}$- and $\hat{y}$-directions, respectively, so that

$$A_o = 4L_x L_y \tag{2.1.78}$$

By allowing L_x and L_y to approach infinity in the above expression, we obtain

$$|\langle I \rangle|^2 = 4\pi^2 A_o \exp\left[-k_{dz}^2 h^2\right] \delta(k_{dx}) \delta(k_{dy}) \tag{2.1.79}$$

where we made use of the identity

$$\lim_{L_x, L_y \to \infty} \frac{L_x\, L_y}{\pi^2} \operatorname{sinc}(k_{dx} L_x) \operatorname{sinc}(k_{dy} L_y) = \delta(k_{dx})\, \delta(k_{dy}) \tag{2.1.80}$$

The integral for $\langle I I^* \rangle$ is given by

$$\langle I I^* \rangle = \int_{A_o} d\overline{r}_\perp \int_{A_o} d\overline{r}'_\perp\, e^{i\overline{k}_{d\perp} \cdot (\overline{r}_\perp - \overline{r}'_\perp)} \langle e^{ik_{dz} (f(\overline{r}_\perp) - f(\overline{r}'_\perp))} \rangle \tag{2.1.81}$$

Using (2.1.73) and making the usual change of variables to the difference and the average of coordinates, we obtain

$$\langle II^* \rangle = \int_{-2L_x}^{2L_x} dx \int_{-2L_y}^{2L_y} dy\, (2L_x - |x|)(2L_y - |y|)$$
$$\times \exp(ik_{dx}x + ik_{dy}y) \exp\Big[-k_{dz}^2\sigma^2\Big(1 - C(\rho)\Big)\Big] \tag{2.1.82}$$

We then have

$$D_I = A_o \int_{-\infty}^{\infty} dx \int_{-\infty}^{\infty} dy \left\{\exp\left[-h^2 k_{dz}^2(1 - C(\rho))\right] - \exp\left[-h^2 k_{dz}^2\right]\right\}$$
$$\cdot \exp\left[ik_{dx}x + ik_{dy}y\right] \tag{2.1.83}$$

We expand the integrand of (2.1.83) in a Taylor series

$$D_I = A_o \exp(-h^2 k_{dz}^2) \int d\overline{r}_\perp \sum_{m=1}^{\infty} \frac{(h^2 k_{dz}^2)^m}{m!} C^m(\rho) \exp(i\overline{k}_{d\perp} \cdot \overline{r}_\perp)$$
$$= 4\pi^2 A_o \exp(-h^2 k_{dz}^2) \sum_{m=1}^{\infty} \frac{k_{dz}^{2m}}{m!} W^{(m)}(|\overline{k}_{d\perp}|) \tag{2.1.84}$$

In a similar manner, the expressions for $|\langle I_t \rangle|^2$ and D_{I_t} may be derived. They are

$$|\langle I_t \rangle|^2 = 4\pi^2 A_o \exp\left[-\sigma^2 k_{tdz}^2\right] \delta(k_{tdx})\delta(k_{tdy}) \tag{2.1.85}$$

and

$$D_{I_t} = 4\pi^2 A_o \exp(-h^2 k_{tdz}^2) \sum_{m=1}^{\infty} \frac{k_{tdz}^{2m}}{m!} W^{(m)}(|\overline{k}_{td}|) \tag{2.1.86}$$

where

$$\overline{k}_{td} = \overline{k}_i - \overline{k}_t = k_{tdx}\hat{x} + k_{tdy}\hat{y} + k_{tdz}\hat{z} \tag{2.1.87}$$

The bistatic scattering coefficients for the reflected intensities are

$$\gamma_{ba}^r(\hat{k}_s, \hat{k}_i) = \frac{4\pi r^2 (S_r)_b}{A_o \cos\theta_i (S_o)_a} \qquad (a, b = v, h) \tag{2.1.88}$$

From (2.1.54) and (2.1.67) we calculate the vertically and horizontally polarized coherent and incoherent scattered intensities for the cases of vertically and horizontally polarized incident fields. Let

$$\overline{F}_b(0,0) = \overline{F}(0,0)\Big|_{\hat{e}_i = \hat{b}_i} \tag{2.1.89}$$

Then

$$\overline{F}_h(0,0) = \cos\theta_i \hat{h}_i(1 - R_{ho}) + \Big(\hat{k}_s \times (\hat{z} \times \hat{h}_i)\Big)(1 + R_{ho}) \tag{2.1.90a}$$

$$\overline{F}_v(0,0) = (\hat{z} \times \hat{h}_i)(1 + R_{vo}) - \cos\theta_i(\hat{k}_s \times \hat{h}_i)(1 - R_{vo}) \tag{2.1.90b}$$

Next we take the dot product of (2.1.90a) and (2.1.90b) with $\hat{v}_s$ and $\hat{h}_s$.

$$\hat{h}_s \cdot \overline{F}_h(0,0) = \Big[(1-R_{ho})\cos\theta_i - (1+R_{ho})\cos\theta_s\Big]\cos(\phi_s - \phi_i) \quad (2.1.91a)$$

$$\hat{v}_s \cdot \overline{F}_h(0,0) = \Big[(1-R_{ho})\cos\theta_i\cos\theta_s - (1+R_{ho})\Big]\sin(\phi_s - \phi_i) \quad (2.1.91b)$$

$$\hat{h}_s \cdot \overline{F}_v(0,0) = \Big[(1+R_{vo}) - (1-R_{vo})\cos\theta_i\cos\theta_s\Big]\sin(\phi_s - \phi_i) \quad (2.1.91c)$$

$$\hat{v}_s \cdot \overline{F}_v(0,0) = \Big[-(1+R_{vo})\cos\theta_s + (1-R_{vo})\cos\theta_i\Big]\cos(\phi_s - \phi_i) \quad (2.1.91d)$$

In view of (2.1.61), the bistatic scattering coefficients γ^r_{ab} can be decomposed into a coherent part $\gamma^r_{c_{ab}}$ and an incoherent part $\gamma^r_{i_{ab}}$.

$$\gamma^r_{ab}(\hat{k}_s, \hat{k}_i) = \gamma^r_{c_{ab}}(\hat{k}_s, \hat{k}_i) + \gamma^r_{i_{ab}}(\hat{k}_s, \hat{k}_i) \quad (2.1.92)$$

where

$$\gamma^r_{c_{ab}}(\hat{k}_s, \hat{k}_i) = \frac{k^2}{4\pi A_o \cos\theta_i}\left|\hat{a}_s \cdot \overline{F}_b(0,0)\right|^2 |\langle I\rangle|^2 \quad (2.1.93a)$$

$$\gamma^r_{i_{ab}}(\hat{k}_s, \hat{k}_i) = \frac{k^2}{4\pi A_o \cos\theta_i}\left|\hat{a}_s \cdot \overline{F}_b(0,0)\right|^2 D_I \quad (2.1.93b)$$

The coherent scattered wave only exists in the specular directions. Using (2.1.79) and (2.1.91) and since

$$\delta(k_{dx})\delta(k_{dy}) = \frac{\delta(\theta_s - \theta_i)\delta(\phi_s - \phi_i)}{(k^2 \sin\theta_i \cos\theta_i)}$$

we find that

$$\gamma^r_{c_{ab}}(\hat{k}_s, \hat{k}_i) = \frac{4\pi\,|R_{bo}|^2}{\sin\theta_i}\exp(-4k^2h^2\cos^2\theta_i)\delta(\theta_s - \theta_i)\delta(\phi_s - \phi_i)\delta_{ab} \quad (2.1.94)$$

where δ_{ab} is the Kronecker delta.

Similarly, we define the bistatic transmission coefficients to be

$$\gamma^t_{ba}(\hat{k}_t, \hat{k}_i) = \frac{4\pi r^2 (S_t)_b}{A_o \cos\theta_i (S_o)_a} \quad (2.1.95)$$

where S_t is the Poynting power density of the transmitted wave. Following the same procedure, we obtain

$$\gamma^t_{ab}(\hat{k}_t, \hat{k}_i) = \gamma^t_{c_{ab}}(\hat{k}_t, \hat{k}_i) + \gamma^t_{i_{ab}}(\hat{k}_t, \hat{k}_i) \quad (2.1.96)$$

where

$$\gamma^t_{c_{ab}}(\hat{k}_t, \hat{k}_i) = \frac{k_1^2}{4\pi A_o \cos\theta_i}\frac{\eta}{\eta_1}\left|\hat{a}_t \cdot \overline{N}_b(0,0)\right|^2 |\langle I_t\rangle|^2 \quad (2.1.97a)$$

$$\gamma^t_{i_{ab}}(\hat{k}_t, \hat{k}_i) = \frac{k_1^2}{4\pi A_o \cos\theta_i}\frac{\eta}{\eta_1}\left|\hat{a}_t \cdot \overline{N}_b(0,0)\right|^2 D_{I_t} \quad (2.1.97b)$$

with

$$\overline{N}_b(0,0) = \overline{N}(0,0)\Big|_{\hat{e}_i=\hat{b}_i} \tag{2.1.98}$$

Thus

$$\hat{h}_t \cdot \overline{N}_h(0,0) = \left[\frac{\eta_1}{\eta}(1 - R_{ho})\cos\theta_i + (1 + R_{ho})\cos\theta_t\right]\cos(\phi_t - \phi_i) \tag{2.1.99a}$$

$$\hat{v}_t \cdot \overline{N}_h(0,0) = \left[-\frac{\eta_1}{\eta}(1 - R_{ho})\cos\theta_i\cos\theta_t - (1 + R_{ho})\right] \cdot \sin(\phi_t - \phi_i) \tag{2.1.99b}$$

$$\hat{h}_t \cdot \overline{N}_v(0,0) = \left[\frac{\eta_1}{\eta}(1 + R_{vo}) + (1 - R_{vo})\cos\theta_i\cos\theta_t\right]\sin(\phi_t - \phi_i) \tag{2.1.99c}$$

$$\hat{v}_t \cdot \overline{N}_v(0,0) = \left[\frac{\eta_1}{\eta}(1 + R_{vo})\cos\theta_t + (1 - R_{vo})\cos\theta_i\right]\cos(\phi_t - \phi_i) \tag{2.1.99d}$$

The coherent component only exists in the specular transmission direction.

$$\gamma^t_{c_{ab}}(\hat{k}_t, \hat{k}_i) = 4\pi\frac{\cos\theta_{1i}}{\cos\theta_i}\frac{|1 + R_{bo}|^2}{\sin\theta_{1i}} g_a \exp\left[-(k_1\cos\theta_{1i} - k\cos\theta_i)^2 h^2\right] \times \delta(\theta_t - \theta_{1i})\delta(\phi_t - \phi_i)\delta_{ab} \tag{2.1.100}$$

where $g_a = \eta_1/\eta$ for $a = v$, $g_a = \eta/\eta_1$ for $a = h$, and θ_{1i} is related to θ_i by Snell's law

$$k_1 \sin\theta_{1i} = k\sin\theta_i \tag{2.1.101}$$

Geometrical Optics Solution

Under the geometrical optics limit as $k \to \infty$, the asymptotic solution to the Kirchhoff-approximated diffraction integrals can be derived using the method of stationary phase. The coherent component of the scattered fields will vanish in this limit, and only the incoherent component will remain. The bistatic scattering coefficients for the reflected and transmitted fields are proportional to the probability of the occurrence of the slopes which will specularly reflect or transmit the incident wave into the observation direction. The bistatic scattering coefficients satisfy reciprocity but violate energy conservation. This is due to the neglect of the effects of multiple scattering and shadowing . The scattering coefficients are next modified to incorporate the shadowing effects. The sum of reflected and transmitted intensities are then shown to be always less than the incident intensity since only single scattering solutions are used.

From (2.1.54) the exponential phase factor is

$$\psi = \overline{k}_d \cdot \overline{r}' = k_{dx}x' + k_{dy}y' + k_{dz}f(x', y') \tag{2.1.102}$$

To determine the stationary phase point, we differentiate (2.1.102) with respect to x' and set it equal to zero.

$$\frac{\partial \psi}{\partial x'} = 0 = k_{dx} + k_{dz}\alpha_o \tag{2.1.103}$$

At the stationary phase point we have

$$\alpha_o = -\frac{k_{dx}}{k_{dz}} \tag{2.1.104}$$

Similarly, by differentiating the phase term ψ with respect to y', we get

$$\beta_o = -\frac{k_{dy}}{k_{dz}} \tag{2.1.105}$$

The slopes α_o and β_o are such that the incident and scattered wave directions form a specular reflection. This can be seen from the fact that from (2.1.52) we have

$$\hat{n}(\alpha_o, \beta_o) = (\overline{k}_s - \overline{k}_i)/|\overline{k}_d| \tag{2.1.106}$$

Replacing the surface slopes α and β by α_o and β_o, we obtain, using (2.1.54),

$$\langle |\overline{E}_s|^2 \rangle = \frac{k^2 \left| E_o \right|^2}{16\pi^2 r^2} \left| (\overline{\overline{I}} - \hat{k}_s \hat{k}_s) \cdot \overline{F}(\alpha_o, \beta_o) \right|^2 \langle II^* \rangle \tag{2.1.107}$$

where

$$\langle II^* \rangle = \left\langle \int_{A_o} d\overline{r}_\perp \int_{A_o} d\overline{r}'_\perp \, e^{i\overline{k}_{d\perp}\cdot(\overline{r}_\perp - \overline{r}'_\perp)} \, e^{ik_{dz}(f(\overline{r}_\perp) - f(\overline{r}'_\perp))} \right\rangle \tag{2.1.108}$$

The above integral can be solved by the asymptotic method. For large k, contributions of the integral come from regions where (x', y') is close to (x, y). Expanding $f(x', y')$ about (x, y),

$$f(x', y') = f(x, y) + \alpha(x' - x) + \beta(y' - y) + \cdots \tag{2.1.109}$$

and replacing the integration variables by

$$u = k(x - x') \tag{2.1.110a}$$

$$v = k(y - y') \tag{2.1.110b}$$

we obtain

$$\langle II^* \rangle = \left\langle \frac{1}{k^2} A_o \iint du \, dv \, \exp\left[iu(q_x + \alpha q_z) + iv(q_y + \beta q_z) + \mathcal{O}\left(\frac{1}{k}\right) \right] \right\rangle \tag{2.1.111}$$

where

$$\overline{q} = \frac{\overline{k}_d}{k} \tag{2.1.112}$$

Ignoring the $O(1/k)$ and higher order terms, we have

$$\langle II^* \rangle = \frac{4\pi^2 A_o}{k^2} \langle \delta(q_x + \alpha q_z)\delta(q_y + \beta q_z) \rangle \tag{2.1.113}$$

Therefore

$$\langle \lim_{k\to\infty} II^* \rangle = \frac{4\pi^2 A_o}{k^2} \int_{-\infty}^{\infty} \int_{-\infty}^{\infty} d\alpha\, d\beta\, \delta(q_x + \alpha q_z)\delta(q_y + \beta q_z)\, p(\alpha, \beta) \tag{2.1.114}$$

where $p(\alpha, \beta)$ is the probability density function for the slopes at the surface. It follows that

$$\langle \lim_{k\to\infty} II^* \rangle = \frac{4\pi^2 A_o}{k_{dz}^2} p\Big(-\frac{k_{dx}}{k_{dz}}, -\frac{k_{dy}}{k_{dz}}\Big) \tag{2.1.115}$$

For the Gaussian random rough surface

$$p(\alpha, \beta) = \frac{1}{2\pi h^2\, |C''(0)|} \exp\left[-\frac{\alpha^2 + \beta^2}{2h^2\, |C''(0)|}\right] \tag{2.1.116}$$

where $C''(0)$ is the double derivative of the correlation function at $\rho = 0$. Thus, $h^2|C''(0)|$ is the mean square surface slope s^2, and for the Gaussian correlation function with correlation length l we have

$$s^2 = h^2\, \big|C''(0)\big| = 2\frac{h^2}{l^2} \tag{2.1.117}$$

Substituting (2.1.116) into (2.1.115) gives

$$\langle II^* \rangle = \frac{2\pi A_o}{k_{dz}^2 h^2\, |C''(0)|} \exp\left[-\frac{k_{dx}^2 + k_{dy}^2}{2k_{dz}^2 h^2\, |C''(0)|}\right] \tag{2.1.118}$$

For an incident field with polarization $\hat{b}_i$, the scattered intensity for polarization $\hat{a}_s$ is given by

$$\langle |E_s(\overline{r})|^2 \rangle = \frac{k^2|E_o|^2}{16\pi^2 r^2} |\hat{a}_s \cdot \overline{F}_b(\alpha_o, \beta_o)|^2 \langle II^* \rangle \tag{2.1.119}$$

where

$$\overline{F}_b(\alpha_o, \beta_o) = \overline{F}(\alpha_o, \beta_o)\Big|_{\hat{e}_i = \hat{b}_i} \tag{2.1.120}$$

and using (2.1.55a), we find that

$$|\hat{a}_s \cdot \overline{F}_b(\alpha_o, \beta_o)|^2 = \frac{|\overline{k}_d|^4}{k^2 |\hat{k}_i \times \hat{k}_s|^4 k_{dz}^2} f_{ba} \tag{2.1.121}$$

with

$$f_{vv} = |(\hat{h}_s \cdot \hat{k}_i)(\hat{h}_i \cdot \hat{k}_s) R_h + (\hat{v}_s \cdot \hat{k}_i)(\hat{v}_i \cdot \hat{k}_s) R_v|^2 \tag{2.1.122a}$$

$$f_{hv} = |(\hat{v}_s \cdot \hat{k}_i)(\hat{h}_i \cdot \hat{k}_s) R_h - (\hat{h}_s \cdot \hat{k}_i)(\hat{v}_i \cdot \hat{k}_s) R_v|^2 \tag{2.1.122b}$$

$$f_{vh} = |(\hat{h}_s \cdot \hat{k}_i)(\hat{v}_i \cdot \hat{k}_s)R_h - (\hat{v}_s \cdot \hat{k}_i)(\hat{h}_i \cdot \hat{k}_s)R_v|^2 \tag{2.1.122c}$$

$$f_{hh} = |(\hat{v}_s \cdot \hat{k}_i)(\hat{v}_i \cdot \hat{k}_s)R_h + (\hat{h}_s \cdot \hat{k}_i)(\hat{h}_i \cdot \hat{k}_s)R_v|^2 \tag{2.1.122d}$$

and R_v and R_h are evaluated at

$$\hat{n} = \frac{k_{dx}/k_{dz}\hat{x} + k_{dy}/k_{dz}\hat{y} + \hat{z}}{\sqrt{k_{dx}^2/k_{dz}^2 + k_{dy}^2/k_{dz}^2 + 1}} \tag{2.1.123}$$

Then, the bistatic scattering coefficients for the reflected intensities are, in view of (2.1.88)

$$\gamma_{ab}^r(\hat{k}_s, \hat{k}_i) = \frac{|\overline{k}_d|^4}{\cos\theta_i |\hat{k}_i \times \hat{k}_s|^4 k_{dz}^4} \frac{1}{2h^2 |C''(0)|} \exp\left[-\frac{k_{dx}^2 + k_{dy}^2}{2k_{dz}^2 h^2 |C''(0)|}\right] f_{ab} \tag{2.1.124}$$

We note that from (2.1.122) the geometrical optics solution does not depend on whether the total or scattered field is used in the diffraction integral because $\overline{F}(\alpha, \beta)$ is evaluated at the stationary phase point.

The backscattering cross sections are defined to be

$$\sigma_{ab}(\hat{k}_i) = \cos\theta_i \, \gamma_{ab}^r(-\hat{k}_i, \hat{k}_i) \tag{2.1.125}$$

From (2.1.125) we obtain

$$\sigma_{hh}(\theta_i) = \sigma_{vv}(\theta_i) = \frac{|R|^2}{\cos^4\theta_i 2h^2 |C''(0)|} \exp\left[-\frac{\tan^2\theta_i}{2h^2 |C''(0)|}\right] \tag{2.1.126a}$$

$$\sigma_{vh}^r(\theta_i) = \sigma_{hv}^r(\theta_i) = 0 \tag{2.1.126b}$$

where R is the reflection coefficient at normal incidence. We note that from (2.1.126b) there is no depolarization in the backscattering direction.

The bistatic scattering coefficients for the transmitted waves can be derived in a similar manner. The stationary phase method is used to evaluate the diffraction integral for the transmitted fields. The stationary phase points are given by

$$\alpha_o = -\frac{k_{tdx}}{k_{tdz}} \tag{2.1.127a}$$

$$\beta_o = -\frac{k_{tdy}}{k_{tdz}} \tag{2.1.127b}$$

where α_o and β_o are the values that the slopes α and β assume at the stationary phase point. At the stationary phase point

$$\hat{n} = \frac{k_{tdx}/k_{tdz}\hat{x} + k_{tdy}/k_{tdz}\hat{y} + \hat{z}}{\sqrt{k_{tdx}^2/k_{tdz}^2 + k_{tdy}^2/k_{tdz}^2 + 1}} \tag{2.1.128}$$

It follows that

$$\overline{k}_i - (\overline{k}_i \cdot \hat{n})\hat{n} = \overline{k}_t - (\overline{k}_t \cdot \hat{n})\hat{n} \tag{2.1.129}$$

which is a statement of Snell's law that the tangential components of the wave vectors $\overline{k}_i$ and $\overline{k}_t$ must be equal. Thus, the slopes α_o and β_o are such that incident and transmitted wave directions form a specular transmission.

The transmitted field is obtained from (2.1.54) by replacing α and β by α_o and β_o, and the associated integrals are evaluated in a manner similar to that of the scattered fields. The bistatic coefficients for the transmitted intensities, in view of (2.1.95), are

$$\gamma^t_{ab}(\hat{k}_t, \hat{k}_i) = \frac{2k_1^2|\overline{k}_{td}|^2(\hat{n}\cdot\hat{k}_t)^2}{\cos\theta_i|\hat{k}_i \times \hat{k}_t|^4 k_{tdz}^4}\frac{\eta}{\eta_1}\frac{1}{h^2|C''(0)|}\exp\left[-\frac{k_{tdx}^2 + k_{tdy}^2}{2k_{tdz}^2 h^2|C''(0)|}\right] W_{ab} \tag{2.1.130}$$

with

$$W_{vv} = \left|(\hat{h}_t\cdot\hat{k}_i)(\hat{h}_i\cdot\hat{k}_t)(1+R_h) + (\hat{v}_t\cdot\hat{k}_i)(\hat{v}_i\cdot\hat{k}_t)\frac{\eta_1}{\eta}(1+R_v)\right|^2 \tag{2.1.131a}$$

$$W_{hv} = \left|-(\hat{v}_t\cdot\hat{k}_i)(\hat{h}_i\cdot\hat{k}_t)(1+R_h) + (\hat{h}_t\cdot\hat{k}_i)(\hat{v}_i\cdot\hat{k}_t)\frac{\eta_1}{\eta}(1+R_v)\right|^2 \tag{2.1.131b}$$

$$W_{vh} = \left|(\hat{h}_t\cdot\hat{k}_i)(\hat{v}_i\cdot\hat{k}_t)(1+R_h) - (\hat{v}_t\cdot\hat{k}_i)(\hat{h}_i\cdot\hat{k}_t)\cdot\frac{\eta_1}{\eta}(1+R_v)\right|^2 \tag{2.1.131c}$$

$$W_{hh} = \left|(\hat{v}_t\cdot\hat{k}_i)(\hat{v}_i\cdot\hat{k}_t)(1+R_h) + (\hat{h}_t\cdot\hat{k}_i)(\hat{h}_i\cdot\hat{k}_t)\cdot\frac{\eta_1}{\eta}(1+R_v)\right|^2 \tag{2.1.131d}$$

The reflection coefficients R_h and R_v are to be taken at the stationary phase point of (2.1.127) so that

$$R_h = \frac{k(\hat{n}\cdot\hat{k}_i) - k_1(\hat{n}\cdot\hat{k}_t)}{k(\hat{n}\cdot\hat{k}_i) + k_1(\hat{n}\cdot\hat{k}_t)} \tag{2.1.132a}$$

$$R_v = \frac{k_1(\hat{n}\cdot\hat{k}_i) - k(\hat{n}\cdot\hat{k}_t)}{k_1(\hat{n}\cdot\hat{k}_i) + k(\hat{n}\cdot\hat{k}_t)} \tag{2.1.132b}$$

Reciprocity and Conservation of Energy

We note that a reciprocity relation exists for the bistatic transmission and reflection coefficients obtained in (2.1.124) and (2.1.130). Consider two media, 1 and 2, with indices of refraction n_1 and n_2 and with wavenumbers k_1 and k_2, respectively. The two media are separated by a rough interface. Then the bistatic transmission coefficient $\gamma^{t_{21}}_{ab}(\hat{k}_2, \hat{k}_1)$ signifies a wave incident from angle (θ_1, ϕ_1) with polarization b onto angle (θ_2, ϕ_2) in medium 2 with polarization a. This is obtained from (2.1.130) by substituting $(\theta_i, \phi_i) = (\theta_1, \phi_1)$, $(\theta_t, \phi_t) = (\theta_2, \phi_2)$, $k_t = k_2$, and $k = k_1$. Similarly, $\gamma_{ba}^{t_{12}}(\hat{k}_1, \hat{k}_2)$ is for an incident wave from medium 2 and can be obtained by substituting the following into (2.1.130), $(\theta_i, \phi_i) = (\theta_2, \phi_2)$, $(\theta_t, \phi_t) = (\theta_1, \phi_1)$, $k_t = k_1$, and $k = k_2$. The following reciprocity relation holds for the bistatic transmission coefficients.

$$n_1^2 \cos\theta_1 \, \gamma_{ab}^{t_{21}}(\hat{k}_2, \hat{k}_1) = n_2^2 \cos\theta_2 \, \gamma_{ba}^{t_{12}}(\hat{k}_1, \hat{k}_2) \tag{2.1.133}$$

For the bistatic reflection coefficients, we similarly obtain the following reciprocity relation

$$\cos\theta_s \, \gamma_{ab}^{r_{jj}}(\hat{k}_i, \hat{k}_s) = \cos\theta_i \, \gamma_{ba}^{r_{jj}}(\hat{k}_s, \hat{k}_i) \tag{2.1.134}$$

with $j = 1, 2$. In (2.1.134), $\gamma_{ab}^{r_{11}}(\hat{k}_s, \hat{k}_i)$ and $\gamma_{ab}^{r_{22}}(\hat{k}_s, \hat{k}_i)$ are the bistatic reflection coefficients in media 1 and 2 between an incident wave with polarization b and a scattered wave with polarization a.

We define the reflectivity

$$r_b(\theta_i) = \frac{1}{4\pi} \sum_a \int_o^{\pi/2} d\theta_s \sin\theta_s \int_o^{2\pi} d\phi_s \, \gamma_{ab}^{r}(\hat{k}_s, \hat{k}_i) \tag{2.1.135a}$$

and the transmissivity

$$t_b(\theta_i) = \frac{1}{4\pi} \sum_a \int_o^{\pi/2} d\theta_t \sin\theta_t \int_o^{2\pi} d\phi_t \, \gamma_{ab}^{t}(\hat{k}_t, \hat{k}_i) \tag{2.1.135b}$$

where the summation a is over both the vertical and horizontal polarizations.

Making use of (2.1.122) through (2.1.124), we find the reflectivities for the vertical and horizontal polarizations to be

$$\begin{aligned} \begin{bmatrix} r_v(\theta_i) \\ r_h(\theta_i) \end{bmatrix} &= \int_o^{\pi/2} d\theta_s \sin\theta_s \int_o^{2\pi} d\phi_s \frac{1}{2\pi h^2 |C''(0)|} \\ &\cdot \exp\left[-\frac{k_{dx}^2 + k_{dy}^2}{2k_{dz}^2 h^2 |C''(0)|} \right] \frac{|\overline{k}_d|^4}{4\cos\theta_i |\hat{k}_i \times \hat{k}_s|^2 k_{dz}^4} \\ &\cdot \begin{bmatrix} (\hat{h}_i \cdot \hat{k}_s)^2 |R_h|^2 + (\hat{v}_i \cdot \hat{k}_s)^2 |R_v|^2 \\ (\hat{v}_i \cdot \hat{k}_s)^2 |R_h|^2 + (\hat{h}_i \cdot \hat{k}_s)^2 |R_v|^2 \end{bmatrix} \end{aligned} \tag{2.1.136}$$

where we made use of the fact that

$$|\hat{k}_i \times \hat{k}_s|^2 = (\hat{v}_i \cdot \hat{k}_s)^2 + (\hat{h}_i \cdot \hat{k}_s)^2 = (\hat{h}_s \cdot \hat{k}_i)^2 + (\hat{v}_s \cdot \hat{k}_i)^2 \tag{2.1.137}$$

Similarly, making use of (2.1.130) and (2.1.131), the transmissivities for the vertical and horizontal polarizations are found to be

$$\begin{bmatrix} t_v(\theta_i) \\ t_h(\theta_i) \end{bmatrix} = \int_o^{\pi/2} d\theta_t \sin\theta_t \int_o^{2\pi} d\phi_t \frac{1}{2\pi h^2 |C''(0)|}$$
$$\cdot \exp\left[-\frac{k_{tdx}^2 + k_{tdy}^2}{2k_{tdz}^2 h^2 |C''(0)|} \right] \frac{k_1^2 |\overline{k}_{td}|^2 (\hat{n}\cdot\hat{k}_i)(\hat{n}\cdot\hat{k}_t)}{\cos\theta_i |\hat{k}_i \times \hat{k}_t|^2 k_{tdz}^4}$$
$$\cdot \begin{bmatrix} (\hat{h}_i\cdot\hat{k}_t)^2(1-|R_h|^2) + (\hat{v}_i\cdot\hat{k}_t)^2(1-|R_v|^2) \\ (\hat{v}_i\cdot\hat{k}_t)^2(1-|R_h|^2) + (\hat{h}_i\cdot\hat{k}_t)^2(1-|R_v|^2) \end{bmatrix} \tag{2.1.138}$$

where we make use of

$$|\hat{k}_i \times \hat{k}_t|^2 = (\hat{h}_t\cdot\hat{k}_i)^2 + (\hat{v}_t\cdot\hat{k}_i)^2 = (\hat{h}_i\cdot\hat{k}_t)^2 + (\hat{v}_i\cdot\hat{k}_t)^2 \tag{2.1.139}$$

and

$$|1+R_h|^2 = (1-|R_h|^2)\frac{(\hat{n}\cdot\hat{k}_i)}{n_t(\hat{n}\cdot\hat{k}_t)} \tag{2.1.140a}$$

$$\frac{|1+R_v|^2}{n_t^2} = (1-|R_v|^2)\frac{(\hat{n}\cdot\hat{k}_i)}{n_t(\hat{n}\cdot\hat{k}_t)} \tag{2.1.140b}$$

with $n_t = k_t/k$.

Conservation of energy relations should also exist for the reflectivity and transmissivity functions. However, as pointed out before, because the shadowing and multiple scattering effects are ignored in Kirchhoff approximation, conservation of energy is not obeyed. To investigate the conservation of energy, we define

$$\Delta_a(\theta_i) = 1 - r_a(\theta_i) - t_a(\theta_i) \tag{2.1.141}$$

with $a = v, h$. In Fig. 2.1.3, $\Delta_h(\theta_i)$ and $\Delta_v(\theta_i)$ are plotted as a function of incident angle. We see that energy loss becomes severe as incident angle increases. However, as incident angle is further increased, the trend reverses and the sum of reflected and transmitted energy becomes greater than unity. Near the normal angle of incidence there is little shadowing and most of the rough surface is illuminated. Therefore, since we only have the single scattering solution and have ignored multiple scattering, there is loss of energy. At higher angle of incidence, the multiple scattering solutions are still left out. However, the shadowing effect that is neglected becomes important. The single scattering solution overestimates scattered energy as incident angle is increased. We shall modify the bistatic scattering coefficients to include the effect of shadowing and show that, in doing so, the sum of reflected and transmitted energy is always less than unity.

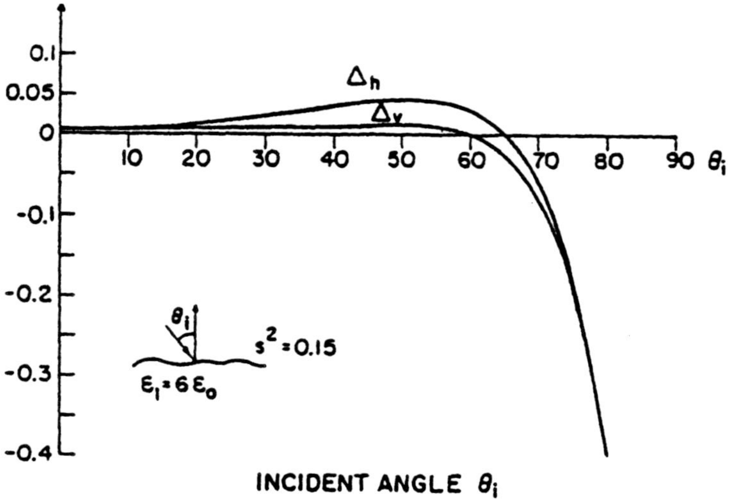

Figure 2.1.3 Energy conservation relations for the geometrical optics solution.

Shadowing Effect

The bistatic reflection coefficients are first considered. The diffraction integral for the reflected field, (2.1.54a), is modified with the addition of an illumination function $L(\hat{k}_i, \hat{k}_s, \overline{r}')$, as follows:

$$\overline{E}_s(\overline{r}) = \frac{ik\, e^{ikr}}{4\pi r} E_o(\overline{\overline{I}} - \hat{k}_s\hat{k}_s) \cdot \int_{A_o} d\overline{r}'_\perp \overline{F}(\alpha, \beta)\, L(\hat{k}_i, \hat{k}_s, \overline{r}')\, e^{i\overline{k}_d \cdot \overline{r}'} \tag{2.1.142}$$

where $L(\hat{k}_i, \hat{k}_s, \overline{r}') = 1$ if a ray having a direction $\hat{k}_i$ is not intersected by the surface and illuminates the point $\overline{r}'$ and if the line drawn from the point $\overline{r}'$ in the direction $\hat{k}_s$ does not strike the surface, and $L(\hat{k}_i, \hat{k}_s, \overline{r}') = 0$ otherwise. The above integral is evaluated by the method of stationary phase. The scattered intensity is, from (2.1.107) and (2.1.108),

$$\langle |\overline{E}_s|^2 \rangle = \frac{k^2 |E_o|^2}{16\pi^2 r^2} \left| (\overline{\overline{I}} - \hat{k}_s\hat{k}_s) \cdot \overline{F}(\alpha_o, \beta_o) \right|^2 \langle II^* \rangle \tag{2.1.143}$$

where $\langle II^* \rangle$ is now modified to include shadowing

$$\begin{aligned} \langle II^* \rangle = \Big\langle \int_{A_o} d\overline{r}_\perp \int_{A_o} d\overline{r}'_\perp \; e^{i\overline{k}_{d\perp} \cdot (\overline{r}_\perp - \overline{r}'_\perp)} \\ \times L(\hat{k}_i, \hat{k}_s, \overline{r}_\perp)\, L(\hat{k}_i, \hat{k}_s, \overline{r}'_\perp)\; e^{ik_{dz}(f(\overline{r}_\perp) - f(\overline{r}'_\perp))} \Big\rangle \end{aligned} \tag{2.1.144}$$

The above integral is solved by the asymptotic method as in (2.1.109) through (2.1.115). We also note that as $k \to \infty$,

$$\lim_{k\to\infty} L(\hat{k}_i, \hat{k}_s, \overline{r}_\perp) L(\hat{k}_i, \hat{k}_s, \overline{r}'_\perp) = \lim_{k\to\infty} L(\hat{k}_i, \hat{k}_s, \overline{r}_\perp) L\Big(\hat{k}_i, \hat{k}_s, x - \frac{u}{k}, y - \frac{v}{k}\Big)$$
$$= L^2(\hat{k}_i, \hat{k}_s, \overline{r}_\perp) \qquad (2.1.145)$$

Hence, we obtain the following expression analogous to (2.1.115).

$$\langle \lim_{k\to\infty} II^* \rangle = \frac{A_o 4\pi^2}{k_{dz}^2} \int dL\ p(\alpha, \beta, L)\Big|_{\alpha = -\frac{k_{dx}}{k_{dz}}, \beta = -\frac{k_{dy}}{k_{dz}}} L^2(\hat{k}_i, \hat{k}_s, \overline{r}_\perp) \quad (2.1.146)$$

where $p(\alpha, \beta, L)$ is the joint probability density function for α, β, and L. Since the process is homogeneous, the result above is independent of $\overline{r}$. The conditional probability density is expressed as

$$p(\alpha, \beta, L) = p(\alpha, \beta)\, p(L|\alpha, \beta) \qquad (2.1.147)$$

where

$$p(L|\alpha, \beta) = P_L(\hat{k}_i, \hat{k}_s|\alpha, \beta)\delta(L-1) + \Big[1 - P_L(\hat{k}_i, \hat{k}_s|\alpha, \beta)\Big]\, \delta(L) \quad (2.1.148)$$

with $P_L(\hat{k}_i, \hat{k}_s|\alpha, \beta)$ is the probability that a point will be illuminated by rays having the directions $\hat{k}_i$ and $-\hat{k}_s$ given the values of the slope α and β at the point. Thus, using (2.1.115), we obtain

$$\langle II^* \rangle = \frac{A_o 4\pi^2}{k_{dz}^2} \frac{1}{2\pi h^2 |C''(0)|} \exp\left[-\frac{k_{dx}^2 + k_{dy}^2}{2k_{dz}^2 h^2 |C''(0)|}\right] P_L\left(\hat{k}_i, \hat{k}_s \Big| -\frac{k_{dx}}{k_{dz}}, -\frac{k_{dy}}{k_{dz}}\right) \quad (2.1.149)$$

The bistatic reflection coefficients are modified to

$$\gamma_{ab}^{m,r}(\hat{k}_s, \hat{k}_i) = \gamma_{ab}^{r}(\hat{k}_s, \hat{k}_i) P_L\left(\hat{k}_i, \hat{k}_s \Big| -\frac{k_{dx}}{k_{dz}}, -\frac{k_{dy}}{k_{dz}}\right) \qquad (2.1.150)$$

We use the shadowing function derived by Smith [1967] and Sancer [1969] to modify the bistatic scattering coefficients.

$$P_L\left(\hat{k}_i, \hat{k}_s \Big| -\frac{k_{dx}}{k_{dz}}, -\frac{k_{dy}}{k_{dz}}\right) = S(\theta_s, \theta_i) \qquad (2.1.151)$$

where

$$S(\hat{k}_s, \hat{k}_i) = \begin{cases} \dfrac{1}{1 + \Lambda(\mu_s)} & \phi_s = \phi_i + \pi,\ \theta_s \geq \theta_i \\ \dfrac{1}{\Lambda(\mu_i) + 1} & \phi_s = \phi_i + \pi,\ \theta_i \geq \theta_s \\ \dfrac{1}{\Lambda(\mu_s) + \Lambda(\mu_i) + 1} & \text{(otherwise)} \end{cases} \qquad (2.1.152)$$

and

$$\mu = \cot\theta \qquad (2.1.153)$$

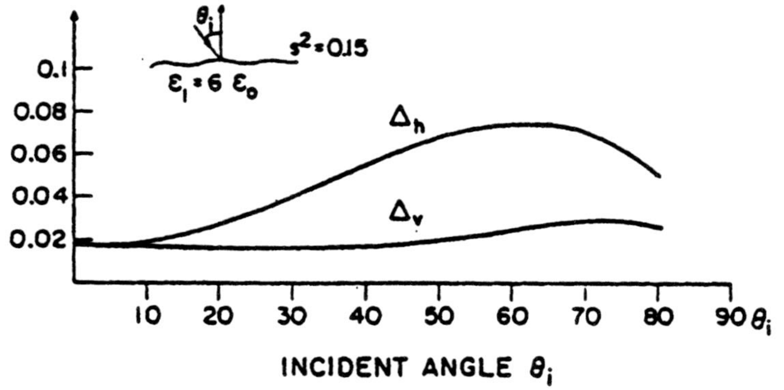

Figure 2.1.4 Energy conservation for the modified geometrical optics solutions.

$$\Lambda(\mu) = \frac{1}{2}\left[\sqrt{\frac{2}{\pi}}\frac{s}{\mu}e^{-\mu^2/2s^2} - \text{erfc}\left(\frac{\mu}{\sqrt{2}s}\right)\right] \tag{2.1.154}$$

s^2 is the mean square surface slope defined in (2.1.117), and erfc is the complementary error function.

The bistatic transmission coefficients can be modified in a similar fashion. Thus,

$$\gamma_{ab}^{m_t}(\hat{k}_t, \hat{k}_i) = \gamma_{ab}^{t}(\hat{k}_t, \hat{k}_i)S_t(\theta_t, \theta_i) \tag{2.1.155}$$

where

$$S_t(\theta_t, \theta_i) = \frac{1}{1 + \Lambda(\mu_t) + \Lambda(\mu_i)} \tag{2.1.156}$$

The modified reflectivity and transmissivity functions $r_b^m(\theta_i)$ and $t_b^m(\theta_i)$ are obtained from (2.1.135) by substituting in the modified bistatic scattering coefficients into the equation. The energy conservation is examined by plotting Δ_v and Δ_h in Fig. 2.1.4. We see that the inclusion of shadowing means that the sum of reflected and transmitted energy is always less than unity. Also we note that $\Delta_h > \Delta_v$.

Emissivities

The emissivity of a rough surface can be calculated from the bistatic reflection coefficient. We have

$$e_a(\theta_i, \phi_i) = 1 - \frac{1}{4\pi}\sum_{b=v,h}\int_o^{\pi/2} d\theta_s \sin\theta_s \int_o^{2\pi} d\phi_s\, \gamma_{ba}^{r}(\theta_s, \phi_s; \theta_i, \phi_i) \tag{2.1.157}$$

The coherent and incoherent bistatic reflection coefficients derived from the Kirchhoff-approximated diffraction integrals in (2.1.93) can be used to calculate the emissivity of the rough surface. In terms of the coherent and incoherent reflectivity functions, the emissivity is given by

$$e_a(\theta_i) = 1 - r_{c_a}(\theta_i) - r_{i_a}(\theta_i) \tag{2.1.158}$$

where, for $\alpha = c$ or i, denoting coherent and incoherent, respectively, we obtain

$$r_{\alpha_a}(\theta_i) = \frac{1}{4\pi} \sum_{b=v,h} \int_o^{\pi/2} d\theta_s \sin\theta_s \int_o^{2\pi} d\phi_s \, \gamma^r_{\alpha_{ba}}(\theta_s, \phi_s; \theta_i, \phi_i) \tag{2.1.159}$$

In the above equation, the dependence on the azimuthal angle of incidence ϕ_i is dropped because the rough surface is isotropic and the results are independent of ϕ_i. After substituting in the explicit expressions for the bistatic scattering coefficients from (2.1.93), carrying out the angular integration and assuming that $C(\rho) = \exp(-\rho^2/l^2)$, we obtain

$$r_{c_a}(\theta_i) = |R_{ao}|^2 \exp\left[-4k^2h^2\cos^2\theta_i\right] \tag{2.1.160}$$

$$\begin{aligned} r_{i_a}(\theta_i) = {} & \frac{k^2l^2}{8\cos\theta_i} \int_o^{\pi/2} d\theta_s \, \sin\theta_s \exp\left[-h^2k^2(\cos\theta_i + \cos\theta_s)^2\right] \\ & \times \sum_{m=1}^{\infty} \frac{(kh(\cos\theta_s + \cos\theta_i))^{2m}}{m!\, m} e^{-y_m} \Bigg\{ \left|(1+R_{ao}) - (1-R_{ao})\cos\theta_i\cos\theta_s\right|^2 \\ & \cdot \frac{I_1(x_m)}{x_m} + \left| -(1+R_{ao})\cos\theta_s + (1-R_{ao})\cos\theta_i \right|^2 \\ & \cdot \left(I_o(x_m) - \frac{I_1(x_m)}{x_m} \right) \Bigg\} \end{aligned} \tag{2.1.161}$$

where $a = v$, h, and

$$y_m = \frac{k^2l^2(\sin^2\theta_i + \sin^2\theta_s)}{4m} \tag{2.1.162}$$

$$x_m = \frac{k^2l^2 \sin\theta_i \sin\theta_s}{2m} \tag{2.1.163}$$

and I_o and I_1 are the zeroth- and first-order modified Bessel functions. We note that if the scattered field is used instead of the total field in the diffraction integral, then for $a = v$ or h we have

$$r_{i_a}(\theta_i) = |R_{ao}|^2 \frac{k^2 l^2}{8\cos\theta_i} \int_o^{\pi/2} d\theta_s \sin\theta_s \exp\left[-h^2k^2(\cos\theta_i + \cos\theta_s)^2\right]$$
$$\times \sum_{m=1}^{\infty} \frac{(kh(\cos\theta_s + \cos\theta_i))^{2m}}{m!\,m} e^{-y_m} \left\{(1 + \cos\theta_i \cos\theta_s)^2 \frac{I_1(x_m)}{x_m}\right.$$
$$\left. +(\cos\theta_s + \cos\theta_i)^2 \left(I_o(x_m) - \frac{I_1(x_m)}{x_m}\right)\right\} \tag{2.1.164}$$

The difference between the incoherent reflectivities obtained using the total and the scattered field in the diffraction integral is due to the approximation made on the integrand $\overline{F}(\alpha, \beta)$.

The emissivity may also be calculated in terms of the bistatic transmission coefficients from medium 1 to medium 0,

$$e_a(\theta_i, \phi_i) = \frac{1}{4\pi} \sum_{b=v,h} \int_o^{\pi/2} d\theta_1 \sin\theta_1 \int_o^{2\pi} d\phi_1 \gamma_{ab}^{t_{01}}(\theta_i, \phi_i; \theta_1, \phi_1) \frac{n_1^2 \cos\theta_1}{n_o^2 \cos\theta_i} \tag{2.1.165}$$

1.3 Second-Order Slope Corrections

We next extend the results of the Kirchhoff approximation up to the second order in slope. The scattered field is

$$\overline{E}_s(\overline{r}) = \int_{S'} dS' \left\{ i\omega\mu_o \overline{\overline{G}}(\overline{r}, \overline{r}') \cdot \hat{n} \times \overline{H}(\overline{r}') + \nabla \times \overline{\overline{G}}(\overline{r}, \overline{r}') \cdot \hat{n} \times \overline{E}(\overline{r}') \right\} \tag{2.1.166}$$

It can readily be shown that

$$\int_{S'} dS' \left\{ i\omega\mu_o \overline{\overline{G}}(\overline{r}, \overline{r}') \cdot \hat{n} \times \overline{H}_i(\overline{r}') + \nabla \times \overline{\overline{G}}(\overline{r}, \overline{r}') \cdot \hat{n} \times \overline{E}_i(\overline{r}') \right\} = 0$$

that is, the surface integral of the incident field is equal to zero. Then

$$\overline{E}_s(\overline{r}) = \int_{S'} dS' \left\{ i\omega\mu_o \overline{\overline{G}}(\overline{r}, \overline{r}') \cdot \hat{n} \times \overline{H}_s(\overline{r}') + \nabla \times \overline{\overline{G}}(\overline{r}, \overline{r}') \cdot \hat{n} \times \overline{E}_s(\overline{r}') \right\} \tag{2.1.167}$$

In this section we will use the scattered field formulation. In the far field we have

$$\overline{E}_s(\overline{r}) = \frac{ike^{ikr}}{4\pi r} (\overline{\overline{I}} - \hat{k}_s \hat{k}_s) \cdot \int_{S'} dS' \left\{ \hat{k}_s \times \left(\hat{n} \times \overline{E}_s(\overline{r}')\right) + \eta \left(\hat{n} \times \overline{H}_s(\overline{r}')\right) \right\} e^{-i\hat{k}_s \cdot \overline{r}'} \tag{2.1.168}$$

The local angle of incidence is θ_{li} such that

$$\cos\theta_{li} = -\hat{n} \cdot \hat{k}_i \tag{2.1.169}$$

$$\hat{n} = \frac{-\alpha\hat{x} - \beta\hat{y} + \hat{z}}{\sqrt{1 + \alpha^2 + \beta^2}} \tag{2.1.170}$$

where α and β are defined in (2.1.53), and

$$\cos\theta_{li} = -\hat{n}\cdot\hat{k}_i = \frac{\alpha\sin\theta_i\cos\phi_i + \beta\sin\theta_i\sin\phi_i + \cos\theta_i}{\sqrt{1+\alpha^2+\beta^2}} \tag{2.1.171}$$

The local perpendicular and parallel polarization reflection coefficients are

$$R_h = \frac{k\cos\theta_{li} - \sqrt{k_1^2 - k^2\sin^2\theta_{li}}}{k\cos\theta_{li} + \sqrt{k_1^2 - k^2\sin^2\theta_{li}}} \tag{2.1.172}$$

$$R_v = \frac{\epsilon_1 k\cos\theta_{li} - \epsilon_o\sqrt{k_1^2 - k^2\sin^2\theta_{li}}}{\epsilon_1 k\cos\theta_{li} + \epsilon_o\sqrt{k_1^2 - k^2\sin^2\theta_{li}}} \tag{2.1.173}$$

Thus

$$\overline{E}_s(\overline{r}) = \frac{ike^{ikr}}{4\pi r}(\overline{\overline{I}} - \hat{k}_s\hat{k}_s)E_o\int_{A_o} d\overline{r}'_\perp \overline{F}(\alpha,\beta)e^{i(\overline{k}_i-\overline{k}_s)\cdot\overline{r}'} \tag{2.1.174}$$

where

$$\begin{aligned}\overline{F}(\alpha,\beta) &= \sqrt{1+\alpha^2+\beta^2}\Big\{\hat{k}_s\times(\hat{n}\times\hat{q}_i)R_h(\hat{e}_i\cdot\hat{q}_i) - (\hat{e}_i\cdot\hat{p}_i)(\hat{n}\cdot\hat{k}_i)(\hat{k}_s\times\hat{q}_i)R_v \\ &\quad + R_h(\overline{e}_i\cdot\hat{q}_i)(\hat{n}\cdot\hat{k}_i)\hat{q}_i + (\hat{e}_i\cdot\hat{p}_i)(\hat{n}\times\hat{q}_i)R_v\Big\}\end{aligned} \tag{2.1.175}$$

Next we expand $\overline{F}(\alpha,\beta)$ about zero slope to the second order in slope [Leader, 1971]. Thus

$$\begin{aligned}\overline{F}(\alpha,\beta) &= \overline{F}(0,0) + \alpha\frac{\partial\overline{F}}{\partial\alpha}\Big|_{\alpha,\beta=0} + \beta\frac{\partial\overline{F}}{\partial\beta}\Big|_{\alpha,\beta=0} \\ &\quad + \frac{\alpha^2}{2}\frac{\partial^2\overline{F}}{\partial\alpha^2}\Big|_{\alpha=0,\beta=0} + \frac{\beta^2}{2}\frac{\partial^2\overline{F}}{\partial\beta^2}\Big|_{\alpha=\beta=0} + \alpha\beta\frac{\partial^2\overline{F}}{\partial\alpha\partial\beta}\Big|_{\alpha=0,\beta=0}\end{aligned} \tag{2.1.176}$$

$$\begin{aligned}\overline{E}_s(\overline{r}) &= \frac{ike^{ikr}}{4\pi r}\left(\overline{\overline{I}} - \hat{k}_s\hat{k}_s\right)E_o\int_{A_o} dx'dy'\Big\{\overline{F}(0,0) + \alpha\overline{F}_\alpha(0,0) + \beta\overline{F}_\beta(0,0) \\ &\quad + \frac{\alpha^2}{2}\overline{F}_{\alpha\alpha}(0,0) + \frac{\beta^2}{2}\overline{F}_{\beta\beta}(0,0) + \alpha\beta\overline{F}_{\alpha\beta}(0,0)\Big\}e^{i\overline{k}_d\cdot\overline{r}'}\end{aligned} \tag{2.1.177}$$

Integrating by parts and ignoring edge effects give

$$\int_{A_o} dx'dy'\frac{\partial f}{\partial x'}e^{ik_{dx}x'+ik_{dy}y'+ik_{dz}f(x',y')} = -\frac{k_{dx}}{k_{dz}}\int_{A_o} dx'dy'e^{i\overline{k}_d\cdot\overline{r}'} \tag{2.1.178}$$

$$\int_{A_o} dx'dy'\beta e^{i\overline{k}_d\cdot\overline{r}'} = -\frac{k_{dy}}{k_{dz}}\int_{A_o} dx'dy'e^{i\overline{k}_d\cdot\overline{r}'} \tag{2.1.179}$$

The higher-order terms can also be recast by integration by parts

$$\iint dx'dy'\alpha^2 e^{i\overline{k}_d\cdot\overline{r}'} = \left(\frac{k_{dx}}{k_{dz}}\right)^2 \iint dx'dy' e^{i\overline{k}_d\cdot\overline{r}'} + \frac{i}{k_{dz}} \iint dx'dy' f_{xx} e^{i\overline{k}_d\cdot\overline{r}'} \quad (2.1.180)$$

$$\iint dx'dy'\beta^2 e^{i\overline{k}_d\cdot\overline{r}'} = \left(\frac{k_{dy}}{k_{dz}}\right)^2 \iint dx'dy' e^{i\overline{k}_d\cdot\overline{r}'} + \frac{i}{k_{dz}} \iint dx'dy' f_{yy} e^{i\overline{k}_d\cdot\overline{r}'} \quad (2.1.181)$$

$$\iint dx'dy'\alpha\beta e^{i\overline{k}_d\cdot\overline{r}'} = \left(\frac{k_{dx}k_{dy}}{k_{dz}^2}\right) \iint dx'dy' e^{i\overline{k}_d\cdot\overline{r}'} + \frac{i}{k_{dz}} \iint dx'dy' f_{xy} e^{i\overline{k}_d\cdot\overline{r}'} \quad (2.1.182)$$

Let

$$\begin{aligned}\overline{\overline{F}}(\overline{k}_{\perp s}, \overline{k}_{\perp i}) &= \overline{F}(0,0) - \frac{k_{dx}}{k_{dz}}\overline{F}_\alpha(0,0) - \frac{k_{dy}}{k_{dz}}\overline{F}_\beta(0,0) + \frac{1}{2}\left(\frac{k_{dx}}{k_{dz}}\right)^2 \overline{F}_{\alpha\alpha}(0,0) \\ &+ \frac{1}{2}\left(\frac{k_{dy}}{k_{dz}}\right)^2 \overline{F}_{\beta\beta}(0,0) + \frac{k_{dx}k_{dy}}{k_{dz}^2}\overline{F}_{\alpha\beta}(0,0) \end{aligned} \quad (2.1.183)$$

Then the scattered field is

$$\begin{aligned}&\overline{E}_s(\overline{r}) \\ &= \frac{ike^{ikr}}{4\pi r}(\overline{\overline{I}} - \hat{k}_s\hat{k}_s)E_o \cdot \int_{A_o} dx'dy' \left\{ \overline{\overline{F}}(\overline{k}_{\perp s}, \overline{k}_{\perp i}) + \frac{i}{2k_{dz}} f_{xx}(x',y')\overline{F}_{\alpha\alpha}(0,0) \right. \\ &\left. + \frac{i}{2k_{dz}} f_{yy}(x',y')\overline{F}_{\beta\beta}(0,0) + \frac{i}{k_{dz}} f_{xy}(x',y')\overline{F}_{\alpha\beta}(0,0) \right\} e^{i\overline{k}_d\cdot\overline{r}'} \end{aligned} \quad (2.1.184)$$

First, we calculate the coherent field. Note that

$$\langle f(x,y)f(x',y')\rangle = h^2C(x-x', y-y') = h^2C(|x-x'|, |y-y'|) \quad (2.1.185)$$

Since $f(x,y)$ is a Gaussian random process, then $f(x,y)$, $\alpha_x(x,y)$, $\beta_y(x,y)$, and $f_{xy}(x,y)$ are joint Gaussian random processes with correlation functions.

$$\langle \alpha_x(x,y)f(x',y')\rangle = h^2\frac{\partial^2 C(x-x', y-y')}{\partial x^2} \quad (2.1.186)$$

$$\langle \beta_y(x,y)f(x',y')\rangle = h^2\frac{\partial^2 C(x-x', y-y')}{\partial y^2} \quad (2.1.187)$$

$$\langle f_{xy}(x,y)f(x',y')\rangle = h^2\frac{\partial^2 C(x-x', y-y')}{\partial x\partial y} \quad (2.1.188)$$

The variances are

$$\langle \alpha_x(x,y)\alpha_x(x,y)\rangle = h^2C_{xxxx}(0,0) \quad (2.1.189)$$

$$\langle \beta_y(x,y)\beta_y(x,y)\rangle = h^2C_{yyyy}(0,0) \quad (2.1.190)$$

$$\langle f_{xy}(x,y)f_{xy}(x,y)\rangle = h^2C_{xyxy}(0,0) \quad (2.1.191)$$

Using the correlation functions and variances, the joint probability density functions of the joint Gaussian processes can be calculated readily. These joint probability density functions can then be used to calculate statistical averages. For coherent field, we have

$$\langle e^{ik_{dz}f(x,y)}\rangle = e^{-\frac{1}{2}h^2k_{dz}^2} \tag{2.1.192}$$

For two zero-mean jointly Gaussian random variables x_1 and x_2, we have

$$\langle x_1x_2\rangle = \sigma_1\sigma_2\rho \tag{2.1.193}$$

where σ_1 and σ_2 are the standard deviations of x_1 and x_2, respectively, and ρ is the correlation coefficient. Then

$$\langle x_2\exp(ivx_1)\rangle = iv\langle x_1x_2\rangle\exp\left(-\frac{v^2\langle x_1^2\rangle}{2}\right) \tag{2.1.194}$$

since

$$\langle f(x,y)\alpha_x(x,y)\rangle = h^2C_{xx}(0,0) \tag{2.1.195a}$$

$$\langle f(x,y)\beta_y(x,y)\rangle = h^2C_{yy}(0,0) \tag{2.1.195b}$$

$$\langle f(x,y)f_{xy}(x,y)\rangle = h^2C_{xy}(0,0) = 0 \tag{2.1.196}$$

assuming that the spectral density $W(k_x,k_y)$ is an even function of k_x and k_y. Then

$$\begin{aligned}\langle\overline{E}_s(\overline{r})\rangle &= \frac{ike^{ikr}}{4\pi r}(\overline{\overline{I}}-\hat{k}_s\hat{k}_s)E_o\exp\left(-\frac{h^2k_{dz}^2}{2}\right)\left\{\overline{\overline{F}}(\overline{k}_{\perp s},\overline{k}_{\perp i})-\frac{1}{2}\overline{F}_{\alpha\alpha}(0,0)h^2C_{xx}(0,0)\right.\\ &\left.-\frac{1}{2}\overline{F}_{\beta\beta}(0,0)h^2C_{yy}(0,0)\right\}\iint dx'dy'e^{i\overline{k}_{d\perp}\cdot\overline{r}'_\perp}\end{aligned} \tag{2.1.197}$$

In the limit of large L_x, L_y, the coherent wave is in the specular direction. Also, from (2.1.183)

$$\overline{\overline{F}}(\overline{k}_{\perp i},\overline{k}_{\perp i}) = \overline{F}(0,0) \tag{2.1.198}$$

Then

$$\begin{aligned}\langle\overline{E}_s(\overline{r})\rangle = \frac{ike^{ikr}}{4\pi r}(\overline{\overline{I}}-\hat{k}_s\hat{k}_s)E_o\cdot&\left\{\overline{F}(0,0)-\frac{\overline{F}_{\alpha\alpha}(0,0)}{2}h^2C_{xx}(0,0)\right.\\ &\left.-\frac{\overline{F}_{\beta\beta}(0,0)}{2}h^2C_{yy}(0,0)\right\}\exp(-2k_{iz}^2h^2)4\pi^2\delta(k_{dx})\delta(k_{dy})\end{aligned} \tag{2.1.199}$$

To calculate incoherent scattering, we take $|\overline{E}_s(\overline{r})|^2$ and keep to second order in slope. Let $E_{sb}=\hat{b}_s\cdot\overline{E}_s$ and $E_{sa}=\hat{a}_s\cdot\overline{E}_s$, where $\hat{b},\hat{a}=\hat{v},\hat{h}$. Then

$$E_{sb}E_{sa}^* = \frac{k^2}{16\pi^2 r^2}|E_o|^2 \iint_{A_o} dx'dy' \iint_{A_o} dx''dy'' e^{ik_{dx}(x'-x'')+ik_{dy}(y'-y'')}$$
$$e^{ik_{dz}(f(x',y')-f(x'',y''))}\Big\{ \left(\hat{b}_s \cdot \overline{\overline{F}}(\overline{k}_{\perp s}, \overline{k}_{\perp i})\right) \left(\hat{a}_s \cdot \overline{\overline{F}}^*(\overline{k}_{\perp s}, \overline{k}_{\perp i})\right)$$
$$+ \frac{i}{2k_{dz}} f_{xx}(x', y') \left(\hat{b}_s \cdot \overline{F}_{\alpha\alpha}(0,0)\right) \left(\hat{a}_s \cdot \overline{\overline{F}}^*(\overline{k}_{\perp s}, \overline{k}_{\perp i})\right)$$
$$+ \frac{i}{2k_{dz}} f_{yy}(x', y') \left(\hat{b}_s \cdot \overline{F}_{\beta\beta}(0,0)\right) \left(\hat{a}_s \cdot \overline{\overline{F}}^*(\overline{k}_{\perp s}, \overline{k}_{\perp i})\right)$$
$$+ \frac{i}{k_{dz}} f_{xy}(x', y') \left(\hat{b}_s \cdot \overline{F}_{\alpha\beta}(0,0)\right) \left(\hat{a}_s \cdot \overline{\overline{F}}^*(\overline{k}_{\perp s}, \overline{k}_{\perp i})\right)$$
$$- \frac{i}{2k_{dz}} f_{xx}(x'', y'') \left(\hat{a}_s \cdot \overline{F}_{\alpha\alpha}^*(0,0)\right) \left(\hat{b}_s \cdot \overline{\overline{F}}(\overline{k}_{\perp s}, \overline{k}_{\perp i})\right)$$
$$- \frac{i}{2k_{dz}} f_{yy}(x'', y'') \left(\hat{a}_s \cdot \overline{F}_{\beta\beta}^*(0,0)\right) \left(\hat{b}_s \cdot \overline{\overline{F}}(\overline{k}_{\perp s}, \overline{k}_{\perp i})\right)$$
$$- \frac{i}{k_{dz}} f_{xy}(x'', y'') \left(\hat{a}_s \cdot \overline{F}_{\alpha\beta}^*(0,0)\right) \left(\hat{b}_s \cdot \overline{\overline{F}}(\overline{k}_{\perp s}, \overline{k}_{\perp i})\right)\Big\} \quad (2.1.200)$$

To calculate $\langle f_{xx}(x', y')e^{iv(f(x',y')-f(x'',y''))}\rangle$, we can define the two zero-mean Gaussian random variables

$$x_1 = f(x', y') - f(x'', y'') \quad (2.1.201)$$
$$x_2 = f_{xx}(x', y') \quad (2.1.202)$$

Then

$$\langle x_1^2\rangle = 2h^2 \left(1 - C(x' - x'', y' - y'')\right) \quad (2.1.203)$$
$$\langle x_1 x_2\rangle = \langle f_{xx}(x', y') \left(f(x', y') - f(x'', y'')\right)\rangle$$
$$= h^2 C_{xx}(0,0) - h^2 C_{xx}(x' - x'', y' - y'') \quad (2.1.204)$$

The slope variances in $\hat{x}$ and $\hat{y}$-directions are respectively

$$\langle \alpha^2\rangle = \sigma_{s_x}^2 = -h^2 C_{xx}(0,0) \quad (2.1.205)$$
$$\langle \beta^2\rangle = \sigma_{s_y}^2 = -h^2 C_{yy}(0,0) \quad (2.1.206)$$

Using (2.1.194), we obtain

$$\langle f_{st}(x', y')e^{ik_{dz}(f(x',y')-f(x'',y''))}\rangle = ik_{dz}h^2 \left(C_{st}(0,0) - C_{st}(x' - x'', y' - y'')\right)$$
$$\exp\Big(-h^2 k_{dz}^2 \left(1 - C(x' - x'', y' - y'')\right)\Big) \quad (2.1.207)$$

where $s, t = x, y$. Also

$$\langle f_{st}(x'', y'')e^{ik_{dz}(f(x',y')-f(x'',y''))}\rangle = -ik_{dz}h^2 \left(C_{st}(0,0) - C_{st}(x' - x'', y' - y'')\right)$$
$$\exp\Big(-h^2 k_{dz}^2 \left(1 - C(x' - x'', y' - y'')\right)\Big) \quad (2.1.208)$$

Using (2.1.207)–(2.1.208) in (2.1.200), we have

$$\langle E_{sb}E_{sa}^*\rangle =$$
$$\frac{k^2|E_o|^2}{16\pi^2 r^2}\iint_{A_o} dx'dy' \iint_{A_o} dx''dy'' \exp\left(ik_{dx}(x'-x'')+ik_{dy}(y'-y'')\right)$$
$$\cdot \exp\Big(-h^2k_{dz}^2\left(1-C(x'-x'',y'-y'')\right)\Big)\Big\{\left(\hat{b}_s\cdot\overline{\overline{F}}(\overline{k}_{\perp s},\overline{k}_{\perp i})\right)$$
$$\cdot\left(\hat{a}_s\cdot\overline{\overline{F}}^*(\overline{k}_{\perp s},\overline{k}_{\perp i})\right)-\frac{h^2}{2}\left(C_{xx}(0,0)-C_{xx}(x'-x'',y'-y'')\right)$$
$$\cdot\Big[\left(\hat{b}_s\cdot\overline{F}_{\alpha\alpha}(0,0)\right)\left(\hat{a}_s\cdot\overline{\overline{F}}^*(\overline{k}_{\perp s},\overline{k}_{\perp i})\right)+\left(\hat{a}_s\cdot\overline{F}^*_{\alpha\alpha}(0,0)\right)\left(\hat{b}_s\cdot\overline{\overline{F}}(\overline{k}_{\perp s},\overline{k}_{\perp i})\right)\Big]$$
$$-\frac{h^2}{2}\left(C_{yy}(0,0)-C_{yy}(x'-x'',y'-y'')\right)$$
$$\cdot\Big[\left(\hat{b}_s\cdot\overline{F}_{\beta\beta}(0,0)\right)\left(\hat{a}_s\cdot\overline{\overline{F}}^*(\overline{k}_{\perp s},\overline{k}_{\perp i})\right)+\left(\hat{a}_s\cdot\overline{F}^*_{\beta\beta}(0,0)\right)\left(\hat{b}_s\cdot\overline{\overline{F}}(\overline{k}_{\perp s},\overline{k}_{\perp i})\right)\Big]$$
$$+h^2C_{xy}(x'-x'',y'-y'')\Big[\left(\hat{b}_s\cdot\overline{F}_{\alpha\beta}(0,0)\right)\left(\hat{a}_s\cdot\overline{\overline{F}}^*(\overline{k}_{\perp s},\overline{k}_{\perp i})\right)$$
$$+\left(\hat{a}_s\cdot\overline{F}^*_{\alpha\beta}(0,0)\right)\left(\hat{b}_s\cdot\overline{\overline{F}}(\overline{k}_{\perp s},\overline{k}_{\perp i})\right)\Big]\Big\} \qquad (2.1.209)$$

with $C_{xy}(0,0)=0$. Letting $x'-x''=x$ and $y'-y''=y$ and subtracting out the coherent field covariance, we have the covariance of incoherent field as

$$\langle E_{sb}E_{sa}^*\rangle-\langle E_{sb}\rangle\langle E_{sa}^*\rangle$$
$$=\frac{k^2|E_o|^2}{16\pi^2 r^2}A_o\iint dxdy\exp(ik_{dx}x+ik_{dy}y-h^2k_{dz}^2)$$
$$\cdot\Big\{\left(\hat{b}_s\cdot\overline{\overline{F}}(\overline{k}_{\perp s},\overline{k}_{\perp i})\right)\left(\hat{a}_s\cdot\overline{\overline{F}}^*(\overline{k}_{\perp s},\overline{k}_{\perp i})\right)\left[\exp\left(h^2k_{dz}^2C(x,y)\right)-1\right]$$
$$-\Big[\left(\hat{b}_s\cdot\overline{F}_{\alpha\alpha}(0,0)\right)\left(\hat{a}_s\cdot\overline{\overline{F}}^*(\overline{k}_{\perp s},\overline{k}_{\perp i})\right)$$
$$+\left(\hat{a}_s\cdot\overline{F}^*_{\alpha\alpha}(0,0)\right)\left(\hat{b}_s\cdot\overline{\overline{F}}(\overline{k}_{\perp s},\overline{k}_{\perp i})\right)\Big]\frac{h^2}{2}$$
$$\times\left[\left(C_{xx}(0,0)-C_{xx}(x,y)\right)\exp\left(h^2k_{dz}^2C(x,y)\right)-C_{xx}(0,0)\right]$$
$$-\Big[\left(\hat{b}_s\cdot\overline{F}_{\beta\beta}(0,0)\right)\left(\hat{a}_s\cdot\overline{\overline{F}}^*(\overline{k}_{\perp s},\overline{k}_{\perp i})\right)$$
$$+\left(\hat{a}_s\cdot\overline{F}^*_{\beta\beta}(0,0)\right)\left(\hat{b}_s\cdot\overline{\overline{F}}(\overline{k}_{\perp s},\overline{k}_{\perp i})\right)\Big]\frac{h^2}{2}$$
$$\times\left[\left(C_{yy}(0,0)-C_{yy}(x,y)\right)\exp\left(h^2k_{dz}^2C(x,y)\right)-C_{yy}(0,0)\right]$$

$$+\exp\left(h^2k_{dz}^2C(x,y)\right)h^2C_{xy}(x,y)\Big[\hat{b}_s\cdot\overline{F}_{\alpha\beta}(0,0)\left(\hat{a}_s\cdot\overline{\overline{F}}^*(\overline{k}_{\perp s},\overline{k}_{\perp i})\right)$$

$$+\left(\hat{a}_s\cdot\overline{F}^*_{\alpha\beta}(0,0)\right)\left(\hat{b}_s\cdot\overline{\overline{F}}(\overline{k}_{\perp s},\overline{k}_{\perp i})\right)\Big]\Big\} \tag{2.1.210}$$

Next, we make Taylor expansions,

$$\exp\left(h^2k_{dz}^2C(x,y)\right)=1+\sum_{m=1}^{\infty}\frac{h^{2m}k_{dz}^{2m}}{m!}C^m(x,y) \tag{2.1.211}$$

We also let

$$-\frac{1}{4\pi^2}h^{2(m+1)}\int_{-\infty}^{\infty}d\overline{r}_\perp e^{i\overline{k}_\perp\cdot\overline{r}_\perp}C_{uv}(x,y)C^m(x,y)=W^{(muv)}(k_x,k_y) \tag{2.1.212}$$

where $u,v=x,y$.

Equations (2.1.211) and (2.1.212) can be used to express the integrals in (2.1.210). For example,

$$\iint dx\,dy\,e^{i\overline{k}_{d\perp}\cdot\overline{r}_\perp}h^2\Big[\left(C_{xx}(0,0)-C_{xx}(x,y)\right)\exp\left(h^2k_{dz}^2C(x,y)\right)-C_{xx}(0,0)\Big]$$

$$=\iint dx\,dy\,e^{i\overline{k}_{d\perp}\cdot\overline{r}_\perp}\Big\{h^2C_{xx}(0,0)\sum_{m=1}^{\infty}\frac{h^{2m}k_{dz}^{2m}}{m!}C^m(x,y)$$

$$-C_{xx}(x,y)h^2\sum_{m=0}^{\infty}\frac{h^{2m}k_{dz}^{2m}}{m!}C^m(x,y)\Big\}$$

$$=4\pi^2\left\{h^2C_{xx}(0,0)\sum_{m=1}^{\infty}\frac{k_{dz}^{2m}}{m!}W^{(m)}(\overline{k}_{d\perp})+\sum_{m=0}^{\infty}\frac{k_{dz}^{2m}}{m!}W^{(mxx)}(\overline{k}_{d\perp})\right\} \tag{2.1.213}$$

Then the bistatic scattering for the incoherent wave is

$$\gamma_{ba}(\hat{k}_s,\hat{k}_i)$$

$$=\frac{\pi k^2}{\cos\theta_i}\exp(-h^2k_{dz}^2)\Big\{\left(\hat{b}_s\cdot\overline{\overline{F}}(\overline{k}_{\perp s},\overline{k}_{\perp i})\right)\left(\hat{a}_s\cdot\overline{\overline{F}}^*(\overline{k}_{\perp s},\overline{k}_{\perp i})\right)$$

$$\times\sum_{m=1}^{\infty}\frac{k_{dz}^{2m}}{m!}W^{(m)}(\overline{k}_{d\perp})-\Big[\left(\hat{b}_s\cdot\overline{F}_{\alpha\alpha}(0,0)\right)\left(\hat{a}_s\cdot\overline{\overline{F}}^*(\overline{k}_{\perp s},\overline{k}_{\perp i})\right)$$

$$+\left(\hat{a}_s\cdot\overline{F}^*_{\alpha\alpha}(0,0)\right)\left(\hat{b}_s\cdot\overline{\overline{F}}(\overline{k}_{\perp s},\overline{k}_{\perp i})\right)\Big]\frac{1}{2}\Big[h^2C_{xx}(0,0)\sum_{m=1}^{\infty}\frac{k_{dz}^{2m}}{m!}W^{(m)}(\overline{k}_{d\perp})$$

$$+\sum_{m=0}^{\infty}\frac{k_{dz}^{2m}}{m!}W^{(mxx)}(\overline{k}_{d\perp})\Big]-\Big[\left(\hat{b}_s\cdot\overline{F}_{\beta\beta}(0,0)\right)\left(\hat{a}_s\cdot\overline{\overline{F}}^*(\overline{k}_{\perp s},\overline{k}_{\perp i})\right)$$

$$+ \left(\hat{a}_s \cdot \overline{F}^*_{\beta\beta}(0,0)\right)\left(\hat{b}_s \cdot \overline{\overline{F}}(\overline{k}_{\perp s}, \overline{k}_{\perp i})\right)\Big]\frac{1}{2}\Big[h^2 C_{yy}(0,0)\sum_{m=1}^{\infty}\frac{k_{dz}^{2m}}{m!}W^{(m)}(\overline{k}_{d\perp})$$

$$+ \sum_{m=0}^{\infty}\frac{k_{dz}^{2m}}{m!}W^{(myy)}(\overline{k}_{d\perp})\Big] - \Big[\left(\hat{b}_s \cdot \overline{F}_{\alpha\beta}(0,0)\right)\left(\hat{a}_s \cdot \overline{\overline{F}}^*(\overline{k}_{\perp s}, \overline{k}_{\perp i})\right)$$

$$+ \left(\hat{a}_s \cdot \overline{F}^*_{\alpha\beta}(0,0)\right)\left(\hat{b}_s \cdot \overline{\overline{F}}(\overline{k}_{\perp s}, \overline{k}_{\perp i})\right)\Big]\sum_{m=0}^{\infty}\frac{k_{dz}^{2m}}{m!}W^{(mxy)}(\overline{k}_{d\perp}) \qquad (2.1.214)$$

2 Phase Perturbation Method

In this section we study the phase perturbation method [Winebrenner and Ishimaru, 1985a,b]. We will study the case of one-dimensional surfaces with the Dirichlet boundary condition.

Consider an incident plane wave with

$$E_{inc} = \exp(ik_{ix}x - ik_{iz}z) \qquad (2.2.1)$$

Then Huygens' principle is, for $z' > f(x')$

$$E(\overline{r}') = E_{inc}(\overline{r}') - \int dx\, g(\overline{r}, \overline{r}')\overline{N} \cdot \nabla E(\overline{r}) \qquad (2.2.2a)$$

and the extinction theorem is, for $z' < f(x)$

$$0 = E_{inc}(\overline{r}') - \int dx\, g(\overline{r}, \overline{r}')\overline{N} \cdot \nabla E(\overline{r}) \qquad (2.2.2b)$$

where $\overline{N} = \frac{\partial f}{\partial x}\hat{x} + \hat{z}$. For two-dimensional problems we have

$$\begin{aligned} g(\overline{r}, \overline{r}') &= \frac{i}{4}H_o^{(1)}\left(k\sqrt{(x-x')^2 + (z-z')^2}\right) \\ &= \frac{i}{4\pi}\int_{-\infty}^{\infty} dk_x \frac{1}{k_z} e^{ik_x(x-x') + ik_z|z-z'|} \end{aligned} \qquad (2.2.3)$$

Using (2.2.3) in (2.2.2a), we obtain

$$E_s(\overline{r}') = -\int dx \frac{i}{4\pi}\int_{-\infty}^{\infty} dk_x \frac{e^{ik_x(x'-x) + ik_z(z' - f(x))}}{k_z}\overline{N} \cdot \nabla E(\overline{r}) \qquad (2.2.4)$$

$$= \int_{-\infty}^{\infty} dk_x\, e^{ik_x x' + ik_z z'} T(k_x, k_{ix}) \qquad (2.2.5)$$

where

$$T(k_x, k_{ix}) = -\frac{i}{4\pi k_z}\int_{-\infty}^{\infty} dx\, e^{-ik_x x - ik_z f(x)}(\overline{N} \cdot \nabla E(\overline{r})) \qquad (2.2.6)$$

is the T-matrix. To solve the surface current through the extinction theorem of (2.2.2*b*), we have for $z' < f(x')$

$$e^{ik_{ix}x'-ik_{iz}z'} = \int_{-\infty}^{\infty} dx \frac{i}{4\pi} \int_{-\infty}^{\infty} \frac{dk_x}{k_z} e^{ik_x(x'-x)-ik_z(z'-f(x))}\overline{N} \cdot \nabla E(\overline{r}) \quad (2.2.7)$$

The solution of (2.2.7) is

$$\delta(k_x - k_{ix}) = \frac{i}{4\pi k_z} \int_{-\infty}^{\infty} dx\, e^{-ik_x x + ik_z f(x)} \left(\overline{N} \cdot \nabla E(\overline{r})\right) \quad (2.2.8)$$

for $\overline{r}$ on S. For a flat surface at $z = 0$, $\overline{N} \cdot \nabla E(\overline{r}) = 2\hat{z} \cdot \nabla E_{inc}(\overline{r}) = -i2k_{iz} \exp(ik_{ix}x)$. To solve (2.2.8), let

$$\overline{N} \cdot \nabla E(\overline{r}) = -i2k_{iz}\, e^{ik_{ix}x} y(x) \quad (2.2.9)$$

Thus, we have

$$\delta(k_x - k_{ix}) = \frac{k_{iz}}{2\pi k_z} \int_{-\infty}^{\infty} dx\, e^{i(k_{ix}-k_x)x}\, e^{ik_z f(x)} y(x) \quad (2.2.10)$$

For the small perturbation method, also known as the field perturbation method, $y(x)$ is expanded in a perturbation series

$$y(x) = y^{(0)}(x) + y^{(1)}(x) + y^{(2)}(x) \quad (2.2.11)$$

Define the Fourier transform of $y(x)$ as

$$\begin{aligned} Y(k_x) &= \frac{1}{2\pi} \int_{-\infty}^{\infty} dx\, e^{-ik_x x} y(x) \\ &= Y^{(0)}(k_x) + Y^{(1)}(k_x) + Y^{(2)}(k_x) + \cdots \end{aligned} \quad (2.2.12)$$

Using (2.2.10), we obtain

$$\begin{aligned} \delta(k_x - k_{ix}) &= \frac{1}{2\pi} \int_{-\infty}^{\infty} dx\, e^{i(k_{ix}-k_x)x} \left(1 + ik_z f(x) - \frac{k_z^2 f^2(x)}{2}\right) \\ &\times \left(y^{(0)}(x) + y^{(1)}(x) + y^{(2)}(x)\right) \end{aligned} \quad (2.2.13)$$

On balancing Eq. (2.2.13) to zeroth order,

$$Y^{(0)}(k_x) = \delta(k_x) \quad (2.2.14)$$

$$y^{(0)}(x) = 1 \quad (2.2.15)$$

Balancing (2.2.13) to the first order gives

$$0 = \int_{-\infty}^{\infty} dx\, e^{i(k_{ix}-k_x)x} ik_z f(x) + \int_{-\infty}^{\infty} dx\, e^{i(k_{ix}-k_x)x} y^{(1)}(x) \quad (2.2.16)$$

Hence

$$Y^{(1)}(k_x - k_{ix}) = -ik_z F(k_x - k_{ix}) \quad (2.2.17)$$

Balancing (2.2.13) to the second order gives

$$0 = Y^{(2)}(k_x - k_{ix}) + \frac{1}{2\pi}\int_{-\infty}^{\infty} dx\, e^{i(k_{ix}-k_x)x} ik_z f(x) y^{(1)}(x)$$
$$- \frac{k_z^2}{2}\frac{1}{2\pi}\int_{-\infty}^{\infty} dx\, e^{i(k_{ix}-k_x)x} f^2(x) y^{(0)}(x) \qquad (2.2.18)$$

Substituting (2.2.17) into (2.2.18), after some algebraic manipulations, we obtain

$$Y^{(2)}(k_x - k_{ix}) = \int_{-\infty}^{\infty} dk'_x F(k_x - k'_x) F(k'_x - k_{ix}) \left(\frac{k_z^2}{2} - k_z k'_z\right) \qquad (2.2.19a)$$

$$y^{(2)}(x) = \int_{-\infty}^{\infty} dk_x e^{i(k_x - k_{ix})x} \int_{-\infty}^{\infty} dk'_x F(k_x - k'_x)$$
$$\cdot F(k'_x - k_{ix}) \left(\frac{k_z^2}{2} - k_z k'_z\right) \qquad (2.2.19b)$$

Using (2.2.9) in (2.2.6) gives

$$T(k_x, k_{ix}) = -\frac{k_{iz}}{2\pi k_z}\int_{-\infty}^{\infty} dx\, e^{i(k_{ix}-k_x)x - ik_z f(x)} y(x) \qquad (2.2.20)$$

In field perturbation theory, we also put $e^{-ik_z f(x)}$ in power series of $f(x)$

$$T(k_x, k_{ix}) = -\frac{k_{iz}}{2\pi k_z}\int_{-\infty}^{\infty} dx\, e^{i(k_{ix}-k_x)x}\left(1 - ik_z f(x) - \frac{k_z^2 f^2(x)}{2}\right)$$
$$\times \left(y^{(0)}(x) + y^{(1)}(x) + y^{(2)}(x)\right) \qquad (2.2.21)$$

Substituting (2.2.19), (2.2.17), and (2.2.15) in (2.2.21) gives

$$T(k_x, k_{ix}) = -\delta(k_x - k_{ix}) + 2ik_{iz} F(k_x - k_{ix})$$
$$+ 2k_{iz}\int_{-\infty}^{\infty} dk'_x\, k'_z F(k_x - k'_x) F(k'_x - k_{ix}) \qquad (2.2.22)$$

Equation (2.2.22) is the solution of $T(k_x, k_{ix})$ based on field perturbation theory. Phase perturbation theory proceeds differently as follows. The perturbation of $y(x)$ is made in the phase rather than in the complex amplitude. Let

$$y(x) = \exp\left(p^{(0)}(x) + p^{(1)}(x) + p^{(2)}(x)\right) \qquad (2.2.23a)$$

so that

$$T(k_x, k_{ix}) = -\frac{k_{iz}}{2\pi k_z}\int_{-\infty}^{\infty} dx e^{i(k_{ix}-k_x)x - ik_z f(x) + p^{(0)}(x) + p^{(1)}(x) + p^{(2)}(x)} \qquad (2.2.23b)$$

Comparing (2.2.23a) with (2.2.11) gives

$$1 + p^{(0)}(x) + p^{(1)}(x) + p^{(2)}(x) + \frac{\left(p^{(0)}(x) + p^{(1)}(x) + p^{(2)}(x)\right)^2}{2}$$
$$= y^{(0)}(x) + y^{(1)}(x) + y^{(2)}(x) \tag{2.2.24}$$

Thus, balancing (2.2.24) to the zeroth, first, and second order gives, respectively,

$$p^{(0)} = 0 \tag{2.2.25}$$
$$p^{(1)} = y^{(1)}(x) \tag{2.2.26}$$
$$p^{(2)} = y^{(2)}(x) - \frac{(y^{(1)}(x))^2}{2} \tag{2.2.27}$$

Putting (2.2.27) in (2.2.23b) and then in (2.2.20) gives

$$T(k_x, k_{ix}) = -\frac{k_{iz}}{2\pi k_z}\int_{-\infty}^{\infty} dx\, e^{i(k_{ix}-k_x)x} \exp(\Lambda(x, k_x)) \tag{2.2.28}$$

where

$$\Lambda(x, k_x) = -ik_z f(x) + y^{(1)}(x) + y^{(2)}(x) - \frac{(y^{(1)}(x))^2}{2} \tag{2.2.29}$$

Equation (2.2.28) is the solution of the T-matrix $T(k_x, k_{ix})$ under the phase perturbation. To calculate statistical averages of fields, we next need to take the average of an exponential function with the random variable in the exponent.

The characteristic function can be written in terms of cumulants

$$\langle \exp(\Lambda) \rangle = \exp(\nu^{(1)} + \nu^{(2)} + \cdots) \tag{2.2.30}$$

Expanding both series and balancing, we get

$$1 + \langle \Lambda \rangle + \frac{\langle \Lambda^2 \rangle}{2} + \cdots = 1 + \nu^{(1)} + \nu^{(2)} + \frac{(\nu^{(1)} + \nu^{(2)})^2}{2!} + \cdots \tag{2.2.31}$$

so that

$$\nu^{(1)} = \langle \Lambda \rangle \tag{2.2.32}$$
$$\nu^{(2)} = \frac{\langle \Lambda^2 \rangle - \langle \Lambda \rangle^2}{2} \tag{2.2.33}$$

Hence, to the second order in cumulants we have

$$\langle \exp(\Lambda(x, k_x)) \rangle = \exp\left(\langle \Lambda \rangle + \frac{\langle \Lambda^2 \rangle - \langle \Lambda \rangle^2}{2}\right) \tag{2.2.34}$$

From (2.2.29), keeping to the second order, we obtain

$$\langle \Lambda(x) \rangle = \langle y^{(2)}(x) \rangle - \frac{\langle (y^{(1)}(x))^2 \rangle}{2} \tag{2.2.35}$$

$$\langle \Lambda^2 \rangle = -k_z^2 \langle f^2(x) \rangle + \langle (y^{(1)}(x))^2 \rangle - 2ik_z \langle f(x) y^{(1)}(x) \rangle \tag{2.2.36}$$

$$\langle \Lambda \rangle + \frac{\langle \Lambda^2 \rangle - \langle \Lambda \rangle^2}{2} = \langle y^{(2)}(x) \rangle - \frac{k_z^2 \langle f^2(x) \rangle}{2} - ik_z \langle f(x) y^{(1)}(x) \rangle \tag{2.2.37}$$

From (2.2.19) and taking the average we get

$$\langle y^{(2)}(x) \rangle = \int_{-\infty}^{\infty} dk_x'' \, W(k_{ix} - k_x'') \left(\frac{k_{iz}^2}{2} - k_{iz} k_z'' \right) \tag{2.2.38}$$

$$\langle f(x) y^{(1)}(x) \rangle = \int_{-\infty}^{\infty} dk_x'' (-ik_z'') W(k_{ix} - k_x'') \tag{2.2.39}$$

Thus

$$\begin{aligned} &\langle \exp(\Lambda(x, k_x)) \rangle \\ &\quad = N(k_x) = \exp\left(-\frac{k_z^2 h^2}{2} + \int_{-\infty}^{\infty} dk_x'' W(k_x'' - k_{ix}) \left(\frac{k_{iz}^2}{2} - k_{iz} k_z'' - k_z k_z'' \right) \right) \\ &\quad = \exp\left(-\frac{(k_z^2 - k_{iz}^2) h^2}{2} - (k_{iz} + k_z) \int_{-\infty}^{\infty} dk_x'' W(k_x'' - k_{ix}) k_z'' \right) \end{aligned} \tag{2.2.40}$$

and is independent of x. In (2.2.40) we have made use of the fact that $\int_{-\infty}^{\infty} dk_x W(k_x) = h^2$.

From (2.2.28) and (2.2.40) we get a Dirac delta function for $\langle T \rangle$

$$\langle T(k_x, k_{ix}) \rangle = -\frac{k_{iz}}{k_z} N(k_x) \delta(k_x - k_{ix}) \tag{2.2.41}$$

Substituting (2.2.41) and (2.2.40) into (2.2.5), we find that the coherent reflected wave is

$$\langle E_s(\overline{r}) \rangle = -e^{ik_{ix}x + ik_{iz}z} \exp\left(-2k_{iz} \int_{-\infty}^{\infty} dk_x k_z W(k_{ix} - k_x) \right) \tag{2.2.42}$$

Thus the phase perturbation theory gives a coherent reflectivity that is different from that of Kirchhoff approximation. Note that the exponent of (2.2.42) can be complex due to the contribution of evanescent waves.

To further compare the phase perturbation theory and the Kirchhoff approximation, one can assume that $kl \gg 1$ so that $(k_{ix} - k_x)$ is sharply peaked at $k_x - k_{ix}$. Then we can make a Taylor expansion of the k_z term about k_{iz}. We get

$$
\begin{aligned}
-2k_{iz}\int_{-\infty}^{\infty} dk_x\, W(k_{ix}-k_x)k_z \\
&\simeq -2k_{iz}\int_{-\infty}^{\infty} dk_x\, W(k_{ix}-k_x)\left\{k_{iz}-\frac{k_{ix}}{k_{iz}}(k_x-k_{ix})-\frac{k^2}{2k_{iz}^3}(k_x-k_{ix})^2\right\} \\
&\simeq -2k_{iz}^2h^2+\frac{k^2}{k_{iz}^2}\int_{-\infty}^{\infty} dk_x\, W(k_x)k_x^2 \qquad (2.2.43a)
\end{aligned}
$$

Thus

$$
\langle E_s(\overline{r})\rangle = -e^{ik_{ix}x+ik_{iz}z}\exp\left(-2k_{iz}^2h^2+\frac{k^2}{k_{iz}^2}\int_{-\infty}^{\infty} dk_x W(k_x)k_x^2\right) \qquad (2.2.43b)
$$

Because of the second term in the parentheses of (2.2.43b) the coherent wave as predicted by phase perturbation theory decays less fast than that of Kirchhoff approximation.

To calculate the incoherent bistatic scattering cross section, we have

$$
\begin{aligned}
&\langle T(k_x,k_{ix})T^*(k_x',k_{ix})\rangle - \langle T(k_x,k_{ix})\rangle\langle T^*(k_x',k_{ix})\rangle = \\
&\frac{k_{iz}^2}{4\pi^2k_zk_z'^*}\int_{-\infty}^{\infty} dx\int_{-\infty}^{\infty} dx'\, e^{i(k_{ix}-k_x)x-i(k_{ix}-k_x')x'} \\
&\cdot\Big\{\langle\exp(\Lambda(x,k_x)+\Lambda^*(x',k_x'))\rangle - \langle\exp(\Lambda(x,k_x))\rangle\langle\exp(\Lambda^*(x',k_x'))\rangle\Big\} \qquad (2.2.44)
\end{aligned}
$$

Using (2.2.34), we have

$$
\begin{aligned}
&\langle\exp(\Lambda(x,k_x)+\Lambda^*(x',k_x'))\rangle - \langle\exp(\Lambda(x,k_x))\rangle\langle\exp(\Lambda^*(x',k_x'))\rangle \\
&\quad = \exp\Big(\langle\Lambda(x,k_x)\rangle + \langle\Lambda^*(x',k_x')\rangle \\
&\qquad + \frac{\langle(\Lambda(x,k_x)+\Lambda^*(x',k_x'))^2\rangle - (\langle\Lambda(x,k_x)+\Lambda^*(x',k_x')\rangle)^2}{2}\Big) \\
&\qquad - \exp\left(\langle\Lambda(x,k_x)\rangle + \frac{\langle\Lambda^2(x,k_x)\rangle - \langle\Lambda(x,k_x)\rangle^2}{2}\right) \\
&\qquad \cdot\exp\left(\langle\Lambda(x',k_x')\rangle + \frac{\langle\Lambda^2(x',k_x')\rangle - \langle\Lambda(x',k_x')\rangle^2}{2}\right) \\
&\quad = N(k_x)N^*(k_x')M(k_x,k_x',x-x') \qquad (2.2.45)
\end{aligned}
$$

where

$$
M(k_x,k_x',x-x') = \exp\Big(\langle\Lambda(x,k_x)\Lambda^*(x',k_x')\rangle - \langle\Lambda(x,k_x)\rangle\langle\Lambda^*(x',k_x')\rangle\Big) - 1 \qquad (2.2.46)
$$

Keeping the exponent to the second order only, we have, using (2.2.29)

$$M(k_x, k_x', x-x') = \exp\Big(k_z k_z'^* \langle f(x)f(x')\rangle + \langle y^{(1)}(x)y^{(1)*}(x')\rangle - ik_z\langle f(x)y^{(1)*}(x')\rangle + ik_z'^*\langle f(x')y^{(1)}(x)\rangle\Big) - 1 \quad (2.2.47)$$

It follows from (2.2.17) that

$$\langle y^{(1)}(x)y^{(1)*}(x')\rangle = \int_{-\infty}^{\infty} dk_x''\, e^{i(k_x''-k_{ix})(x-x')}|k_z''|^2 W(k_x''-k_{ix}) \quad (2.2.48)$$

$$\langle f(x)y^{(1)*}(x')\rangle = \int_{-\infty}^{\infty} dk_x'' ik_z''^*\, e^{i(k_x''-k_{ix})(x-x')} W(k_x''-k_{ix}) \quad (2.2.49)$$

Putting (2.2.48) and (2.2.49) into (2.2.47),

$$\begin{aligned} M(k_x,&k_x', x-x') \\ &= \exp\Bigg(k_z k_z'^* h^2 C(x-x') + \int_{-\infty}^{\infty} dk_x''\, e^{i(k_x''-k_{ix})(x-x')} W(k_x''-k_{ix}) \\ &\quad \cdot \Big[|k_z''|^2 + k_z k_z''^* + k_z'^* k_z''\Big]\Bigg) - 1 \end{aligned} \quad (2.2.50)$$

Substituting (2.2.50) and (2.2.45) into (2.2.44) gives

$$\begin{aligned} &\langle T(k_x, k_{ix})T^*(k_x', k_{ix})\rangle - \langle T(k_x, k_{ix})\rangle\langle T^*(k_x', k_{ix})\rangle \\ &\quad = \frac{k_{iz}^2}{4\pi^2 k_z k_z'^*}\int_{-\infty}^{\infty} dx \int_{-\infty}^{\infty} dx'\, e^{i(k_{ix}-k_x)x - i(k_{ix}-k_x')x'} \\ &\quad \cdot N(k_x)N^*(k_x')M(k_x, k_x', x-x') \end{aligned} \quad (2.2.51)$$

Because M depends on $x - x'$ only, we can make change of variables of $x_a = (x+x')/2$ and $x_d = x - x'$. The integration of $x_a = (x+x')/2$ gives $\delta(k_x - k_x')$. Hence

$$\begin{aligned} &\langle T(k_x, k_{ix})T^*(k_x', k_{ix})\rangle - \langle T(k_x, k_{ix})\rangle\langle T^*(k_x', k_{ix})\rangle \\ &= \delta(k_x - k_x')\frac{k_{iz}^2}{2\pi|k_z|^2}\int_{-\infty}^{\infty} dx\, e^{i(k_{ix}-k_x)x}|N(k_x)|^2 M(k_x, k_x, x) \\ &= \delta(k_x - k_x')\frac{k_{iz}^2}{2\pi|k_z|^2}\int_{-\infty}^{\infty} dx\, e^{i(k_{ix}-k_x)x}|N(k_x)|^2\Big(e^{N_{11}(x,k_x)} - 1\Big) \end{aligned} \quad (2.2.52)$$

where

$$\begin{aligned} N_{11}(x, k_x) &= |k_z|^2 h^2 C(x) \\ &\quad + \int_{-\infty}^{\infty} dk_x''\, e^{i(k_x''-k_{ix})x} W(k_x''-k_{ix})\Big(|k_z''|^2 + 2\,\mathrm{Re}(k_z k_z'')\Big) \\ &= \int_{-\infty}^{\infty} dk_x''\, e^{i(k_x''-k_{ix})x} W(k_x''-k_{ix})|k_z + k_z''|^2 \end{aligned} \quad (2.2.53)$$

The Poynting's vector associated with the incoherent wave is from (9.3.34) of Volume I. We have

$$\langle \overline{S}_s \cdot \hat{z} \rangle = \frac{1}{2\omega\mu} \int_{-k}^{k} dk_x \, k_z \mathcal{I}(k_x) \tag{2.2.54}$$

where

$$\mathcal{I}(k_x) = \frac{k_{iz}^2}{2\pi |k_z|^2} \int_{-\infty}^{\infty} dx \, e^{i(k_{ix}-k_x)x} |N(k_x)|^2 \left(e^{N_{11}(x,k_x)} - 1 \right) \tag{2.2.55}$$

Changing to angular variables by letting $k_x = k \sin\theta_s$ in (2.2.54), we have,

$$\langle \overline{S}_s \cdot \hat{z} \rangle = \frac{1}{2\eta} \int_{-\pi/2}^{\pi/2} d\theta_s k \cos^2\theta_s \mathcal{I}(k_x = k \sin\theta_s) \tag{2.2.56}$$

Note that $|\langle \overline{S}_i \cdot \hat{z} \rangle| = \cos\theta_i / 2\eta$, we define $\sigma(\theta_s)$ so that

$$\langle \overline{S}_s \cdot \hat{z} \rangle = |\langle \overline{S}_i \cdot \hat{z} \rangle| \int_{-\pi/2}^{\pi/2} d\theta_s \sigma(\theta_s) \tag{2.2.57}$$

where $\sigma(\theta_s)$ is the incoherent bistatic scattering coefficient and is given by

$$\sigma(\theta_s) = \frac{k \cos^2\theta_s}{\cos\theta_i} \mathcal{I}(k_x = k \sin\theta_s) \tag{2.2.58}$$

Equations (2.2.55) and (2.2.58) together gives the bistatic scattering coefficient under the phase perturbation theory.

3 Emissivity Based on Composite Surface Model

In this section, we illustrate the emissivity of ocean surface based on a composite surface model. As discussed in Chapter 4, Section 8 of Volume I, ocean surfaces consist of large scale roughness with small scale roughness superimposed on it. In a composite surface model, the ocean spectrum is divided into two parts with the dividing line dependent on the microwave frequency. The large scale roughness spectrum is used to calculate the local slope of the small roughness spectrum.

Consider an ocean spectrum

$$W(k_x, k_y) = W(k_\rho, \phi_k).$$

In the two-scale model, this is decomposed by using a two-scale cutoff k_d.

$$W_l(k_\rho, \phi_k) = \begin{cases} W(k_\rho, \phi_k) & \text{if } k_\rho < k_d \\ 0 & \text{otherwise} \end{cases} \tag{2.3.1}$$

$$W_s(k_\rho, \phi_k) = \begin{cases} 0 & \text{if } k_\rho < k_d \\ W(k_\rho, \phi_k) & \text{otherwise} \end{cases} \tag{2.3.2}$$

where W_l and W_s denote the large-scale and small-scale spectrum, respectively. The choice of k_d depends on the frequency of electromagnetic wave because large scale or small scale is relative to the operating frequency of the electromagnetic wave. The large scale spectrum (large scale ocean waves) will be used to create a Gaussian probability density function of slope distribution. Let $\hat{x}$ be the direction of the wind. Then the probability density function of slope is

$$p(s_x, s_y) = \frac{1}{2\pi s_u s_c} \exp\left[-\frac{1}{2}\left(\frac{s_x^2}{s_u^2} + \frac{s_y^2}{s_c^2}\right)\right] \tag{2.3.3}$$

where s_x and s_y are the respective slopes in $\hat{x}$ and $\hat{y}$-directions of large scale ocean waves, and

$$s_u^2 = \int_0^\infty dk_\rho \int_0^{2\pi} d\phi_k k_\rho^3 \cos^2\phi_k W_l(k_\rho, \phi_k) \tag{2.3.4a}$$

$$s_c^2 = \int_0^\infty dk_\rho \int_0^{2\pi} d\phi_k k_\rho^3 \sin^2\phi_k W_l(k_\rho, \phi_k) \tag{2.3.4b}$$

are the upwind and crosswind surface slopes, respectively.

It is also known that hydrodynamic modulation makes the small scale roughness appear more on the leeward faces of large scale ocean waves. This is modeled by a hydrodynamic modulation parameter H [Yueh, 1997]. Thus the small scale roughness is further modified to be

$$W_s(k_\rho, \phi_k, s_x) = H W_s(k_\rho, \phi_k) \tag{2.3.5}$$

where

$$H = \begin{cases} 1 - 0.5\operatorname{sgn}(s_x) & \text{if } |s_x/s_u| > 1.25 \\ 1 - 0.4 s_x/s_u & \text{if } |s_x/s_u| \le 1.25 \end{cases} \tag{2.3.6}$$

$\operatorname{sgn}(s_x) = 1$ if s_x is positive and $\operatorname{sgn}(s_x) = -1$ if s_x is negative. Thus H is larger than 1 for negative s_x and less than 1 for positive s_x.

The large scale roughness affects the local incident angles. For a given slope s_x and s_y at a point based on the large scale roughness, we let $\hat{z}_l$ point in the direction of local normal, then

$$\hat{z}_l = \frac{-s_x\hat{x} - s_y\hat{y} + \hat{z}}{\sqrt{1 + s_x^2 + s_y^2}} \tag{2.3.7}$$

Let the incident direction be

$$\hat{k}_i = \sin\theta_i \cos\phi_i \hat{x} + \sin\theta_i \sin\phi_i \hat{y} - \cos\theta_i \hat{z} \tag{2.3.8}$$

Note that the emission direction is $\hat{k}_e = -\hat{k}_i$. However, in using the small perturbation methods on the small scale roughness, we need to use the local in-

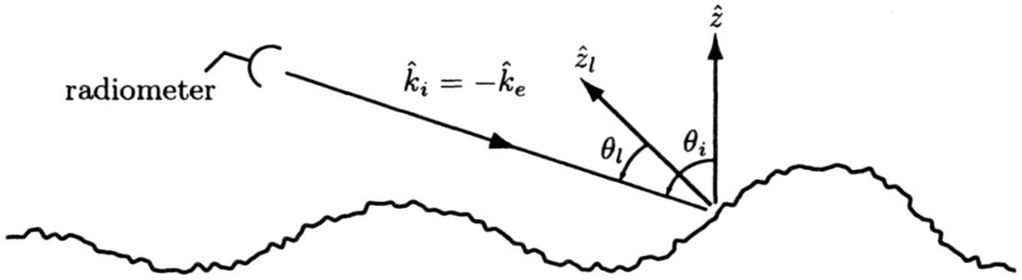

Figure 2.3.1 Local angle of incidence. $\hat{k}_i$ is the incident wave direction, $\hat{k}_e$ is the emission direction, and $\hat{k}_e = -\hat{k}_i$.

cident angle. Let the local zenith and azimuthal angles be (θ_l, ϕ_l) (Fig. 2.3.1). Thus

$$\hat{k}_i = \sin\theta_l \cos\phi_l \hat{x}_l + \sin\theta_l \sin\phi_l \hat{y}_l - \cos\theta_l \hat{z}_l = -\hat{k}_e \tag{2.3.9a}$$

$$\hat{k}_e = \sin\theta_l \cos(\pi - \phi_l)\hat{x}_l + \sin\theta_l \sin(\pi - \phi_l)\hat{y}_l + \cos\theta_l \hat{z}_l \tag{2.3.9b}$$

We will use (θ_l, ϕ_l) to replace (θ_i, ϕ_i) in the emissivity formulas derived by small perturbation method in Eqs. (1.3.29), (1.3.31), (1.3.50), and (1.3.57) of Chapter 1, Section 3.2. To determine (θ_l, ϕ_l), we first choose $\hat{x}_l$ to be in the $(\hat{x}, \hat{z})$ plane. Note that $\hat{x}$ is in the direction of the wind vector.

Let β be the angle between x and $\hat{x}_l$ such that

$$\hat{x}_l = \cos\beta\hat{x} - \sin\beta\hat{z} \tag{2.3.10}$$

$$\hat{y}_l = \hat{z}_l \times \hat{x}_l \tag{2.3.11}$$

To calculate β, we note that $\hat{x}_l$ is perpendicular to both $\hat{y}$ and $\hat{z}_l$. Using (2.3.7)

$$\hat{x}_l = \frac{\hat{y} \times \hat{z}_l}{|\hat{y} \times \hat{z}_l|} = \frac{\hat{x} + s_x\hat{z}}{\sqrt{1 + s_x^2}} \tag{2.3.12}$$

$$\hat{y}_l = \frac{-s_x s_y \hat{x} + (1 + s_x^2)\hat{y} + s_y\hat{z}}{\sqrt{(1 + s_x^2)(1 + s_x^2 + s_y^2)}} \tag{2.3.13}$$

Equations (2.3.7), (2.3.12), and (2.3.13) express $(\hat{x}_l, \hat{y}_l, \hat{z}_l)$ in terms of $(\hat{x}, \hat{y}, \hat{z})$. Comparing (2.3.10) and (2.3.12), we have

$$\cos\beta = \frac{1}{\sqrt{1 + s_x^2}} \tag{2.3.14}$$

$$\sin\beta = -\frac{s_x}{\sqrt{1 + s_x^2}} \tag{2.3.15}$$

To find (θ_l, ϕ_l), we equate $\hat{k}_i$ in the two systems of $(\hat{x}, \hat{y}, \hat{z})$ and $(\hat{x}_l, \hat{y}_l, \hat{z}_l)$

$$\begin{aligned}\hat{k}_i \cdot \hat{z}_l &= -\cos\theta_l \\ &= \sin\theta_i \cos\phi_i(\hat{x}\cdot\hat{z}_l) + \sin\theta_i \sin\phi_i(\hat{y}\cdot\hat{z}_l) - \cos\theta_i(\hat{z}\cdot\hat{z}_l)\end{aligned} \tag{2.3.16}$$

Thus, taking the dot product as indicated in (2.3.16) and using (2.3.7)

$$\cos\theta_l = \frac{1}{\sqrt{1+s_x^2+s_y^2}}\{\sin\theta_i\cos\phi_i s_x + \sin\theta_i \sin\phi_i s_y + \cos\theta_i\} \tag{2.3.17}$$

$$\sin\theta_l \cos\phi_l = \hat{k}_i \cdot \hat{x}_l \tag{2.3.18}$$

which gives

$$\sin\theta_l\cos\phi_l = \frac{1}{\sqrt{1+s_x^2}}\{\sin\theta_i\cos\phi_i - \cos\theta_i s_x\} \tag{2.3.19}$$

$$\begin{aligned}\sin\theta_l \sin\phi_l = \hat{k}_i\cdot\hat{y}_l = &\frac{1}{\sqrt{(1+s_x^2)(1+s_x^2+s_y^2)}} \\ &\cdot\{-\sin\theta_i\cos\phi_i s_x s_y + \sin\theta_i\sin\phi_i(1+s_x^2) - s_y\cos\theta_i\}\end{aligned} \tag{2.3.20}$$

We can determine θ_l and ϕ_l from (2.3.19) and (2.3.20).

Besides local zenith and azimuth angles, we also need to make a polarization transformation. However, we know that from the emission point of view, the emission direction is opposite to that of the incident direction, $\hat{k}_e = -\hat{k}_i$. Thus $\hat{k}_i$ is in direction $(\pi-\theta_i, \phi_i)$ and $(\pi-\theta_l, \phi_l)$ while $\hat{k}_e$ is in direction $(\theta_i, \pi+\phi_i)$ and $(\theta_l, \pi+\phi_l)$. Thus

$$\hat{h}(\theta_i, \pi+\phi_i) = \sin\phi_i\hat{x} - \cos\phi_i\hat{y} \tag{2.3.21}$$

$$\hat{v}(\theta_i, \pi+\phi_i) = -\cos\theta_i\cos\phi_i\hat{x} - \cos\theta_i\sin\phi_i\hat{y} - \sin\theta_i\hat{z} \tag{2.3.22}$$

$$\hat{h}_l(\theta_l, \pi+\phi_l) = \sin\phi_l\hat{x}_l - \cos\phi_l\hat{y}_l \tag{2.3.23}$$

$$\hat{v}_l(\theta_l, \pi+\phi_l) = -\cos\theta_l\cos\phi_l\hat{x}_l - \cos\theta_l\sin\phi_l\hat{y}_l - \sin\theta_l\hat{z}_l \tag{2.3.24}$$

Let α be the angle between the two polarization systems $(\hat{h}_l, \hat{v}_l)$ and $(\hat{h}, \hat{v})$

$$\begin{aligned}\cos\alpha &= \hat{h}_l(\theta_l, \pi+\phi_l)\cdot\hat{h}(\theta_i, \pi+\phi_i) = \hat{v}_l(\theta_l, \pi+\phi_l)\cdot\hat{v}(\theta_i, \pi+\phi_i) \\ &= \sin\phi_i\sin\phi_l(\hat{x}\cdot\hat{x}_l) + \cos\phi_i\cos\phi_l(\hat{y}\cdot\hat{y}_l) \\ &\quad - \sin\phi_i\cos\phi_l(\hat{x}\cdot\hat{y}_l) - \cos\phi_i\sin\phi_l(\hat{x}_l\cdot\hat{y}) \\ &= \sin\phi_i\sin\phi_l\frac{1}{\sqrt{1+s_x^2}} + \cos\phi_i\cos\phi_l\sqrt{\frac{1+s_x^2}{1+s_x^2+s_y^2}} \\ &\quad + \frac{\sin\phi_i\cos\phi_l s_x s_y}{\sqrt{(1+s_x^2)(1+s_x^2+s_y^2)}}\end{aligned} \tag{2.3.25}$$

Also

$$\sin\alpha = \hat{v}(\theta_i, \pi+\phi_i)\cdot\hat{h}_l(\theta_l, \pi+\phi_l) = -\hat{h}(\theta_i, \pi+\phi_i)\cdot\hat{v}_l(\theta_l, \pi+\phi_l)$$
$$= -\frac{\cos\phi_i\cos\theta_i\sin\phi_l}{\sqrt{1+s_x^2}} - \cos\phi_i\cos\theta_i\cos\phi_l\frac{s_x s_y}{\sqrt{1+s_x^2}\sqrt{1+s_x^2+s_y^2}}$$
$$+ \sin\phi_i\cos\theta_i\cos\phi_l\sqrt{\frac{1+s_x^2}{1+s_x^2+s_y^2}} - \sin\theta_i\sin\phi_l\frac{s_x}{\sqrt{1+s_x^2}}$$
$$+ \sin\theta_i\cos\phi_l\frac{s_y}{\sqrt{1+s_x^2}\sqrt{1+s_x^2+s_y^2}} \tag{2.3.26}$$

Then the transformation of the polarizations is

$$E_v = E_{v_l}\cos\alpha + E_{h_l}\sin\alpha \tag{2.3.27a}$$
$$E_h = -E_{v_l}\sin\alpha + E_{h_l}\cos\alpha \tag{2.3.27b}$$

which implies that

$$T_v(s_x, s_y) = T_{v_l}\cos^2\alpha + T_{h_l}\sin^2\alpha + U_l\cos\alpha\sin\alpha \tag{2.3.28a}$$
$$T_h(s_x, s_y) = T_{v_l}\sin^2\alpha + T_{h_l}\cos^2\alpha - U_l\cos\alpha\sin\alpha \tag{2.3.28b}$$
$$U(s_x, s_y) = -(T_{v_l} - T_{h_l})\sin 2\alpha + U_l(\cos^2\alpha - \sin^2\alpha) \tag{2.3.28c}$$
$$V(s_x, s_y) = V_l \tag{2.3.28d}$$

Note that

$$U = 2\mathrm{Re}(E_v E_h^*) \tag{2.3.29a}$$
$$V = 2\mathrm{Im}(E_v E_h^*) \tag{2.3.29b}$$
$$U_l = 2\mathrm{Re}(E_{v_l} E_{h_l}^*) \tag{2.3.29c}$$
$$V_l = 2\mathrm{Im}(E_{v_l} E_{h_l}^*) \tag{2.3.29d}$$

The emissivity formulas of SPM of Chapter 1 Section 3.2 are to be used for T_{v_l}, T_{h_l}, U_l, and V_l. Thus

$$T_v(s_x, s_y) = T\left\{e_v(\theta_l, \pi+\phi_l)\cos^2\alpha + e_h(\theta_l, \pi+\phi_l)\sin^2\alpha \right.$$
$$\left. + \, e_3(\theta_l, \pi+\phi_l)\cos\alpha\sin\alpha\right\} \tag{2.3.30}$$

where T is the physical temperature and e_v, e_h, and e_3 are given by (1.3.31), (1.3.29), and (1.3.50), respectively with the incident and scattered angles in the formulas all changed to θ_l, ϕ_l, $\hat{x}_l$, and $\hat{y}_l$. For example, in (1.3.29), we need to substitute $\overline{k}_{\perp i}$ as follows

$$\overline{k}_{\perp i} \rightarrow k\sin\theta_l\cos\phi_l\hat{x}_l + k\sin\theta_l\sin\phi_l\hat{y}_l$$

On the other hand, we substitute $\overline{k}_\perp$ by

$$\overline{k}_\perp \rightarrow k\sin\theta_k\cos\phi_k\hat{x}_l + k\sin\theta_k\sin\phi_k\hat{y}_l \tag{2.3.31}$$

The same applies to equations (1.3.31) and (1.3.50). Similarly

$$\begin{aligned} T_h =& T\left\{e_v(\theta_l, \pi + \phi_l)\sin^2\alpha + e_h(\theta_l, \pi + \phi_l)\cos^2\alpha \right. \\ &\left. - e_3(\theta_l, \pi + \phi_l)\cos\alpha\sin\alpha\right\} \end{aligned} \tag{2.3.32}$$

$$\begin{aligned} U =& T\left\{-\left[e_v(\theta_l, \pi + \phi_l) - e_h(\theta_l, \pi + \phi_l)\right]\sin 2\alpha \right. \\ &\left. + e_3(\theta_l, \pi + \phi_l)(\cos^2\alpha - \sin^2\alpha)\right\} \end{aligned} \tag{2.3.33}$$

$$V = e_4(\theta_l, \pi + \phi_l) \tag{2.3.34}$$

where e_4 is given by (1.3.57) with the same substitution of (2.3.31).

The brightness temperatures are to be weighted by the probability density function of the slope distribution. The tilting angles of large scale roughness also affect the surface area that is projected along the line of sight.

To study the projection, consider an incident wave in direction $\hat{k}_i$ on x-z plane of a one-dimensional surface. Let the surface be composed of facets, each of slope s_j, length l_j, and normal $\hat{n}_j$, $j = 1, \ldots, N$ where N is the number of facets. Then the probability of occurrence of slope s_j as viewed by the radiometer is

$$p = \frac{-l_j\hat{n}_j \cdot \hat{k}_i}{\sum\limits_{j=1}^{N}(-l_j\hat{n}_j \cdot \hat{k}_i)} p(s_x) \tag{2.3.35}$$

where $p(s_x)$ is the probability density function of s_x. In 1-D rough surface

$$\hat{k}_i = \sin\theta_i\hat{x} - \cos\theta_i\hat{z},$$
$$\hat{n} = (-s_x\hat{x} + \hat{z})/\sqrt{1 + s_x^2}.$$

Also $l \sim \sqrt{1 + s_x^2}$. Thus $-l\hat{n} \cdot \hat{k} \sim s_x \sin\theta_i + \cos\theta_i$. Substitute in (2.3.35) gives

$$p = A(s_x\sin\theta_i + \cos\theta_i)p(s_x) \tag{2.3.36}$$

where A is a normalization factor such that the probability is equal to 1. Therefore $A = 1/\cos\theta_i$ and

$$p = (1 + s_x\tan\theta_i)p(s_x) \tag{2.3.37}$$

Also for $s_x < -\cot\theta_i$, the surface is shadowed.

For the two-dimensional case, the radiometer is viewing at angle $\phi_{id} = \pi + \phi_i$ where ϕ_{id} is the emission azimuthal angle while ϕ_i is the incident azimuthal angle. We make the transformation between (s_x, s_y) and (s'_x, s'_y)

$$s'_x = s_x\cos\phi_i + s_y\sin\phi_i \tag{2.3.38a}$$

$$s'_y = -s_x\sin\phi_i + s_y\cos\phi_i \tag{2.3.38b}$$

$$s_x = s'_x \cos\phi_i - s'_y \sin\phi_i \tag{2.3.39a}$$

$$s_y = s'_x \sin\phi_i + s'_y \cos\phi_i \tag{2.3.39b}$$

Then the four Stokes parameters are

$$T_\gamma = \int_{-\infty}^{\infty} ds'_y \int_{-\cot\theta_i}^{\infty} ds'_x (1 + s'_x \tan\theta_i) p(s_x, s_y) T_\gamma(s_x, s_y) \tag{2.3.40}$$

where $T_\gamma(s_x, s_y)$ are given by (2.3.28) for the four Stokes parameters and the probability density function $p(s_x, s_y)$ is given by (2.3.3).

Foam effects have been modeled by Stogryn [1972] empirically. Let ν be the microwave frequency in GHz, W is the wind speed in m/s, $\theta_{ld} = \theta_l(180/\pi)$ is the local incidence angle in degrees, T_s is the sea surface temperature in Kelvin. Then the foam coverage function F is

$$F = b_0 + b_1 W + b_2 W^2 \tag{2.3.41}$$

where the coefficients b_0, b_1, and b_2 are frequency dependent and are given by

$$b_0 = 1.707 \times 10^{-2} + 8.560 \times 10^{-4}\nu + 1.120 \times 10^{-5}\nu^2 \tag{2.3.42a}$$

$$b_1 = -1.501 \times 10^{-2} + 1.821 \times 10^{-3}\nu - 4.634 \times 10^{-5}\nu^2 \tag{2.3.42b}$$

$$b_2 = 2.442 \times 10^{-4} - 2.282 \times 10^{-6}\nu + 4.194 \times 10^{-7}\nu^2 \tag{2.3.42c}$$

The emissivity vectors from the foam can be computed using an empirical formula

$$\begin{aligned} \epsilon_{fh}^{(l)} =& 0.005\nu + \frac{208 + 1.29\nu}{T_s}(1 - 7.48 \times 10^{-3}\theta_{ld} \\ & - 7.336 \times 10^{-5}\theta_{ld}^2 + 1.044 \times 10^{-7}\theta_{ld}^3) \end{aligned} \tag{2.3.43a}$$

and

$$\begin{aligned} \epsilon_{fv}^{(l)} =& 0.005\nu + \frac{208 + 1.29\nu}{T_s}(1 - 9.946 \times 10^{-4}\theta_{ld} + 3.218 \times 10^{-5}\theta_{ld}^2 \\ & - 1.187 \times 10^{-6}\theta_{ld}^3 + 7.0 \times 10^{-20}\theta_{ld}^{10}). \end{aligned} \tag{2.3.43b}$$

Transforming to the natural frame and putting in the foam factor give

$$T_{fv}(s_x, s_y) = TF\left(\epsilon_{fh}^{(l)} \sin^2\alpha + \epsilon_{fv}^{(l)} \cos^2\alpha\right) \tag{2.3.44a}$$

$$T_{fh}(s_x, s_y) = TF\left(\epsilon_{fv}^{(l)} \sin^2\alpha + \epsilon_{fh}^{(l)} \cos^2\alpha\right) \tag{2.3.44b}$$

$$U_f(s_x, s_y) = V_f(s_x, s_y) = 0 \tag{2.3.44c}$$

where the subscript f refers to surface with the foam correction. Thus the brightness temperature for the foam-covered ocean surface is

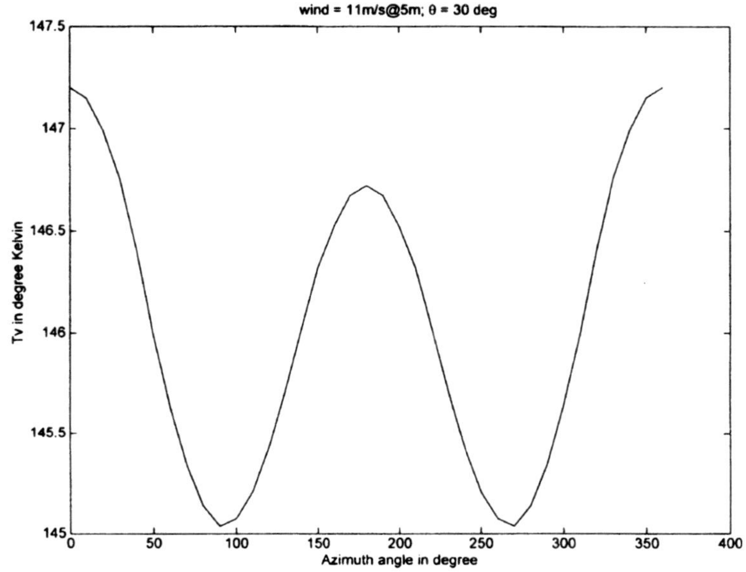

Figure 2.3.2 Brightness temperature for T_v as a function of azimuthal angles.

$$T_\gamma = \int_{-\infty}^{\infty} ds'_y \int_{-\infty}^{\infty} ds'_x (1 + s'_x \tan\theta_i) p(s_x, s_y) \cdot [(1-F) T_\gamma(s_x, s_y) + F T_{f\gamma}(s_x, s_y)] \tag{2.3.45}$$

We next illustrate numerically the final results of the brightness temperature as given by (2.3.45).

Numerical Illustrations

In Figs. 2.3.2, 2.3.3, 2.3.4, and 2.3.5, we plot respectively the results of T_v, T_h, U, and V as a function of azimuthal angle $\phi_{id} = \pi + \phi_i$. The parameters are $\theta_i = 30°$, wind speed is 11 m/sec at elevation of 5 m. The ocean temperature is 285 K, frequency is 19.39 GHz, and the ocean relative permittivity is $28.59 + i36.89$. The choice of k_d is $120\,\mathrm{m}^{-1}$. The ocean spectrum used is that of Durden-Vesecky as described in Chapter 4 of Volume I.

Following Yueh et al. [1994a] and Yueh [1997], we also used spectrum amplitude $a_0 = 0.008$. We see that the four Stokes parameters exhibit azimuthal variation. Because of the hydrodynamic modulation, they also exhibit inversion asymmetry in the sense that results at ϕ_{id} are different from that at $\phi_{id} + \pi$.

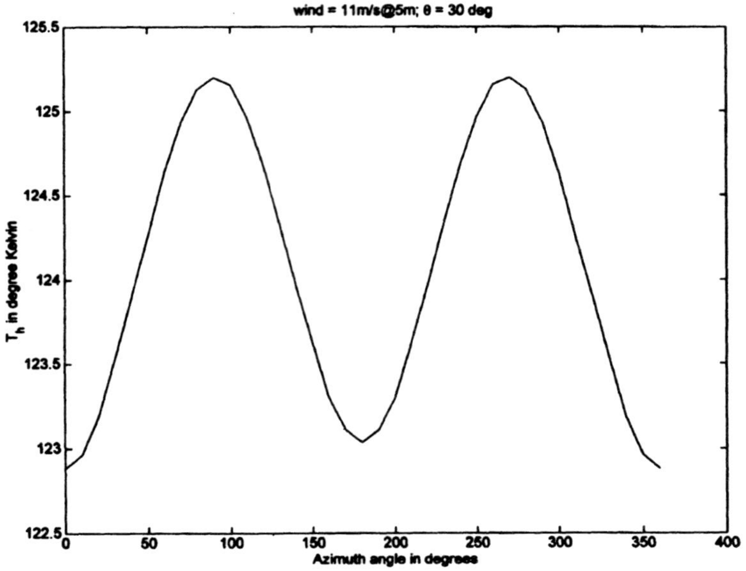

Figure 2.3.3 Brightness temperature for T_h as a function of azimuthal angles.

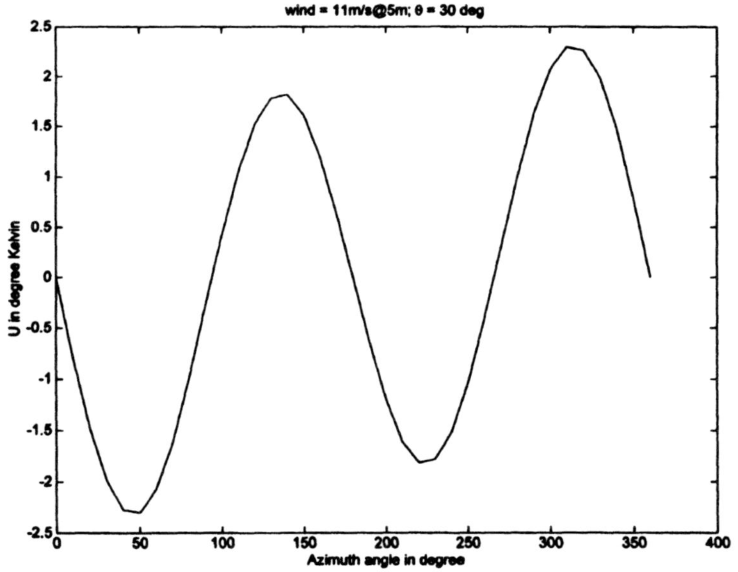

Figure 2.3.4 Brightness temperature for U as a function of azimuthal angles.

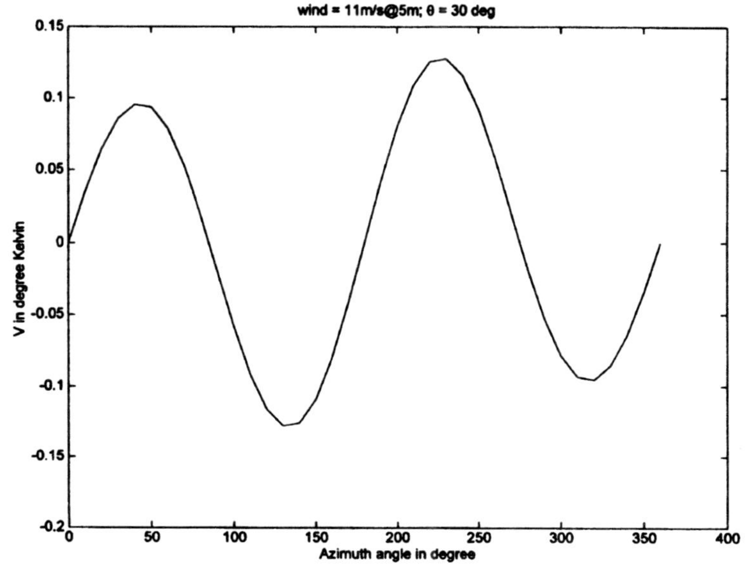

Figure 2.3.5 Brightness temperature for V as a function of azimuthal angles.

REFERENCES AND ADDITIONAL READINGS

Bahar, E. and M. A. Fitzwater (1989), Depolarization and backscatter enhancement in light scattering from random rough surfaces: Comparison of full-wave theory with experiment, *J. Opt. Soc. Am. A*, **6**, 33–43.

Bass, F. G. and I. M. Fuks (1979), *Wave Scattering from Statistically Rough Surfaces*, translated by C. B. Vesecky and J. F. Vesecky, Pergamon Press, Oxford, UK.

Beckmann, P. and A. Spizzichino (1963), *The Scattering of Electromagnetic Waves from Rough Surfaces*, Pergamon Press, Oxford.

Berman, D. H. (1985), The Kirchhoff approximation and first-order perturbation theory for rough surface scattering, *J. Acous. Soc. Am.*, **78**, 1045–1051.

Broschat, S. L. (1993), The small slope approximation reflection coefficient for scattering from a "Pierson-Moskowitz" sea surface, *IEEE Trans. Geosci. Remote Sens.*, **31(5)**, 1112–1114.

Broschat, S. L., L. Tsang, A. Ishimaru, and E. I. Thorsos (1988), A numerical comparison of the phase perturbation technique with the classical field perturbation and Kirchhoff approximations for random rough surface scattering problems, *J. Electromag. Waves and Appl.*, **2(1)**, 85–102.

Chen, J. S. and A. Ishimaru (1990), Numerical simulation of the second order Kirchhoff approximation from very rough surfaces and study of backscattering enhancement, *J. Acous. Soc. Am.*, **88**, 1846–1850.

Chen, K. S. and A. K. Fung (1995), A comparison of backscattering models for rough surfaces, *IEEE Trans. Geosci. Remote Sens.*, **33(1)**, 195–199.

Chen, K. S., T. D. Wu, M. K. Tsay, and A. K. Fung (2000), A note on the multiple scattering of the IEM model, *IEEE Trans. Geosci. Remote Sens.*, **38(1)**, 249–256.

Chen, M. F. and A. K. Fung (1988), A numerical study of the regions of validity of the Kirchhoff and small-perturbation rough surface scattering models, *Radio Sci.*, **23**, 163–170.

Collin, R. E. (1994), Full wave theories for rough surface scattering: An updated assessment, *Radio Sci.*, **29**, 1237–1254.

DeSanto, J. A. (1974), Green's function for electromagnetic scattering from a random rough surfaces, *J. Math. Phys.*, **15**, 283–288.

DeSanto, J. A. and O. Shisha (1974), Numerical solution of a singular integral equation in random rough surface scattering theory, *J. of Comp. Phys.*, **15**, 286–292.

DeSanto, J. A. and R. J. Wombell (1991), Rough surface scattering, *Waves in Random Media*, **1**, 41–56.

Durden, S. L. and J. F. Vesecky (1985), A physical radar cross-section model for a wind driven sea with swell, *IEEE J. Ocean. Eng.*, **10**, 445–451.

Durden, S. L. and J. F. Vesecky (1990), A numerical study of the separation wavenumber in the two scale scattering approximation, *IEEE Trans. Geosci. Remote Sens.*, **28**, 271–272.

Fung, A. K. (1994), *Microwave Scattering and Emission Models and Their Applications*, Artech House, Norwood, Massachusetts.

Fung, A. K. and M. F. Chen (1985), Numerical simulation of scattering from simple and composite random surfaces, *J. Opt. Soc. Am. A*, **2**, 2274–2284.

Fung, A. K. and K. K. Lee (1982), A semi-empirical sea-spectrum model for scattering coefficient estimation, *IEEE J. Ocean. Eng.*, **7**, 166–176.

Fung, A. K. and G. W. Pan (1987), A scatter model for perfectly conducting random surface: I. Model development. II. Range of validity, *Int. J. Remote Sens.*, **8(11)**, 1579–1605.

Greffet, J. J. and M. Nieto-Vesperinas (1998), Field theory for generalized bidirectional reflectivity: derivation of Helmholtz's reciprocity principle and Kirchhoff's law, *J. Opt. Soc. Am. A*, **15(10)**, 2735–2744.

Irisov, V. G. (1994), Small slope expansion for electromagnetic-wave diffraction on a rough surface, *Waves in Random Media*, **4(4)**, 441–452.

Irisov, V. G. (1997), Small-slope expansion for thermal and reflected radiation from a rough surface, *Waves in Random Media*, **7(1)**, 1–10.

Ishimaru, A. (1978), *Wave Propagation and Scattering in Random Media*, **1** and **2**, Academic Press, New York.

Ishimaru, A. and J. S. Chen (1990), Scattering from very rough surfaces based on the modified second order Kirchhoff approximation with angular and propagation shadowing, *J. Acous. Soc. Am. A*, **88**, 1877–1883.

Ishimaru, A. and J. S. Chen (1991), Scattering from very rough metallic and dielectric surfaces: a theory based on the modified Kirchhoff approximation, *Waves in Random Media*, **1(1)**, 21–34.

Ito, S. (1985), Analysis of scalar wave scattering from slightly rough random surface: A multiple scattering theory, *Radio Sci.*, **20**, 1–12.

Jackson, D. R., D. P. Winebrenner, and A. Ishimaru (1988), Comparison of perturbation theories for rough-surface scattering, *J. Acous. Soc. Am.*, **83(3)**, 961–969.

Leader, J. C. (1971), Bidirectional scattering of electromagnetic waves from rough surfaces, *J. Appl. Phys.*, **42**, 4808–4816.

Nieto-Vesperinas, M. and N. Garcia (1981), A detailed study of the scattering of scalar waves from random rough surfaces, *Opt. Acta*, **28**, 1651–1672.

Ogilvy, J. A. (1991), *Theory of Wave Scattering from Random Rough Surfaces*, Adam Hilger, Bristol.

Ogura, H. and N. Takahashi (1995), Wave scattering from a random rough surface: reciprocal theorem and backscattering enhancement, *Waves in Random Media*, **5(2)**, 223–242.

Plant, W. J. (1986), A two-scale model of short wind generated waves and scatterometry, *J. Geophys. Res.*, **91(C9)**, 10735–10749.

Rice, S. O. (1951), Reflection of electromagnetic waves from slightly rough surfaces, *Comm. Pure Appl. Math.*, **4**, 351–378.

Rice, S. O. (1963), Reflection of EM waves by slightly rough surfaces, *The Theory of Electromagnetic Waves*, M. Kline, Ed., Interscience, New York.

Rodriguez, E. (1989), Beyond the Kirchhoff approximation, *Radio Sci.*, **24**, 681–693.

Rodriguez, E. (1991), Beyond the Kirchhoff approximation II electromagnetic scattering, *Radio Sci.*, **26**, 121–132.

Sancer, M. I. (1969), Shadow-corrected electromagnetic scattering from a randomly rough surface, *IEEE Trans. Antennas Propagat.*, **17**, 577–585.

Sentenac, A. and J.-J. Greffet (1998), Mean-field theory of light scattering by one-dimensional rough surfaces, *J. Opt. Soc. Am. A*, **15**, 528–532.

Shin, R. T. (1984), Theoretical models for microwave remote sensing of earth terrain, Ph.D. thesis, Department of Electrical Engineering and Computer Science, Massachusetts Institute of Technology.

Smith, B. G. (1967), Geometrical shadowing of a random rough surface, *IEEE Trans. Antennas Propagat.*, **15**, 668–671.

Stogryn, A. (1972), The emissivity of sea foam at microwave frequency, *J. Geophys. Res.*, **77 (9)**, 1658–1666.

Tatarskii, V. I. and V. V. Tatarskii (1994), Statistical description of rough-surface scattering using the quasi-small-slope approximation for random surfaces with a Gaussian multivariate probability distribution, *Waves in Random Media*, **4(2)**, 191–214.

Thorsos, E. I. (1988), The validity of the Kirchhoff approximation for rough surface scattering using a Gaussian roughness spectrum, *J. Acous. Soc. Am.*, **83(1)**, 78–92.

Thorsos, E. I. and S. L. Broschat (1995), An investigation of the small slope approximation for scattering from rough surfaces. Part I. Theory, *J. Acous. Soc. Am.*, **97(4)**, 2082–2093.

Thorsos, E. I. and D. R. Jackson (1989), The validity of the perturbation approximation for rough surface scattering using a Gaussian roughness spectrum, *J. Acous. Soc. Am.*, **86(1)**, 261–272.

Tsang, L., J. A. Kong, and R. T. Shin (1985), *Theory of Microwave Remote Sensing*, Wiley-Interscience, New York.

Ulaby, F. T., R. K. Moore, and A. K. Fung (1981), *Microwave Remote Sensing: Active and Passive*, **1** and **2**, Addison-Wesley, Reading, MA.

Ulaby, F. T., R. K. Moore, and A. K. Fung (1982), *Microwave Remote Sensing: Active and Passive*, **3**, Artech House, Norwood, Massachusetts.

Valenzuela, G. R. (1967), Depolarization of EM waves by slightly rough surfaces, *IEEE Trans. Antennas Propagat.*, **15**, 552–557.

Valenzuela, G. R. (1968), Scattering of electromagnetic waves from a tilted slightly rough surface, *Radio Sci.*, **3**, 1057–1066.

Voronovich, A. G. (1994a), Small-slope approximation for electromagnetic wave scattering at a rough interface of two dielectric half-spaces, *Waves in Random Media*, **4(3)**, 337–367.

Voronovich, A. G. (1994b), *Wave Scattering from Rough Surfaces*, Springer-Verlag, Berlin.

Voronovich, A. G. (1996), A two-scale model from the point of view of the small-slope approximation, *Waves in Random Media*, **6(1)**, 73–83.

Winebrenner, D. P. and A. Ishimaru (1985a), Application of the phase-perturbation technique to randomly rough surface, *J. Opt. Soc. Am. A*, **2**, 2285–2294.

Winebrenner, D. P. and A. Ishimaru (1985b), Investigation of a surface field perturbation technique for scattering from rough surface, *Radio Sci.*, **20**, 161–170.

Yueh, S. H. (1997), Modeling of wind direction signal in polarimetric sea surface brightness temperatures, *IEEE Trans. Geosci. Remote Sens.*, **35(6)**, 1400–1418.

Yueh, S. H., R. Kwok, F. K. Li, S. V. Nghiem, W. J. Wilson, and J. A. Kong (1994a), Polarimetric passive remote sensing of ocean wind vector, *Radio Sci.*, **29**, 799–814.

Yueh, S. H., S. V. Nghiem, W. Wilson, F. K. Li, J. T. Johnson, and J. A. Kong (1994b), Polarimetric passive remote sensing of periodic water surfaces, *Radio Sci.*, **29**, 87–96.

Yueh, S. H., W. J. Wilson, F. K. Li, S. V. Nghiem, and W. B. Ricketts (1995), Polarimetric measurements of sea surface brightness temperatures using an aircraft K-band radiometer, *IEEE Trans. Geosci. Remote Sens.*, **33**, 85–92.

Yueh, S. H., W. J. Wilson, F. K. Li, S. V. Nghiem, and W. B. Ricketts (1997), Polarimetric brightness temperatures of sea surface measured with aircraft K- and Ka-Band radiometer, *IEEE Trans. Geosci. Remote Sens.*, **35**, 1177–1187.

Chapter 3

VOLUME SCATTERING: CASCADE OF LAYERS

The theory of multiple scattering can be treated from several viewpoints. In this chapter, we take the viewpoint of cascade of layers. The basic idea is to first use a thin layer that can only have single scattering. The precise nature of "thinness" is discussed. We then find the bistatic scattering results of such a thin layer. To simplify the problem, we use point scatterers.

The scattering of a thin layer is important because a thick layer can be considered as a cascading of many thin layers together with single scattering occurring in each thin layer. We first study the propagation of coherent wave with this cascading approach. The results give the effective propagation constant that agrees with Foldy's approximation. Then we study the second-order solution of a thin layer. The radiative transfer equations can then be developed from the point of view of a result of cascading of thin layers. The cascading approach illustrates, in a more intuitive way, the role of multiple scattering in a thick layer. It also gives results such as Foldy's approximation and radiative transfer theory that are usually derived from more formal multiple scattering theory. We also study the effects of clustering taking into account the effects of pair distribution function between clusters and within a cluster. When compared with independent scattering, the pair distribution between different clusters and the pair distribution function within a cluster can have opposite effects on the extinction.

1 Single Scattering Solution of a Thin Layer, Coherent Wave, and Effective Propagation Constant

Consider a plane wave impinging upon a slab of point scatterers of thickness D. The background permittivities of regions 0, 1, and 2 are all equal to ϵ which is not absorptive so that the boundaries at $z = 0$ and $z = -D$ are nonreflective (Fig. 3.1.1). The incident wave is

$$\psi_{inc} = e^{ik_{ix}x + ik_{iy}y - ik_{iz}z} \tag{3.1.1}$$

where

$$k_{ix} = k \sin\theta_i \cos\phi_i \tag{3.1.2}$$

$$k_{iy} = k \sin\theta_i \sin\phi_i \tag{3.1.3}$$

$$k_{iz} = k \cos\theta_i \tag{3.1.4}$$

The first-order scattered field is

$$\psi^{(s)(1)}(\overline{r}) = \sum_{l=1}^{N} \psi_l^{(s)(1)}(\overline{r}) \tag{3.1.5}$$

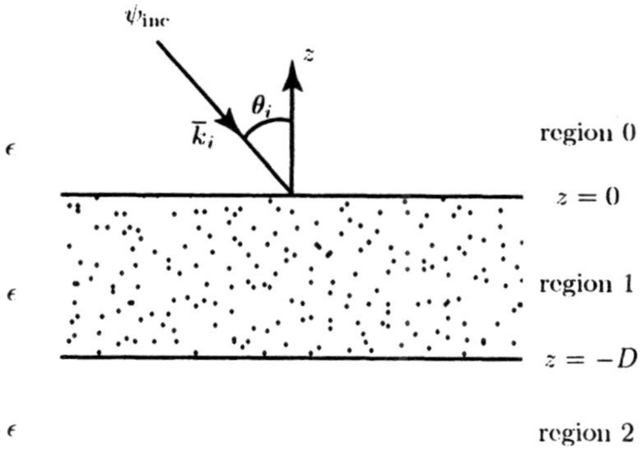

Figure 3.1.1 Plane wave impinging upon a slab of scatterers.

where $\psi_l^{(s)(1)}(\overline{r})$ is the first-order scattered field from the point scatterer located at $\overline{r} = \overline{r}_l = \overline{\rho}_l + \hat{z}z_l$:

$$\psi_l^{(s)(1)}(\overline{r}) = f\frac{e^{ik|\overline{r}-\overline{r}_l|}}{|\overline{r}-\overline{r}_l|}e^{i\overline{k}_{i\rho}\cdot\overline{\rho}_l - ik_{iz}z_l} \tag{3.1.6}$$

where $\overline{k}_{i\rho} = k_{ix}\hat{x} + k_{iy}\hat{y}$, and f is an isotropic scattering amplitude. The plane wave representation of spherical waves is, from (2.1.14) of Volume I, Chapter 2.

$$\frac{e^{ik|\overline{r}-\overline{r}_l|}}{|\overline{r}-\overline{r}_l|} = \frac{i}{2\pi}\int_{-\infty}^{\infty}\frac{d\overline{k}_\rho}{k_z}e^{i\overline{k}_\rho\cdot(\overline{\rho}-\overline{\rho}_l)}e^{ik_z|z-z_l|} \tag{3.1.7}$$

Using the plane wave representation of Green's function, we obtain

$$\begin{aligned}\psi_l^{(s)(1)}(\overline{r}) &= \frac{if}{2\pi}\int d\overline{k}_\rho\frac{1}{k_z}e^{i\overline{k}_\rho\cdot(\overline{\rho}-\overline{\rho}_l)}\\ &\cdot\exp(i\overline{k}_{i\rho}\cdot\overline{\rho}_l - ik_{iz}z_l)\begin{cases}e^{ik_z(z-z_l)} & \overline{r}\text{ in region 0}\\ e^{-ik_z(z-z_l)} & \overline{r}\text{ in region 2}\end{cases}\end{aligned} \tag{3.1.8}$$

The positions of the point scatterers $\overline{r}_l$'s are random. We see from (3.1.5)–(3.1.6) that the random positions of $\overline{r}_l$ will create a randomly fluctuating contribution from each term of $\psi^{(s)(1)}(\overline{r})$ which has summation over l as well as integration over k_ρ. As a function of $\overline{k}_\rho$, the integrand is randomly fluctuating, except when $\overline{k}_\rho = \overline{k}_{i\rho}$ and observation is in region 2. Thus in Monte Carlo simulation, for a single realization, the scattered waves are in all possible directions as represented by the integration over $d\overline{k}_\rho$. However, for the transmitted wave in region 2, there is a sharp peak in the direction of $\overline{k}_\rho = \overline{k}_{i\rho}$ because in that direction the phase is not fluctuating when summed over all particles $\sum_{l=1}^{N}$.

Let the random medium be extended over an area $A = L_x \times L_y$ and thickness D and the single-particle position be given by the probability density function

$$p(\overline{r}_l) = \frac{1}{V} \qquad \text{for } \overline{r}_l \text{ in region 1} \tag{3.1.9}$$

where $V = AD$.

The averaging of (3.1.5) is performed as follows:

$$\left\langle \sum_{l=1}^{N} h(\overline{r}_l) \right\rangle = \sum_{l=1}^{N} \langle h(\overline{r}_l) \rangle = \sum_{l=1}^{N} \int d\overline{r}_l \, h(\overline{r}_l) \, p(\overline{r}_l)$$
$$= N \int d\overline{r}_l \, h(\overline{r}_l) \, p(\overline{r}_l) = n_o \int d\overline{r}_l \, h(\overline{r}_l) \tag{3.1.10}$$

Before averaging, $\overline{r}_l$ is a deterministic variable. However, on averaging, each $\overline{r}_l$ becomes a dummy variable, so that the summation becomes just a product of N. Note that this is a result of averaging while for a single realization, each $\overline{r}_l$, that is randomly generated, is distinct. The last equality in (3.1.10) is a result of using (3.1.9) and $n_o = N/V$ as the number of particles per unit volume.

Thus,

$$\langle \psi^{s(1)}(\overline{r}) \rangle = n_o \int d\overline{r}_l \, \psi_l^{s(1)}(\overline{r}) \tag{3.1.11}$$

Using (3.1.8), we get the integral

$$\int_A d\overline{\rho}_l e^{i(\overline{k}_{i\rho} - \overline{k}_\rho)\cdot\overline{\rho}_l} = L_x L_y \, \text{sinc}\left((k_{ix} - k_x) \frac{L_x}{2} \right) \text{sinc}\left((k_{iy} - k_y) \frac{L_y}{2} \right) \tag{3.1.12}$$

where sinc $x = \sin x / x$. In the limit of $L_x, L_y \gg \lambda$, we obtain Dirac delta functions.

$$\lim_{L_x, L_y \to \infty} \frac{L_x L_y}{4\pi^2} \, \text{sinc}\left((k_{ix} - k_x) \frac{L_x}{2} \right) \text{sinc}\left((k_{iy} - k_y) \frac{L_y}{2} \right)$$
$$= \delta(k_{ix} - k_x) \delta(k_{iy} - k_y) \tag{3.1.13}$$

We also note that for statistical averaging to be applicable, the thickness of the slab must be much larger than the correlation length l of particle positions. Particle positions can be correlated within the correlation distance or length denoted by l. To have "sufficient amount of randomness" in the thin layer, we have the first condition.

$$(i) \quad D \gg l, \lambda \tag{3.1.14}$$

Using (3.1.11)–(3.1.13) in (3.1.8) and integration of $\int_{-D}^{0} dz_l$ gives

$$\lim_{A\to\infty} \langle \psi^{s(1)}(\overline{r}) \rangle = if2\pi n_o \begin{cases} \dfrac{1}{k_{iz}} \dfrac{e^{i\overline{k}_{i\rho}\cdot\overline{\rho}}\left(e^{2ik_{iz}D}-1\right)e^{ik_{iz}z}}{2ik_{iz}} & \overline{r} \text{ in region 0} \\ \dfrac{De^{i\overline{k}_{i\rho}\cdot\overline{\rho}-ik_{iz}z}}{k_{iz}} & \overline{r} \text{ in region 2} \end{cases} \tag{3.1.15}$$

For $D \gg \lambda$, the coherent transmitted wave is much larger than the coherent reflected wave. As is clear from (3.1.15), for $\overline{r}$ in region 2, the solution diverges if $D \to \infty$. In order for the first-order solution of coherent field to be valid, the first-order transmitted wave in (3.1.15) must be much smaller than the original incident wave. Thus we have a second condition.

$$(ii) \quad \frac{f2\pi n_o}{k_{iz}} D \ll 1 \tag{3.1.16}$$

which as shown later in equivalent to $D \ll l_{mfp}$ where l_{mfp} is the mean free path. Note that condition (i) gives a lower bound for D, while condition (ii) gives an upper bound on D. Condition (ii) actually gives a condition for the convergence of the Born series or the Neumann expansion as given by the iterative solution of the Foldy-Lax equation. From (3.1.1) and (3.1.15) we obtain

$$\langle \psi^{(1)}(\overline{r}) \rangle = \begin{cases} e^{i\overline{k}_{i\rho}\cdot\overline{\rho}-ik_{iz}z} + \dfrac{if2\pi n_o D}{k_{iz}} e^{i\overline{k}_{i\rho}\cdot\overline{\rho}-ik_{iz}z} & \text{for } z < -D \\ 0 & \text{for } z > 0 \end{cases} \tag{3.1.17}$$

We can express (3.1.17) as an input and output relation with the incident wave as the input and the transmitted wave as the output:

$$\langle \psi \rangle_{output} - \langle \psi \rangle_{input} e^{ik_{iz}D} = \frac{if2\pi n_o D}{k_{iz}} \langle \psi \rangle_{input} e^{ik_{iz}D} \tag{3.1.18}$$

where $e^{ik_{iz}D}$ is the phase shift of the incident wave as it passes through the layer of thickness D. The solution of a thick layer can be solved by cascading layers of thicknesses equal to D (Fig. 3.1.2).

Suppose the total thickness of the layer is d with $d \gg D$. Then we can divide d into many layers with each layer of thickness D that obeys both conditions (i) and (ii). That is, each layer is "thin" enough such that single scattering applies. Nevertheless, D still has to be thick enough to obey condition (i). Then we have the recurrence relation

$$\langle \psi \rangle_{n+1} = \left(1 + \frac{if2\pi n_o D}{k_{iz}}\right) \langle \psi \rangle_n e^{ik_{iz}D} \tag{3.1.19}$$

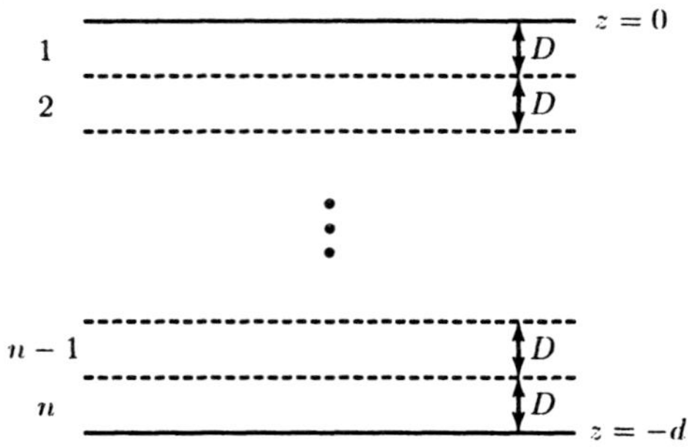

Figure 3.1.2 Studying large thickness problem by cascading: Total thickness is d and each layer thickness is D.

where the subscript n denotes the nth layer solution. Thus letting the zeroth layer solution be $\langle\psi\rangle_0 = \psi_{inc}$, we have, from (3.1.19),

$$\langle\psi\rangle_n = \left(1 + \frac{if2\pi n_o D}{k_{iz}}\right)^n e^{ik_{iz}nD}\psi_{inc}(z=0)$$

Using $D = d/n$ and taking the limit of large n with fixed d, we have

$$\begin{aligned}\lim_{n\to\infty}\langle\psi\rangle_n &= \lim_{n\to\infty}\left(1 + \frac{if2\pi n_o d}{k_{iz}n}\right)^n \psi_{inc}(z=0)e^{ik_{iz}nD} \\ &= \exp\left(\frac{i2\pi f n_o d}{k_{iz}} + i\overline{k}_{i\rho}\cdot\overline{\rho} + ik_{iz}d\right)\end{aligned} \quad (3.1.20)$$

The net effect of cascading for the coherent wave is that the net change occurs in the exponent. The result of (3.1.20) is an effective propagation constant K of the medium such that

$$K_{iz} = (K^2 - k_{i\rho}^2)^{1/2} \quad (3.1.21)$$

so that

$$\lim_{n\to\infty}\langle\psi\rangle_n = e^{i\overline{k}_{i\rho}\cdot\overline{\rho}}e^{-iK_{iz}z} \quad (3.1.22)$$

From (3.1.20) and (3.1.22) we have

$$K_{iz} = \frac{2\pi f n_o}{k_{iz}} + k_{iz} \quad (3.1.23)$$

Note that $\overline{k}_{i\rho}$ stays the same because of phase matching. Using (3.1.21) and (3.1.23) gives an expression for K with $K^2 = k_{i\rho}^2 + K_{iz}^2$. We assume that the effective propagation constant is such that $|K - k|/|k| \ll 1$.

$$K^2 = k_{i\rho}^2 + \left(k_{iz} + \frac{2\pi f n_o}{k_{iz}}\right)^2 \simeq k_{i\rho}^2 + k_{iz}^2 + 4\pi f n_o = k^2 + 4\pi f n_o \quad (3.1.24)$$

Thus

$$K = k + \frac{2\pi f n_o}{k} \tag{3.1.25}$$

Equation (3.1.25) is the same result as Foldy's approximation for the coherent wave. Note that since f has an imaginary part, K has a positive imaginary part even though k is real. The extinction rate κ_e relates to power taken away from the forward direction and is defined as

$$\kappa_e = 2\,\mathrm{Im}(K) \tag{3.1.26}$$

The mean free path is defined as

$$l_{mfp} = \frac{1}{\kappa_e} \tag{3.1.27}$$

The optical thickness of a layer is $\tau = \kappa_e d = d/l_{mfp}$ is the number of mean free path in the layer thickness. Thus in the random media problem, we have at least three distance scales: λ, l, and l_{mfp} with l_{mfp} usually a much larger distance than λ and l. When $l_{mfp} \simeq \lambda$, it is known as the Ioffe-Reggi criterion signifying the onset of strong localization [Sheng, 1990]. In the remote sensing problem, at microwave frequencies, l_{mfp} ranges from 10λ to thousands of wavelengths. The Foldy's approximation can be regarded as a cumulative result of many layers with single scattering in each layer to yield an effective propagation constant for the coherent wave. We also have the following observation. Long distance multiple scattering can occur over distance $d \gg l_{mfp}$. Such multiple scattering over a long distance can be studied by cascading "thin" layers with $D \ll l_{mfp}$. This is also the idea of invariant imbedding. Invariant imbedding in radiative transfer theory is described in Chapter 8, Section 3 of Volume I.

In the next section we will generalize the scalar point scatterer case to vector electromagnetic case. However, it is to be noted that we have assumed in this section that the point scatterers have negligible coherent mutual interaction. Nevertheless, there are cases when the point scatterers are in the neighborhood of each other so that there is strong coherent neighborhood mutual interaction. For example, the scatterer can be in clusters (locally dense) or densely distributed (uniformly dense as in dense media). In such cases there can be important coherent multiple scattering and correlated scattering among scatterers in the neighborhood of each other. Such effects have been simulated in Volume II. Obviously this is a different kind of multiple scattering from the long-distance multiple scattering. Thus we shall distinguish these two kinds of multiple scattering: (1) long distance or large thickness ($d > l_{mfp}$) multiple scattering and (2) "neighborhood" coherent multiple scattering. The latter kind can be important for discrete scatterers because of the moderate to large dielectric contrast between the background permittivity and the scatterer permittivity. However, the neighborhood multiple scattering can be studied by a single "thin layer" with $D > \lambda$ and $D \gg l$. The neighborhood multiple scattering can have significant effect on the effective propagation constants. It also depends heavily on the joint probability density functions of particle positions.

If $l_{mfp} \gg \lambda, l$ the two kinds of multiple scattering are distinct and can also be treated "separately."

A differential equation can also be found from (3.1.18) and (3.1.19). Let the coherent field be written in terms of an envelope function $\psi_A(z)$ so that

$$\langle \psi(\overline{r}) \rangle = \psi_A(z) e^{i\overline{k}_{id}\cdot\overline{r}} \tag{3.1.28}$$

Then from (3.1.19) we have

$$\psi_A(z - D) = \left(1 + \frac{if 2\pi n_o D}{k_{iz}}\right) \psi_A(z) \tag{3.1.29}$$

so that

$$-\frac{d\psi_A(z)}{dz} = \frac{if 2\pi n_o}{k_{iz}} \psi_A(z) \tag{3.1.30}$$

where the derivative with respect to z must be taken in the sense that every step size is of order D that obeys conditions (i) and (ii). We note that $k_{iz} = k\cos\theta_i$ and $-\cos\theta_i d\psi_A(z)/dz = d\psi_A/ds$, where $\hat{s}$ is the direction of propagation. Equation (3.1.30) can be generalized to the case of general propagation direction $\hat{s}$ with

$$\frac{d\psi_A}{ds} = \frac{if 2\pi n_o}{k} \psi_A \tag{3.1.31}$$

2 Transition Operator

In the far field, the scattering from a single particle centered at the origin is given by

$$\overline{E}_s(\overline{r}) = \frac{e^{ikr}}{r} \overline{\overline{F}}(\hat{k}_s, \hat{k}_i) \cdot \hat{e}_i \tag{3.2.1}$$

where $\hat{k}_s$ and $\hat{k}_i$ are the scattered and incident direction, respectively, and $\hat{e}_i$ is the polarization of the incident wave. We can write the scattering in a more general form. Using a transition operator $\overline{\overline{T}}$ for the scatterer centered at the origin, we obtain

$$\overline{E}_s(\overline{r}) = \int d\overline{r}' \int d\overline{r}'' \overline{\overline{G}}_o(\overline{r}, \overline{r}') \cdot \overline{\overline{T}}(\overline{r}', \overline{r}'') \cdot \overline{E}_{inc}(\overline{r}'') \tag{3.2.2}$$

Equation (3.2.2) includes near- and far-field effects, and the incident field $\overline{E}_{inc}$ can also be of general excitation instead of just plane wave. It is also convenient to use the Fourier transform of the transition operator. This is known as momentum representation:

$$\overline{\overline{T}}_p(\overline{p}', \overline{p}'') = \int d\overline{r}' \int d\overline{r}'' e^{-i\overline{p}'\cdot\overline{r}'} \overline{\overline{T}}(\overline{r}', \overline{r}'') e^{i\overline{p}''\cdot\overline{r}''} \tag{3.2.3}$$

The inverse transformation is

$$\overline{\overline{T}}(\overline{r}', \overline{r}'') = \frac{1}{(2\pi)^6} \int d\overline{p}' \int d\overline{p}'' e^{i\overline{p}'\cdot\overline{r}'} \overline{\overline{T}}_p(\overline{p}', \overline{p}'') e^{-i\overline{p}''\cdot\overline{r}''} \tag{3.2.4}$$

To derive a relation between $\overline{\overline{T}}$ and $\overline{\overline{F}}$, we consider an incident wave of

$$\overline{E}_{inc}(\overline{r}) = \hat{e}_i e^{i\overline{k}_i\cdot\overline{r}} \tag{3.2.5}$$

Substituting (3.2.5) and (3.2.4) into (3.2.2) and using the momentum representation of $\overline{\overline{T}}$, we have

$$\overline{E}_s(\overline{r}) = \int d\overline{r}' \overline{\overline{G}}_o(\overline{r}, \overline{r}') \int \frac{d\overline{p}'}{(2\pi)^3} e^{i\overline{p}'\cdot\overline{r}'} \overline{\overline{T}}_p(\overline{p}', \overline{k}_i) \cdot \hat{e}_i \tag{3.2.6}$$

In the far field in the observation direction of $\hat{k}_s = \hat{r}(\theta_s, \phi_s)$ we have

$$\overline{\overline{G}}_o(\overline{r}, \overline{r}') = \frac{e^{ikr}}{4\pi r}(\hat{\theta}_s\hat{\theta}_s + \hat{\phi}_s\hat{\phi}_s) e^{-i\overline{k}_s\cdot\overline{r}'} \tag{3.2.7}$$

Substituting (3.2.7) into (3.2.6) gives

$$\overline{E}_s(\overline{r}) = (\hat{\theta}_s\hat{\theta}_s + \hat{\phi}_s\hat{\phi}_s) \cdot \frac{e^{ikr}}{4\pi r} \overline{\overline{T}}_p(k\hat{k}_s, k\hat{k}_i) \cdot \hat{e}_i \tag{3.2.8}$$

Comparing (3.2.8) with (3.2.1) gives

$$\overline{\overline{F}}(\hat{k}_s, \hat{k}_i) \cdot \hat{e}_i = \frac{1}{4\pi}(\hat{\theta}_s\hat{\theta}_s + \hat{\phi}_s\hat{\phi}_s) \cdot \overline{\overline{T}}_p(k\hat{k}_s, k\hat{k}_i) \cdot \hat{e}_i \tag{3.2.9}$$

Thus, we note that the scattering amplitude is related to the $\overline{\overline{T}}$ operator when the two arguments of $\overline{\overline{T}}$ in the momentum representation have magnitude equal to the wavenumber k of the background medium. However, $\overline{\overline{T}}_p$ can have arguments of arbitrary magnitude and thus is more general than the far-field scattering dyad $\overline{\overline{F}}$.

If the scatterer is centered at $\overline{r}_j$ instead of at the origin, then the transition operator is given by, in coordinate and momentum representations,

$$\overline{\overline{T}}^j(\overline{r}',\overline{r}'') = \overline{\overline{T}}(\overline{r}' - \overline{r}_j, \overline{r}'' - \overline{r}_j) \tag{3.2.10}$$

$$\overline{\overline{T}}^j_p(\overline{p}',\overline{p}'') = e^{-i\overline{p}'\cdot\overline{r}_j}\overline{\overline{T}}_p(\overline{p}',\overline{p}'')e^{i\overline{p}''\cdot\overline{r}_j} \tag{3.2.11}$$

For the case of point scatterers for the scalar wave equation, the scalar transition operator is

$$T(\overline{r}',\overline{r}'') = 4\pi f\delta(\overline{r}')\delta(\overline{r}'') \tag{3.2.12}$$

3 Electromagnetic Wave Case of a Thin Layer and Extinction Matrix

The results of Section 1 can be readily generalized to the vector electromagnetic wave case to calculate the effective propagation constants for sparsely distributed scatterers. The line of attack is first to find the coherent wave solution of a thin layer of thickness D with $D \gg l, \lambda$ and $D \ll l_{mfp}$. Then we use cascade of layers to find the coherent wave propagation constant. Using the cascading of layer approach, we derive the effective propagation constant equations that are identical to that of Foldy approximation. We consider the thin layer of Fig. 3.1.1. Let the incident wave that impinges on the layer be given by

$$\overline{E}_{inc}(\overline{r}) = (E_{vi}\hat{v}_i + E_{hi}\hat{h}_i)e^{i\overline{k}_{i\rho}\cdot\overline{\rho}}e^{-ik_{iz}z} \tag{3.3.1}$$

where $\hat{v}_i = \hat{\theta}(\pi - \theta_i, \phi_i)$ and $\hat{h}_i = \hat{\phi}(\pi - \theta_i, \phi_i)$. The incident direction is $\hat{k}_{id} = \hat{r}(\pi - \theta_i, \phi_i)$, where the subscript d denotes downward going. The N scatterers are located at $\overline{r}_1, \ldots, \overline{r}_N$.

In this case we allow the scatterers to have different scattering properties from each other. For instance, the scatterers can be different in size, permittivity, shape, and orientation. We use an index s_j to denote the species of scatterer j.

The first-order solution is given by

$$\overline{E}_s^{(1)}(\overline{r}) = \sum_{j=1}^{N} \int d\overline{r}' \int d\overline{r}''\, \overline{\overline{G}}_o(\overline{r},\overline{r}') \cdot \overline{\overline{T}}^{s_j}(\overline{r}' - \overline{r}_j, \overline{r}'' - \overline{r}_j) \cdot \overline{E}_{inc}(\overline{r}'') \tag{3.3.2}$$

Using (3.3.1) in (3.3.2) and (3.2.11) gives

$$\overline{E}_s^{(1)}(\overline{r}) = \sum_{j=1}^{N} \int d\overline{r}'\, \overline{\overline{G}}_o(\overline{r},\overline{r}') \cdot \int \frac{d\overline{p}'}{(2\pi)^3} e^{i\overline{p}'\cdot(\overline{r}'-\overline{r}_j)} e^{i\overline{k}_{i\rho}\cdot\overline{\rho}_j}$$

$$e^{-ik_{iz}z_j}\overline{\overline{T}}^{s_j}_p(\overline{p}', k\hat{k}_{id}) \cdot (\hat{v}_i E_{vi} + \hat{h}_i E_{hi}) \tag{3.3.3}$$

Assuming that the species are independent of their positions (i.e., the scatterer size, shape, etc. are independent of their positions), we will have, using the uniform probability density function of (3.1.9),

$$\langle \overline{E}_s^{(1)}(\overline{r})\rangle = \frac{N}{V}\int_{-\infty}^{\infty} d\overline{\rho}_j \int_{-D}^{0} dz_j \int d\overline{r}'\, \overline{\overline{G}}_o(\overline{r},\overline{r}') \int \frac{d\overline{p}}{(2\pi)^3} e^{i\overline{p}\cdot(\overline{r}'-\overline{r}_j)} \cdot e^{i\overline{k}_{i\rho}\cdot\overline{\rho}_j} e^{-ik_{iz}z_j} \langle \overline{\overline{T}}_p^{s_j}(\overline{p}, k\hat{k}_{id})\rangle_s \cdot (\hat{v}_i E_{vi} + \hat{h}_i E_{hi}) \tag{3.3.4}$$

where $\langle \overline{\overline{T}}^{s_j}\rangle_s$ denotes the average of the transition operator averaged over the species properties.

Integrating over $d\overline{\rho}_j$ in (3.3.4) gives a Dirac delta function of $\delta(\overline{p}_\rho - \overline{k}_{i\rho})$. Note that for $\overline{p} = \overline{p}_\rho + p_z\hat{z}$ we have

$$\langle \overline{E}_s^{(1)}(\overline{r})\rangle = n_o \int d\overline{r}'\, \overline{\overline{G}}_o(\overline{r},\overline{r}') \int_{-D}^{0} dz_j \int_{-\infty}^{\infty} \frac{dp_z}{2\pi} e^{i\overline{k}_{i\rho}\cdot\overline{\rho}'} e^{ip_z z'} e^{-ip_z z_j - ik_{iz}z_j} \cdot \langle \overline{\overline{T}}_p(\overline{k}_{i\rho} + p_z\hat{z}, k\hat{k}_{id})\rangle_s \cdot (\hat{v}_i E_{vi} + \hat{h}_i E_{hi}) \tag{3.3.5}$$

If $D \gg \lambda$, then the integration of dz_j in (3.3.5) gives a delta function of $\delta(p_z + k_{iz})$, so that

$$\langle \overline{E}_s^{(1)}(\overline{r})\rangle = n_o \int_{-\infty}^{\infty} d\overline{\rho}' \int_{-D}^{0} dz' \overline{\overline{G}}_o(\overline{r},\overline{r}') e^{i\overline{k}_{i\rho}\cdot\overline{\rho}' - ik_{iz}z'} \cdot \langle \overline{\overline{T}}_p(k\hat{k}_{id}, k\hat{k}_{id})\rangle_s \cdot (\hat{v}_i E_{vi} + \hat{h}_i E_{hi}) \tag{3.3.6}$$

Using the plane wave representation of dyadic Green's function from (2.1.16) in Chapter 2 of Volume I, we have for z in region 2 and $z < z'$

$$\overline{\overline{G}}_o(\overline{r},\overline{r}') = \frac{i}{8\pi^2}\int_{-\infty}^{\infty} d\overline{k}_\rho \frac{1}{k_z}(\overline{\overline{I}} - \hat{k}_d\hat{k}_d) e^{i\overline{k}_\rho\cdot(\overline{\rho}-\overline{\rho}')} e^{-ik_z(z-z')} \tag{3.3.7}$$

Substituting (3.3.7) in (3.3.6), carrying out $d\overline{\rho}'$ integration gives $\delta(\overline{k}_\rho - \overline{k}_{i\rho})$, and after some algebra, we get

$$\langle \overline{E}_s^{(1)}(\overline{r})\rangle = \frac{n_o i D}{2k_{iz}}(\overline{\overline{I}} - \hat{k}_{id}\hat{k}_{id}) e^{i\overline{k}_{i\rho}\cdot\overline{\rho} - ik_{iz}z} \cdot \langle \overline{\overline{T}}_p(k\hat{k}_{id}, k\hat{k}_{id})\rangle_s \cdot (\hat{v}_i E_{vi} + \hat{h}_i E_{hi}) \tag{3.3.8}$$

Using the relation between scattering function matrix and transition operator as given by (3.2.9), we recognize that (3.3.8) is proportional to the forward scattering amplitude. Hence

$$\langle \overline{E}_s^{(1)}(\overline{r})\rangle = \frac{2\pi i n_o D}{k_{iz}} \langle \overline{\overline{F}}(\hat{k}_{id}, \hat{k}_{id})\rangle_s \cdot (\hat{v}_i E_{vi} + \hat{h}_i E_{hi}) e^{i\overline{k}_{id}\cdot\overline{r}} \tag{3.3.9}$$

Thus the coherent wave propagation depends only on the average values of the forward scattering amplitudes.

If we use the concept of cascade of layers to obtain the result for the large layer thickness case from the thin layer case (Fig. 3.1.2), then in a manner analogous to (3.1.19) we obtain

$$\langle \overline{E}(\overline{r}) \rangle_{n+1} = \left(\overline{\overline{I}} + \frac{2\pi i n_o D}{k_{iz}} \langle \overline{\overline{F}}(\hat{k}_{id}, \hat{k}_{id}) \rangle_s \right) \cdot \langle \overline{E}(\overline{r}) \rangle_n e^{ik_{iz}D} \tag{3.3.10a}$$

where the subscript n denotes the solution in the nth layer. A differential equation similar to (3.1.29)–(3.1.31) with the derivative to be taken in the sense that the step size D obeys conditions (i) and (ii) of Section 1.

$$\langle \overline{E}(\overline{r}) \rangle_n = \left(\overline{\overline{I}} + \frac{2\pi i n_o D}{k_{iz}} \langle \overline{\overline{F}}(\hat{k}_{id}, \hat{k}_{id}) \rangle_s \right)^n \cdot \langle \overline{E}(z=0) \rangle e^{ik_{iz}nD} \tag{3.3.11}$$

Since

$$nD = -z \tag{3.3.12}$$

in the limit of large n, we obtain

$$\lim_{n\to\infty} \left(\overline{\overline{I}} - \frac{2\pi i n_o z}{k_{iz} n} \langle \overline{\overline{F}}(\hat{k}_{id}, \hat{k}_{id}) \rangle_s \right)^n = \exp\left(\overline{\overline{I}} - \frac{2\pi i n_o z}{k_{iz}} \langle \overline{\overline{F}}(\hat{k}_{id}, \hat{k}_{id}) \rangle_s \right)$$

$$\lim_{n\to\infty} \langle \overline{E}(\overline{r}) \rangle_n = \langle \overline{E}(\overline{r}) \rangle = \exp\left(\overline{\overline{I}} - \frac{2\pi i n_o z}{k_{iz}} \langle \overline{\overline{F}}(\hat{k}_{id}, \hat{k}_{id}) \rangle_s \right) \cdot \langle \overline{E}(z=0) \rangle \exp(-ik_{iz}z) \tag{3.3.10b}$$

Let $\overline{E}^A(z)$ be the amplitude envelope

$$\langle \overline{E}(\overline{r}) \rangle = \overline{E}^A(z) e^{i\overline{k}_{id}\cdot\overline{r}} \tag{3.3.13}$$

Then, from (3.3.10b) we have

$$\overline{E}^A(z) e^{i\overline{k}_{id}\cdot\overline{r}} = \exp\left(\overline{\overline{I}} - \frac{2\pi i n_o z}{k_{iz}} \langle \overline{\overline{F}}(\hat{k}_{id}, \hat{k}_{id}) \rangle_s \right) \cdot \overline{E}^A(z=0) \exp(i\overline{k}_{i\rho}\cdot\overline{\rho} - ik_{iz}z) \tag{3.3.13a}$$

giving

$$\overline{E}^A(z) = \exp\left(\overline{\overline{I}} - \frac{2\pi i n_o z}{k_{iz}} \langle \overline{\overline{F}}(\hat{k}_{id}, \hat{k}_{id}) \rangle_s \right) \cdot \overline{E}^A(z=0) \tag{3.3.13b}$$

Equation (3.3.13b) can be put into a differential equation

$$-\frac{d}{dz} \overline{E}^A(z) = \frac{i2\pi n_o}{k_{iz}} \langle \overline{\overline{F}}(\hat{k}_{id}, \hat{k}_{id}) \rangle_s \cdot \overline{E}^A(z) \tag{3.3.14}$$

for this case. For a general direction of propagation $\hat{s}$, we have

$$\frac{d}{ds} \overline{E}^A = \frac{i2\pi n_o}{k} \langle \overline{\overline{F}}(\hat{s}, \hat{s}) \rangle_s \cdot \overline{E}^A \tag{3.3.15}$$

In (3.3.15) if we use vertical and horizontal polarization components E_v^A and E_h^A as basis vectors, then

$$\frac{dE_v^A}{ds} = M_{vv}E_v^A + M_{vh}E_h^A \tag{3.3.16}$$

$$\frac{dE_h^A}{ds} = M_{hv}E_v^A + M_{hh}E_h^A \tag{3.3.17}$$

where

$$M_{jl} = \frac{i2\pi n_o}{k}\langle f_{jl}(\hat{s},\hat{s})\rangle_s \tag{3.3.18}$$

with $j, l = v, h$. Equations (3.3.16)–(3.3.17) are known as the coherent wave propagation equations under Foldy's approximation. Foldy approximation is often derived using diagrammatic methods or operator methods as shown in Chapter 5. However, the derivation in this chapter is based on cascading of thin layers and a single scattering within each thin layer.

Multiplying (3.3.16) by E_v^{A*} and adding to its complex conjugate give

$$\frac{d}{ds}\{|E_v^A|^2\} = 2\text{Re}(M_{vv})\left|E_v^A\right|^2 + 2\text{Re}(M_{vh}E_v^{A*}E_h^A) \tag{3.3.19}$$

Using the definition of Stokes parameters I_v, I_h, U, and V, (3.3.19) takes the form

$$\frac{dI_v}{ds} = 2\text{Re}(M_{vv})I_v + \text{Re}(M_{vh})U + \text{Im}(M_{vh})V \tag{3.3.20}$$

Similar differential equations can be derived for dI_h/ds, dU/ds, and dV/ds by making use of (3.3.16) and (3.3.17)

$$\frac{dI_h}{ds} = 2\text{Re}(M_{hh})I_h + \text{Re}(M_{hv})U - \text{Im}(M_{hv})V \tag{3.3.21}$$

$$\begin{aligned}\frac{dU}{ds} &= 2\text{Re}(M_{hv})I_v + 2\text{Re}(M_{vh})I_h + [\text{Re}(M_{vv}) + \text{Re}(M_{hh})]U \\ &\quad - [\text{Im}(M_{vv}) - \text{Im}(M_{hh})]V\end{aligned} \tag{3.3.22}$$

$$\begin{aligned}\frac{dV}{ds} &= -2\text{Im}(M_{hv})I_v + 2\text{Im}(M_{vh})I_h + [\text{Im}(M_{vv}) - \text{Im}(M_{hh})]U \\ &\quad + [\text{Re}(M_{vv}) + \text{Re}(M_{hh})]V\end{aligned} \tag{3.3.23}$$

If we write for the coherent wave

$$\frac{d\overline{I}}{ds} = -\overline{\overline{\kappa}}_e \cdot \overline{I} \tag{3.3.24}$$

where $\overline{I}$ stands for the four-component Stokes vector and $\overline{\overline{\kappa}}_e$ is the extinction matrix, then

$$\overline{\overline{\kappa}}_e = \begin{bmatrix} -2\mathrm{Re}M_{vv} & 0 & -\mathrm{Re}(M_{vh}) & -\mathrm{Im}(M_{vh}) \\ 0 & -2\mathrm{Re}M_{hh} & -\mathrm{Re}(M_{hv}) & \mathrm{Im}(M_{hv}) \\ -2\mathrm{Re}M_{hv} & -2\mathrm{Re}M_{vh} & -(\mathrm{Re}M_{vv} + \mathrm{Re}M_{hh}) & (\mathrm{Im}M_{vv} - \mathrm{Im}M_{hh}) \\ 2\mathrm{Im}M_{hv} & -2\mathrm{Im}M_{vh} & -(\mathrm{Im}M_{vv} - \mathrm{Im}M_{hh}) & -(\mathrm{Re}M_{vv} + \mathrm{Re}M_{hh}) \end{bmatrix} \tag{3.3.25}$$

Equation (3.3.16) can also be solved by an eigenvalue and eigenvector approach.

Let the envelope functions be given by

$$\begin{bmatrix} E_v^A \\ E_h^A \end{bmatrix} = \begin{bmatrix} E_{vo}^A \\ E_{ho}^A \end{bmatrix} e^{i(K-k)s} \tag{3.3.26}$$

where K is the effective propagation constant. Substituting (3.3.26) in (3.3.16) and (3.3.17) gives an eigenfunction problem of

$$\begin{bmatrix} i(K-k) - M_{vv} & -M_{vh} \\ -M_{hv} & i(K-k) - M_{hh} \end{bmatrix} \begin{bmatrix} E_{vo}^A \\ E_{ho}^A \end{bmatrix} = 0 \tag{3.3.27}$$

The eigenvalues are K_1 and K_2, where

$$K_1 = k - \frac{i}{2}[M_{vv} + M_{hh} + r] \tag{3.3.28}$$

$$K_2 = k - \frac{i}{2}[M_{vv} + M_{hh} - r] \tag{3.3.29}$$

$$r = \left\{(M_{vv} - M_{hh})^2 + 4M_{hv}M_{vh}\right\}^{1/2} \tag{3.3.30}$$

and the corresponding eigenvectors are

$$\begin{bmatrix} E_{vo}^A \\ E_{ho}^A \end{bmatrix} = \begin{bmatrix} 1 \\ b_1 \end{bmatrix}, \quad \begin{bmatrix} b_2 \\ 1 \end{bmatrix} \tag{3.3.31}$$

where

$$b_1 = \frac{2M_{hv}}{M_{vv} - M_{hh} + r} \tag{3.3.32}$$

$$b_2 = \frac{2M_{vh}}{-M_{vv} + M_{hh} - r} \tag{3.3.33}$$

The eigenvalues and eigenvectors represent, respectively, the effective propagation constants and the characteristic polarizations of the characteristic waves. The characteristic waves do not have to be vertically polarized or horizontally polarized. The characteristic polarizations generally can be linear combinations of vertical and horizontal polarizations. We have used the results of coherent wave propagation in Chapter 7, Section 2.2 of Volume I in conjunction with vector radiative transfer theory with various applications in remote sensing.

4 First- and Second-Order Solutions: Incoherent Waves

Consider the case of point scatterers as studied in Section 1.1. Use the geometry of Fig. 3.1.1 and the incident plane wave as given by (3.1.1)–(3.1.4). We note that to the second order we have

$$\psi = \psi_{inc} + \psi^{s(1)} + \psi^{s(2)} \tag{3.4.1}$$

If we study the second moment, then

$$\begin{aligned}\langle \psi\psi^* \rangle &= \psi_{inc}\psi^*_{inc} + \psi_{inc}\langle \psi^{s(1)*} \rangle + \langle \psi^{s(1)} \rangle \psi^*_{inc} \\ &\quad + \langle \psi^{s(1)}\psi^{s(1)*} \rangle + \psi_{inc}\langle \psi^{s(2)*} \rangle + \langle \psi^{s(2)} \rangle \psi^*_{inc} + \cdots\end{aligned} \tag{3.4.2}$$

We will calculate the intensity contained in the first-order scattered field, $\langle \psi^{s(1)}\psi^{s(1)*} \rangle$. However, such a term is of the same order as $\psi_{inc}\langle \psi^{s(2)*} \rangle$. Thus, one has to consider $\langle \psi^{s(2)} \rangle$ also in the second order corrections to the second moment.

If we separate into coherent and incoherent waves

$$\psi = \langle \psi \rangle + \psi_f \tag{3.4.3}$$

then writing the coherent field to the second order gives

$$\langle \psi \rangle = \psi_{inc} + \langle \psi^{s(1)} \rangle + \langle \psi^{s(2)} \rangle \tag{3.4.4}$$

The coherent field intensity is

$$\begin{aligned}\langle \psi \rangle \langle \psi^* \rangle &= \psi_{inc}\psi^*_{inc} + \psi_{inc}\langle \psi^{s(1)*} \rangle + \langle \psi^{s(1)} \rangle \psi^*_{inc} \\ &\quad + \langle \psi^{s(1)} \rangle \langle \psi^{s(1)*} \rangle + \psi_{inc}\langle \psi^{s(2)*} \rangle + \langle \psi^{s(2)} \rangle \psi^*_{inc}\end{aligned}$$

To the same order as in (3.4.2), then

$$\langle \psi\psi^* \rangle = \langle \psi \rangle \langle \psi^* \rangle + \langle \psi_f \psi_f^* \rangle \tag{3.4.5}$$

where the incoherent intensity is

$$\langle \psi_f \psi_f^* \rangle = \langle \psi^{s(1)}\psi^{s(1)*} \rangle - \langle \psi^{s(1)} \rangle \langle \psi^{s(1)*} \rangle \tag{3.4.6}$$

For scalar wave equation and nonabsorptive scatterers, the energy conservation relation is

$$\nabla \cdot \left(\psi^*(\overline{r}) \nabla \psi(\overline{r}) - \psi(\overline{r}) \nabla \psi^*(\overline{r}) \right) = 0 \tag{3.4.7}$$

We define the Poynting vector

$$\overline{S} = \mathrm{Im}\Big(\psi^*(\overline{r}) \nabla \psi(\overline{r}) \Big) \tag{3.4.8}$$

Since

$$\nabla\Big(\psi(\overline{r})\psi^*(\overline{r}')\Big) = \psi^*(\overline{r}')\nabla\psi(\overline{r}) \tag{3.4.9}$$

we can write

$$\overline{S}(\overline{r}) = \lim_{\overline{r}'\to\overline{r}} \operatorname{Im}\Big(\nabla\left(\psi(\overline{r})\psi^*(\overline{r}')\right)\Big) \tag{3.4.10a}$$

$$\langle\overline{S}(\overline{r})\rangle = \lim_{\overline{r}'\to\overline{r}} \operatorname{Im}\Big(\nabla\langle\psi(\overline{r})\psi^*(\overline{r}')\rangle\Big) \tag{3.4.10b}$$

In (3.4.10b), the average Poynting's vector is expressed in terms of field correlation $\langle\psi(\overline{r})\psi^*(\overline{r}')\rangle$. The conservation theorem is

$$\nabla\cdot\overline{S} = 0 \tag{3.4.11a}$$

which has the integral form

$$\oint dA\hat{n}\cdot\overline{S} = 0 \tag{3.4.11b}$$

The incoming power on the slab is

$$-\hat{z}\cdot\operatorname{Im}\Big(\psi^*_{inc}(\overline{r})\nabla\psi_{inc}(\overline{r})\Big)A = k_{iz}A \tag{3.4.11c}$$

where A is the area of the slab. The power flux is

$$\hat{z}\cdot\langle\overline{S}(\overline{r})\rangle = \lim_{\overline{r}'\to\overline{r}} \hat{z}\cdot\operatorname{Im}\Big(\nabla\langle\psi(\overline{r})\psi^*(\overline{r}')\rangle\Big) \tag{3.4.12}$$

In the calculation of the intensity of the first-order scattered fields, we have

$$\psi^{s(1)}(\overline{r})\psi^{s(1)*}(\overline{r}') = \sum_{l=1}^{N}\sum_{j=1}^{N}\psi_j^{s(1)}(\overline{r})\psi_l^{s(1)*}(\overline{r}') \tag{3.4.13}$$

Thus the product of scattered fields can come from different particles of $j \neq l$ in (3.1.6). Instead of specifying only the single-particle position probability density function as given by $p(\overline{r}_l) = 1/V$, we need to specify also the two-particle joint probability density function. The joint probability density function of two particles is, from (4.5.21) and (4.5.23) in Chapter 4 of Volume I,

$$p(\overline{r}_1,\overline{r}_2) = \frac{g(\overline{r}_1,\overline{r}_2)}{V^2}\left(\frac{N}{N-1}\right) \tag{3.4.14}$$

where g is the pair distribution function. The case of independent particle positions is given by $g(\overline{r}_1,\overline{r}_2) = 1$. However, such is an idealized case because particles cannot penetrate each other so that $g(\overline{r}_1,\overline{r}_2) = 0$ for $|\overline{r}_1 - \overline{r}_2| <$

d_{min}, where d_{min} is a minimum distance. In the case that $|\overline{r}_1 - \overline{r}_2| \to \infty$, the particle positions must be independent. Hence

$$\lim_{|\overline{r}_1-\overline{r}_2|\to\infty} g(\overline{r}_1, \overline{r}_2) = 1 \tag{3.4.15}$$

The pair function can be assumed to be translational invariant so that $g(\overline{r}_1, \overline{r}_2) = g(\overline{r}_1 - \overline{r}_2)$. The summation in (3.4.13) is next split into two parts, $j = l$ and $j \neq l$:

$$\psi^{s(1)}(\overline{r})\psi^{s(1)*}(\overline{r}') = \sum_{l=1}^{N} \psi_l^{s(1)}(\overline{r})\psi_l^{s(1)*}(\overline{r}') + \sum_{l=1}^{N}\sum_{\substack{j=1\\ l\neq j}}^{N} \psi_j^{s(1)}(\overline{r})\psi_l^{s(1)*}(\overline{r}') \tag{3.4.16}$$

Hence

$$\langle\psi^{s(1)}(\overline{r})\psi^{s(1)*}(\overline{r}')\rangle = N\int d\overline{r}_l\psi_l^{s(1)}(\overline{r})\psi_l^{s(1)*}(\overline{r}')p(\overline{r}_l)$$
$$+ N(N-1)\int d\overline{r}_l\int d\overline{r}_j\psi_j^{s(1)}(\overline{r})\psi_l^{s(1)*}(\overline{r}')\frac{\left[\left(g(\overline{r}_l-\overline{r}_j)-1\right)+1\right]}{V^2} \tag{3.4.17}$$

For integration purposes, we write $g = (g-1)+1$ because the asymptotic value of g is unity and that makes $g-1$ integrable. We recognize, from (3.4.17), that the $+1$ term in (3.4.17) is actually the products of the coherent field and its complex conjugate. Hence

$$\langle\psi^{s(1)}(\overline{r})\psi^{s(1)*}(\overline{r}')\rangle = n_o\int d\overline{r}_l\psi_l^{s(1)}(\overline{r})\psi_l^{s(1)*}(\overline{r}')$$
$$+ n_o^2\int d\overline{r}_l\int d\overline{r}_j\psi_j^{s(1)}(\overline{r})\psi_l^{s(1)*}(\overline{r}')[g(\overline{r}_l-\overline{r}_j)-1]$$
$$+ \langle\psi^{s(1)}(\overline{r})\rangle\langle\psi^{s(1)*}(\overline{r}')\rangle \tag{3.4.18}$$

The pair distribution functions representing the correlation of particle scattering arise in the intensity of the first-order scattered fields because the field and field conjugates can come from different particles. The pair distribution function term does not occur in $\psi_{inc}\langle\psi^{s(1)*}\rangle$. Thus the perturbation series expansion of (3.4.2) indicates that we need to include $\psi_{inc}\langle\psi^{s(2)*}\rangle$, which must depend on the pair distribution function.

The second-order scattered field is

$$\psi^{s(2)}(\overline{r}) = \sum_{l=1}^{N}\sum_{\substack{j=1\\ j\neq l}}^{N} f^2\frac{e^{ik|\overline{r}-\overline{r}_l|}}{|\overline{r}-\overline{r}_l|}\frac{e^{ik|\overline{r}_l-\overline{r}_j|}}{|\overline{r}_l-\overline{r}_j|}\psi_{inc}(\overline{r}_j) \tag{3.4.19}$$

so that

$$\langle \psi^{s(2)}(\overline{r}) \rangle = n_o^2 f^2 \int d\overline{r}_j \int d\overline{r}_l \frac{e^{ik|\overline{r}-\overline{r}_l|}}{|\overline{r}-\overline{r}_l|} \frac{e^{ik|\overline{r}_l-\overline{r}_j|}}{|\overline{r}_l-\overline{r}_j|} e^{i\overline{k}_{id}\cdot\overline{r}_j} \Big(g(\overline{r}_l - \overline{r}_j) - 1 + 1 \Big) \tag{3.4.20}$$

Using plane wave expansion of the spherical waves, the first term of (3.4.18) becomes

$$\begin{aligned} n_o \int_V & d\overline{r}_j \psi_j^{s(1)}(\overline{r}) \psi_j^{s(1)*}(\overline{r}') \\ &= n_o \frac{|f|^2}{4\pi^2} \int_{-\infty}^{\infty} d\overline{\rho}_j \int_{-D}^{0} dz_j \int_{-\infty}^{\infty} d\overline{k}_{1\rho} \int_{-\infty}^{\infty} d\overline{k}_{2\rho} \\ &\quad \times \frac{e^{i\overline{k}_{1\rho}\cdot(\overline{\rho}-\overline{\rho}_j)} e^{-i\overline{k}_{2\rho}\cdot(\overline{\rho}'-\overline{\rho}_j)}}{k_{1z}k_{2z}^*} e^{\pm ik_{1z}(z-z_j)\mp ik_{2z}^*(z'-z_j)} \end{aligned} \tag{3.4.21}$$

where the upper sign is for z and z' in region 0 and the lower sign is for z and z' in region 2. This convention will be adopted in this section. In (3.4.21) $k_{1z} = (k^2 - k_{1\rho}^2)^{1/2}$ and $k_{2z} = (k^2 - k_{2\rho}^2)^{1/2}$. Integrating over $d\overline{\rho}_j$ gives $\delta(\overline{k}_{1\rho} - \overline{k}_{2\rho})$. Hence we have

$$\begin{aligned} n_o \int_V & d\overline{r}_j \psi_j^{s(1)}(\overline{r}) \psi_j^{s(1)*}(\overline{r}') \\ &= n_o |f|^2 \int_{-D}^{0} dz_j \int_{-\infty}^{\infty} d\overline{k}_\rho \frac{e^{i\overline{k}_\rho\cdot(\overline{\rho}-\overline{\rho}')}}{|k_z|^2} e^{\pm ik_z z \mp ik_z^* z' \pm 2\mathrm{Im}\,(k_z) z_j} \end{aligned} \tag{3.4.22}$$

Next we assume that only radiating waves in (3.4.21) contribute and that evanescent waves are negligible. Let $k_x = k\sin\theta\cos\phi$, $k_y = k\sin\theta\sin\phi$, $k_\rho = k\sin\theta$, and $k_z = k\cos\theta$, yielding

$$\begin{aligned} \int_{-\infty}^{\infty} d\overline{k}_\rho &= \int_0^\infty dk_\rho k_\rho \int_0^{2\pi} d\phi \cong \int_0^k dk_\rho k_\rho \int_0^{2\pi} d\phi \\ &= k^2 \int_0^{\pi/2} d\theta \sin\theta\cos\theta \int_0^{2\pi} d\phi \end{aligned} \tag{3.4.23}$$

Also the layer D is thin, so that $D \ll l_{mfp}$. This means that $\mathrm{Im}(k_z)D \ll 1$. Hence, we can ignore the $e^{\pm 2\mathrm{Im}(k_z)z_j}$ term in (3.4.22). Then,

$$\begin{aligned} n_o \int_V & d\overline{r}_j \psi_j^{s(1)}(\overline{r}) \psi_j^{s(1)*}(\overline{r}') \\ &= n_o |f|^2 D \int_0^{\pi/2} d\theta \frac{\sin\theta}{\cos\theta} \int_0^{2\pi} d\phi \; e^{i\overline{k}_\rho\cdot(\overline{\rho}-\overline{\rho}')} e^{\pm ik\cos\theta(z-z')} \end{aligned} \tag{3.4.24}$$

From (3.4.24), it can be seen that the field correlation $\langle \psi^{s(1)}(\overline{r}) \cdot \psi^{s(1)*}(\overline{r}') \rangle$ contains a spectrum of plane waves with phase dependence of $\exp(i\overline{k}_\rho \cdot (\overline{\rho} -$

$\overline{\rho}') \pm ik\cos\theta(z - z'))$. This is unlike the coherent wave, which is only in the original incident direction. The $\cos\theta$ in the denominator will disappear when $\partial\psi/\partial z$ is to be taken to get $\hat{z} \cdot \langle \overline{S} \rangle$.

Next we evaluate the second term of (3.4.18). We note that $g - 1$ is only nonzero over the correlation length of the pair distribution function l. We change the integration variable to average and difference coordinates

$$\overline{r}_A = \frac{\overline{r}_j + \overline{r}_l}{2} \tag{3.4.25}$$

$$\overline{r}_d = \overline{r}_j - \overline{r}_l \tag{3.4.26}$$

The integration limits of $\overline{r}_d$ can be from $-\infty$ to ∞ since the slab thickness D is assumed to be much larger than the correlation length l and $g - 1$ vanishes for distance larger than l. Thus $\int d\overline{r}_l \int d\overline{r}_j = \int_{-\infty}^{\infty} d\overline{\rho}_A \int_{-D}^{0} dz_A \int_{-\infty}^{\infty} d\overline{r}_d$. Substituting in (3.4.18) and carrying out the $d\overline{\rho}_A$ integration, we have

$$\begin{aligned} n_o^2 \int d\overline{r}_l \int d\overline{r}_j \psi_j^{s(1)}(\overline{r}) \psi_l^{s(1)*}(\overline{r}')[g(\overline{r}_l - \overline{r}_j) - 1] \\ = n_o^2 |f|^2 \int_{-D}^{0} dz_A \int_{-\infty}^{\infty} d\overline{r}_d [g(\overline{r}_d) - 1] \int_{-\infty}^{\infty} d\overline{k}_\rho \, e^{i\overline{k}_\rho \cdot (\overline{\rho} - \overline{\rho}' - \overline{\rho}_d)} \\ \cdot e^{i\overline{k}_{i\rho} \cdot \overline{\rho}_d - ik_{iz} z_d} e^{\pm ik_z (z - z_A - z_d/2)} \frac{e^{\mp ik_z^* (z' - z_A + z_d/2)}}{|k_z|^2} \end{aligned} \tag{3.4.27a}$$

Since $D \ll l_{mfp}$, we can ignore the attenuation. Then (3.4.27a) simplifies to

$$\begin{aligned} n_o^2 \int d\overline{r}_l \int d\overline{r}_j \psi_j^{s(1)}(\overline{r}) \psi_l^{s(1)*}(\overline{r}')[g(\overline{r}_l - \overline{r}_j) - 1] \\ = n_o^2 |f|^2 D \int_0^{\pi/2} d\theta \frac{\sin\theta}{\cos\theta} \int_0^{2\pi} d\phi \, e^{i\overline{k}_\rho \cdot (\overline{\rho} - \overline{\rho}')} e^{\pm ik\cos\theta(z - z')} \\ \int_{-\infty}^{\infty} d\overline{r}_d [g(\overline{r}_d) - 1] e^{i(\overline{k}_{i\rho} - \overline{k}_\rho) \cdot \overline{\rho}_d} e^{\mp ik_z z_d} e^{-ik_{iz} z_d} \end{aligned} \tag{3.4.27b}$$

where we have assumed that only radiating waves contribute. We note that the scattering in (3.4.27*b*) is proportional to the Fourier transform of the difference of pair function and unity.

Let

$$h(\overline{r}) = g(\overline{r}) - 1 \tag{3.4.28}$$

and

$$H(\overline{p}) = \frac{1}{(2\pi)^3} \int_{-\infty}^{\infty} d\overline{r} h(\overline{r}) e^{-i\overline{p} \cdot \overline{r}} \tag{3.4.29}$$

is the structure factor. Because $g(\overline{r})$ is symmetric in x, y, and z, another useful expression for $H(\overline{p})$ is

$$H(\overline{p}) = \frac{2}{(2\pi)^3} \int_{-\infty}^{\infty} d\overline{\rho} \int_0^{\infty} dz h(\overline{r}) \cos(\overline{p}_\rho \cdot \overline{\rho} + p_z z) \tag{3.4.30}$$

Hence

$$n_o^2 \int d\overline{r}_l \int d\overline{r}_j \psi_j^{s(1)}(\overline{r}) \psi_l^{s(1)*}(\overline{r}')[g(\overline{r}_l - \overline{r}_j) - 1]$$

$$= (2\pi)^3 n_o^2 |f|^2 D \int_0^{\pi/2} d\theta \frac{\sin\theta}{\cos\theta} \int_0^{2\pi} d\phi \, e^{i\overline{k}_\rho \cdot (\overline{\rho} - \overline{\rho}')}$$

$$e^{\pm ik\cos\theta(z-z')} H(\overline{p}_\rho = \overline{k}_\rho - \overline{k}_{i\rho}, p_z = \pm k_z + k_{iz}) \tag{3.4.31}$$

Comparing (3.4.24) and (3.4.31), we see that the first term of (3.4.18) can be interpreted as independent scattering while the second term of (3.4.18) is interpreted as correlated scattering. We note that the correlated scattering is proportional to H which can be of order l^3, where l is the correlation length. Thus the ratio of correlated scattering to independent scattering is of the order $n_o l^3$, which can be appreciable. Equation (3.4.18) can be expressed as

$$\langle \psi^{s(1)}(\overline{r}) \psi^{(s)(1)*}(\overline{r}') \rangle = \langle \psi_f(\overline{r}) \psi_f^*(\overline{r}') \rangle + \langle \psi^{s(1)}(\overline{r}) \rangle \langle \psi^{s(1)*}(\overline{r}') \rangle \tag{3.4.31a}$$

where

$$\langle \psi_f^{s(1)}(\overline{r}) \psi_f^{s(1)*}(\overline{r}') \rangle = n_o |f|^2 D \int_0^{\pi/2} d\theta \frac{\sin\theta}{\cos\theta} \int_0^{2\pi} d\phi e^{i\overline{k}_\rho \cdot (\overline{\rho} - \overline{\rho}')} e^{\pm ik\cos\theta(z-z')}$$

$$\cdot \left[1 + n_o (2\pi)^3 H(\overline{p}_\rho = \overline{k}_\rho - \overline{k}_{i\rho}, p_z = \pm k_z + k_{iz})\right] \tag{3.4.31b}$$

We next compute outgoing intensity. We will include later the $\langle \psi^{s(2)} \rangle$, that is, the second-order coherent field terms. Also from (3.1.17) we have

$$\langle \psi_s^{(1)}(\overline{r}) \rangle = \begin{cases} \dfrac{if 2\pi n_o D}{k_{iz}} e^{i\overline{k}_{i\rho} \cdot \overline{\rho} - ik_{iz} z} & \text{at } z = -D \\ 0 & \text{at } z = 0 \end{cases} \tag{3.4.32}$$

Note that as shown in Section 1, the coherent first-order scattered field is negligible at $z = 0$. The first-order coherent scattered field only gives nonzero correlation with the transmitted wave.

$$\oint dS\hat{n} \cdot \text{Im} \lim_{\overline{r}' \to \overline{r}} \nabla \Big[\psi_{inc}(\overline{r}) \psi_{inc}^*(\overline{r}') + \psi_{inc}(\overline{r}) \langle \psi^{s(1)*}(\overline{r}') \rangle$$

$$+ \langle \psi^{s(1)}(\overline{r}) \rangle \psi_{inc}^*(\overline{r}') + \langle \psi^{s(1)}(\overline{r}) \psi^{s(1)*}(\overline{r}') \rangle \Big]$$

$$= \hat{z} \cdot \text{Im} \lim_{\overline{r}' \to \overline{r}} \nabla \Big[\psi_{inc}(\overline{r}) \psi_{inc}^*(\overline{r}') + \psi_{inc}(\overline{r}) \langle \psi^{s(1)*}(\overline{r}') \rangle$$

$$+ \langle \psi^{s(1)}(\overline{r}) \rangle \psi_{inc}^*(\overline{r}') + \langle \psi^{s(1)}(\overline{r}) \psi^{s(1)*}(\overline{r}') \rangle \Big]_{z=0}$$

$$
\begin{aligned}
&- \hat{z} \cdot \operatorname{Im} \lim_{\overline{r}' \to \overline{r}} \nabla \Big[\psi_{inc}(\overline{r}) \psi^*_{inc}(\overline{r}') + \psi_{inc}(\overline{r}) \langle \psi^{s(1)*}(\overline{r}') \rangle \\
&\qquad + \langle \psi^{s(1)}(\overline{r}) \rangle \psi^*_{inc}(\overline{r}') + \langle \psi^{s(1)}(\overline{r}) \psi^{s(1)*}(\overline{r}') \rangle \Big]_{z=-D} \\
&= A \Big\{ -\operatorname{Im}(f) 4\pi n_o D + n_o |f|^2 D k \int_0^{\pi/2} d\theta \sin\theta \int_0^{2\pi} d\phi \\
&\times \left[1 + (2\pi)^3 n_o H(\overline{p}_\rho = \overline{k}_\rho - \overline{k}_{i\rho}, p_z = -k_z + k_{iz}) \right] \\
&+ |f|^2 \frac{(2\pi n_o D)^2}{k_{iz}} + n_o |f|^2 D k \int_0^{\pi/2} d\theta \sin\theta \int_0^{2\pi} d\phi \\
&\times \left[1 + (2\pi)^3 n_o H(\overline{p}_\rho = \overline{k}_\rho - \overline{k}_{i\rho}, p_z = k_z + k_{iz}) \right] \Big\}
\end{aligned} \tag{3.4.33}
$$

where A is the area of the slab. In (3.4.33), the first term is from $\psi_{inc} \langle \psi^{s(1)*} \rangle + \psi^*_{inc} \langle \psi^{s(1)} \rangle$. The second term is from the outgoing intensity of the incoherent wave at $z = -D$. The third term is from the correlation function of the first-order coherent field $\langle \psi^{s(1)} \rangle \langle \psi^{s(1)*} \rangle$. The last term is the outgoing intensity at top boundary $z = 0$. We note that (3.4.33) is not equal to 0. We also note that the coherent field is in the original direction of propagation, while the incoherent intensity is in all directions as exhibited by angular integration of $\int d\theta \sin\theta \int d\phi$.

To obey energy conservation, we need to calculate $\langle \psi^{s(2)} \rangle$ next. The second-order term of the coherent wave is as given by (3.4.19). We substitute in the plane wave representations $\int d\overline{k}_{1\rho}$ and $\int d\overline{k}_{2\rho}$ of the two spherical waves. We also let $\overline{\rho}_A = (\overline{\rho}_j + \overline{\rho}_l)/2$ and $\overline{\rho}_d = (\overline{\rho}_j - \overline{\rho}_l)$. Integrating over $\int d\overline{\rho}_A$ then gives $\delta(\overline{k}_{1\rho} - \overline{k}_{i\rho})$. Since the coherent wave is only appreciable in the forward direction, we consider z in region 2 for $\langle \psi^{s(2)}(\overline{r}) \rangle$. Thus

$$
\begin{aligned}
&\langle \psi^{s(2)}(\overline{r}) \rangle \\
&= -\frac{n_o^2 f^2}{4\pi^2} \int_{-\infty}^{\infty} d\overline{\rho}_l \int_{-\infty}^{\infty} d\overline{\rho}_j \int_{-D}^{0} dz_j \int_{-D}^{0} dz_l \int_{-\infty}^{\infty} d\overline{k}_{1\rho} \frac{e^{i\overline{k}_{1\rho} \cdot (\overline{\rho} - \overline{\rho}_l)} e^{-ik_{1z}(z - z_l)}}{k_{1z}} \\
&\int_{-\infty}^{\infty} d\overline{k}_{2\rho} \frac{e^{i\overline{k}_{2\rho} \cdot (\overline{\rho}_l - \overline{\rho}_j)}}{k_{2z}} e^{+ik_{2z}|z_l - z_j|} e^{i\overline{k}_{i\rho} \cdot \overline{\rho}_j} e^{-ik_{iz} \cdot z_j} g\left(|\overline{\rho}_l - \overline{\rho}_j|, |z_l - z_j| \right) \\
&= -n_o^2 f^2 \int_{-\infty}^{\infty} d\overline{\rho}_d \int_{-D}^{0} dz_j \int_{-D}^{0} dz_l \frac{1}{k_{iz}} e^{i\overline{k}_{i\rho} \cdot \overline{\rho}} e^{-ik_{iz} z} e^{i\overline{k}_{i\rho} \cdot \overline{\rho}_d} \\
&\cdot \left[1 + \left(g(\overline{\rho}_d, |z_l - z_j|) - 1 \right) \right] \int d\overline{k}_\rho e^{i\overline{k}_\rho \cdot \overline{\rho}_d} \frac{1}{k_z} e^{ik_z |z_j - z_l|} e^{-ik_{iz}(z_j - z_l)}
\end{aligned} \tag{3.4.34}
$$

We note that

$$\int_{-\infty}^{\infty} d\overline{\rho}_d \int_{-D}^{0} dz_j \int_{-D}^{0} dz_l e^{i\overline{k}_{i\rho}\cdot\overline{\rho}_d} \int d\overline{k}_\rho e^{i\overline{k}_\rho\cdot\overline{\rho}_d} \frac{1}{k_z} e^{ik_z|z_j-z_l|-ik_{iz}(z_j-z_l)}$$

$$= 4\pi^2 \int_{-D}^{0} dz_j \int_{-D}^{0} dz_l \frac{1}{k_{iz}} e^{ik_{iz}|z_j-z_l|-ik_{iz}(z_j-z_l)} \simeq 2\pi^2 D^2/k_{iz}$$

for $D \gg \lambda$. For the $g-1$ term, we set $z_j - z_l = z_d$ and $(z_j + z_l)/2 = z_A$. The dz_A can be integrated since the integrand is independent of z_A.

$$\int_{-D}^{0} dz_j \int_{-D}^{0} dz_l(\cdot)(g-1) \simeq D \int dz_d(\cdot)(g-1) \tag{3.4.35}$$

where $(\cdot)$ denotes the expression in (3.4.34). The last equality is due to the fact that $D \gg l$, where l, the correlation length, is the distance over which $g-1$ is nonzero. Thus, in region 2 we have

$$\langle\psi^{s(2)}(\overline{r})\rangle = -n_o^2 f^2 4\pi^2 D^2 \frac{e^{i\overline{k}_{i\rho}\cdot\overline{\rho}-ik_{iz}z}}{2k_{iz}^2} - n_o^2 f^2 D \frac{e^{i\overline{k}_{i\rho}\cdot\overline{\rho}-ik_{iz}z}}{k_{iz}} \cdot$$

$$\int_{-\infty}^{\infty} d\overline{\rho}_d \int d\overline{k}_\rho \int_{-\infty}^{\infty} dz_d \Big(g(\overline{\rho}_d, |z_d|) - 1\Big) e^{i(\overline{k}_{i\rho}-\overline{k}_\rho)\cdot\overline{\rho}_d} \frac{e^{ik_z|z_d|}e^{-ik_{iz}z_d}}{k_z} \tag{3.4.36}$$

and $\langle\psi^{s(2)}(\overline{r})\rangle = 0$ in region 0. Both terms in (3.4.36) exhibit the phase dependence of $\exp(i\overline{k}_{i\rho}\cdot\overline{\rho} - ik_{iz}z)$, indicating that the second-order coherent field is also in the forward direction. Since $\mathrm{Re}\, f \gg \mathrm{Im}\, f$, f^2 can be approximated by $|f|^2$. We calculate the outgoing power at $z = -D$ due to $\psi_{inc}\langle\psi^{s(2)*}\rangle + \langle\psi^{s(2)}\rangle\psi_{inc}^*$. The second-order coherent wave is a transmitted wave.

$$A\Big\{-\hat{z}\cdot \mathrm{Im}\Big[\nabla\psi_{inc}(\overline{r})\langle\psi^{s(2)*}(\overline{r}')\rangle + \nabla\langle\psi^{s(2)}(\overline{r})\rangle\psi_{inc}^*(\overline{r}')\Big]\Big\}_{\overline{r}'=\overline{r},\ z=-D}$$

$$= 2k_{iz} A\, \mathrm{Re}\Big\{-n_o^2|f|^2 \frac{4\pi^2 D^2}{2k_{iz}^2} - \frac{n_o^2|f|^2 D}{k_{iz}} \int_{-\infty}^{\infty} d\overline{\rho}_d \int d\overline{k}_\rho$$

$$\int_{-\infty}^{\infty} dz_d \Big(g(\overline{\rho}_d, |z_d|) - 1\Big) e^{i(\overline{k}_{i\rho}-\overline{k}_\rho)\cdot\overline{\rho}_d} \frac{e^{ik_z|z_d|}e^{-ik_{iz}z_d}}{k_z}\Big\} \tag{3.4.37}$$

Taking the real part in (3.4.37) gives cosine terms for the complex exponentials in (3.4.37). Thus, including radiating waves only in (3.4.37), we have

$$A\Big\{-\hat{z}\cdot \mathrm{Im}\Big[\nabla\psi_{inc}(\overline{r})\langle\psi^{s(2)*}(\overline{r}')\rangle + \nabla\langle\psi^{s(2)}(\overline{r})\rangle\psi_{inc}^*(\overline{r}')\Big]\Big\}_{\overline{r}'=\overline{r},\ z=-D}$$

$$= A\Big\{-n_o^2|f|^2 \frac{4\pi^2 D^2}{k_{iz}} - 2n_o^2|f|^2 Dk \int_0^{\pi/2} d\theta \sin\theta \int_0^{2\pi} d\phi \int_{-\infty}^{\infty} d\overline{\rho}_d$$

$$
\cdot \left[\int_0^{\infty} dz_d \Big(g(\overline{\rho}_d, |z_d|) - 1 \Big) \cos\Big((\overline{k}_{i\rho} - \overline{k}_\rho) \cdot \overline{\rho}_d + (k_z - k_{iz}) z_d \Big) \right.
$$

$$
\left. + \int_{-\infty}^{0} dz_d \Big(g(\overline{\rho}_d, |z_d|) - 1 \Big) \cos\Big((\overline{k}_{i\rho} - \overline{k}_\rho) \cdot \overline{\rho}_d + (k_z + k_{iz}) z_d \Big) \right] \Bigg\}
$$

$$
= A \Bigg\{ -n_o^2 |f|^2 \frac{4\pi^2 D^2}{k_{iz}} - n_o^2 |f|^2 D k (2\pi)^3 \int_0^{\pi/2} d\theta \sin\theta \int_0^{2\pi} d\phi
$$

$$
\times \Big[H(\overline{p}_\rho = \overline{k}_\rho - \overline{k}_{i\rho}, p_z = k_{iz} - k_z)
$$

$$
+ H(\overline{p}_\rho = \overline{k}_\rho - \overline{k}_{i\rho}, p_z = k_z + k_{iz}) \Big] \Bigg\} \tag{3.4.38}
$$

The second equality in (3.4.38) is from (3.4.30). Note that (3.4.38) are identical to the terms in (3.4.33) and will cancel on adding. Thus adding (3.4.38) to (3.4.33) gives

$$
\oint dS \hat{n} \cdot \mathrm{Im} \lim_{\overline{r}' \to \overline{r}} \nabla \Big[\psi_{inc}(\overline{r}) \psi_{inc}^*(\overline{r}') + \psi_{inc}(\overline{r}) \langle \psi^{s(1)*}(\overline{r}') \rangle + \langle \psi^{s(1)}(\overline{r}) \rangle \psi_{inc}^*(\overline{r}')
$$

$$
+ \langle \psi^{s(1)}(\overline{r}) \psi^{s(1)*}(\overline{r}') \rangle + \psi_{inc}(\overline{r}) \langle \psi^{s(2)*}(\overline{r}') \rangle + \langle \psi^{s(2)}(\overline{r}) \rangle \psi_{inc}^*(\overline{r}') \Big]
$$

$$
= A \Bigg\{ -\mathrm{Im}(f) 4\pi n_o D + 2 n_o |f|^2 D k \int_0^{\pi/2} d\theta \sin\theta \int_0^{2\pi} d\phi \Bigg\}
$$

$$
= A \left\{ -\mathrm{Im}(f) 4\pi n_o D + 4\pi n_o D k |f|^2 \right\} = 0 \tag{3.4.39}
$$

The last equality is a result of the forward scattering theorem for the scattering amplitude f. That is, $\mathrm{Im}\,(f) = k|f|^2$, the conservation of energy condition for a single particle. Equation (3.4.39) indicates that when carried out to the second-order scattering for a thin layer, energy conservation is obeyed.

Note that in the derivations we have made use of plane wave representations, which is particularly convenient to represent the bistatic scattering of a thin slab. One can also think in terms of discrete propagation directions when the values of k_x and k_y are discretized.

5 Cascading of Layers: From First- and Second-Order Wave Solutions to Radiative Transfer Equation

In Sections 1 and 2 we note that by the process of cascading of layers we can use the first-order coherent wave solution of thickness D to derive the effective propagation constant and the Foldy's approximation.

In Section 4 we showed that to obey energy conservation, we need to calculate pair distribution functions and also the second-order coherent wave.

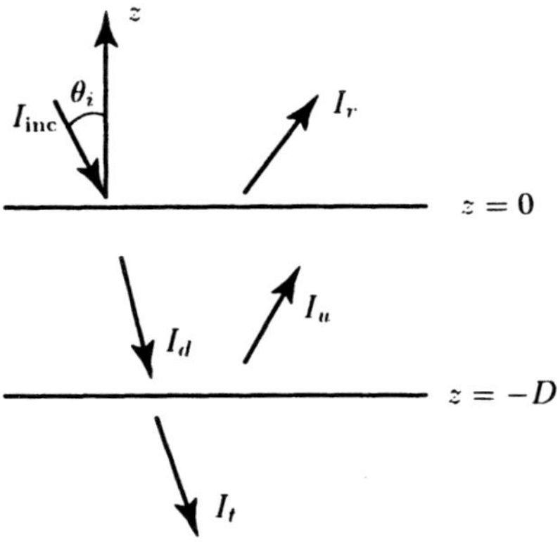

Figure 3.5.1 An illustration of incident, reflected, upward-going, downward-going and transmitted intensities for a single layer geometry.

The intensity is also decomposed into coherent and incoherent intensities. In this section we will show the results of cascading of the single-layer solutions of coherent wave and incoherent wave. By single layer, it is meant that if D obeys the conditions of $l \ll D$ and $\lambda < D \ll l_{mfp}$, then the first- and second-order solutions are sufficient to represent the input and output of the thin layer. By cascading the layers of thickness D to give a much thicker layer of d, we solve a large thickness or long-distance problem $d \gg D$, where multiple scattering solutions are represented by cumulative single-layer solutions. The single-layer solution consists of the first- and second-order solutions in each layer. We will show that the cascading of the first- and second-order solutions of a single layer to many layers will lead to the radiative transfer equation in difference form. The derivation of the radiative transfer equation by this manner of cascading also answers the question of when radiative transfer equation is valid.

First we will rewrite the solution of Section 4 in terms of input and output solutions for a layer (Fig. 3.5.1). We define the intensity in direction (θ, ϕ) as

$$I_u(\theta, \phi) = \frac{\hat{z} \cdot \mathrm{Im}\langle \psi^*(\overline{r}) \nabla \psi(\overline{r}) \rangle_{(\theta,\phi)}}{k \cos\theta} \tag{3.5.1}$$

where $\langle \psi^*(\overline{r}) \nabla \psi(\overline{r}) \rangle_{(\theta,\phi)}$ denotes the wave equation solution associated with direction (θ, ϕ), and subscript u denotes upward-going. Similarly, the downward-going intensity has

$$I_d(\theta, \phi) = \frac{-\hat{z} \cdot \mathrm{Im}\,\langle \psi^*(\overline{r}) \nabla \psi(\overline{r}) \rangle_{(\theta,\phi)}}{k \cos\theta} \tag{3.5.2}$$

The first- and second-order solutions of Section 4 state that when the incident wave is impinging upon the layer, the wave is transmitted and reflected. The transmitted wave has two parts: One is in the original direction of propagation known as the coherent wave, and the other is in all directions (θ, ϕ) due to an incoherent wave. The reflected wave has the incoherent wave in all directions θ and ϕ. The incoherent wave part is due to $\langle \psi_s^{(1)} \nabla \psi_s^{(1)*} \rangle - \langle \psi_s^{(1)} \rangle \langle \nabla \psi_s^{(1)*} \rangle$ in Section 4.

The intensity of the incident wave is

$$I_{inc}(\theta_i, \phi_i) = 1 \tag{3.5.3}$$

For the transmitted wave, the intensity that is associated with the original direction of propagation is

$$\begin{aligned} I_t(\theta_i, \phi_i) = \frac{1}{k \cos \theta_i} (-\hat{z}) \cdot \operatorname{Im} \Big\{ & \langle \psi_{inc}^*(\overline{r}) \nabla \psi_{inc}(\overline{r}) \rangle \\ & + \langle \psi^{s(1)*}(\overline{r}) \rangle \nabla \psi_{inc}(\overline{r}) + \psi_{inc}^*(\overline{r}) \langle \nabla \psi^{s(1)}(\overline{r}) \rangle \\ & + \langle \psi^{s(1)*}(\overline{r}) \rangle \nabla \langle \psi^{s(1)}(\overline{r}) \rangle + \psi_{inc}^*(\overline{r}) \langle \nabla \psi^{s(2)}(\overline{r}) \rangle \\ & + \langle \psi^{s(2)*}(\overline{r}) \rangle \nabla \psi_{inc}(\overline{r}) \Big] \Big\}_{z=-D} \end{aligned} \tag{3.5.4}$$

which represents the coherent field intensity to the second order. Using (3.4.33) and (3.4.38) in (3.5.4) gives

$$\begin{aligned} I_t(\theta_i, \phi_i) = & \frac{1}{k \cos \theta_i} \Big\{ k_{iz} - \operatorname{Im}(f) 4\pi n_o D + |f|^2 \frac{(2\pi n_o D)^2}{k_{iz}} - \frac{n_o^2 |f|^2 4\pi^2 D^2}{k_{iz}} \\ & - n_o^2 |f|^2 D k (2\pi)^3 \int_0^{\pi/2} d\theta \sin\theta \int_0^{2\pi} d\phi [H(\overline{p}_\rho = \overline{k}_\rho - \overline{k}_{i\rho}, p_z = k_z + k_{iz}) \\ & + H(\overline{p}_\rho = \overline{k}_\rho - \overline{k}_{i\rho}, p_z = -k_z + k_{iz})] \Big\} I_{inc} \end{aligned} \tag{3.5.5a}$$

Inside the curly brackets, the third term arises from $\langle \psi^{s(1)} \rangle \langle \psi^{s(1)*} \rangle$ while the fourth and last terms arise from $\psi^{inc} \langle \psi^{s(2)*} \rangle + \langle \psi^{s(2)} \rangle \psi^{inc*}$. The third and fourth terms cancel. We recognize that

$$4\pi \operatorname{Im} f = k|f|^2 4\pi = \int_0^{\pi/2} d\theta \sin\theta \int_0^{2\pi} d\phi 2k|f|^2$$

Thus

$$I_t(\theta_i,\phi_i) = \frac{1}{k\cos\theta_i}\Big\{k\cos\theta_i - |f|^2 n_o Dk \int_0^{\pi/2} d\theta \sin\theta \int_0^{2\pi} d\phi\, 2$$
$$- n_o^2 |f|^2 Dk(2\pi)^3 \int_0^{\pi/2} d\theta \sin\theta \int_0^{2\pi} d\phi\, \big[H(\overline{p}_\rho = \overline{k}_\rho - \overline{k}_{i\rho}, p_z = k_{iz} + k_z)$$
$$+ H(\overline{p}_\rho = \overline{k}_\rho - \overline{k}_{i\rho}, p_z = k_{iz} - k_z)\big]\Big\} \tag{3.5.5b}$$

To put the notations consistent with radiative transfer theory, we define phase functions as

$$P(\theta,\phi;\pi-\theta_i,\phi_i) = n_o|f|^2\Big[1 + n_o(2\pi)^3 H(\overline{p}_\rho = \overline{k}_\rho - \overline{k}_{i\rho}, p_z = k_z + k_{iz})\Big] \tag{3.5.6}$$

where $\overline{k}_\rho = k\sin\theta\cos\phi\hat{x} + k\sin\theta\sin\phi\hat{y}$ and $k_z = k\cos\theta$. Hence from (3.5.5b) the transmitted intensity in the original direction is

$$I_t(\theta_i,\phi_i) = \frac{1}{k\cos\theta_i}\Big\{k\cos\theta_i - kD\int_0^{\pi/2} d\theta\sin\theta \int_0^{2\pi} d\phi$$
$$\Big[P(\theta,\phi;\pi-\theta_i,\phi_i) + P(\pi-\theta,\phi;\pi-\theta_i,\phi_i)\Big]\Big\} I_{inc}$$
$$= \Big\{1 - \frac{\kappa_e}{\cos\theta_i} D\Big\} I_{inc} \tag{3.5.7}$$

where κ_e (the extinction rate) is, from (3.5.7),

$$\kappa_e = \int_0^{\pi/2} d\theta\sin\theta \int_0^{2\pi} d\phi \Big[P(\theta,\phi;\pi-\theta_i,\phi_i) + P(\pi-\theta,\phi;\pi-\theta_i,\phi_i)\Big] \tag{3.5.8}$$

The transmitted region also contains the incoherent intensity, which is, from (3.4.31) and (3.4.33),

$$-\hat{z}\cdot \mathrm{Im}\langle\psi_f^*(\overline{r})\nabla\psi_f(\overline{r})\rangle\Big|_{z=-D}$$
$$= n_o|f|^2 Dk \int_0^{\pi/2} d\theta\sin\theta \int_0^{2\pi} d\phi$$
$$\Big[1 + (2\pi)^3 n_o H(\overline{p}_\rho = \overline{k}_\rho - \overline{k}_{i\rho}, p_z = -k_z + k_{iz})\Big]$$
$$= Dk \int_0^{\pi/2} d\theta\sin\theta \int_0^{2\pi} d\phi P(\pi-\theta,\phi;\pi-\theta_i,\phi_i) \tag{3.5.9}$$

which consists of transmitted waves in various directions (θ,ϕ) as represented by the integration over $\sin\theta d\theta d\phi$. Thus if we write

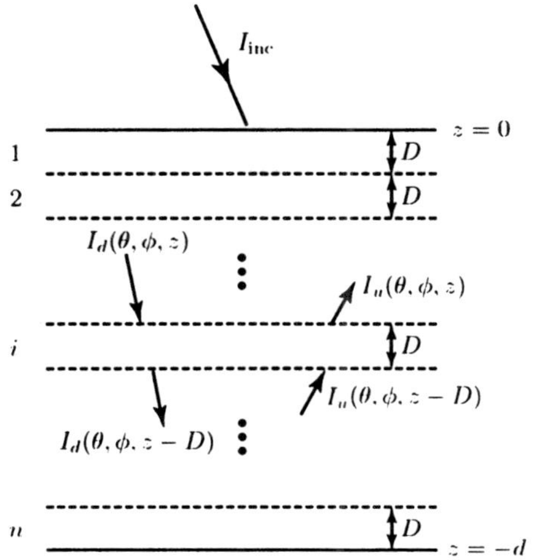

Figure 3.5.2 An illustration of the cascading of many layers of thickness D to form a thick layer of thickness d.

$$-\hat{z} \cdot \mathrm{Im}\langle \psi_f^*(\overline{r}) \nabla \psi_f(\overline{r}) \rangle \Big|_{z=-D}$$
$$= \int_0^{\pi/2} d\theta \sin\theta \int_0^{2\pi} d\phi \left[-\hat{z} \cdot \mathrm{Im}\langle \psi_f^*(\overline{r}) \nabla \psi_f(\overline{r}) \rangle_{(\theta,\phi)} \right] \quad (3.5.10)$$

where (θ, ϕ) represents the incoherent intensity $\langle \psi_f^* \nabla \psi_f \rangle$ in direction (θ, ϕ), then $I_d(\theta, \phi)$, the downward diffuse intensity due to scattering is, from (3.5.10), (3.5.2), and (3.5.9),

$$I_d(\theta, \phi) = \frac{D}{\cos\theta} P(\pi - \theta, \phi; \pi - \theta_i, \phi_i) I_{inc} \quad (3.5.11)$$

The angular variables in the argument of the phase function have the meaning of scattering from the $(\pi - \theta_i, \phi_i)$ direction into the $(\pi - \theta, \phi)$ direction. Similarly, by using (3.4.31) and (3.4.33), the upward diffuse intensity due to scattering of I_{inc} is, using $\hat{z} \cdot \mathrm{Im}\,\langle \psi_f^*(\overline{r}) \nabla \psi_f(\overline{r}) \rangle|_{z=0}$,

$$I_u(\theta, \phi) = \frac{D}{\cos\theta} P(\theta, \phi; \pi - \theta_i, \phi_i) I_{inc} \quad (3.5.12)$$

Next we consider the cascading of many layers with thickness D to form a much thicker layer of thickness d (Fig. 3.5.2). Consider a typical layer which will now have intensities from many directions impinging upon it. A general direction intensity is denoted by $I_u(\theta', \phi')$ and $I_d(\pi - \theta', \phi')$ with $0 \le \theta' \le \pi/2$ and $0 \le \phi' \le 2\pi$.

Let us focus on the direction $(\pi - \theta, \phi)$. The input intensity is $I_d(\theta, \phi)$, which, according to previous results, will generate a transmitted intensity in the same direction and diffuse intensities in other directions. The "output" transmitted intensity in direction $(\pi - \theta, \phi)$ is, from (3.5.7),

$$\left[1 - \frac{\kappa_e}{\cos\theta} D\right] I_d(\theta, \phi) \tag{3.5.13}$$

We examine the diffuse intensity from the scattering of waves from $I_u(\theta', \phi')$ and $I_d(\theta', \phi')$ into the direction $(\pi - \theta, \phi)$. Using (3.5.11) and (3.5.12), these will be

$$\frac{D}{\cos\theta} P(\pi - \theta, \phi; \pi - \theta', \phi') I_d(\theta', \phi') + \frac{D}{\cos\theta} P(\pi - \theta, \phi; \theta', \phi') I_u(\theta', \phi') \tag{3.5.14}$$

Representing (3.5.13) and (3.5.14) by input and output relations and summing contributions from all directions of $I_d(\theta', \phi')$ and $I_u(\theta', \phi')$ to the $(\pi - \theta, \phi)$ direction, we have

Output in direction $(\pi - \theta, \phi)$

$$\begin{aligned} &= \left(1 - \frac{\kappa_e D}{\cos\theta}\right) I_d(\theta, \phi) + \frac{D}{\cos\theta} \left\{ \int_0^{\pi/2} d\theta' \sin\theta' \int_0^{2\pi} d\phi' \right. \\ &\qquad \left. \Big[P(\pi - \theta, \phi; \pi - \theta', \phi') I_d(\theta', \phi') + P(\pi - \theta, \phi; \theta', \phi') I_u(\theta', \phi') \Big] \right\} \end{aligned} \tag{3.5.15}$$

Thus we have the following difference equation for input and output:

$$\begin{aligned} I_d(\theta, \phi, z - D) - I_d(\theta, \phi, z) &= (\text{Output} - \text{Input}) \text{ in direction } (\pi - \theta, \phi) \\ &= \frac{D}{\cos\theta} \left\{ -\kappa_e I_d(\theta, \phi, z) + \int_0^{\pi/2} d\theta' \sin\theta' \int_0^{2\pi} d\phi' \right. \\ &\qquad \Big[P(\pi - \theta, \phi; \pi - \theta', \phi') I_d(\theta', \phi', z) \\ &\qquad \left. + P(\pi - \theta, \phi; \theta', \phi') I_u(\theta', \phi', z) \Big] \right\} \end{aligned} \tag{3.5.16}$$

Based on (3.5.16), one can write a differential equation:

$$\begin{aligned} &- \cos\theta \frac{d}{dz} I_d(\theta, \phi, z) \\ &= -\kappa_e I_d(\theta, \phi, z) + \int_0^{\pi/2} d\theta' \sin\theta' \int_0^{2\pi} d\phi' \Big[P(\pi - \theta, \phi; \pi - \theta', \phi') I_d(\theta', \phi', z) \\ &\quad + P(\pi - \theta, \phi; \theta', \phi') I_u(\theta', \phi', z) \Big] \end{aligned} \tag{3.5.17}$$

By considering $I_u(\theta, \phi, z)$, difference equations and differential equations can be obtained similar to (3.5.16) and (3.5.17):

$$I_u(\theta,\phi, z) - I_u(\theta, \phi, z - D) = \frac{D}{\cos\theta}\Big\{-\kappa_e I_u(\theta, \phi, z) + \int_0^{\pi/2} d\theta' \sin\theta' \int_0^{2\pi} d\phi' \Big[P(\theta, \phi; \theta', \phi')I_u(\theta', \phi', z) + P(\theta, \phi; \pi - \theta', \phi')I_d(\theta', \phi', z)\Big]\Big\} \tag{3.5.18}$$

$$\cos\theta \frac{d}{dz} I_u(\theta, \phi, z) = -\kappa_e I_u(\theta, \phi, z) + \int_0^{\pi/2} d\theta' \sin\theta' \int_0^{2\pi} d\phi' \Big[P(\theta, \phi; \theta', \phi')I_u(\theta', \phi', z) + P(\theta, \phi; \pi - \theta', \phi')I_d(\theta', \phi', z)\Big] \tag{3.5.19}$$

Equations (3.5.16) and (3.5.18) are the difference forms of the radiative transfer equations, while (3.5.17) and (3.5.19) are the differential forms.

Remarks

(1) The differential form of radiative transfer equations in (3.5.17) and (3.5.19) can be misleading since it suggests an equation of intensity changes at a point in space. However, as we see from previous sections, it is the wave equation which governs wave propagation in space. The result of the difference equation in (3.5.16) is obtained by assuming $l \ll D$ and $\lambda < D \ll l_{mfp}$. The conditions $\lambda < D$ and $l \ll D$ are necessary because in such a limit of a layer D, averaging can be taken over random particle positions and over a spatial extent of several to many wavelengths. The condition $D \ll l_{mfp}$ is needed because it means that one can represent input and output relations of a thin layer D by the first- and second-order scattering solutions. Thus (3.5.17) has to be interpreted in the sense that the distance has been calibrated with the distance scale D.

(2) In microwave remote sensing problems, usually $\lambda, l \ll l_{mfp}$, so that it is possible to find a distance scale D such that $\lambda, l < D \ll l_{mfp}$ with the appropriate radiative transfer equation. However, in problems of strong localization, $\lambda \leq l_{mfp}$, the distance scale D such that $\lambda, l < D \ll l_{mfp}$ does not exist.

(3) We are able to derive simple expressions for the phase function because of the isotropic scattering of the point particle models. In problems of complicated particle shapes and correlation of particle positions, it is generally difficult to analytically determine the phase function. In Volume II, Monte Carlo methods have been employed to determine solutions

of Maxwell's equations. However, to use Maxwell's equations to solve a problem of layer thickness $d \gg l_{mfp}$ is a computationally formidable problem. The development in this section justifies the useful alternative that has been adopted in Volume II. Divide the problem into "thin layers" of thickness D with $\lambda, l < D \ll l_{mfp}$. Use Maxwell's equations to calculate the input and output relations of layer of thickness D problem. Then use the cascading of layers approach to find the overall solution to include the multiple scattering for the large thickness d problem. By using D such that $\lambda,\ l,\ < D \ll l_{mfp}$, we have carried out extensive Monte Carlo simulation in Volume II to calculate the extinction coefficient and the phase matrix.

(4) Based on (3.5.16), we can interpret the phase function as the *averaged* bistatic coefficients of a thin layer.

6 Effects of Clustering

In Section 3 we have seen the importance of the pair distribution function in calculating the lowest-order scattered intensity because the field and field conjugates originate from different scatterers. In some problems the scatterers occur in clusters. For example, in a vegetation canopy the branches and leaves can occur in clusters, so that they are locally densely distributed and not uniformly dense. In such a case the pair distribution function depends on whether the two scatterers are in the same cluster or in different clusters. In Chapter 7 of Volume II, we have performed Monte Carlo simulations for collections of clusters of scatterers. In the following, we consider the scattering of a layer of thickness D of point scatterers when the scatterers can be in clusters. We assume that the particles may occur in clusters with N_s particles in each cluster. Note that N_s may not be a large number. For the case of nonclustering, $N_s = 1$. The number of clusters in the layer is N_c with

$$N_c = \frac{N}{N_s} \tag{3.6.1a}$$

and $N, N_c \rightarrow \infty$. The center of each cluster is at $\overline{r}_\alpha$, $\alpha = 1, 2, \ldots, N_c$ (Fig. 3.6.1). The N_s scatterers in each cluster α are labeled as secondary scatterers, with indices j_α, $j_\alpha = 1, \ldots, N_s$. The location of the j_α scatterer is at

$$\overline{r}_\alpha + \overline{r}_{j_\alpha} \tag{3.6.1b}$$

with $j_\alpha = 1, \ldots N_s$. The volume occupied by the clusters is V_c, where V_c does not have to be large compared with wavelength. Thus the single-particle

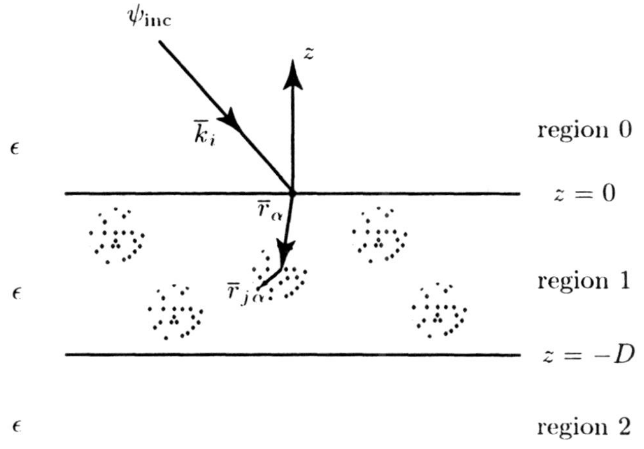

Figure 3.6.1 An illustration of wave imping on a slab of point particles in clusters.

probability density functions for primary and secondary scatterers are

$$p(\overline{r}_\alpha) = \frac{1}{V} \qquad \text{for } \overline{r}_\alpha \text{ in } V \tag{3.6.2}$$

$$p(\overline{r}_{j_\alpha}) = \frac{1}{V_c} \qquad \text{for } \overline{r}_{j_\alpha} \text{ in } V_c \tag{3.6.3}$$

For the sake of simplicity, we have assumed that the primary cluster location and the secondary scatterer locations are independent. In taking averaging, we note that N, N_c and $V \to \infty$, while N_s and V_c, the number of particles in each cluster and the volume occupied by each cluster are finite. The layer D is such that $\lambda, l \ll D \ll l_{mfp}$. The cluster size V_c is related also to l and the layer thickness D must contain many clusters so that $D \gg V_c^{1/3}$. Furthermore, we let

$$n_c = \frac{N_c}{V} \tag{3.6.4}$$

$$n_o = \frac{N}{V} \tag{3.6.5}$$

to denote number of clusters per unit volume and number of particles per unit volume, respectively. The joint probability density function of two different clusters is

$$p(\overline{r}_\alpha, \overline{r}_\beta) = \frac{g_p(\overline{r}_\alpha, \overline{r}_\beta)}{V^2} \frac{N_c}{N_c - 1} \tag{3.6.6}$$

where the subscript p denotes primary scatterers (clusters). The joint probability density function of two different scatterers in the same cluster is

$$p(\overline{r}_{j_\alpha}, \overline{r}_{l_\alpha}) = \frac{g_s(\overline{r}_{j_\alpha}, \overline{r}_{l_\alpha})}{V_c^2} \frac{N_s}{N_s - 1} \tag{3.6.7}$$

where the subscript s stands for secondary scatterers. We have assumed that the pair distribution function for primary scatterers is translational invariant while the translational invariant property may not apply to g_s because of the much smaller cluster volume V_c. We also assume that the joint probability density functions of two secondary scatterers in two different clusters are independent.

The single scattering is given by

$$\begin{aligned}\psi^{s(1)}(\overline{r}) &= \sum_{l=1}^{N} \psi_l^{s(1)}(\overline{r}) = \sum_{\alpha=1}^{N_c} \sum_{l_\alpha=1}^{N_s} \psi_{l_\alpha}^{s(1)}(\overline{r}) \\ &= \sum_{\alpha=1}^{N_c} \sum_{l_\alpha=1}^{N_s} f \frac{e^{ik|\overline{r} - \overline{r}_\alpha - \overline{r}_{l_\alpha}|}}{|\overline{r} - \overline{r}_\alpha - \overline{r}_{l_\alpha}|} e^{i\overline{k}_{id} \cdot (\overline{r}_\alpha + \overline{r}_{l_\alpha})} \end{aligned} \tag{3.6.8}$$

$$\langle \psi^{s(1)}(\overline{r}) \rangle = N_c N_s \int_V d\overline{r}_\alpha \frac{1}{V} \int_{V_c} \frac{d\overline{r}_{l_\alpha}}{V_c} \psi_{l_\alpha}^{s(1)}(\overline{r}) \tag{3.6.9}$$

For the lth scatterer, it is the l_α scatterer of the αth cluster so that its location is at $\overline{r}_l = \overline{r}_\alpha + \overline{r}_{l_\alpha}$. The probability density functions for $\overline{r}_l$ and $\overline{r}_\alpha$ are uniform in V while probability density function for $\overline{r}_{l_\alpha}$ is uniform over a smaller cluster volume V_c. In (3.6.9) the integrand $\psi_{l_\alpha}^{s(1)}(\overline{r})$ depends only on $\overline{r}_\alpha + \overline{r}_{l_\alpha}$. Thus with the transformation $\overline{r}_\alpha + \overline{r}_{l_\alpha} = \overline{r}_l$ we have

$$\langle \psi^{s(1)}(\overline{r}) \rangle = \frac{N}{V} \int_V d\overline{r}_l \psi_l^{s(1)}(\overline{r}) \int_{V_c} \frac{d\overline{r}_{l\alpha}}{V_c} = n_o \int_V d\overline{r}_l \psi_l^{s(1)}(\overline{r}) \tag{3.6.10}$$

which is the same result as the nonclustered case of Section 1. The coherent scattered field, as indicated in Section 1, is only appreciable in the transmitted region 2 and is, for z in region 2,

$$\langle \psi^{s(1)}(\overline{r}) \rangle = \frac{n_o 2\pi i f D}{k_{iz}} \exp(i\overline{k}_{i\rho} \cdot \overline{\rho} - ik_{iz} z) \tag{3.6.11}$$

To calculate intensity, we use the same approach as in Section 4. The double summation of scatterers over l and j is decomposed as follows:

$$\sum_{l=1}^{N} \sum_{j=1}^{N} = \sum_{\alpha=1}^{N_c} \sum_{j_\alpha=1}^{N_s} + \sum_{\alpha=1}^{N_c} \sum_{\substack{j_\alpha=1 \\ j_\alpha \neq l_\alpha}}^{N_s} \sum_{l_\alpha=1}^{N_s} + \sum_{\alpha=1}^{N_c} \sum_{j_\alpha=1}^{N_s} \sum_{\substack{\beta=1 \\ \beta \neq \alpha}}^{N_c} \sum_{l_\beta=1}^{N_s} \tag{3.6.12}$$

Thus the first term in (3.6.12) is from the same scatterer, the second term is from different scatterers in the same cluster, and the last term is from two scatterers in different clusters. Averaging of the first-order scattered intensity is then

$$\langle\psi^{s(1)}(\overline{r})\psi^{s(1)*}(\overline{r}')\rangle = N_c N_s \int_V d\overline{r}_\alpha \frac{1}{V} \int_{V_c} d\overline{r}_{j_\alpha} \frac{1}{V_c} \psi_{j_\alpha}^{s(1)}(\overline{r})\psi_{j_\alpha}^{s(1)*}(\overline{r}')$$
$$+ N_c N_s^2 \int_V d\overline{r}_\alpha \frac{1}{V} \int_{V_c} d\overline{r}_{j_\alpha} \int_{V_c} d\overline{r}_{l_\alpha} \psi_{j_\alpha}^{s(1)}(\overline{r})\psi_{l_\alpha}^{s(1)*}(\overline{r}') \frac{g_s(\overline{r}_{j_\alpha}, \overline{r}_{l_\alpha})}{V_c^2}$$
$$+ N_c^2 N_s^2 \int_V d\overline{r}_\alpha \int_V d\overline{r}_\beta \int_{V_c} d\overline{r}_{j_\alpha} \int_{V_c} d\overline{r}_{l_\beta} \psi_{j_\alpha}^{s(1)}(\overline{r})\psi_{l_\beta}^{s(1)*}(\overline{r}') \frac{g_p(\overline{r}_\alpha, \overline{r}_\beta)}{V^2} \frac{1}{V_c^2} \tag{3.6.13}$$

For the first term in (3.6.13), since the integrand depends only on $\overline{r}_j = \overline{r}_\alpha + \overline{r}_{j_\alpha}$, transformation can be made to the $\overline{r}_j$-coordinate as done between (3.6.9) and (3.6.10). Integration can then be carried out for $\overline{r}_{j_\alpha}$. We also rewrite g_p as $(g_p - 1) + 1$ since g_p asymptotically approaches unity. On comparing with (3.6.9), it follows that the unity term gives $\langle\psi^{s(1)}(\overline{r})\rangle\langle\psi^{s(1)*}(\overline{r}')\rangle$:

$$\langle\psi^{s(1)}(\overline{r})\psi^{s(1)*}(\overline{r}')\rangle = n_o \int d\overline{r}_j \psi_j^{s(1)}(\overline{r})\psi_j^{s(1)*}(\overline{r}')$$
$$+ n_c N_s^2 \int_V d\overline{r}_\alpha \int_{V_c} d\overline{r}_{j_\alpha} \int_{V_c} d\overline{r}_{l_\alpha} \psi_{j_\alpha}^{s(1)}(\overline{r})\psi_{l_\alpha}^{s(1)*}(\overline{r}') \frac{g_s(\overline{r}_{j_\alpha}, \overline{r}_{l_\alpha})}{V_c^2}$$
$$+ n_c^2 N_s^2 \int_V d\overline{r}_\alpha \int_V d\overline{r}_\beta \int_{V_c} d\overline{r}_{j_\alpha} \int_{V_c} d\overline{r}_{l_\beta} \psi_{j_\alpha}^{s(1)}(\overline{r})\psi_{l_\beta}^{s(1)*}(\overline{r}') \frac{[g_p(\overline{r}_\alpha - \overline{r}_\beta) - 1]}{V_c^2}$$
$$+ \langle\psi^{s(1)}(\overline{r})\rangle\langle\psi^{s(1)*}(\overline{r}')\rangle \tag{3.6.14}$$

To facilitate the calculation of (3.6.14), we can define scattering amplitude for a cluster of N_s scatterers. For real $\overline{k}_\rho$ and k_z, let

$$f^{(P)}(\overline{k}_\rho, k_z) = \sum_{j_\alpha=1}^{N_s} f e^{-i\overline{k}_\rho\cdot\overline{\rho}_{j_\alpha} - ik_z z_{j_\alpha}} \tag{3.6.15}$$

where the superscript (P) denotes a primary scatterer. Then the first moment of $f^{(P)}$ is

$$\langle f^{(P)}(\overline{k}_\rho, k_z)\rangle = N_s f \int_{V_c} \frac{d\overline{r}_{j\alpha}}{V_c} e^{-i\overline{k}_\rho\cdot\overline{\rho}_{j_\alpha} - ik_z z_{j_\alpha}} \tag{3.6.16}$$

If we consider the integration of the first-order scattered field over the cluster volume, then we obtain

$$\begin{aligned} N_s \int_{V_c} \frac{d\overline{r}_{j_\alpha}}{V_c} \psi_{j_\alpha}^{s(1)}(\overline{r}) &= N_s \int_{V_c} \frac{d\overline{r}_{j_\alpha}}{V_c} f \frac{e^{ik|\overline{r}-\overline{r}_\alpha-\overline{r}_{j_\alpha}|}}{|\overline{r}-\overline{r}_\alpha-\overline{r}_{j_\alpha}|} e^{i\overline{k}_{id}\cdot(\overline{r}_\alpha+\overline{r}_{j_\alpha})} \\ &= \frac{if}{2\pi} N_s \int d\overline{k}_\rho \int \frac{d\overline{r}_{j_\alpha}}{V_c} \frac{e^{i\overline{k}_\rho\cdot(\overline{\rho}-\overline{\rho}_\alpha-\overline{\rho}_{j_\alpha})}}{k_z} \\ &\quad \cdot e^{i\overline{k}_{i\rho}\cdot(\overline{\rho}_\alpha+\overline{\rho}_{j_\alpha})} e^{-ik_{iz}(z_\alpha+z_{j_\alpha})\pm ik_z(z-z_\alpha-z_{j_\alpha})} \\ &= \frac{i}{2\pi} \int \frac{d\overline{k}_\rho}{k_z} e^{i\overline{k}_\rho\cdot(\overline{\rho}-\overline{\rho}_\alpha)} e^{i\overline{k}_{i\rho}\cdot\overline{\rho}_\alpha} e^{-ik_{iz}z_\alpha} \\ &\quad \cdot e^{\pm ik_z(z-z_\alpha)} f^{(P)}(\overline{k}_\rho - \overline{k}_{i\rho}, \pm k_z + k_{iz}) \end{aligned} \tag{3.6.17}$$

If we calculate the second moment of $f^{(P)}$, then

$$|f^{(P)}(\overline{k}_\rho, k_z)|^2 = \sum_{j_\alpha=1}^{N_s} |f|^2 + |f|^2 \sum_{\substack{j_\alpha=1 \\ j_\alpha\neq l_\alpha}}^{N_s} \sum_{l_\alpha=1}^{N_s} e^{-i\overline{k}_\rho\cdot(\overline{\rho}_{j_\alpha}-\overline{\rho}_{l_\alpha})-ik_z(z_{j_\alpha}-z_{l_\alpha})} \tag{3.6.18}$$

Thus

$$\begin{aligned} \langle |f^{(P)}(\overline{k}_\rho, k_z)|^2 \rangle &= N_s|f|^2 + N_s^2|f|^2 \int_{V_c} d\overline{r}_{j_\alpha} \int_{V_c} d\overline{r}_{l_\alpha} \frac{g_s(\overline{r}_{j_\alpha}, \overline{r}_{l_\alpha})}{V_c^2} \\ &\quad \times e^{-i\overline{k}_\rho\cdot(\overline{\rho}_{j_\alpha}-\overline{\rho}_{l_\alpha})-ik_z(z_{j_\alpha}-z_{l_\alpha})} \end{aligned} \tag{3.6.19}$$

$$\begin{aligned} &= N_s|f|^2 + N_s^2|f|^2 \int_{V_c} d\overline{r}_{j_\alpha} \int_{V_c} d\overline{r}_{l_\alpha} \frac{g_s(\overline{r}_{j_\alpha}, \overline{r}_{l_\alpha})}{V_c^2} \\ &\quad \times \cos\Big(\overline{k}_\rho \cdot (\overline{\rho}_{j_\alpha} - \overline{\rho}_{l_\alpha}) + k_z(z_{j_\alpha} - z_{l_\alpha})\Big) \end{aligned} \tag{3.6.20}$$

This is to be distinguished from the product of the averages

$$|\langle f^{(P)}(k_\rho, k_z)\rangle|^2 = N_s^2|f|^2 \int \frac{d\overline{r}_{j_\alpha}}{V_c} \int \frac{d\overline{r}_{l_\alpha}}{V_c} \cos\Big(\overline{k}_\rho \cdot (\overline{\rho}_{j_\alpha} - \overline{\rho}_{l_\alpha}) + k_z(z_{j_\alpha} - z_{l_\alpha})\Big) \tag{3.6.21}$$

The equality of (3.6.20) is a result of assuming that g_s is symmetric, that is, $g_s(\overline{r}_{j_\alpha}, \overline{r}_{l_\alpha}) = g_s(\overline{r}_{l_\alpha}, \overline{r}_{j_\alpha})$. For the special case that $g_s = 1$, then

$$\langle |f^{(P)}(\overline{k}_\rho, k_z)|^2 \rangle = N_s|f|^2 + |\langle f^{(P)}(\overline{k}_\rho, k_z)\rangle|^2 \tag{3.6.22}$$

On comparing $(1/V_c)\int_{V_c} d\overline{r}_{j_\alpha} \psi_{j_\alpha}^{s(1)}(\overline{r})$ with $\psi_j^{(1)}(\overline{r})$ of (3.1.6), it follows that the integration of a scattered field over clustering volume gives an expression of a scattered field that is centered at the cluster center $\overline{r}_\alpha$ and with the point scatterer amplitude f replaced by a directional-dependent scattering amplitude of the primary scatterer $f^{(P)}$. Then the second to the last term of (3.6.14) can be calculated in a manner similar to (3.4.27)–(3.4.31)—for example, letting $\overline{\rho}_\alpha - \overline{\rho}_\beta = \overline{\rho}_d$, $(\overline{\rho}_\alpha + \overline{\rho}_\beta)/2 = \overline{\rho}_A$, and so on. Thus

$$n_c^2 N_s^2 \int_V d\overline{r}_\alpha \int_V d\overline{r}_\beta \int_{V_c} \frac{d\overline{r}_{j_\alpha}}{V_c} \int_{V_c} \frac{d\overline{r}_{l_\beta}}{V_c} \psi_{j_\alpha}^{(s)(1)}(\overline{r}) \psi_{l_\beta}^{s(1)*}(\overline{r}') \Big[g_p(\overline{r}_\alpha - \overline{r}_\beta) - 1 \Big]$$

$$= (2\pi)^3 n_c^2 D \int_0^{\pi/2} d\theta \frac{\sin\theta}{\cos\theta} \int_0^{2\pi} d\phi \, e^{i\overline{k}_\rho \cdot (\overline{\rho} - \overline{\rho}')} e^{\pm ik\cos\theta(z-z')}$$

$$\times H_p(\overline{p}_\rho = \overline{k}_\rho - \overline{k}_{i\rho}, p_z = \pm k_z + k_{iz}) \Big| \langle f^{(P)}(\overline{k}_\rho - \overline{k}_{i\rho}, \pm k_z + k_{iz}) \rangle \Big|^2 \quad (3.6.23)$$

where

$$h_p(\overline{r}) = g_p(\overline{r}) - 1 \quad (3.6.24)$$

$$H_p(\overline{p}) = \frac{1}{(2\pi)^3} \int_{-\infty}^{\infty} d\overline{r} \, h_p(\overline{r}) \, e^{-i\overline{p}\cdot\overline{r}} \quad (3.6.25)$$

are now the corresponding quantities for primary scatterers analogous to (3.4.28) and (3.4.29). The quantity $H_p(\overline{p})$ is the structure factor that relates different primary scatterers.

The sum of the first two terms of (3.6.14) gives, in a manner similar to (3.4.18),

$$n_o \int d\overline{r}_j \psi_j^{s(1)}(\overline{r}) \, \psi_j^{s(1)*}(\overline{r}')$$

$$+ n_c N_s^2 \int_V d\overline{r}_\alpha \int_{V_c} d\overline{r}_{j_\alpha} \int_{V_c} d\overline{r}_{l_\alpha} \psi_{j_\alpha}^{s(1)}(\overline{r}) \psi_{l_\alpha}^{s(1)*}(\overline{r}') \frac{g_s(\overline{r}_{j_\alpha}, \overline{r}_{l_\alpha})}{V_c^2}$$

$$= D \int d\overline{k}_\rho \frac{e^{i\overline{k}_\rho \cdot (\overline{\rho} - \overline{\rho}') \pm ik_z(z-z')}}{|k_z|^2} \Big\{ n_o |f|^2 + n_c N_s^2 |f|^2 \times$$

$$\int \frac{d\overline{r}_{j_\alpha}}{V_c} \int \frac{d\overline{r}_{l_\alpha}}{V_c} g_s(\overline{r}_{j_\alpha}, \overline{r}_{l_\alpha}) e^{-i(\overline{k}_\rho - \overline{k}_{i\rho}) \cdot (\overline{\rho}_{j_\alpha} - \overline{\rho}_{l_\alpha})} e^{-i(\pm k_z + k_{iz})(z_{j_\alpha} - z_{l_\alpha})} \Big\}$$

$$= D n_c \int_0^{\pi/2} d\theta \frac{\sin\theta}{\cos\theta} \int_0^{2\pi} d\phi \, e^{i\overline{k}_\rho \cdot (\overline{\rho} - \overline{\rho}') \pm ik\cos\theta(z-z')}$$

$$\cdot \langle |f^{(P)}(\overline{k}_\rho - \overline{k}_{i\rho}, \pm k_z + k_{iz})|^2 \rangle \quad (3.6.26)$$

The last equality in (3.6.26) is a result of using (3.6.19) and (3.4.23).

The sum of (3.6.23) and (3.6.26) gives the intensity of the incoherent first-order scattered fields $\langle \psi_f^{(1)}(\overline{r}) \psi_f^{(1)*}(\overline{r}') \rangle$. We note from (3.6.24) and (3.6.25) that $H_p(\overline{p})$ is usually negative because $g_p(\overline{r}) = 0$ for $|\overline{r}| < d_{min}$ so that the correlation of scattering between different clusters decreases incoherent scattering. The decrease of incoherent scattering is associated with an increase in the coherent forward scattering which, as we saw earlier, does not contribute to extinction. Thus the scattering effect of g_p is to decrease extinction. From (3.6.16), we note that g_s is positive so that $\langle|f^{(p)}(\overline{k}_\rho, k_z)|^2\rangle$ is usually larger than $N_s|f|^2$. This means that clustering due to g_s generally increases the scattering cross section above that of independent scattering. Thus g_p and g_s have opposite effects on extinction. The overall scattering of a random collection of scatterers depends on the interplay between the two pair functions g_p and g_s, namely, the pair function between different clusters and the pair function within a cluster. Without the second-order coherent field, the outgoing power from the slab is analogous to (3.4.33). From (3.6.11), (3.6.14), (3.6.23), and (3.6.26) we have

$$\begin{aligned}
&\oint ds\hat{n}\cdot \text{Im} \lim_{\overline{r}'\to\overline{r}} \nabla \Big[\psi_{inc}(\overline{r})\psi^*_{inc}(\overline{r}') + \psi_{inc}(\overline{r})\langle\psi^{s(1)*}(\overline{r}')\rangle + \langle\psi^{s(1)}(\overline{r})\rangle\psi^*_{inc}(\overline{r}') \\
&\qquad + \langle\psi^{s(1)}(\overline{r})\psi^{s(1)*}(\overline{r}')\rangle\Big]_{\substack{\hat{n}=-\hat{z} \text{ at } z=-0\\ \hat{n}=\hat{z} \text{ at } z=0}} \\
&= A\Big\{-\text{Im}(f)4\pi n_o D + |f|^2\frac{(2\pi n_o D)^2}{k_{iz}} \\
&\quad + Dn_c k\int_0^{\pi/2} d\theta\sin\theta\int_0^{2\pi} d\phi\langle|f^{(P)}(\overline{k}_\rho - \overline{k}_{i\rho}, -k_z + k_{iz})|^2\rangle \\
&\quad + Dkn_c^2(2\pi)^3\int_0^{\pi/2} d\theta\sin\theta\int_0^{2\pi} d\phi H_p(\overline{k}_\rho - \overline{k}_{i\rho}, -k_z + k_{iz}) \\
&\quad \cdot|\langle f^{(P)}(\overline{k}_\rho - \overline{k}_{i\rho}, -k_z + k_{iz})\rangle|^2 \\
&\quad + Dn_c k\int_0^{\pi/2} d\theta\sin\theta\int_0^{2\pi} d\phi\langle|f^{(P)}(\overline{k}_\rho - \overline{k}_{i\rho}, k_z + k_{iz})|^2\rangle \\
&\quad + Dkn_c^2(2\pi)^3\int_0^{\pi/2} d\theta\sin\theta\int_0^{2\pi} d\phi H_p(\overline{k}_\rho - \overline{k}_{i\rho}, k_z + k_{iz}) \\
&\quad \cdot|\langle f^{(P)}(\overline{k}_\rho - \overline{k}_{i\rho}, k_z + k_{iz})\rangle|^2\Big\}
\end{aligned} \tag{3.6.27}$$

We can see that, in (3.6.27), the second term is from $\langle\psi^{s(1)}\rangle\langle\psi^{s(1)*}\rangle$, the third and fourth terms are the outgoing incoherent intensities from the boundary, and the fifth and sixth terms are the outgoing incoherent intensities from the top boundary.

The second-order coherent scattered field can be calculated as follows. We use (3.4.19). The double summation of $\sum_l \sum_{j\neq l}$ is decomposed into the sum of scattering from two different scatterers in the same cluster and from two scatterers in two different clusters. Thus taking the average of (3.4.19) gives

$$\begin{aligned}\langle\psi^{s(2)}(\overline{r})\rangle &= N_c N_s(N_s-1)\int\frac{d\overline{r}_\alpha}{V}\int d\overline{r}_{j_\alpha}\int d\overline{r}_{l_\alpha}\frac{g_s(\overline{r}_{j_\alpha},\overline{r}_{l_\alpha})}{V_c^2}\\
&\times f^2\frac{e^{ik|\overline{r}-\overline{r}_\alpha-\overline{r}_{l\alpha}|}}{|\overline{r}-\overline{r}_\alpha-\overline{r}_{l\alpha}|}\frac{e^{ik|\overline{r}_{l\alpha}-\overline{r}_{j\alpha}|}}{\overline{r}_{l\alpha}-\overline{r}_{j\alpha}|}e^{i\overline{k}_{i\rho}\cdot(\overline{\rho}_\alpha+\overline{\rho}_{j\alpha})-ik_{iz}(z_\alpha+z_{j\alpha})}\\
&+N_c(N_c-1)N_s^2\int d\overline{r}_\alpha\int d\overline{r}_\beta\frac{g_p(\overline{r}_\alpha,\overline{r}_\beta)}{V^2}\\
&\times\int\frac{d\overline{r}_{l\alpha}}{V_c}\int\frac{d\overline{r}_{j\beta}}{V_c}f^2\frac{e^{ik|\overline{r}-\overline{r}_\alpha-\overline{r}_{l\alpha}|}}{|\overline{r}-\overline{r}_\alpha-\overline{r}_{l\alpha}|}\frac{e^{ik|\overline{r}_\alpha+\overline{r}_{l\alpha}-\overline{r}_\beta-\overline{r}_{j\beta}|}}{|\overline{r}_\alpha+\overline{r}_{l\alpha}-\overline{r}_\beta-\overline{r}_{j\beta}|}\\
&\times e^{i\overline{k}_{i\rho}\cdot(\overline{\rho}_\beta+\overline{\rho}_{j\beta})-ik_{iz}(z_\beta+z_{j\beta})}\end{aligned}\tag{3.6.28}$$

By using the plane wave representation of Green's functions in (3.6.28), one can show that the coherent scattered field is only appreciable in region 2. By considering the transmitted intensity associated with the second-order coherent field, one gets

$$\begin{aligned}&-\hat{z}\cdot\lim_{\overline{r}'\to\overline{r}}\text{Im}\Big[\nabla\psi_{inc}(\overline{r})\langle\psi^{s(2)}(\overline{r}')\rangle+\nabla\langle\psi^{s(2)}(\overline{r})\rangle\psi^*_{inc}(\overline{r}')\Big]\\
&=-n_c kD\int_0^{\pi/2}d\theta\sin\theta\int_0^{2\pi}d\phi\Big\{|f^{(P)}(\overline{k}_\rho-\overline{k}_{i\rho},k_z+k_{iz})|^2\\
&\quad-2N_s|f|^2+|f^{(P)}(\overline{k}_\rho-\overline{k}_{i\rho},-k_z+k_{iz})|^2\Big\}\\
&\quad-n_c^2kD\int_0^{\pi/2}d\theta\sin\theta\int_0^{2\pi}d\phi(2\pi)^3\Big\{H(\overline{k}_\rho-\overline{k}_{i\rho},k_z+k_{iz})\\
&\quad\cdot|\langle f^{(P)}(\overline{k}_\rho-\overline{k}_{i\rho},k_z+k_{iz})\rangle|^2\\
&\quad+H(\overline{k}_\rho-\overline{k}_{i\rho},-k_z+k_{iz})|\langle f^{(P)}(\overline{k}_\rho-\overline{k}_{i\rho},-k_z+k_{iz})\rangle|^2\Big\}\\
&\quad-|f|^2\frac{(2\pi n_o D)^2}{k_{iz}}\end{aligned}\tag{3.6.29}$$

By combining (3.6.27) and (3.6.29), energy conservation is exactly obeyed.

For thick-layer problems with layer thickness d, one can divide the problem into many layers of thickness D and perform cascading as done in Section 5. A difference form of radiative transfer equation can be obtained as in (3.5.16), with extinction coefficients and phase matrices.

The extinction coefficients can be obtained from the transmitted intensity of the coherent wave:

$$\kappa_e = \frac{1}{(-kD\,|\psi_{inc}(\overline{r})|^2)}\Big\{-\hat{z}\cdot\lim_{\overline{r}'\to\overline{r}}\mathrm{Im}\Big[\psi_{inc}(\overline{r})\langle\psi^{s(1)*}(\overline{r}')\rangle + \langle\psi^{s(1)}(\overline{r})\rangle\psi^*_{inc}(\overline{r}')$$
$$+ \langle\psi^{s(1)}(\overline{r})\rangle\langle\psi^{s(1)*}(\overline{r}')\rangle + \nabla\psi_{inc}(\overline{r})\langle\psi^{s(2)}(\overline{r}')\rangle + \nabla\langle\psi^{s(2)}(\overline{r})\rangle\psi^*_{inc}(\overline{r}')\Big]\Big\} \tag{3.6.30}$$

Note that $\langle\psi^{s(1)}\rangle\langle\psi^{s(1)*}\rangle$ is used in (3.6.30), which contains the coherent component.

Substituting (3.6.27), (3.6.28), and (3.6.29) into (3.6.30) gives the following equations for this case of clustered point scatterers with two pair distribution functions g_s and g_p

$$\kappa_e = \frac{4\pi n_o \mathrm{Im}(f)}{k} - 4\pi n_o|f|^2 + n_c\int_0^{\pi/2} d\theta\sin\theta\int_0^{2\pi} d\phi$$
$$\times\Big\{\langle|f^{(P)}(\overline{k}_\rho - \overline{k}_{i\rho}, k_z + k_{iz})|^2\rangle + \langle|f^{(P)}(\overline{k}_\rho - \overline{k}_{i\rho}, -k_z + k_{iz})|^2\rangle\Big\}$$
$$+ n_c^2\int_0^{\pi/2} d\theta\sin\theta\int_0^{2\pi} d\phi(2\pi)^3\Big\{H(\overline{k}_\rho - \overline{k}_{i\rho}, k_z + k_{iz})$$
$$\cdot|\langle f^{(P)}(\overline{k}_\rho - \overline{k}_{i\rho}, k_z + k_{iz})\rangle|^2$$
$$+ H(\overline{k}_\rho - \overline{k}_{i\rho}, -k_z + k_{iz})|\langle f^{(P)}(\overline{k}_\rho - \overline{k}_{i\rho}, -k_z + k_{iz})\rangle|^2\Big\} \tag{3.6.31a}$$

The first two terms in (3.6.31) cancel by invoking the optical theorem. Hence

$$\kappa_e = n_c\int_0^{\pi/2} d\theta\sin\theta\int_0^{2\pi} d\phi\Big\{\langle|f^{(P)}(\overline{k}_\rho - \overline{k}_{i\rho}, k_z + k_{iz})|^2\rangle$$
$$+ \langle|f^{(P)}(\overline{k}_\rho - \overline{k}_{i\rho}, -k_z + k_{iz})|^2\rangle\Big\}$$
$$+ n_c^2\int_0^{\pi/2} d\theta\sin\theta\int_0^{2\pi} d\phi(2\pi)^3\Big\{H(\overline{k}_\rho - \overline{k}_{i\rho}, k_z + k_{iz})$$
$$\cdot|\langle f^{(P)}(\overline{k}_\rho - \overline{k}_{i\rho}, k_z + k_{iz})\rangle|^2$$
$$+ H(\overline{k}_\rho - \overline{k}_{i\rho}, -k_z + k_{iz})|\langle f^{(P)}(\overline{k}_\rho - \overline{k}_{i\rho}, -k_z + k_{iz})\rangle|^2\Big\} \tag{3.6.31b}$$

The phase function is obtained from the angular distribution of the incoherent intensity term in (3.6.28), $\langle\psi_f(\overline{r})\psi_f^*(\overline{r}')\rangle$, in a manner similar to that performed in Section 5. Thus the phase function for this case of clustering point scatterers is

$$P(\theta,\phi;\pi-\theta_i,\phi_i) = n_c\langle|f^{(P)}(\overline{k}_\rho - \overline{k}_{i\rho}, k_z + k_{iz})|^2\rangle$$
$$+ n_c^2(2\pi)^3 H_p(\overline{k}_\rho - \overline{k}_{i\rho}, k_z + k_{iz})\left|\langle f^{(P)}(\overline{k}_\rho - \overline{k}_{i\rho}, k_z + k_{iz})\rangle\right|^2 \tag{3.6.32}$$

This is quite similar to (3.5.6), of the nonclustering case with $|f|^2$ replaced by $\langle|f^{(P)}|^2\rangle$ and $|\langle f^{(P)}\rangle|^2$, respectively, and n_o replaced by n_c. When we use the phase function as defined in (3.6.32), the extinction coefficient of (3.6.31) can be written as

$$\kappa_e = \int_0^{\pi/2} d\theta \sin\theta \int_0^{2\pi} d\phi \Big[P(\theta,\phi;\pi-\theta_i,\phi_i) + P(\pi-\theta,\phi;\pi-\theta_i,\phi_i) \Big] \quad (3.6.33)$$

Note that depending on the two pair functions g_s and g_p, the extinction coefficient κ_e can be directionally dependent on (θ_i, ϕ_i).

The results of (3.6.32) and (3.6.33) have the following physical interpretation. If we view the random media as consisting of primary scatterers, then the scattering is a result of correlated scattering of primary scatterers similar to dense media theory. The result of scattering can be smaller than the independent scattering of primary scatterers, an effect due to g_p. However, each primary scatterer consists of a cluster of small scatterers. Thus the scattering by a primary scatterer can be larger than the sum of the scattering cross sections of the small scatterers in the cluster, an effect due to g_s.

REFERENCES AND ADDITIONAL READINGS

Chandrasekhar, S. (1960), *Radiative Transfer*, Dover, New York.

Fante, R. I. (1981), Relation between radiative transport theory and Maxwell's equations in dielectric media, *J. Opt. Soc. Am.*, **71**, 460–468.

Karam, M. and A. K. Fung (1988), Electromagnetic scattering from a layer of finite length, randomly oriented dielectric cylinders over a rough interface with application to vegetation, *Int. J. Remote Sens.*, **9**, 1109–1134.

Lang, R. H. (1981), Electromagnetic backscattering from a sparse distribution of lossy dielectric scatterers, *Radio Sci.*, **16**, 15–30.

Lang, R. H. and J. S. Sidhu (1983), Electromagnetic backscattering from a layer of vegetation: a discrete approach, *IEEE Trans. Geosci. Remote Sens.*, **21**, 62–71.

Sheng, P., Ed. (1990), *Scattering and Localization of Classical Waves in Random Media*, World Scientific, Singapore.

Tsang, L. and K. H. Ding (1991), Polarimetric signatures of a layer of random nonspherical discrete scatterers overlying a homogeneous half space based on the first and second order vector radiative transfer theory, *IEEE Trans. Geosci. Remote Sens.*, **29**, 242–253.

Tsang, L. and J. A. Kong (1977), Thermal microwave emission from a random homogeneous layer over a homogeneous medium using the method of invariant imbedding, *Radio Sci.*, **12**, 185–195.

Tsang, L. and J. A. Kong (1982), Effective propagation constants for coherent electromagnetic wave propagation in media embedded with dielectric scatterers, *J. Appl. Phys.*, **53**, 7162–7173.

Tsang, L., J. A. Kong, Z. Chen, K. Pak, and C. Hsu (1995), Theory of microwave scattering from vegetation based on the collective scattering effects of discrete scatterers, *Passive Microwave Remote Sensing of Land-Atmosphere Interaction*, B. J. Choudhury, Y. H. Kerr, E. G. Njoku, and P. Pampaloni, Eds., 117–154.

Tsang, L., J. A. Kong, and R. T. Shin (1985), *Theory of Microwave Remote Sensing*, Wiley-Interscience, New York.

Twersky, V. (1977), Coherent scalar field in pair-correlated random distributions of aligned scatterers, *J. Math. Phys.*, **18**, 2468–2486.

Yueh, S. H., J. A. Kong, J. K. Jao, R. T. Shin, and T. Le Toan (1992), Branching model for vegetation, *IEEE Trans. Geosci. Remote Sens.*, **30**, 390–402.

Chapter 4

ANALYTIC WAVE THEORY FOR A MEDIUM WITH PERMITTIVITY FLUCTUATIONS

In this chapter we will use the diagrammatic approach to study multiple scattering in random media. The diagrammatic approach leads to Dyson's equation for the mean field and the Bethe-Salpeter equation for the covariance of the field. Both Dyson's equation and Bethe-Salpeter equation are exact. Approximations are made on the Dyson's equation. The bilocal and nonlinear approximations will be examined. For the Bethe-Salpeter equation, the ladder approximation is used. We also study the case when the variance of the permittivity fluctuation is appreciable, leading to the analytic theory of strong permittivity fluctuations.

1 Dyson's Equation for the Mean Field

Consider an unbounded random medium with permittivity $\epsilon(\overline{r}) = \langle\epsilon\rangle + \epsilon_f(\overline{r})$. The dyadic Green's function $\overline{\overline{G}}(\overline{r}, \overline{r}_o)$ in the medium satisfies the vector wave equation

$$\nabla \times \nabla \times \overline{\overline{G}}(\overline{r}, \overline{r}_o) - \omega^2 \mu\langle\epsilon\rangle \overline{\overline{G}}(\overline{r}, \overline{r}_o) = Q(\overline{r})\overline{\overline{G}}(\overline{r}, \overline{r}_o) + \overline{\overline{I}}\delta(\overline{r} - \overline{r}_o) \quad (4.1.1)$$

with $Q(\overline{r}) = \omega^2 \mu \epsilon_f(\overline{r})$. In (4.1.1), we have put the $\langle\epsilon\rangle$ part on the left-hand side of the equation while putting the ϵ_f part on the right-hand side.

The solution to (4.1.1) can be expressed as an integral equation by using the homogeneous dyadic Green's function $\overline{\overline{G}}^{(0)}(\overline{r}, \overline{r}_o)$, which satisfies the wave equation with mean wavenumber $k_m = \omega\sqrt{\mu\langle\epsilon\rangle}$:

$$\nabla \times \nabla \times \overline{\overline{G}}^{(0)}(\overline{r}, \overline{r}_o) - k_m^2 \overline{\overline{G}}^{(0)}(\overline{r}, \overline{r}_o) = \overline{\overline{I}}\delta(\overline{r} - \overline{r}_o) \quad (4.1.2)$$

Then

$$\overline{\overline{G}}(\overline{r}, \overline{r}_o) = \overline{\overline{G}}^{(0)}(\overline{r}, \overline{r}_o) + \int d\overline{r}_1 \, \overline{\overline{G}}^{(0)}(\overline{r}, \overline{r}_1) \cdot Q(\overline{r}_1)\overline{\overline{G}}(\overline{r}_1, \overline{r}_o) \quad (4.1.3)$$

Equation (4.1.3) may be solved formally by iteration to generate the Neumann series for the dyadic Green's function

$$\begin{aligned}
\overline{\overline{G}}(\overline{r}, \overline{r}_o) =& \overline{\overline{G}}^{(0)}(\overline{r}, \overline{r}_o) + \int d\overline{r}_1 \, \overline{\overline{G}}^{(0)}(\overline{r}, \overline{r}_1) \cdot Q(\overline{r}_1)\overline{\overline{G}}^{(0)}(\overline{r}_1, \overline{r}_o) \\
&+ \int d\overline{r}_1 \, d\overline{r}_2 \, \overline{\overline{G}}^{(0)}(\overline{r}, \overline{r}_1) \cdot Q(\overline{r}_1)\overline{\overline{G}}^{(0)}(\overline{r}_1, \overline{r}_2) Q(\overline{r}_2) \cdot \overline{\overline{G}}^{(0)}(\overline{r}_2, \overline{r}_o) \\
&+ \cdots + \int d\overline{r}_1 d\overline{r}_2 \cdots d\overline{r}_n \overline{\overline{G}}^{(0)}(\overline{r}, \overline{r}_1) \cdot Q(\overline{r}_1)\overline{\overline{G}}^{(0)}(\overline{r}_1, \overline{r}_2) Q(\overline{r}_2) \\
&\cdot \overline{\overline{G}}^{(0)}(\overline{r}_2, \overline{r}_3) \cdots Q(\overline{r}_n)\overline{\overline{G}}^{(0)}(\overline{r}_n, \overline{r}_o) + \cdots
\end{aligned} \quad (4.1.4)$$

The first term in the perturbation series (4.1.4) is the zeroth-order approximation and represents the solution to the homogeneous medium in which there is no scattering. The next term is the first-order Born approximation. It has the physical interpretation of single scattering of the incident wave as shown in the figure below:

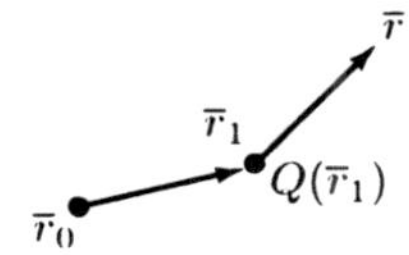

In general, the nth-order Born approximation represents the n-tuply scattered incident wave as represented below:

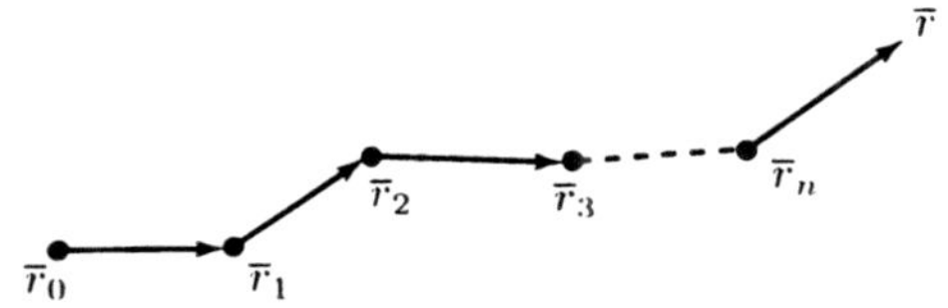

For the statistics of the random fluctuations $\epsilon_f(\overline{r})$, a common assumption made is that $\epsilon_f(\overline{r})$ is a Gaussian stationary random process so that the first and second moments of $\epsilon_f(\overline{r})$ are sufficient to describe the statistics. Therefore, we have

$$\langle \epsilon_f(\overline{r}) \rangle = 0 \tag{4.1.5}$$

and, in general, for integer n, the odd and even moments of ϵ_f are respectively

$$\langle \epsilon_f(\overline{r}_1) \cdots \epsilon_f(\overline{r}_{2n+1}) \rangle = 0 \tag{4.1.6}$$

and

$$\langle \epsilon_f(\overline{r}_1) \cdots \epsilon_f(\overline{r}_{2n}) \rangle = \sum_{\substack{\text{distinct} \\ \text{pairs}}} \langle \epsilon_f(\overline{r}_i)\epsilon_f(\overline{r}_j) \rangle \cdots \langle \epsilon_f(\overline{r}_p)\epsilon_f(\overline{r}_s) \rangle \tag{4.1.7}$$

where the summation extends over all possible distinct pairs of arguments to give $(2n-1)!!$ terms. For example, if $n = 2$, then we have

$$\begin{aligned} &\langle \epsilon_f(\overline{r}_1)\epsilon_f(\overline{r}_2)\epsilon_f(\overline{r}_3)\epsilon_f(\overline{r}_4) \rangle \\ &\quad = \langle \epsilon_f(\overline{r}_1)\epsilon_f(\overline{r}_2) \rangle \langle \epsilon_f(\overline{r}_3)\epsilon_f(\overline{r}_4) \rangle + \langle \epsilon_f(\overline{r}_1)\epsilon_f(\overline{r}_3) \rangle \langle \epsilon_f(\overline{r}_2)\epsilon_f(\overline{r}_4) \rangle \\ &\qquad + \langle \epsilon_f(\overline{r}_1)\epsilon_f(\overline{r}_4) \rangle \langle \epsilon_f(\overline{r}_2)\epsilon_f(\overline{r}_3) \rangle \end{aligned} \tag{4.1.8}$$

In the case of non-Gaussian statistics, the moments of $\epsilon_f(\overline{r})$ may still be cluster-expanded but not exclusively in terms of the two-point correlation function [Frisch, 1968]. Assuming Gaussian statistics and taking the ensemble average of (4.1.4), the perturbation series for the mean dyadic Green's function is obtained

$$
\begin{aligned}
\langle\overline{\overline{G}}(\overline{r},\overline{r}_o)\rangle &= \overline{\overline{G}}^{(0)}(\overline{r},\overline{r}_o) \\
&+ \int d\overline{r}_1 d\overline{r}_2 \overline{\overline{G}}^{(0)}(\overline{r},\overline{r}_1)\cdot\overline{\overline{G}}^{(0)}(\overline{r}_1,\overline{r}_2)\cdot\overline{\overline{G}}^{(0)}(\overline{r}_2,\overline{r}_o)\langle Q(\overline{r}_1)Q(\overline{r}_2)\rangle \\
&+ \int d\overline{r}_1\, d\overline{r}_2\, d\overline{r}_3\, d\overline{r}_4\, \overline{\overline{G}}^{(0)}(\overline{r},\overline{r}_1) \\
&\cdot \overline{\overline{G}}^{(0)}(\overline{r}_1,\overline{r}_2)\cdot\overline{\overline{G}}^{(0)}(\overline{r}_2,\overline{r}_3)\cdot\overline{\overline{G}}^{(0)}(\overline{r}_3,\overline{r}_4)\overline{\overline{G}}^{(0)}(\overline{r}_4,\overline{r}_o) \\
&\times [\langle Q(\overline{r}_1)Q(\overline{r}_2)\rangle\langle Q(\overline{r}_3)Q(\overline{r}_4)\rangle + \langle Q(\overline{r}_1)Q(\overline{r}_4)\rangle\langle Q(\overline{r}_2)Q(\overline{r}_3)\rangle \\
&+ \langle Q(\overline{r}_1)Q(\overline{r}_3)\rangle\langle Q(\overline{r}_2)Q(\overline{r}_4)\rangle] + \cdots \qquad (4.1.9)
\end{aligned}
$$

If the iteration of (4.1.9) is continued, the expressions will become increasingly cumbersome to handle. Therefore, Feynman diagrams [Frisch, 1968] are introduced as follows:

• $\equiv Q(\overline{r}_1)$, vertex over which spatial integration and contraction of dyadic indices are implied

—— $\equiv \overline{\overline{G}}^{(0)}(\overline{r},\overline{r}_o)$

══ $\equiv \langle\overline{\overline{G}}(\overline{r},\overline{r}_o)\rangle$

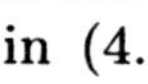 $\equiv \langle Q(\overline{r}_1)Q(\overline{r}_2)\rangle$

These rules set up a one-to-one correspondence between the analytical expressions in (4.1.9) and their graphical representations. Thus, (4.1.9) becomes

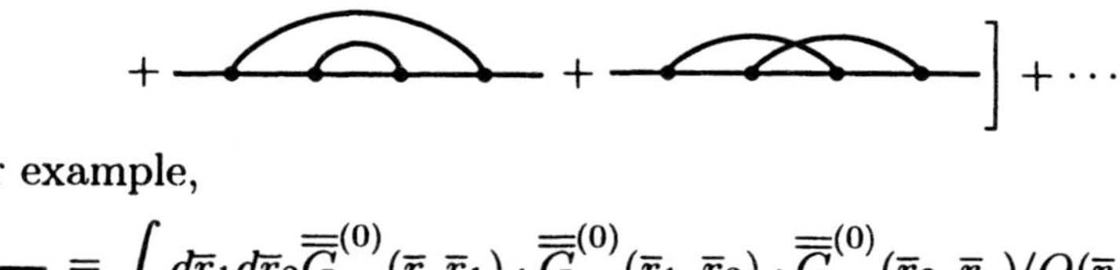 (4.1.10)

where, for example,

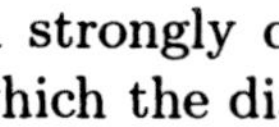 $\equiv \int d\overline{r}_1 d\overline{r}_2 \overline{\overline{G}}^{(0)}(\overline{r},\overline{r}_1)\cdot\overline{\overline{G}}^{(0)}(\overline{r}_1,\overline{r}_2)\cdot\overline{\overline{G}}^{(0)}(\overline{r}_2,\overline{r}_o)\langle Q(\overline{r}_1)Q(\overline{r}_2)\rangle$

A strongly connected diagram or irreducible diagram is defined as one in which the diagram cannot be divided without breaking the correlation connections. For example, the following diagrams are strongly connected:

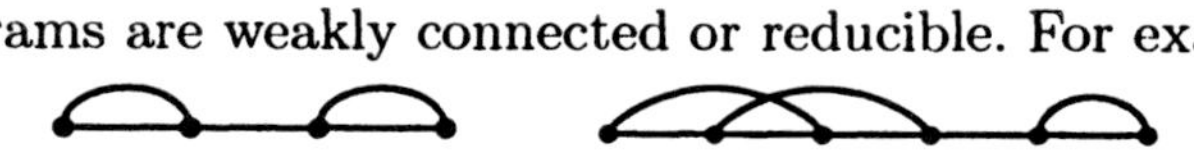

All other diagrams are weakly connected or reducible. For example

Note that weakly connected diagrams can be obtained by connecting strongly connected diagrams. For example,

 can be connected to give 

This means that on iteration of an operator equation or an integral equation where the operator is a strongly connected diagram, the weakly connected diagrams will be reproduced from those iterations. Thus when one sets up an operator equation or an integral equation, only strongly connected diagrams need to be included. The mass operator is defined as the sum of all strongly connected diagrams, minus end connectors:

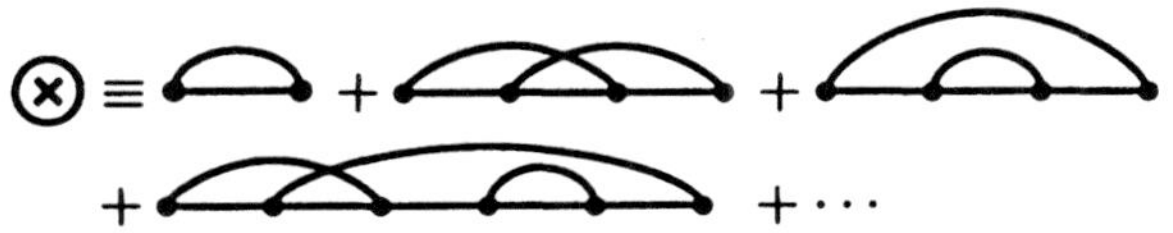

(4.1.11)

With the mass operator of (4.1.11), we have

(4.1.12)

The weakly connected diagrams in (4.1.10) which contain two strongly connected elements may be summed as

(4.1.13)

Weakly connected diagrams are reproduced by cascading of mass operators. Continuing with this process, the Neumann series for the mean dyadic Green's function may be written in the form:

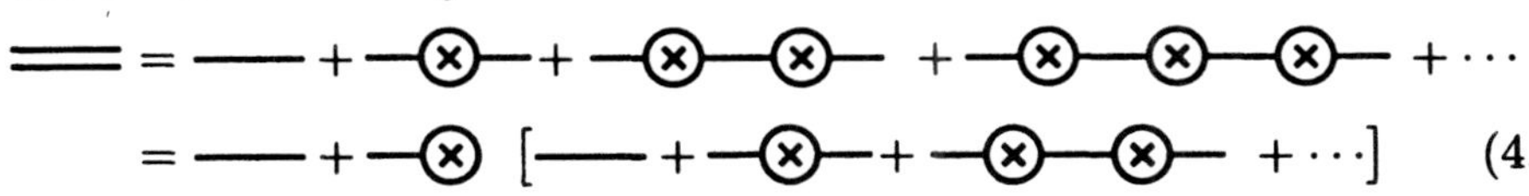

(4.1.14)

or

(4.1.15)

Equation (4.1.15) is the diagrammatic representation of the Dyson's equation. One can verify that all the diagrams, both weak and strong, are included in the Dyson's equation. The Dyson's equation is an exact equation for the mean field. In analytical form, Dyson's equation becomes

$$\langle \overline{\overline{G}}(\overline{r}, \overline{r}_o) \rangle = \overline{\overline{G}}^{(0)}(\overline{r}, \overline{r}_o) + \int d\overline{r}_1 \, d\overline{r}_2 \, \overline{\overline{G}}^{(0)}(\overline{r}, \overline{r}_1) \cdot \overline{\overline{Q}}(\overline{r}_1, \overline{r}_2) \cdot \langle \overline{\overline{G}}(\overline{r}_2, \overline{r}_o) \rangle \quad (4.1.16)$$

where $\overline{\overline{Q}}(\overline{r}_1, \overline{r}_2)$ is the dyadic mass operator. For a statistical homogeneous medium, the mass operator is statistically translational invariant so that the mean field is also translational invariant.

$$\langle \overline{\overline{Q}}(\overline{r}_1, \overline{r}_2) \rangle = \overline{\overline{Q}}(\overline{r}_1 - \overline{r}_2) \tag{4.1.17a}$$

$$\langle \overline{\overline{G}}(\overline{r}, \overline{r}_o) \rangle = \overline{\overline{G}}_c(\overline{r} - \overline{r}_o) \tag{4.1.17b}$$

We define the Fourier transform by

$$\overline{\overline{G}}_c(\overline{k}) = \int d\overline{r}\, \overline{\overline{G}}_c(\overline{r}) e^{-i\overline{k}\cdot\overline{r}} \tag{4.1.18a}$$

$$\overline{\overline{Q}}(\overline{k}) = \int d\overline{r}\, \overline{\overline{Q}}(\overline{r}) e^{-i\overline{k}\cdot\overline{r}} \tag{4.1.18b}$$

The inverse Fourier transform of $\overline{\overline{G}}_c(\overline{k})$ is

$$\overline{\overline{G}}_c(\overline{r}) = \frac{1}{(2\pi)^3} \int d\overline{k}\; \overline{\overline{G}}_c(\overline{k}) e^{i\overline{k}\cdot\overline{r}} \tag{4.1.18c}$$

The Fourier transform of $\overline{\overline{G}}^{(0)}$ is

$$\overline{\overline{G}}^{(0)}(\overline{k}) = \left(\overline{\overline{I}} - \frac{\overline{k}\,\overline{k}}{k_m^2} \right) \frac{1}{k^2 - k_m^2}$$

Taking (4.1.16) in the transform domain, we obtain

$$\overline{\overline{G}}_c(\overline{k}) = \overline{\overline{G}}^{(0)}(\overline{k}) + \overline{\overline{G}}^{(0)}(\overline{k}) \overline{\overline{Q}}(\overline{k}) \overline{\overline{G}}_c(\overline{k}) \tag{4.1.19}$$

Thus

$$\overline{\overline{G}}_c(\overline{k}) = \left(\overline{\overline{I}} - \overline{\overline{G}}^{(0)}(\overline{k}) \overline{\overline{Q}}(\overline{k}) \right)^{-1} \overline{\overline{G}}^{(0)}(\overline{k}) \tag{4.1.20}$$

For the scalar wave case, the analogous result is

$$G^{(0)}(\overline{k}) = \frac{1}{(k^2 - k_m^2)} \tag{4.1.21}$$

Then

$$G_c(k) = \frac{\dfrac{1}{(k^2 - k_m^2)}}{1 - \dfrac{Q(\overline{k})}{(k^2 - k_m^2)}} = \frac{1}{k^2 - k_m^2 - Q(\overline{k})} \tag{4.1.22}$$

The poles of the mean Green's function in the spectral domain are the effective propagation constants of the coherent wave. Thus from (4.1.22) we can show that the effective propagation constant of the coherent field is K where

$$K^2 = k_m^2 + Q(\overline{K}) \tag{4.1.23}$$

Dyson's equation is an exact equation for the mean dyadic Green's function. However, the mass operator as given in (4.1.11) is in the form of an infinite series containing all the strongly connected diagrams. It must be approximated to solve (4.1.16). Approximations to the mass operator will next be examined.

1.1 Bilocal Approximation

The mass operator is approximated by retaining only the first term. This is called the bilocal or first-order smoothing approximation:

$\overline{\overline{Q}}(\overline{r}_1, \overline{r}_2) \equiv$  $= C(\overline{r}_1 - \overline{r}_2)\,\overline{\overline{G}}^{(0)}(\overline{r}_1, \overline{r}_2)$ (4.1.24)

Under this approximation, Dyson's equation becomes

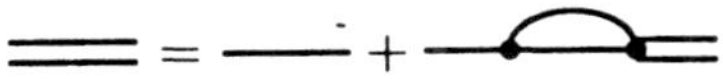 (4.1.25)

In equation form,

$$\langle \overline{\overline{G}}(\overline{r}, \overline{r}_o) \rangle = \overline{\overline{G}}^{(0)}(\overline{r}, \overline{r}_o) + \int d\overline{r}_1 d\overline{r}_2 \overline{\overline{G}}^{(0)}(\overline{r}, \overline{r}_1) \cdot \overline{\overline{G}}^{(0)}(\overline{r}_1, \overline{r}_2) \cdot \langle \overline{\overline{G}}(\overline{r}_2, \overline{r}_o) \rangle C(\overline{r}_1 - \overline{r}_2) \tag{4.1.26}$$

To visualize the terms of the Neumann series that are included in the bilocal approximation to Dyson's equation, (4.1.25) can be iterated to obtain

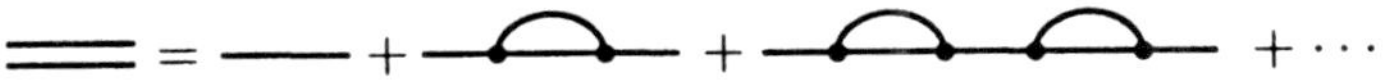 (4.1.27)

Thus, the bilocal approximation accounts for those weakly connected diagrams that are formed by products of bilocal diagrams (i.e., ).

Bilocal Approximation Applied to the Scalar Wave Equation

The bilocally approximated Dyson's equation for scalar wave is given by

$$\langle G(\overline{r}, \overline{r}_o) \rangle = G^{(0)}(\overline{r}, \overline{r}_o) + \int d\overline{r}_1 d\overline{r}_2 G^{(0)}(\overline{r}, \overline{r}_1) G^{(0)}(\overline{r}_1, \overline{r}_2) \cdot \langle G(\overline{r}_2, \overline{r}_o) \rangle C(\overline{r}_1 - \overline{r}_2) \tag{4.1.28}$$

where we have used

$$Q(\overline{r}) \cong C(\overline{r}) G^{(0)}(\overline{r}) \tag{4.1.29}$$

and $G^{(0)}(r) = \exp(ik_m r)/4\pi r$. For the case of

$$C(\overline{r}) = \delta k_m'^4 e^{-r/a} \tag{4.1.30}$$

where δ is the variance of permittivity fluctuations and k'_m is the real part of k_m,

$$Q(\overline{k}) = \int d\overline{r} \delta k'^4_m e^{-r/a} \frac{1}{4\pi r} e^{ik_m r} e^{-i\overline{k}\cdot\overline{r}} = \frac{\delta k'^4_m}{k} \int_0^\infty dr \; e^{-r/a} e^{ik_m r} \sin kr$$
$$= \frac{\delta k'^4_m a^2}{k^2 a^2 + (1 - ik_m a)^2} \tag{4.1.31}$$

Thus the effective propagation constant is using (4.1.23) and (4.1.31)

$$K^2 = k_m^2 + \frac{\delta k'^4_m a^2}{K^2 a^2 + (1 - ik_m a)^2} \tag{4.1.32}$$

It is important to keep both the leading real part and the leading imaginary part even though the leading imaginary part can be much smaller than the leading real part. The leading imaginary part represents attenuation due to scattering. If we assume $K \simeq k_m$, then

$$K^2 \simeq k_m^2 + \frac{\delta k'^4_m a^2}{k_m^2 a^2 + (1 - ik_m a)^2} = k_m^2 + \frac{\delta k'^4_m a^2}{1 - 2ik_m a} \tag{4.1.33}$$

For the case of a small correlation length of $k_m a \ll 1$ we have

$$K^2 \simeq k_m^2 + \delta k'^4_m a^2 (1 + 2ik_m a) \tag{4.1.34}$$

The first term in (4.1.34) can be regarded as the zeroth-order solution, and the second term in (4.1.34) can be regarded as the first-order solution. Thus, for (4.1.34) to be valid, the first-order solution has to be smaller than the zeroth-order solution, giving the condition

$$\delta k_m^2 a^2 \ll 1 \tag{4.1.35}$$

Since $k_m a \ll 1$, we may further simplify (4.1.34) to get

$$K \simeq k_m \left[1 + \frac{\delta k_m^2 a^2}{2} (1 + i2k_m a) \right] \tag{4.1.36}$$

We see that the randomness has introduced an effective loss term that causes the coherent wave to decay as it propagates in the random medium.

Bilocal Approximation Applied to the Vector Wave Equation

For the vector wave equation we have

$$\overline{\overline{Q}}(\overline{k}) = \int d\overline{r} \; e^{-i\overline{k}\cdot\overline{r}} \; \overline{\overline{G}}^{(0)}(\overline{r}) C(\overline{r}) \tag{4.1.37}$$

The dyadic Green's function $\overline{\overline{G}}^{(0)}(\overline{r})$ is

$$\overline{\overline{G}}^{(0)}(\overline{r}) = PS\overline{\overline{G}}^{(0)}(\overline{r}) - \frac{\overline{\overline{I}}}{3k_m^2} \delta(\overline{r}) \quad \text{(sphere)} \tag{4.1.38}$$

where PS stands for principal value with an infinitesimal volume of defined shape excluded about the point $\overline{r} = 0$. The expression (sphere) at the end of (4.1.38) indicates that a spherical infinitesimal volume has been chosen for the shape of the exclusion. The coefficient in front of the Dirac delta function of the second term depends on the shape of the exclusion volume. If shapes other than spherical are chosen, then the coefficient will be different from that of (4.1.38).

For a spherical exclusion we have

$$PS\,\overline{\overline{G}}^{(0)}(\overline{r}) = PS\,G_1(r)\overline{\overline{I}} + PS\,G_2(r)\hat{r}\hat{r} \tag{4.1.39}$$

where

$$G_1(r) = (-1 + ik_m r + k_m^2 r^2)\frac{e^{ik_m r}}{4\pi k_m^2 r^3} \tag{4.1.40}$$

$$G_2(r) = (3 - 3ik_m r - k_m^2 r^2)\frac{e^{ik_m r}}{4\pi k_m^2 r^3} \tag{4.1.41}$$

Using (4.1.39) through (4.1.41) and the spherically symmetric correlation functions,

$$\int d\overline{r}\, C(\overline{r})\, PS\overline{\overline{G}}^{(0)}(\overline{r}) = \frac{2}{3}\overline{\overline{I}}\int_0^\infty dr\, C(r)\, r\, e^{ik_m r} \tag{4.1.42}$$

Using the exponential correlation function as in the scalar case, we obtain the following in the low-frequency limit of $k_m a \ll 1$

$$K^2 = k_m^2 - \frac{\delta}{3}k_m^2 + \frac{2}{3}\delta k_m^4 a^2(1 + 2ik_m a) \tag{4.1.43}$$

The result in (4.1.43) is valid if the zeroth-order solution [the first term in (4.1.43)] is much larger than the first-order solution. Thus the validity conditions of (4.1.43) are

$$\delta \ll 1 \tag{4.1.44}$$

and

$$\delta k_m^2 a^2 \ll 1 \tag{4.1.45}$$

Note that (4.1.45) is identical to (4.1.35) of the scalar case. However, the condition (4.1.44) is new and is not in the scalar wave case. Thus the vector wave case has one more condition than the scalar wave case. Equation (4.1.44) assumes that the variance of permittivity fluctuations is small. The weak permittivity fluctuation criterion of (4.1.44) is restrictive in applications because many media are characterized by large permittivity fluctuations. Physically, in cases of large permittivity fluctuations, the internal field inside an inhomogeneity can be quite different from that of the background medium. The restriction in (4.1.44) can be eliminated by taking into account the singularity of the dyadic Green's function. The incorporation of the singularity will be done in Section 3 on strong permittivity fluctuations.

1.2 Nonlinear Approximation

The terms in the mass operator can be rearranged to form a new renormalized mass operator.

Consider the diagram: 

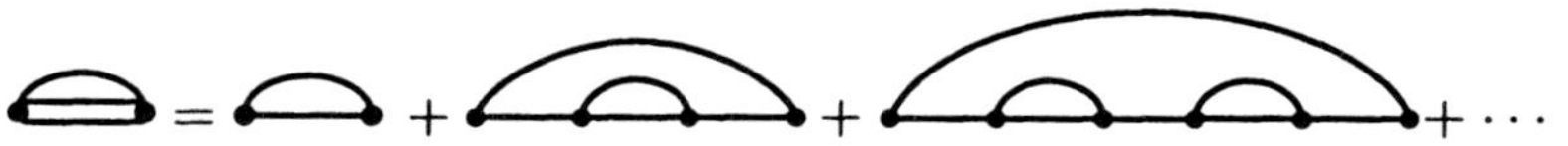 (4.1.46)

Equation (4.1.46) reproduces all strongly connected diagrams in the Neumann series for the mean field which have one outer correlation and both strongly and weakly connected diagrams inside. Next, observe that all the strongly connected diagrams with two-crossed outer correlations may be summed by the following

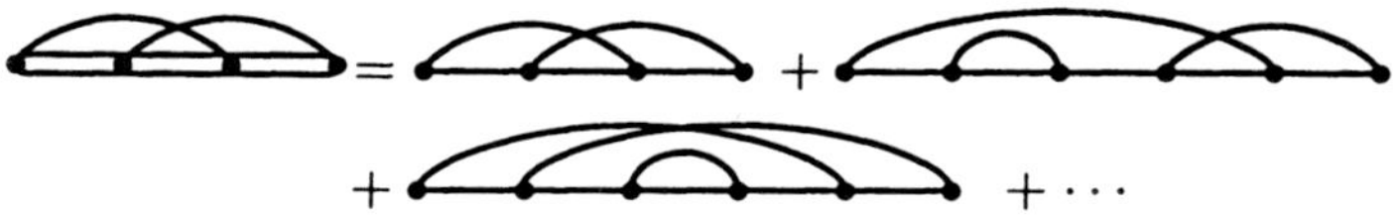 (4.1.47)

By continuing this process of resummation, all the strongly connected diagrams can be reproduced. This yields the renormalized mass operator:

 (4.1.48)

If only the first term of the new mass operator is retained, the nonlinear approximation to Dyson's equation is obtained. In the nonlinear approximation,

 (4.1.49)

Then

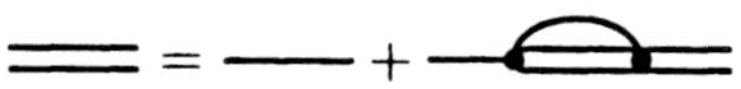 (4.1.50)

In equation form,

$$\langle\overline{\overline{G}}(\overline{r},\overline{r}_o)\rangle = \overline{\overline{G}}^{(0)}(\overline{r},\overline{r}_o) + \int d\overline{r}_1 d\overline{r}_2 \overline{\overline{G}}^{(0)}(\overline{r},\overline{r}_1)\cdot\langle\overline{\overline{G}}(\overline{r}_1,\overline{r}_2)\rangle \cdot\langle\overline{\overline{G}}(\overline{r}_2,\overline{r}_o)\rangle C(\overline{r}_1-\overline{r}_2) \quad (4.1.51)$$

The nonlinear Dyson's equation may be solved by Fourier transforms. We consider the scalar wave case:

$$(\nabla^2 + k_m^2)\langle G(\overline{r},\overline{r}_o)\rangle = -\delta(\overline{r}-\overline{r}_o) - \int d\overline{r}_1 \langle G(\overline{r},\overline{r}_1)\rangle\langle G(\overline{r}_1,\overline{r}_o)\rangle C(\overline{r}-\overline{r}_1) \tag{4.1.52}$$

For the case $k_m a \ll 1$, the correlation function $C(\overline{r}-\overline{r}_1)$ is sharply peaked at $\overline{r}=\overline{r}_1$. Hence, for the nonlinear approximation we have

$$K^2 = k_m^2 + \int d\overline{r}\langle G(\overline{r})\rangle C(\overline{r}) \tag{4.1.53}$$

Assuming $\langle G(\overline{r})\rangle = \exp(iKr)/4\pi r$ and $C(\overline{r}) = \delta k_m'^4 \exp(-r/a)$, we have

$$K^2 \simeq k_m^2 + \delta k_m'^4 a^2(1+2iKa) \tag{4.1.54}$$

Next we compare the result of bilocal approximation of (4.1.34) with that of nonlinear approximation of (4.1.54). Suppose k_m is real, then $\mathrm{Re}(K^2)$ are the same for both approximations. The imaginary part of K^2 is related to the attenuation rate due to scattering. The two results for the imaginary part are different since one is $2\delta k_m'^5 a^3$ while the other is $2\delta k_m'^4 K a^3$. Since $\mathrm{Re}(K)$ is much greater than $\mathrm{Im}(K)$, the attenuation rates based on the two approximations will be essentially the same if $\mathrm{Re}(K)$ is approximately equal to k_m. This is usually the case for a tenuous medium with small δ. Under the circumstances where $\delta \ll 1$, the results of bilocal approximation and nonlinear approximation are roughly the same.

2 Second Moment of the Field

2.1 Bethe-Salpeter Equation

In Section 1, Dyson's equation for the mean field is derived by renormalization, i.e., re-summing the averaged Neumann series for the field. In a similar fashion, an equation governing the second moment of the field can be derived by using the diagrammatic method.

We consider a random medium with permittivity

$$\epsilon(\overline{r}) = \langle\epsilon\rangle + \epsilon_f(\overline{r}) \tag{4.2.1}$$

The solution to the scattering of electromagnetic waves in terms of dyadic Green's functions is given in the previous section.

$$\overline{\overline{G}}(\overline{r},\overline{r}_o) = \overline{\overline{G}}^{(0)}(\overline{r},\overline{r}_o) + \int d\overline{r}_1 \overline{\overline{G}}^{(0)}(\overline{r},\overline{r}_1) Q(\overline{r}_1)\cdot\overline{\overline{G}}(\overline{r}_1,\overline{r}_o) \tag{4.2.2}$$

where $Q(\overline{r}) = \omega^2\mu\epsilon_f(\overline{r})$. Equation (4.2.2) may be solved formally by iteration to yield the Neumann series solution for the random dyadic Green's function.

$$\begin{aligned}\overline{\overline{G}}(\overline{r},\overline{r}_o) &= \overline{\overline{G}}^{(0)}(\overline{r},\overline{r}_o) + \int d\overline{r}_1 \overline{\overline{G}}^{(0)}(\overline{r},\overline{r}_1)Q(\overline{r}_1)\cdot\overline{\overline{G}}^{(0)}(\overline{r}_1,\overline{r}_o) \\ &+ \int d\overline{r}_1 \int d\overline{r}_2 \overline{\overline{G}}^{(0)}(\overline{r},\overline{r}_1)Q(\overline{r}_1)\cdot\overline{\overline{G}}^{(0)}(\overline{r}_1,\overline{r}_2)Q(\overline{r}_2)\cdot\overline{\overline{G}}^{(0)}(\overline{r}_2,\overline{r}_o) \\ &+ \int d\overline{r}_1 \int d\overline{r}_2 \int d\overline{r}_3 \overline{\overline{G}}^{(0)}(\overline{r},\overline{r}_1)Q(\overline{r}_1)\cdot\overline{\overline{G}}^{(0)}(\overline{r}_1,\overline{r}_2)Q(\overline{r}_2)\cdot \\ &\quad \overline{\overline{G}}^{(0)}(\overline{r}_2,\overline{r}_3)Q(\overline{r}_3)\cdot\overline{\overline{G}}^{(0)}(\overline{r}_3,\overline{r}_o) + \cdots \end{aligned} \tag{4.2.3}$$

To calculate intensity, we perform the tensor product of the Green's function and ensemble average the resulting expression. The Neumann series solution for the covariance of the field is obtained.

$$\begin{aligned}&\langle\overline{\overline{G}}(\overline{r},\overline{r}_o)\overline{\overline{G}}^*(\overline{r}',\overline{r}_o')\rangle = \overline{\overline{G}}^{(0)}(\overline{r},\overline{r}_o)\overline{\overline{G}}^{(0)*}(\overline{r}',\overline{r}_o') \\ &+ \overline{\overline{G}}^{(0)}(\overline{r},\overline{r}_o) \int d\overline{r}_1' \int d\overline{r}_2' \overline{\overline{G}}^{(0)*}(\overline{r}',\overline{r}_1')\cdot\overline{\overline{G}}^{(0)*}(\overline{r}_1',\overline{r}_2')\cdot\overline{\overline{G}}^{(0)*}(\overline{r}_2',\overline{r}_o') \\ &\quad \langle Q^*(\overline{r}_1')Q^*(\overline{r}_2')\rangle \\ &+ \int d\overline{r}_1 \int d\overline{r}_2 \overline{\overline{G}}^{(0)}(\overline{r},\overline{r}_1)\cdot\overline{\overline{G}}^{(0)}(\overline{r}_1,\overline{r}_2)\cdot\overline{\overline{G}}^{(0)}(\overline{r}_2,\overline{r}_o)\langle Q(\overline{r}_1)Q(\overline{r}_2)\rangle\overline{\overline{G}}^{(0)*}(\overline{r}',\overline{r}_o') \\ &+ \overline{\overline{G}}^{(0)}(\overline{r},\overline{r}_o) \int d\overline{r}_1' \int d\overline{r}_2' \int d\overline{r}_3' \overline{\overline{G}}^{(0)*}(\overline{r}',\overline{r}_1')\cdot\overline{\overline{G}}^{(0)*}(\overline{r}_1',\overline{r}_2')\cdot\overline{\overline{G}}^{(0)*}(\overline{r}_2',\overline{r}_3')\cdot \\ &\quad \overline{\overline{G}}^{(0)*}(\overline{r}_3',\overline{r}_o')\langle Q^*(\overline{r}_1')Q^*(\overline{r}_2')Q^*(\overline{r}_3')\rangle \\ &+ \int d\overline{r}_1 \int d\overline{r}_2 \int d\overline{r}_3 \overline{\overline{G}}^{(0)}(\overline{r},\overline{r}_1)\cdot\overline{\overline{G}}^{(0)}(\overline{r}_1,\overline{r}_2)\cdot\overline{\overline{G}}^{(0)}(\overline{r}_2,\overline{r}_3)\cdot\overline{\overline{G}}^{(0)}(\overline{r}_3,\overline{r}_o) \\ &\quad \langle Q(\overline{r}_1)Q(\overline{r}_2)Q(\overline{r}_3)\rangle\overline{\overline{G}}^{(0)*}(\overline{r}',\overline{r}_o') \\ &+ \int d\overline{r}_1 \int d\overline{r}_1' \int d\overline{r}_2' \overline{\overline{G}}^{(0)}(\overline{r},\overline{r}_1)\cdot\overline{\overline{G}}^{(0)}(\overline{r}_1,\overline{r}_o)\overline{\overline{G}}^{(0)*}(\overline{r}',\overline{r}_1')\cdot\overline{\overline{G}}^{(0)*}(\overline{r}_1',\overline{r}_2')\cdot \\ &\quad \overline{\overline{G}}^{(0)*}(\overline{r}_2',\overline{r}_o')\langle Q(\overline{r}_1)Q^*(\overline{r}_1')Q^*(\overline{r}_2')\rangle \\ &+ \int d\overline{r}_1 \int d\overline{r}_2 \int d\overline{r}_1' \overline{\overline{G}}^{(0)}(\overline{r},\overline{r}_1)\cdot\overline{\overline{G}}^{(0)}(\overline{r}_1,\overline{r}_2)\cdot\overline{\overline{G}}^{(0)}(\overline{r}_2,\overline{r}_o)\overline{\overline{G}}^{(0)*}(\overline{r}',\overline{r}_1')\cdot \\ &\quad \overline{\overline{G}}^{(0)*}(\overline{r}_1',\overline{r}_o')\langle Q(\overline{r}_1)Q(\overline{r}_2)Q^*(\overline{r}_1')\rangle + \cdots \end{aligned} \tag{4.2.4}$$

Feynman diagrams are introduced in conjunction with those of the previous section:

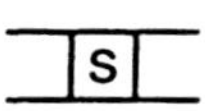

$$\equiv \langle\overline{\overline{G}}(\overline{r},\overline{r}_o)\overline{\overline{G}}^*(\overline{r}',\overline{r}_o')\rangle \tag{4.2.5}$$

is the averaged tensor product of the Green's function

 $\equiv \overline{\overline{G}}(\bar{r}, \bar{r}_o)$ (4.2.6)

is the exact Green's function of the random medium problem without average, and

$\equiv \langle Q(\bar{r}_1)Q(\bar{r}_2)\rangle$ (4.2.7)

The expressions for the mean Green's function $\langle\overline{\overline{G}}\rangle$ is given in (4.1.9) with diagrammatic representation given in (4.1.10). Using diagrams, the equation for the Green's function (4.2.3) can be represented as

 (4.2.8)

Iterating (4.2.8) yields the graphical representation of (4.2.3)

 (4.2.9)

Two-level diagrams can be introduced where the upper level represents a vector space and the lower level represents the complex conjugate vector space. For example,

 $\equiv \overline{\overline{G}}(\bar{r}, \bar{r}_o)\overline{\overline{G}}^*(\bar{r}', \bar{r}_o')$ (4.2.10)

which is the exact second moment of the Green's function without average. The average is defined by

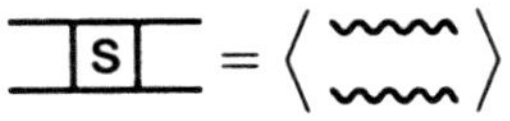

$= \langle\overline{\overline{G}}(\bar{r}, \bar{r}_o)\overline{\overline{G}}^*(\bar{r}', \bar{r}_o')\rangle$ (4.2.11)

Upon forming the tensor product of the Neumann series (4.2.9), we obtain

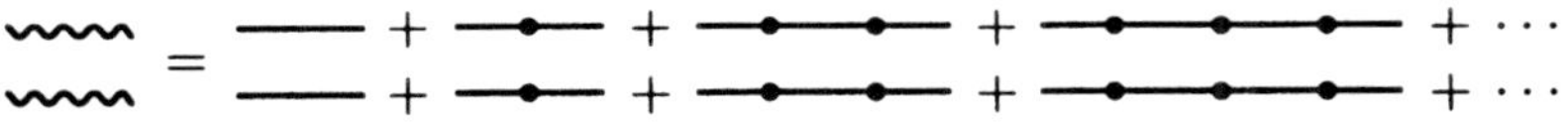 (4.2.12)

we obtain the ensemble average of (4.2.11),

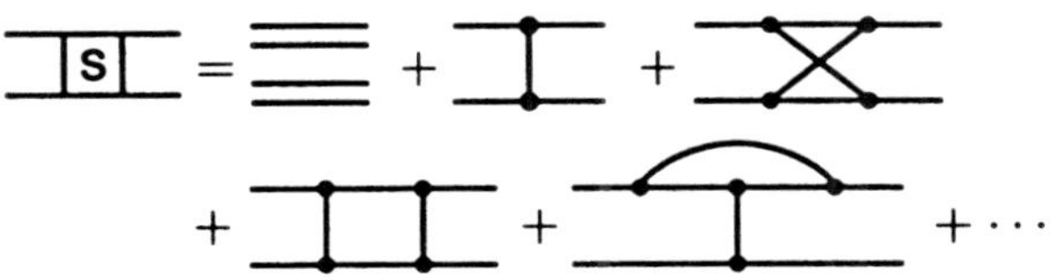

(4.2.13)

where the first term on the right-hand side of (4.2.13) follows from (4.1.10):

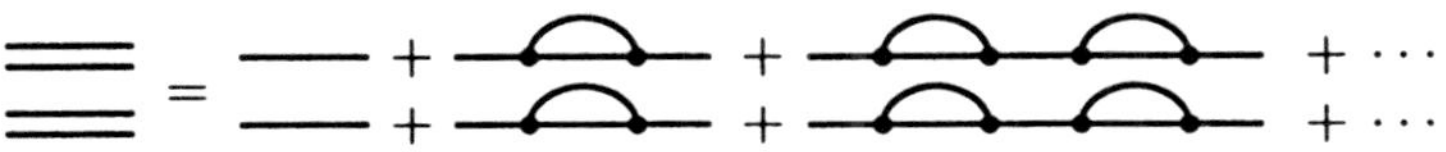 (4.2.14)

Weakly connected or reducible diagrams are those diagrams that can be reduced without breaking a correlation connection.

Weakly Connected Examples Are:

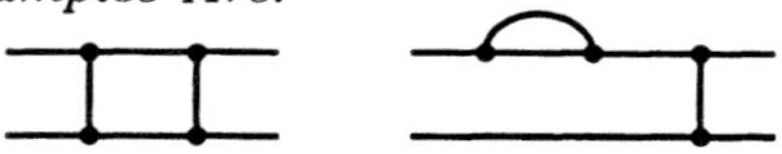

All other diagrams are strongly connected or irreducible.

Strongly Connected Examples Are:

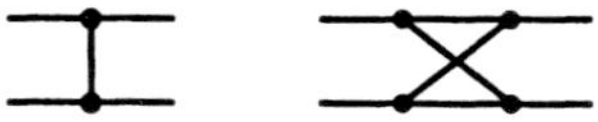

The intensity operator is the sum of all strongly connected diagrams (minus end connectors):

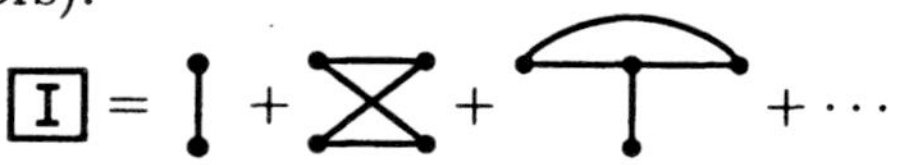
(4.2.15)

The sum of all strongly connected diagrams in (4.2.13) may be written in terms of the intensity operator as

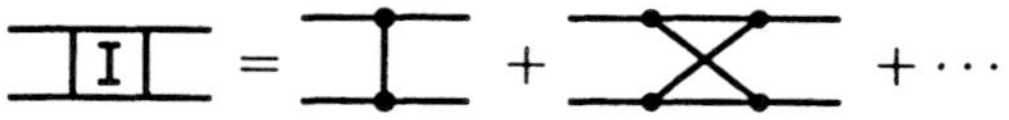

The sum of all strongly and weakly connected diagrams in (4.2.13) containing one intensity operator is given by

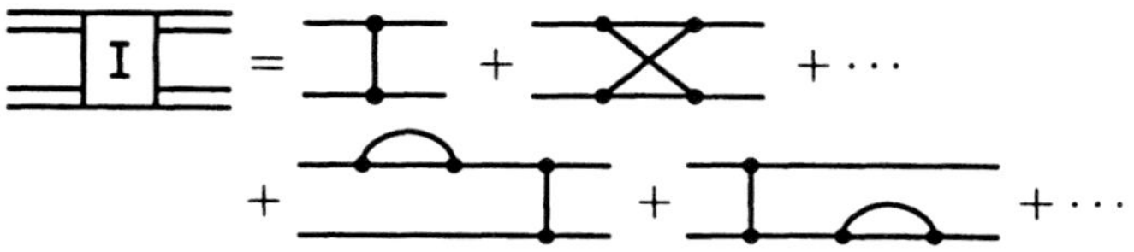

The sum of all weakly connected diagrams in (4.2.13) containing two intensity operators in cascade is given by

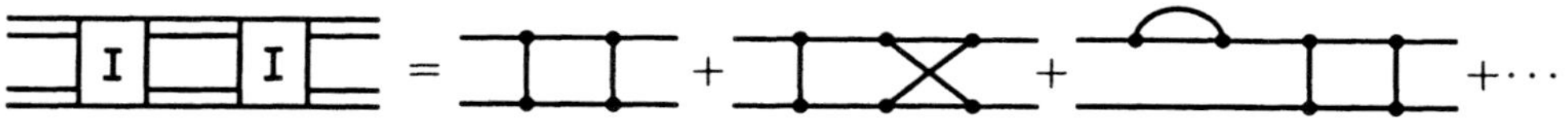

Continuing with this process, the series for the field correlation can be rewritten in the form

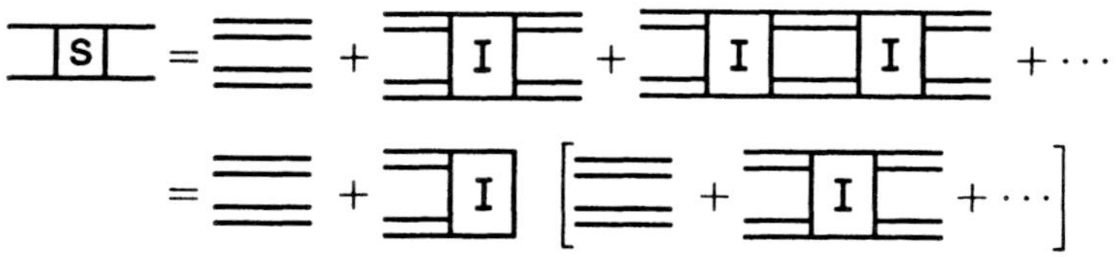
(4.2.16)

The key point is that weakly connected diagrams can be reproduced from strongly connected diagrams on iteration. On summation of (4.2.16), we have

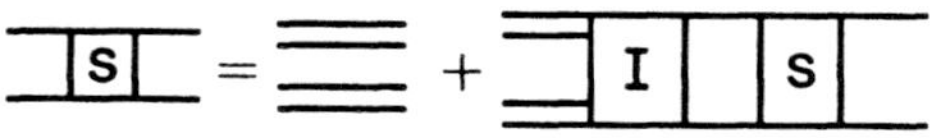
(4.2.17)

Equation (4.2.17) is the diagrammatic representation of the Bethe-Salpeter equation for the correlation of the field. In analytical form, the Bethe-Salpeter equation reads

$$\begin{aligned}\langle G_{ij}(\overline{r},\overline{r}_o)G^*_{kl}(\overline{r}',\overline{r}'_o)\rangle &= \langle G_{ij}(\overline{r},\overline{r}_o)\rangle\langle G^*_{kl}(\overline{r}',\overline{r}'_o)\rangle \\ &+ \sum_{m=1}^{3}\sum_{n=1}^{3}\sum_{g=1}^{3}\sum_{r=1}^{3}\int d\overline{r}_1\, d\overline{r}_2\, d\overline{r}'_1\, d\overline{r}'_2 \langle G_{im}(\overline{r},\overline{r}_1)\rangle\langle G^*_{kg}(\overline{r}',\overline{r}'_1)\rangle \\ &I_{mn,gr}(\overline{r}_1,\overline{r}_2;\overline{r}'_1,\overline{r}'_2)\langle G_{nj}(\overline{r}_2,\overline{r}_o)G^*_{rl}(\overline{r}'_2,\overline{r}'_o)\rangle \end{aligned} \tag{4.2.18}$$

where $I_{mn,gr}$ is the intensity operator as given by (4.2.15). Note that in (4.2.18), we have integration over intermediate variables and summation over intermediate dyadic vector components.

Quite often the Bethe-Salpeter equation is written in terms of the field covariance rather than the field correlation. The field covariance is defined by the following diagram

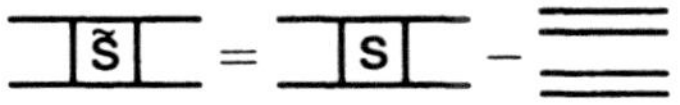

(4.2.19)

Eliminating 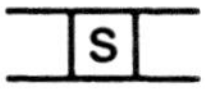 in (4.2.17) yields

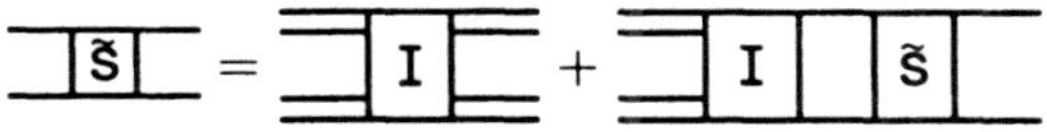

(4.2.20)

The Bethe-Salpeter equation is an exact equation for the second moment of the field. However, the intensity operator as given by (4.2.15) is in the form of an infinite series. It must be approximated to give tractable solutions. A common approximation known as the ladder approximation is to retain only the first term of the series expansion in (4.2.15):

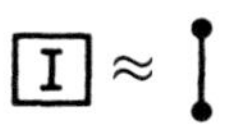 $= \delta(\overline{r}_1 - \overline{r}_2)\delta(\overline{r}'_1 - \overline{r}'_2)C(\overline{r}_1 - \overline{r}'_1)$ (4.2.21)

The Bethe-Salpeter equation then becomes

(4.2.22)

To see which terms of the Neumann series for the field covariance are included in the ladder approximation, we iterate (4.2.22) to obtain

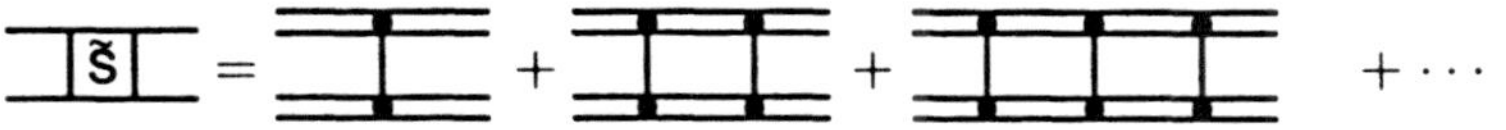

(4.2.23)

Thus under the ladder approximation, the field covariance contains all the "ladder-like" terms. On the other hand, the distorted Born approximation consists of only the first term in (4.2.23).

2.2 Energy Conservation

Dyson's equation for the mean field and the Bethe-Salpeter equation for the covariance have been derived and are exact equations for the respective statistical quantities. The exact field equations satisfy energy conservation. However, when approximations are imposed upon the Dyson and Bethe-Salpeter equations to the mass and intensity operators respectively, the results may be inconsistent with the requirement of the conservation of energy. Therefore, the question of having approximations on Dyson and Bethe-Salpeter equations that are consistent with energy conservation is an important one. In this section it is shown that the ladder approximation and the nonlinear approximation satisfy energy conservation exactly. The general relation between mass and intensity operators for energy conservation is also established. The consistency in conservation law is derived for the case of real permittivity; that is, both $\langle\epsilon\rangle$ and ϵ_f are real.

Consider scalar wave in a continuous random medium. The wave equation is

$$(\nabla^2 + k^2)\psi(\overline{r}) = -\omega^2\mu\epsilon_f(\overline{r})\psi(\overline{r}) \tag{4.2.24}$$

where $k^2 \equiv \omega^2\mu\langle\epsilon\rangle$. It follows that

$$\nabla \cdot [\,\psi^*(\overline{r})\nabla\psi(\overline{r}) - \psi(\overline{r})\nabla\psi^*(\overline{r})\,] + (k^2 - k^{*2})\psi(\overline{r})\psi^*(\overline{r}) = 0 \tag{4.2.25}$$

Equation (4.2.25) is an energy conservation law where the quantity in square brackets defines an energy flux vector, $\overline{S}_\psi$:

$$\overline{S}_\psi \equiv \psi^*(\overline{r})\nabla\psi(\overline{r}) - \psi(\overline{r})\nabla\psi^*(\overline{r}) \tag{4.2.26}$$

The Dyson and Bethe-Salpeter equations are

$$-(\nabla^2 + k^2)\langle\psi(0)\rangle = Q(0,1)\langle\psi(1)\rangle \tag{4.2.27}$$

$$-(\nabla^2 + k^2)\langle G(0,1)\rangle = \delta(0-1) + Q(0,2)\langle G(2,1)\rangle \tag{4.2.28}$$

$$\begin{aligned}\langle\psi^*(0')\psi(0)\rangle &= \langle\psi^*(0')\rangle\langle\psi(0)\rangle \\ &\quad + \langle G(0,1)\rangle\langle G^*(0',1')\rangle I(1,2;1',2')\langle\psi(2)\psi^*(2')\rangle\end{aligned} \tag{4.2.29}$$

where we have used a shorthand notation: $0 \equiv \overline{r}_o, 1 \equiv \overline{r}_1, 2 \equiv \overline{r}_2$, and so on, and the integration is implied over internal arguments. In (4.2.27) through (4.2.29), $Q(0,1)$ and $I(1,2;1',2')$ are the mass and intensity operators, respectively.

By using (4.2.27) through (4.2.29), we have

$$-\Big\{\langle\nabla\cdot\overline{S}_\psi(0)\rangle + (k^2 - k^{*2})\langle\psi^*(0)\psi(0)\rangle\Big\}$$

$$
\begin{aligned}
&= \lim_{0'\to 0}[(\nabla^2 + k^2)\langle\psi(0)\psi^*(0')\rangle - (\nabla'^2 + k^{*2})\langle\psi(0)\psi^*(0')\rangle] \\
&= \langle\psi^*(0)\rangle Q(0,1)\langle\psi(1)\rangle + \langle G^*(0,1')\rangle I(0,2;1',2')\langle\psi(2)\psi^*(2')\rangle \\
&\quad + Q(0,3)\langle G(3,1)\rangle\langle G^*(0,1')\rangle I(1,2;1',2')\langle\psi(2)\psi^*(2')\rangle \\
&\quad - Q^*(0,1)\langle\psi^*(1)\rangle\langle\psi(0)\rangle - \langle G(0,1)\rangle I(1,2;0,2')\langle\psi(2)\psi^*(2')\rangle \\
&\quad - \langle G(0,1)\rangle Q^*(0,3)\langle G^*(3,1')\rangle I(1,2;1',2')\langle\psi(2)\psi^*(2')\rangle \qquad (4.2.30)
\end{aligned}
$$

Next we apply the Bethe-Salpeter equation of (4.2.29) to the second and the fifth terms on the right-hand side of (4.2.30)

RHS of (4.2.30)

$$
\begin{aligned}
&= \langle\psi^*(0)\rangle Q(0,1)\langle\psi(1)\rangle + \langle G^*(0,1')\rangle I(0,2;1',2')\langle\psi(2)\rangle\langle\psi^*(2')\rangle \\
&\quad + \langle G^*(0,1')\rangle I(0,2;1',2')\langle G(2,3)\rangle\langle G^*(2',3')\rangle I(3,4;3',4')\langle\psi(4)\psi^*(4')\rangle \\
&\quad + Q(0,3)\langle G(3,1)\rangle\langle G^*(0,1')\rangle I(1,2;1',2')\langle\psi(2)\psi^*(2')\rangle \\
&\quad - Q^*(0,1)\langle\psi^*(1)\rangle\langle\psi(0)\rangle - \langle G(0,1)\rangle I(1,2;0,2')\langle\psi(2)\rangle\langle\psi^*(2')\rangle \\
&\quad - \langle G(0,1)\rangle I(1,2;0,2')\langle G(2,3)\rangle\langle G^*(2',3')\rangle I(3,4;3',4')\langle\psi(4)\psi^*(4')\rangle \\
&\quad - \langle G(0,1)\rangle Q^*(0,3)\langle G^*(3,1')\rangle I(1,2;1',2')\langle\psi(2)\psi^*(2')\rangle \qquad (4.2.31)
\end{aligned}
$$

Next we examine the *integrated optical relation* which is a result of integrating over the 0 coordinate in (4.2.30) and (4.2.31). Upon this integration, all the arguments on the right-hand side of the equation are integration dummy variables. These dummy integration variables are then interchanged. By keeping the arguments of the second and third terms unchanged and interchanging arguments for the rest of the terms in the integration of (4.2.31), we have

$$
\begin{aligned}
&-\int d0\Big\{\langle\nabla\cdot\overline{S}_\psi(0)\rangle + (k^2 - k^{*2})\langle\psi^*(0)\psi(0)\rangle\Big\} \\
&= \langle\psi^*(2')\rangle Q(2',2)\langle\psi(2)\rangle + \langle G^*(0,1')\rangle I(0,2;1',2')\langle\psi(2)\rangle\langle\psi^*(2')\rangle \\
&\quad + \langle G^*(0,1')\rangle I(0,2;1',2')\langle G(2,3)\rangle\langle G^*(2',3')\rangle I(3,4;3',4')\langle\psi(4)\psi^*(4')\rangle \\
&\quad + Q(2',2)\langle G(2,3)\rangle\langle G^*(2',3')\rangle I(3,4;3',4')\langle\psi(4)\psi^*(4')\rangle \\
&\quad - Q^*(2,2')\langle\psi^*(2')\rangle\langle\psi(2)\rangle - \langle G(1',0)\rangle I(0,2;1',2')\langle\psi(2)\rangle\langle\psi^*(2')\rangle \\
&\quad - \langle G(1',0)\rangle I(0,2;1',2')\langle G(2,3)\rangle\langle G^*(2',3')\rangle I(3,4;3',4')\langle\psi(4)\psi^*(4')\rangle \\
&\quad - \langle G(2,3)\rangle Q^*(2,2')\langle G^*(2',3')\rangle I(3,4;3',4')\langle\psi(4)\psi^*(4')\rangle \qquad (4.2.32)
\end{aligned}
$$

Integration is performed over all repeated arguments on the right-hand side of (4.2.32).

We next rearrange the terms in (4.2.32). Then

$$-\int d0\left\{\langle\overline{\nabla}\cdot\overline{S}_\psi(0)\rangle+(k^2-k^{*2})\langle\psi^*(0)\psi(0)\rangle\right\}=\Big\{Q(2',2)-Q^*(2,2')$$
$$+\langle G^*(0,1')\rangle I(0,2;1',2')-\langle G(1',0)\rangle I(0,2;1',2')\Big\}\langle\psi(2)\rangle\langle\psi^*(2')\rangle$$
$$+\Big\{\langle G^*(0,1')\rangle I(0,2;1',2')+Q(2',2)$$
$$-\langle G(1',0)\rangle I(0,2;1',2')-Q^*(2,2')\Big\}$$
$$\cdot\langle G(2,3)\rangle\langle G^*(2',3')\rangle I(3,4;3',4')\langle\psi(4)\psi^*(4')\rangle \tag{4.2.33}$$

For the case of non-absorptive random medium $k=k^*$. The left-hand side of (4.2.33) is zero. For the right-hand side to be zero, we set the terms in the curly brackets in (4.2.33) to be zero. This gives the condition

$$Q(2',2)-Q^*(2,2')+\left[\langle G^*(0,1')\rangle-\langle G(1',0)\rangle\right]I(0,2;1',2')=0 \tag{4.2.34}$$

with integration over repeated arguments. Equation (4.2.34) is the general condition between mass and intensity operators in order to obey the integrated optical relation. The integrated optical relation is

$$-\int d0\left\{\langle\overline{\nabla}\cdot\overline{S}_\psi(0)\rangle+(k^2-k^{*2})\langle\psi^*(0)\psi(0)\rangle\right\}=0 \tag{4.2.35}$$

The ladder approximation and the nonlinear approximation are, respectively,

$$I(1,2;1',2')=C(1-1')\delta(1-2)\delta(1'-2') \tag{4.2.36}$$
$$Q(0,1)=\langle G(0,1)\rangle C(0-1) \tag{4.2.37}$$

These two approximations together obey the condition in (4.2.34). Thus ladder approximation and nonlinear approximation are consistent with each other from the point of view of energy conservation. On the other hand, ladder approximation and bilocal approximation are not consistent with each other in terms of energy conservation.

3 Strong Permittivity Fluctuations

It was noted in Section 1 that when the bilocal approximation is applied in a straightforward manner to vector electromagnetic wave equations, the results are only valid when the permittivity fluctuation variance δ is much less than unity. The results for weak permittivity fluctuations are in Eqs. (4.1.37)–(4.1.45). The small δ criterion severely restricts the usefulness of the random medium model. Geophysical media and composite materials are often mixtures of substances with very different dielectric properties.

In the bilocal-approximated Dyson's equation, both the observation point and the source point of the dyadic Green's function are within the random medium and can coincide with each other within the domain of integration. Physically, this corresponds to the fact that the scatterers act as sources of radiation. By taking into account the singularity of the dyadic Green's function, a strong permittivity fluctuation theory of electromagnetic propagation can be derived that is applicable for larger variances of the permittivity.

In this section we apply the strong permittivity fluctuation theory (SPF) to study scattering from a layer of random medium with spherical correlation function. The first moment of the field is calculated by using the bilocal approximation, and the second moment of the field is calculated by using the distorted Born approximation. The correlation function is obtained by using simple physical arguments and is expressed in terms of the fractional volumes and the particle sizes of the constituents of the dielectric mixture. The low-frequency assumption is made in the derivations. The low-frequency effective propagation constant has an imaginary part that is due to a combination of absorption and scattering. The scattering part is dependent on particle size and can be ignored in the very low frequency limit. It is also shown that the derived result of the effective permittivity in the very low frequency limit is identical to the Polder and van Santern mixing formula. It is also identical to the result of the effective medium theory. In this section we only study media with spherically symmetric correlation functions. Results for anisotropic correlation functions can be found in the references.

3.1 Random Medium with Spherically Symmetric Correlation Function

Consider a medium with permittivity $\epsilon(\overline{r})$ which is a random function of position. The vector wave equation is

$$\nabla \times \nabla \times \overline{E} - k_o^2 \frac{\epsilon(\overline{r})}{\epsilon_o} \overline{E} = 0 \tag{4.3.1}$$

where $k_o = \omega\sqrt{\mu_o \epsilon_o}$ is the free-space wavenumber. We rewrite (4.3.1) as follows:

$$\nabla \times \nabla \times \overline{E} - k_o^2 \frac{\epsilon_g}{\epsilon_o} \overline{E} = k_o^2 \left(\frac{\epsilon(\overline{r}) - \epsilon_g}{\epsilon_o} \right) \overline{E} \tag{4.3.2}$$

The permittivity ϵ_g is chosen such that the effective permittivity calculated by the bilocal approximation differs from ϵ_g by a term that is required to be small. That small difference is dependent on particle size. The particle-size-dependent term vanishes as $k_o \to 0$ so that ϵ_g is the effective permittivity at the very low frequency limit.

Let $\overline{\overline{G}}_g(\overline{r}, \overline{r}')$ be the dyadic Green's function that satisfies a vector wave equation with wavenumber k_g with

$$k_g = \omega\sqrt{\mu_o \epsilon_g} \tag{4.3.3}$$

Thus,

$$\nabla \times \nabla \times \overline{\overline{G}}_g(\overline{r}, \overline{r}') - k_g^2 \overline{\overline{G}}_g(\overline{r}, \overline{r}') = \overline{\overline{I}}\delta(\overline{r} - \overline{r}') \tag{4.3.4}$$

Note that both ϵ_g and $\overline{\overline{G}}_g$ are non-random quantities and are chosen to approximate well the effective background through conditions that are to be imposed later. Using (4.3.4) in (4.3.2), we have

$$\overline{E}(\overline{r}) = \overline{E}_o(\overline{r}) + k_o^2 \int d\overline{r}' \, \overline{\overline{G}}_g(\overline{r}, \overline{r}') \frac{\epsilon(\overline{r}') - \epsilon_g}{\epsilon_o} \cdot \overline{E}(\overline{r}') \tag{4.3.5}$$

where $\overline{E}_o(\overline{r})$ is the field that satisfies the homogeneous wave equation with wavenumber k_g. We note that in (4.3.5) both $\overline{r}$, the observation point, and $\overline{r}'$, the source point of the dyadic Green's function $\overline{\overline{G}}_g(\overline{r}, \overline{r}')$, are inside the random medium. The two points can coincide with each other within the domain of integration of $d\overline{r}'$. The singular nature of the dyadic Green's function at $\overline{r} = \overline{r}'$ must be taken into consideration.

The singularity of the dyadic Green's function depends on the shape of the infinitesimal exclusion volume as discussed in Chapter 2, Section 1.3 of Volume II. For permittivity fluctuations with spherical symmetric correlation functions, a spherical exclusion volume is chosen for the singularity of $\overline{\overline{G}}_g$. We decompose the dyadic Green's function $\overline{\overline{G}}_g(\overline{r}, \overline{r}')$ as follows:

$$\overline{\overline{G}}_g(\overline{r}, \overline{r}') = PS\,\overline{\overline{G}}_g(\overline{r}, \overline{r}') - \frac{\overline{\overline{I}}}{3k_g^2}\delta(\overline{r} - \overline{r}') \tag{4.3.6}$$

where PS stands for principal value. For nonspherical correlation functions, other shapes of exclusion volumes are to be chosen. Substituting (4.3.6) into (4.3.5) gives

$$\overline{F}(\overline{r}) = \overline{E}_o(\overline{r}) + k_o^2 \int d\overline{r}' PS\,\overline{\overline{G}}_g(\overline{r}, \overline{r}')\xi(\overline{r}')\overline{F}(\overline{r}') \tag{4.3.7}$$

where

$$\overline{F}(\overline{r}) = \frac{\epsilon(\overline{r}) + 2\epsilon_g}{3\epsilon_g} \overline{E}(\overline{r}) \tag{4.3.8}$$

$$\xi(\overline{r}) = 3\frac{\epsilon_g}{\epsilon_o}\left(\frac{\epsilon(\overline{r}) - \epsilon_g}{\epsilon(\overline{r}) + 2\epsilon_g}\right) \tag{4.3.9}$$

The diagrammatic method of random medium of Section 1 can be applied to (4.3.7) using $\overline{F}(\overline{r})$ as the field quantity, $PS\overline{\overline{G}}_g(\overline{r},\overline{r}')$ as the propagator, and $k_0^2\xi(\overline{r})$ as the "scatterer."

The expression $\overline{E}(\overline{r}) = \frac{3\epsilon_g}{\epsilon(\overline{r})+2\epsilon_g}\overline{F}(\overline{r})$ can be understood from electrostatics. If $\overline{E}$ is the incident field for a spherical particle with permittivity $\epsilon(\overline{r})$ embedded in a background medium of permittivity ϵ_g, then $3\epsilon_g/(\epsilon(\overline{r})+2\epsilon_g)$ is the factor that relates the internal field to the incident field. Thus $\overline{F}(\overline{r})$ and $\overline{E}(\overline{r})$ play, respectively, the roles of external and internal fields.

For the required small corrections in the bilocal approximation, we choose ϵ_g such that

$$\langle \xi(\overline{r})\rangle = 0 \tag{4.3.10}$$

Then, applying the bilocal approximation to (4.3.7) gives an equation that is analogous to (4.1.26),

$$\langle \overline{F}(\overline{r})\rangle = \overline{E}_o(\overline{r}) + k_o^2 \int d\overline{r}'' \int d\overline{r}' PS\overline{\overline{G}}_g(\overline{r},\overline{r}')\cdot \overline{\overline{\xi}}_{\text{eff}}(\overline{r}',\overline{r}'')\cdot \langle \overline{F}(\overline{r}'')\rangle \tag{4.3.11}$$

with

$$\overline{\overline{\xi}}_{\text{eff}}(\overline{r}',\overline{r}'') = k_o^2\, PS\overline{\overline{G}}_g(\overline{r}',\overline{r}'')R_\xi(|\overline{r}'-\overline{r}''|) \tag{4.3.12}$$

where

$$R_\xi(|\overline{r}'-\overline{r}''|) = \langle \xi(\overline{r}')\xi(\overline{r}'')\rangle \tag{4.3.13}$$

is the spherically symmetric correlation function of $\xi(\overline{r})$. The Fourier transform of $\overline{\overline{\xi}}_{\text{eff}}(\overline{r})$ can be readily determined from (4.3.12).

$$\overline{\overline{\xi}}_{\text{eff}}(\overline{k}) = \int d\overline{r} e^{-i\overline{k}\cdot\overline{r}} k_o^2\, PS\overline{\overline{G}}_g(\overline{r})R_\xi(|\overline{r}|) \tag{4.3.14}$$

3.2 Very Low Frequency Effective Permittivity

At very low frequencies, the scattering attenuation rate can be neglected and the effective permittivity is given by ϵ_g, which obeys the equation from (4.3.9) and (4.3.10):

$$\left\langle \frac{\epsilon(\overline{r})-\epsilon_g}{\epsilon(\overline{r})+2\epsilon_g}\right\rangle = 0 \tag{4.3.15}$$

We shall show that ϵ_g, as given by (4.3.15), when applied to a dielectric mixture, is identical to that of the mixing formula of Polder and van Santern [1946].

Suppose there are n constituents in the mixture with ϵ_p and f_p ($p = 1,2,3,4,...,n$) as the permittivity and the fractional volume, respectively, of the pth constituent. Thus, $Pr\,[\epsilon(r) = \epsilon_p] = f_p$, with Pr standing for probability. Then

$$\sum_{p=1}^{n} f_p = 1 \tag{4.3.16}$$

and (4.3.15) takes the form

$$\sum_{p=1}^{n} \frac{\epsilon_p - \epsilon_g}{\epsilon_p + 2\epsilon_g} f_p = 0 \tag{4.3.17}$$

From (4.3.17) we obtain

$$\sum_{p=1}^{n} \left[1 - \frac{3\epsilon_g}{\epsilon_p + 2\epsilon_g}\right] f_p = 0 \tag{4.3.18a}$$

Using (4.3.16) in (4.3.18a) gives

$$\frac{1}{3\epsilon_g} = \sum_{p=1}^{n} \frac{f_p}{\epsilon_p + 2\epsilon_g} \tag{4.3.18b}$$

Next, we rearrange terms in (4.3.17)

$$\sum_{p=1}^{n} \frac{\epsilon_p - \epsilon_o}{\epsilon_p + 2\epsilon_g} f_p = (\epsilon_g - \epsilon_o) \sum_{p=1}^{n} \frac{f_p}{\epsilon_p + 2\epsilon_g} \tag{4.3.19}$$

where ϵ_o is the free-space permittivity. Using (4.3.18) in (4.3.19), we obtain

$$\sum_{p=1}^{n} \frac{\epsilon_p - \epsilon_o}{\epsilon_p + 2\epsilon_g} f_p = \frac{\epsilon_g - \epsilon_o}{3\epsilon_g} \tag{4.3.20}$$

which is identical to Polder and van Santern's mixing formula. In Polder and van Santern's original derivation, the inhomogeneities are regarded as dipoles. This assumption is consistent with the low-frequency limit.

3.3 Effective Permittivity Under the Bilocal Approximation

To calculate the effective permittivity under the bilocal approximation, we have from (4.3.7) and (4.3.8)

$$\xi(\overline{r})\overline{F}(\overline{r}) = \frac{\overline{D}(\overline{r})}{\epsilon_o} - \frac{\epsilon_g}{\epsilon_o}\overline{E}(\overline{r}) \tag{4.3.21}$$

where $\overline{D}(\overline{r}) = \epsilon(\overline{r})\overline{E}(\overline{r})$ is the electric displacement. Taking the ensemble average of (4.3.7), we obtain

$$\langle\overline{F}(\overline{r})\rangle = \overline{E}_o(\overline{r}) + k_o^2 \int d\overline{r}' P S \overline{\overline{G}}_g(\overline{r}, \overline{r}') \langle \xi(\overline{r}')\overline{F}(\overline{r}')\rangle \tag{4.3.22}$$

Comparing (4.3.22) with (4.3.11) gives

$$\langle \xi(\overline{r})\overline{F}(\overline{r})\rangle = \int d\overline{r}'\ \overline{\overline{\xi}}_{\text{eff}}(\overline{r}-\overline{r}')\cdot\langle\overline{F}(\overline{r}')\rangle \tag{4.3.23}$$

where $\overline{\overline{\xi}}_{\text{eff}}(\overline{r})$ is given in (4.3.12). From (4.3.8) we have

$$\langle\overline{F}(\overline{r})\rangle = \frac{\langle\overline{D}(\overline{r})\rangle}{3\epsilon_g} + \frac{2}{3}\langle\overline{E}(\overline{r})\rangle \tag{4.3.24}$$

Analogous to (4.3.23), we define $\overline{\overline{\epsilon}}_{\text{eff}}(\overline{r})$ such that

$$\langle\overline{D}(\overline{r})\rangle = \int d\overline{r}'\,\overline{\overline{\epsilon}}_{\text{eff}}(\overline{r}-\overline{r}')\cdot\langle\overline{E}(\overline{r}')\rangle \tag{4.3.25}$$

Let $\overline{\overline{\epsilon}}_{\text{eff}}(\overline{k})$, $\overline{\overline{\xi}}_{\text{eff}}(\overline{k})$, $\overline{E}(\overline{k})$, $\overline{D}(\overline{k})$, and $\overline{F}(\overline{k})$ be, respectively, the three-dimensional Fourier transforms of $\overline{\overline{\epsilon}}_{\text{eff}}(\overline{r})$, $\overline{\overline{\xi}}_{\text{eff}}(\overline{r})$, $\overline{E}(\overline{r})$, $\overline{D}(\overline{r})$, and $\overline{F}(\overline{r})$. For example,

$$\overline{\overline{\epsilon}}_{\text{eff}}(\overline{k}) = \int d\overline{r}\,\overline{\overline{\epsilon}}_{\text{eff}}(\overline{r})\,e^{-i\overline{k}\cdot\overline{r}} \tag{4.3.26}$$

The $\overline{k}$ dependence of the effective permittivity $\overline{\overline{\epsilon}}_{\text{eff}}(\overline{k})$ represents spatial dispersion effects. We can now express $\overline{\overline{\epsilon}}_{\text{eff}}(\overline{k})$ in terms of $\overline{\overline{\xi}}_{\text{eff}}(\overline{k})$ as follows. Substituting (4.3.21) into (4.3.23) and taking Fourier transform of the resultant equation yield

$$\frac{\langle\overline{D}(\overline{k})\rangle}{\epsilon_o} = \frac{\epsilon_g}{\epsilon_o}\langle\overline{E}(\overline{k})\rangle + \overline{\overline{\xi}}_{\text{eff}}(\overline{k})\cdot\langle\overline{F}(\overline{k})\rangle \tag{4.3.27}$$

Taking Fourier transform of (4.3.24) and then substituting into (4.3.27) we have

$$\frac{\langle\overline{D}(\overline{k})\rangle}{\epsilon_o} = \frac{\epsilon_g}{\epsilon_o}\langle\overline{E}(\overline{k})\rangle + \overline{\overline{\xi}}_{\text{eff}}(\overline{k})\cdot\left[\frac{\langle\overline{D}(\overline{k})\rangle}{3\epsilon_g} + \frac{2}{3}\langle\overline{E}(\overline{k})\rangle\right]$$

Simplifying gives

$$\begin{aligned}\left(\overline{\overline{I}} - \frac{\epsilon_o}{3\epsilon_g}\overline{\overline{\xi}}_{\text{eff}}(\overline{k})\right)\cdot\frac{\langle\overline{D}(\overline{k})\rangle}{\epsilon_o} &= \frac{\epsilon_g}{\epsilon_o}\langle\overline{E}(\overline{k})\rangle + \frac{2}{3}\overline{\overline{\xi}}_{\text{eff}}(\overline{k})\cdot\langle\overline{E}(\overline{k})\rangle \\ &= \frac{\epsilon_g}{\epsilon_o}\left(\overline{\overline{I}} - \frac{\epsilon_o}{3\epsilon_g}\overline{\overline{\xi}}_{\text{eff}}(\overline{k})\right)\langle\overline{E}(\overline{k})\rangle + \overline{\overline{\xi}}_{\text{eff}}(\overline{k})\cdot\langle\overline{E}(\overline{k})\rangle\end{aligned}$$

Since

$$\langle\overline{D}(\overline{k})\rangle = \overline{\overline{\epsilon}}_{\text{eff}}(\overline{k})\cdot\langle\overline{E}(\overline{k})\rangle \tag{4.3.28a}$$

we then have the effective permittivity tensor in the spectral domain

$$\overline{\overline{\epsilon}}_{\text{eff}}(\overline{k}) = \epsilon_g\overline{\overline{I}} + \epsilon_o\left(\overline{\overline{I}} - \frac{\epsilon_o}{3\epsilon_g}\overline{\overline{\xi}}_{\text{eff}}(\overline{k})\right)^{-1}\overline{\overline{\xi}}_{\text{eff}}(\overline{k}) \tag{4.3.28b}$$

Thus, the second term in (4.3.28b) is the correction provided by the bilocal approximation to $\overline{\overline{\epsilon}}_{\text{eff}}(\overline{k})$. The validity of the bilocal approximation requires

$$|\overline{\overline{\xi}}_{\text{eff}}(\overline{k})| \ll 1 \tag{4.3.29}$$

Making the approximation of (4.3.29) in (4.3.28b) gives

$$\overline{\overline{\epsilon}}_{\text{eff}}(\overline{k}) \simeq \epsilon_g \overline{\overline{I}} + \epsilon_o \overline{\overline{\xi}}_{\text{eff}}(\overline{k}) \tag{4.3.30}$$

For low frequencies, the spatial dispersion effects can be ignored so that the $\overline{k}$ argument in (4.3.30) can be replaced by zero. Thus

$$\overline{\overline{\epsilon}}_{\text{eff}} = \epsilon_g \overline{\overline{I}} + \epsilon_o \overline{\overline{\xi}}^{(0)}_{\text{eff}} \tag{4.3.31}$$

where

$$\overline{\overline{\xi}}^{(0)}_{\text{eff}} = \int d\overline{r}\, \overline{\overline{\xi}}_{\text{eff}}(\overline{r}) \tag{4.3.32}$$

From (4.3.32) and (4.3.12) we have

$$\overline{\overline{\xi}}^{(0)}_{\text{eff}} = k_o^2 \int d\overline{r}\ PS\ \overline{\overline{G}}_g(\overline{r}) R_\xi(|\overline{r}|) \tag{4.3.33a}$$

Thus, in the low-frequency limit we obtain

$$\overline{\overline{\epsilon}}_{\text{eff}} = \epsilon_g \overline{\overline{I}} + k_o^2 \epsilon_o \int d\overline{r}\, PS\, \overline{\overline{G}}_g(\overline{r}) R_\xi(|\overline{r}|) \tag{4.3.33b}$$

Note that

$$\int d\overline{r}\ PS\ \overline{\overline{G}}_g(\overline{r}) R_\xi(|\overline{r}|) = \int d\overline{r} \Big(PSG_1(r)\overline{\overline{I}} + PSG_2(r)\hat{r}\hat{r} \Big) R_\xi(r)$$

where $G_1(r)$ and $G_2(r)$ are given in (4.1.40) and (4.1.41) with k_m replaced by k_g. Since

$$\int_{4\pi} d\Omega \hat{r}\hat{r} = \frac{4\pi}{3}\overline{\overline{I}}$$

we have

$$\begin{aligned}
\int d\overline{r}\ PS\ \overline{\overline{G}}_g(\overline{r}) R_\xi(|\overline{r}|) &= 4\pi \overline{\overline{I}} \int_0^\infty dr\ r^2 \Big(PSG_1(r) + \frac{1}{3} PSG_2(r) \Big) R_\xi(r) \\
&= \frac{2}{3}\overline{\overline{I}} \int_0^\infty dr\ r e^{ik_g r} R_\xi(r) \\
&= \frac{2}{3}\overline{\overline{I}} \int_0^\infty dr\ r(1 + ik_g r) R_\xi(r)
\end{aligned} \tag{4.3.34a}$$

In (4.3.34a) we keep both the leading term in the real part and the leading term in the imaginary part. Thus

$$\epsilon_{\text{eff}} = \epsilon_g + \epsilon_o \frac{2}{3} k_o^2 \int_0^\infty dr\, r\, R_\xi(r) + i\frac{2}{3} k_o^2 k_g \epsilon_o U \tag{4.3.34b}$$

where

$$U = \int_0^{\infty} dr\, r^2\, R_\xi(r) \tag{4.3.35}$$

In (4.3.34*b*) the real and imaginary parts of the second term are always much smaller than the real and imaginary parts, respectively, of the first term. Thus the second term can be neglected. The third term, however, cannot be ignored because its imaginary part can be much larger than the imaginary part of the first term. Hence, (4.3.34*b*) simplifies to

$$\epsilon_{\text{eff}} = \epsilon_g + i\frac{2}{3}k_o^2 k_g \epsilon_o U \tag{4.3.36}$$

We note from (4.3.36) that for the real part, $\text{Re}(\epsilon_{\text{eff}}) \simeq \text{Re}(\epsilon_g)$. However, for the imaginary part, both the absorption attenuation and scattering attenuation are included in the imaginary part of ϵ_{eff}, whereas ϵ_g contains only the absorption effect in its imaginary part.

3.4 Backscattering Coefficients

Consider a plane wave incident on a half-space dielectric mixture with effective permittivity ϵ_{eff}. The field $\overline{F}(\overline{r})$ is decomposed into a coherent part $\langle \overline{F}(\overline{r}) \rangle$ and an incoherent part $\overline{\mathcal{F}}(\overline{r})$, $\overline{F}(\overline{r}) = \langle \overline{F}(\overline{r}) \rangle + \overline{\mathcal{F}}(\overline{r})$. In Chapter 6, Section 2.3 of Volume I, we applied the Born approximation. In this section, the distorted Born approximation will be used to calculate the incoherent scattered field. The distorted Born approximation corresponds physically to the single scattering of the coherent field. It is also known as first-order multiple scattering [Ishimaru, 1978]. The scattered intensity $\langle |\overline{\mathcal{F}}(\overline{r})|^2 \rangle$ in the upper half-space is, under the distorted Born approximation,

$$\begin{aligned} \langle |\overline{\mathcal{F}}(\overline{r})|^2 \rangle = k_o^4 \int\!\!\int_A dx_1\, dy_1 \int_{-\infty}^{0} dz_1 \int\!\!\int_A dx_2\, dy_2 \int_{-\infty}^{0} dz_2 \\ \times C_\xi(|\overline{r}_1 - \overline{r}_2|)(\langle \overline{\overline{G}}_{01}(\overline{r}, \overline{r}_1) \rangle \cdot \langle \overline{F}(\overline{r}_1) \rangle) \cdot (\langle \overline{\overline{G}}{}^*_{01}(\overline{r}, \overline{r}_2) \rangle \cdot \langle \overline{F}^*(\overline{r}_2) \rangle) \end{aligned} \tag{4.3.37}$$

where

$$C_\xi(|\overline{r}_1 - \overline{r}_2|) = \langle \xi(\overline{r}_1) \xi^*(\overline{r}_2) \rangle \tag{4.3.38}$$

A denotes the area of observation by the radar, and $\langle \overline{\overline{G}}_{01}(\overline{r}, \overline{r}_1) \rangle$ is the mean Green's function with $\overline{r}$ in the upper half-space and $\overline{r}_1$ in the lower half-space. The coherent reflection and transmission coefficients for random scattering in the low-frequency limit are the classical Fresnel reflection and transmission coefficients with the permittivity of the lower region replaced by the effective permittivity of the random medium. Thus, we shall make this assumption in calculating the half-space mean Green's function and the mean field in the lower half-space.

For incident horizontally polarized wave with amplitude E_o, the mean field in the random medium is

$$\langle \overline{F}(\overline{r}) \rangle = \hat{e}_{1i}\, E_o\, X_{01i}\, e^{i\overline{K}_i \cdot \overline{r}} \tag{4.3.39}$$

For incident vertically polarized wave, the mean field is

$$\langle \overline{F}(\overline{r}) \rangle = \hat{h}_1(-K_{iz})\, E_o\, Y_{01i}\, \frac{k_o}{K}\, e^{i\overline{K}_i \cdot \overline{r}} \tag{4.3.40}$$

where the subscript i is used to denote incident direction. In (4.3.39) and (4.3.40), $\overline{K}_i = \hat{x}k_{ix} + \hat{y}k_{iy} - \hat{z}K_{iz}$ denotes the transmitted wave vector of the coherent field in the effective medium of the lower half-space, and K is the effective propagation constant $K = \omega\sqrt{\mu_o \epsilon_{\text{eff}}}$. Since the observation point $\overline{r}$ of $\langle \overline{\overline{G}}_{01}(\overline{r}, \overline{r}') \rangle$ is in the far field in region 0, we obtain

$$\langle \overline{\overline{G}}_{01}(\overline{r}, \overline{r}') \rangle = \frac{k_{sz}}{K_{sz}} \left[X_{10s}\hat{e}_s\hat{e}_{1s} + \frac{K}{k_o} Y_{10s}\hat{h}(k_{sz})\hat{h}_1(K_{sz}) \right] \frac{e^{ik_o r}}{4\pi r} \exp(-i\overline{K}_s \cdot \overline{r}') \tag{4.3.41}$$

where the subscript s denotes scattered direction. The expression in (4.3.41) is similar to the unperturbed Green's function $G_{01}^{(0)}(\overline{r}, \overline{r}')$, with the key difference being that the mean permittivity ϵ_{1m} and mean wavenumber k_{1m} are replaced, respectively, by the effective permittivity ϵ_{eff} and the effective wavenumber K. In (4.3.39) through (4.3.41), $\overline{K}_s = \hat{x}k_{sx} + \hat{y}k_{sy} + \hat{z}K_{sz}$, with $k_x = k\sin\theta\cos\phi$, $k_y = k\sin\theta\sin\phi$, $K_z = (K^2 - k_x^2 - k_y^2)^{1/2}$, and $k_z = k\cos\theta$. In the incident direction, $(\theta, \phi) = (\theta_i, \phi_i)$; in the scattered direction $(\theta, \phi) = (\theta_s, \phi_s)$.

In (4.3.39) through (4.3.41), X_{01}, Y_{01}, X_{10}, and Y_{10} are Fresnel transmission coefficients with k_{1z} and ϵ_1 replaced, respectively, by ϵ_{eff} and K_z. For example,

$$Y_{01} = \frac{2\epsilon_{\text{eff}} k_z}{\epsilon_{\text{eff}} k_z + \epsilon_o K_z}$$

In (4.3.39) and (4.3.41), $\hat{e}$ and $\hat{h}$ are, respectively, transverse electric (TE) and transverse magnetic (TM) polarization vectors in the upper region, and $\hat{e}_1$ and $\hat{h}_1$ are the TE and TM polarization vectors in the lower half-space with medium 1 parameters replaced by the effective parameters of ϵ_{eff}, K, and K_z. The unit polarization vectors $\hat{e}$ and $\hat{h}$ are as defined in (2.1.18*a*) and (2.1.18*b*) of Volume I.

In the low-frequency limit, when the correlation lengths are short compared with the wavelength, $C_\xi(|\overline{r}_1 - \overline{r}_2|)$ is peaked at $\overline{r}_1 = \overline{r}_2$ and (4.3.37) can be approximated by

$$\langle|\overline{\mathcal{F}}(\overline{r})|^2\rangle = 4\pi k_o^4 W \int\int_A dx_1\, dy_1 \int_{-\infty}^{0} dz_1 \left|\langle\overline{\overline{G}}_{01}(\overline{r},\overline{r}_1)\rangle \cdot \langle\overline{F}(\overline{r}_1)\rangle\right|^2 \quad (4.3.42)$$

where

$$W = \int_0^\infty dr\, r^2\, C_\xi(r) \quad (4.3.43)$$

To calculate backscattering coefficients, we use the backscattering direction in the expression for $\langle\overline{\overline{G}}_{01}\rangle$ with $k_{xs} = -k_{xi}$ and $k_{ys} = -k_{yi}$. Thus,

$$\sigma_{vv} = \frac{k_o^4 W k_{iz}^2}{4K_{iz}''\,|K_{iz}|^2}\,|Y_{10i}Y_{01i}|^2 \quad (4.3.44)$$

$$\sigma_{hh} = \frac{k_o^4 W k_{iz}^2}{4K_{iz}''\,|K_{iz}|^2}\,|X_{10i}X_{01i}|^2 \quad (4.3.45)$$

$$\sigma_{vh} = \sigma_{hv} = 0 \quad (4.3.46)$$

with $K_{iz}'' = \mathrm{Im}(K_{iz})$.

As discussed in Section 2.1, the distorted Born approximation consists of the first term of (4.2.23). It is valid for small albedo. The depolarization return for a random medium with a spherically symmetric correlation function is proportional to the square of the albedo. Calculation of the depolarization return requires a second-order solution for a spherically symmetric correlation function. The depolarization calculation can be obtained by using the second term of (4.2.23).

3.5 Results of Effective Permittivity and Bistatic Coefficients

In this section we illustrate effective permittivity and backscattering coefficients using typical parameters encountered in microwave remote sensing of dry and wet snow. Dry snow is a mixture of two constituents, air and ice, with the fractional volume of ice between 10% and 40%. Wet snow is a mixture of three constituents, air, ice, and water, with the percentage of water between 0% and 10%. We use ϵ_b, ϵ_{s1}, and ϵ_{s2} to denote permittivities of air, ice, and water, respectively, and use f_b, f_{s1}, and f_{s2} to denote fractional volumes of air, ice, and water, respectively. The permittivity of ice, ϵ_{s1}, is chosen to be $3.2(1 + i0.001)\epsilon_o$; and the permittivity of water, ϵ_{s2}, obeys the relation

$$\frac{\epsilon_{s2}}{\epsilon_o} = 5.5 + \frac{82.5}{1 - i3.59/\lambda} \quad (4.3.47)$$

where λ is wavelength in centimeters.

Mixture of Two Constituents

The parameters are ϵ_b, f_b, ϵ_s, and f_s with $f_b + f_s = 1$. The quantity ϵ_g is first determined by using (4.3.20). We use the simple correlation functions

$$R_\xi(r) = \begin{cases} \langle \xi^2 \rangle & \text{for } \; 0 \le r \le a \\ 0 & \text{otherwise} \end{cases} \tag{4.3.48}$$

$$C_\xi(r) = \begin{cases} \langle |\xi|^2 \rangle & \text{for } \; 0 \le r \le a \\ 0 & \text{otherwise} \end{cases} \tag{4.3.49}$$

Taking averages, $\langle \xi^2 \rangle = f_s \xi_s^2 + f_b \xi_b^2$ and $\langle |\xi|^2 \rangle = f_s |\xi_s|^2 + f_b |\xi_b|^2$. From (4.3.9) we have $\xi_s = 3r_s \epsilon_g / \epsilon_o$, $\xi_b = 3r_b \epsilon_g / \epsilon_o$ and

$$r_s = \frac{\epsilon_s - \epsilon_g}{\epsilon_s + 2\epsilon_g} \tag{4.3.50}$$

$$r_b = \frac{\epsilon_b - \epsilon_g}{\epsilon_b + 2\epsilon_g} \tag{4.3.51}$$

Using (4.3.36), (4.3.44), and (4.3.45) gives

$$\epsilon_{\text{eff}} = \epsilon_g [1 + i2k_g^3 a^3 (f_s r_s^2 + f_b r_b^2)] \tag{4.3.52}$$

$$\sigma_{hh} = 3 |k_g|^4 a^3 \left[f_s |r_s|^2 + f_b |r_b|^2 \right] \frac{k_{iz}^2}{|K_{iz}|^2} |X_{01i} X_{10i}|^2 \frac{1}{4K_{iz}''} \tag{4.3.53}$$

$$\sigma_{vv} = 3 |k_g|^4 a^3 \left[f_s |r_s|^2 + f_b |r_b|^2 \right] \frac{k_{iz}^2}{|K_{iz}|^2} |Y_{01i} Y_{10i}|^2 \frac{1}{4K_{iz}''} \tag{4.3.54}$$

In Figs. 4.3.1 and 4.3.2 the real and imaginary parts of ϵ_g / ϵ_o and $\epsilon_{\text{eff}} / \epsilon_o$ are plotted as a function of the fractional volume of scatterers. The limit of $f_s = 1$ corresponds to the entire volume being occupied by scatterers. This is not attainable for particles of the same size and is purely of theoretical interest. In Fig. 4.3.3 the backscattering coefficients are plotted as a function of incidence angle θ_i at a frequency of 8.6 GHz and for a half-space with a fractional volume of scatterers equal to 30%. The calculated effective permittivity is $(1.47 + i0.663 \times 10^{-3})\epsilon_o$. The results of the figure indicate that the polarization difference is small.

Mixture of Three Constituents

For three constituents we have a background medium with permittivity ϵ_b, two kinds of scatterers with permittivities ϵ_{s1} and ϵ_{s2}, and fractional volumes f_{s1} and f_{s2}. The scatterers are assumed to be spherical with radii a_1 and

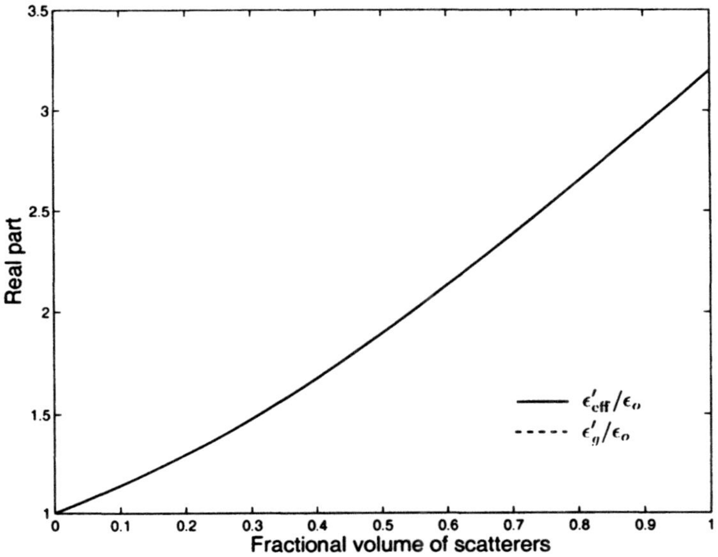

Figure 4.3.1 Real part of ϵ_g/ϵ_o and $\epsilon_{\rm eff}/\epsilon_o$ as a function of fractional volume of scatterers for $\epsilon_b = \epsilon_o$, $\epsilon_s = 3.2(1 + i0.001)\epsilon_o$, frequency = 8.6 GHz, and $a = 0.5$ mm.

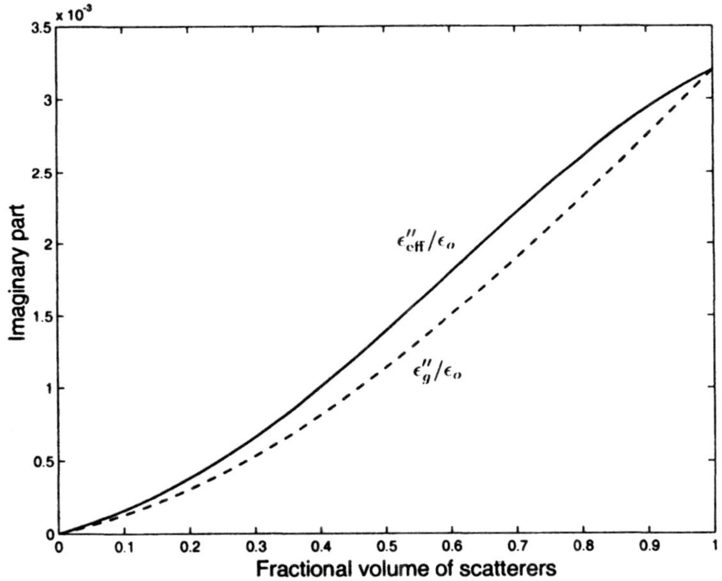

Figure 4.3.2 Imaginary part of ϵ_g/ϵ_o and $\epsilon_{\rm eff}/\epsilon_o$ as a function of fractional volume of scatterers for $\epsilon_b = \epsilon_o$, $\epsilon_s = 3.2(1 + i0.001)\epsilon_o$, frequency = 8.6 GHz, and $a = 0.5$ mm.

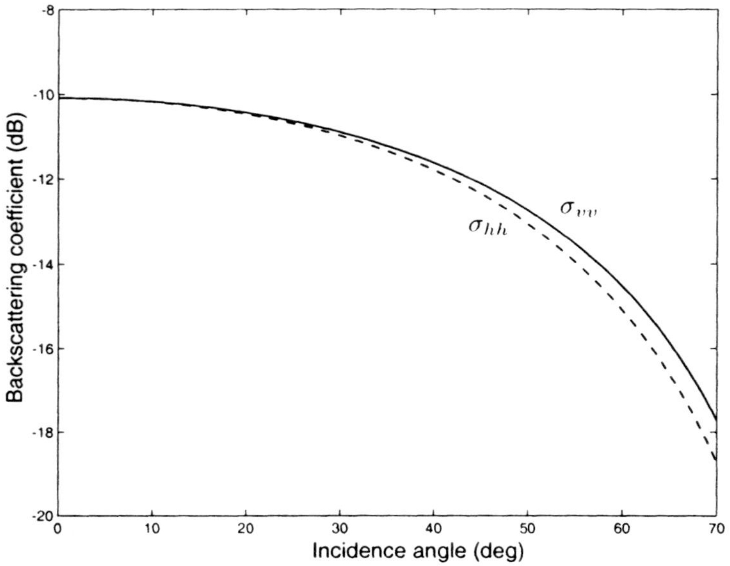

Figure 4.3.3 Backscattering coefficients as a function of incidence angle θ_i for $\epsilon_b = \epsilon_o$, $\epsilon_s = 3.2(1 + i0.001)\epsilon_o$, $f_s = 30\%$, $a = 0.5$ mm, and frequency = 8.6 GHz. Calculated value of ϵ_{eff} is $(1.47 + i0.663 \times 10^{-3})\epsilon_o$.

a_2. We have $f_b + f_{s1} + f_{s2} = 1$. The quantity ϵ_g is first calculated by using (4.3.20). Assuming, without loss of generality, that $a_2 \geq a_1$, the correlation functions are

$$R_\xi(r) = \begin{cases} \langle \xi^2 \rangle & \text{for } 0 \leq r \leq a_1 \\ \xi_{s2}^2 f_{s2}(1 + f_{s2}) & \text{for } a_1 < r \leq a_2 \\ 0 & \text{otherwise} \end{cases} \tag{4.3.55}$$

$$C_\xi(r) = \begin{cases} \langle |\xi|^2 \rangle & \text{for } 0 \leq r \leq a_1 \\ |\xi_{s2}|^2 f_{s2}(1 + f_{s2}) & \text{for } a_1 < r \leq a_2 \\ 0 & \text{otherwise} \end{cases} \tag{4.3.56}$$

Using (4.3.35) and (4.3.43) to calculate U and W respectively, we have

$$U = \frac{3\epsilon_g^2}{\epsilon_o^2}\left\{(f_{s1}r_{s1}^2 + f_{s2}r_{s2}^2 + f_b r_b^2)a_1^3 + r_{s2}^2 f_{s2}(1 + f_{s2})(a_2^3 - a_1^3)\right\} \tag{4.3.57}$$

$$W = 3\frac{|\epsilon_g|^2}{\epsilon_o^2}\left\{(f_{s1}|r_{s1}|^2 + f_{s2}|r_{s2}|^2 + f_b|r_b|^2)a_1^3 + |r_{s2}|^2 f_{s2}(1 + f_{s2})(a_2^3 - a_1^3)\right\} \tag{4.3.58}$$

with $r_{s1} = (\epsilon_{s1} - \epsilon_g)/(\epsilon_{s1} + 2\epsilon_g)$, $r_{s2} = (\epsilon_{s2} - \epsilon_g)/(\epsilon_{s2} + 2\epsilon_g)$, and r_b as given by (4.3.51).

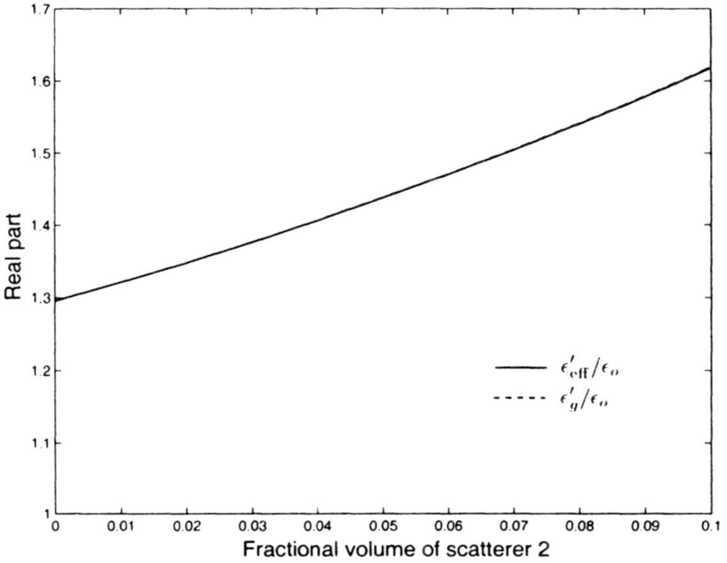

Figure 4.3.4 Real part of ϵ_g/ϵ_o and $\epsilon_{\rm eff}/\epsilon_o$ as a function of fractional volume of scatterer 2 for $\epsilon_b = \epsilon_o$, $\epsilon_{s1} = 3.2(1 + i0.001)\epsilon_o$, $f_b = 80\%$, $a_1 = 0.5$ mm, $a_2 = 0.5$ mm, and frequency = 17 GHz.

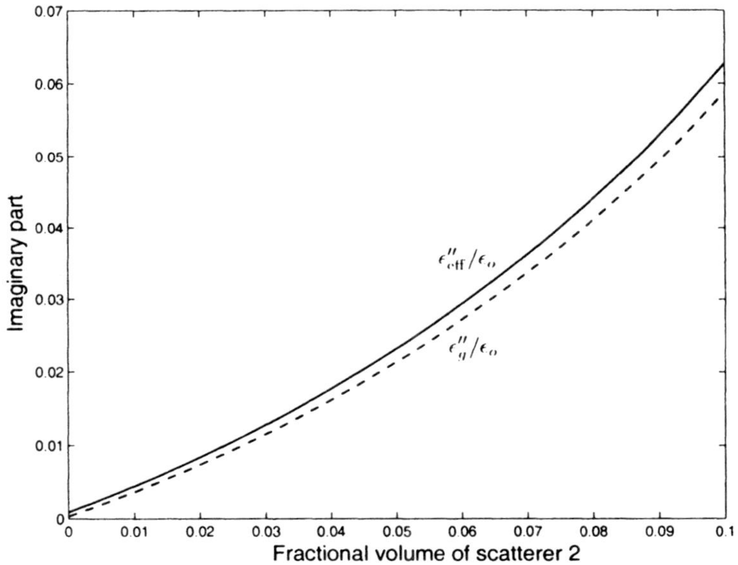

Figure 4.3.5 Imaginary part of ϵ_g/ϵ_o and $\epsilon_{\rm eff}/\epsilon_o$ as a function of fractional volume of scatterer 2 for $\epsilon_b = \epsilon_o$, $\epsilon_{s1} = 3.2(1 + i0.001)\epsilon_o$, $f_b = 80\%$, $a_1 = 0.5$ mm, $a_2 = 0.5$ mm, and frequency = 17 GHz.

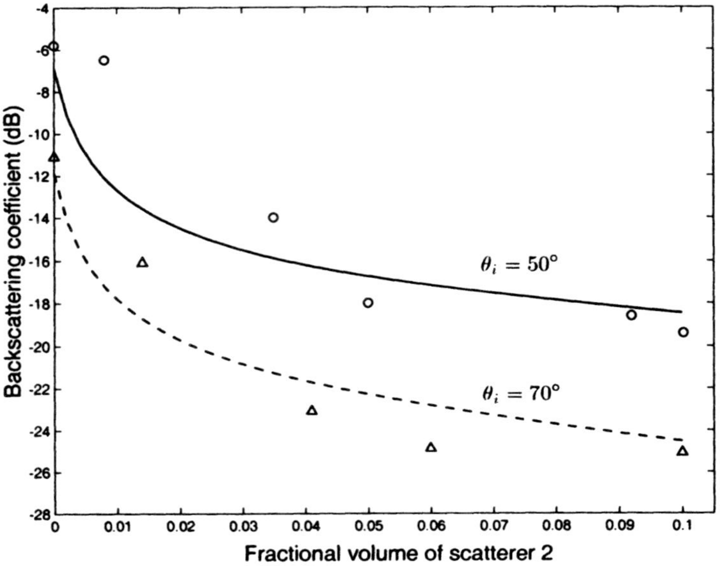

Figure 4.3.6 Backscattering coefficients σ_{hh} at $\theta_i = 50°$ and $\theta_i = 70°$ as a function of fractional volume of scatterer 2 for $\epsilon_b = \epsilon_o$, $\epsilon_{s1} = 3.2(1 + i0.001)\epsilon_o$, $f_b = 80\%$, $a_1 = 0.5$ mm, $a_2 = 0.5$ mm, and frequency = 17 GHz. Comparison with experimental data.

We use the theory for a mixture of three constituents to illustrate the application to wet snow. In Figs. 4.3.4, 4.3.5, and 4.3.6, we plot, respectively, the real part of the effective dielectric constant, the imaginary part of the effective dielectric constant, and backscattering coefficients as a function of fractional volume of scatterer 2 from 0% to 10%. The permittivity of scatterer 2 is taken to be that of water and is given by (4.3.47). The fractional volume of the background medium is fixed at 80%. Therefore, as f_{s2} increases from 0 to 10%, f_{s1} decreases from 20% to 10%. From Fig. 4.3.6 we note that the backscattering coefficients decrease as the wetness of the snow is increased. The radii of both scatterers 1 and 2 are chosen to be 0.5 mm, which falls within the grain size of snow.

In Fig. 4.3.6 the theoretical result is also used to match experimental data taken from wet snow [Stiles and Ulaby, 1980]. The frequency of operation for the data was 17 GHz and was for HH polarization. The data were plotted as a function of volumetric snow wetness, M_v, which we have equated with f_{s2}. The incident angles were 50° and 70°, and the snow depth d was 30 cm with water equivalent W equal to 5 cm. Since ice has a smaller density than water, the quantity obtained by dividing the water equivalent W by snow depth d is generally about 5% to 20% smaller than the fractional volume of ice. Thus in Fig. 4.3.6 we take the fractional volume of air to be 80%. There is good match between theory and experimental data.

In recent years, extensive studies have been made on strong permittivity fluctuation theory with extensions and improvements. Applications have been made to dry and wet snow. Relations were also established between the dense medium discrete scatterer model and the strong permittivity fluctuation model [Veysoglu, 1994]. Interested readers can refer to the references.

REFERENCES AND ADDITIONAL READINGS

Chuah, H. T. and H. S. Tan (1989), A high order renormalization method for radar backscatter from a random medium, *IEEE Trans. Geosci. Remote Sens.*, **27**, 79–85.

Frisch, V. (1968), Wave propagation in random medium, *Probabilistic Methods in Applied Mathematics*, **1**, Bharuch-Reid, Ed., Academic Press.

Fung, A. K. (1979), Scattering from a vegetation layer, *IEEE Trans. Geosci. Electronics*, **17**, 1–5.

Fung, A. K. and H. S. Fung (1979), Application of first-order renormalization method to scattering from a vegetation-like half-space, *IEEE Trans. Geosci. Electronics*, **15**, 189–195.

Ishimaru, A. (1978), *Wave Propagation and Scattering in Random Media*, **1** and **2**, Academic Press, New York.

Jin, Y. Q. (1989), The radiative transfer equation for strongly-fluctuation continuous random media., *J. Quant. Spectrosc. Radiat. Transfer.*, Vol. 42, 529–537.

Jin, Y. Q. (1985), Strong fluctuation theory for a bounded layer of random discrete scatterers, Ph.D. thesis, Massachusetts Institute of Technology.

Jin, Y. Q. and J. A. Kong (1984), Strong fluctuation theory for electromagnetic wave scattering by a layer of random discrete scatters, *J. Appl. Phys.*, **55**, 1364–1369.

Jin, Y. Q. and J. A. Kong (1985), Strong fluctuation theory for scattering, attenuation, and transmission of microwave through snowfall, *IEEE Trans. Geosci. Remote Sens.*, **23**, 754–760.

Karal, F. C. and J. B. Keller (1960), Elastic electromagnetic and other waves in random medium, *J. Math. Phys.*, **5**, 537–547.

Keller, J. B. (1964), Stochastic equations and wave propagation in random media, *Proc. Symp. Appl. Math.*, **16**, 145–170 (Am Math. Soc., Providence, RI).

Keller, J. B. and F. C. Karal (1966), Effective dielectric constant, permeability and conductivity of a random medium and the velocity and attenuation coefficient of coherent waves, *J. Math. Phys.*, **7**, 661–670.

Lim, H. H., M. E. Veysoglu, S. H. Yueh, R. T. Shin, and J. A. Kong (1994), Random medium model approach to scattering from a random collection of discrete scatterers, *J. Electromag. Waves and Appl.*, **8**, 801–817.

Mätzler, C. (1996), Microwave permittivity of dry snow, *IEEE Trans. Geosci. Remote Sens.*, **34(2)**, 573-581.

Mätzler, C. (1998), Improved Born approximation for scattering of radiation in a granular medium, *J. Appl. Phys.*, **83(11)**, 6111–6117.

Nghiem, S. V. (1991), Electromagnetic wave models for polarimetric remote sensing of geophysical media, Ph.D. thesis, Massachusetts Institute of Technology.

Nghiem, S. V., R. Kwok, J. A. Kong, R. T. Shin (1993), A model with ellipsoidal scatterers for polarimetric remote sensing of anisotropic layered media, *Radio Sci.*, **28(5)**, 687–703.

Nghiem, S. V., T. Le Toan, J. A. Kong, H. C. Han, M. Borgeaud (1993), Layer model with random spheroidal scatterers for remote sensing of vegetation canopy, *J. Electromag. Waves and Appl.*, **7(1)**, 49–75.

Polder, D. and J. H. van Santern (1946), The effective permeability of mixture of solids, *Physica*, **12**, 257–271.

Ryzhov, Y. A. and V. V. Tamoikin (1970), Radiation and propagation of electromagnetic waves in randomly inhomogeneous media, *Radiophysics Quantum Electron.*, **13**, 273–300.

Sihvola, A. H. (1999), *Electromagnetic Mixing Formulas and Applications*, The Institute of Electrical Engineering, London, UK.

Sihvola, A. H. and J. A. Kong (1988), Effective permittivity of dielectric mixtures, *IEEE Trans. Geosci. Remote Sens.*, **26(4)**, 420–429; *see also* "Correction" in *IEEE Trans. Geosci. Remote Sens.*, **27(1)**, 101–102, 1999.

Stiles, W.H. and F. T. Ulaby (1980), Radar observation of snowpacks, in NASA Conference Publication 2153 Microwave Remote Sensing of Snowpack Properties, Workshop at Fort Collins, Colorado, May 20–22,1980.

Stogryn, A. (1974), Electromagnetic scattering by random dielectric constant fluctuations in a bounded medium, *Radio Sci.*, **9**, 509–518.

Stogryn, A. (1983), A note on the singular part of the dyadic Green's function in strong fluctuation theory, *Radio Sci.*, **18**, 1283–1286.

Stogryn, A. (1984a), The bilocal approximation for the effective dielectric constant of an isotropic random medium, *IEEE Trans. Antennas Propagat.*, **32**, 517–520.

Stogryn, A. (1984b), Correlation functions for random granular medium in strong fluctuation theory, *IEEE Trans. Geosci. Remote Sens.*, **22**, 150–154.

Stogryn, A. (1990), Strong fluctuation theory equations for electric field second moments in anisotropic media, *IEEE Trans. Antennas Propagat.*, **38(7)**, 1099–1101.

Tamoikin, V. V. (1971),The average field in a medium having strong anisotropic inhomogeneities, *Radiophysics Quantum Electron.*, **14**, 228–233.

Tan, H. S. and A. K. Fung (1979), The mean Green's dyadic for a half-space random medium: A non-linear approximation, *IEEE Trans. Antennas Propagat.*, **27**, 517–523.

Tan, H. S., A. K. Fung, and H. Eom (1980), A second-order renormalization theory for cross-polarized backscatter from a half-space random medium, *Radio Sci.*, **15**, 1059–1065.

Tatarskii, V. I. (1961), *Wave Propagation in a Turbulent Medium*, translated from Russian by R. Silverman, McGraw-Hill, New York.

Tatarskii, V. I. (1964), Propagation of electromagnetic waves in a medium with strong dielectric fluctuations, *Sov. Phys. JETP*, **19**, 946–953.

Tatarskii, V. I. and M. E. Gertsenshtein (1963), Propagation of waves in a medium with strong fluctuations of the refractive index, *Sov. Phys. JETP*, **17**, 458–463.

Tatarskii, V. I. (1971), The effects of turbulent atmosphere on wave propagation, National Tech. Information Service, 472, Springfield, VA.

Tsang, L. and J. A. Kong (1980), Thermal microwave emission from a three-layer random medium with three-dimensional variations, *IEEE Trans. Geosci. Remote Sens.*, **18**, 212–216.

Tsang, L. and J. A. Kong (1981a), Scattering of electromagnetic waves from random media with strong permittivity fluctuations, *Radio Sci.*, **16**, 303–320.

Tsang, L. and J. A. Kong (1981b), Application of strong fluctuation random medium theory to scattering from vegetation-like half-space, *IEEE Trans. Geosci. Remote Sens.*, **19**, 62–69.

Tsang, L., J. A. Kong, and R. W. Newton (1982), Application of strong fluctuation random medium theory to scattering of electromagnetic waves from a half-space of dielectric mixture, *IEEE Trans. Antennas Propagat.*, **30**, 292–302.

Veysoglu, M. E. (1994), Direct and inverse scattering models for random media and rough surfaces, Ph.D. thesis, Massachusetts Institute of Technology.

Wang, H., J. Pulliainen, and M. Hallikainen (1998), Correlation functions and correlation lengths for dry snow, *J. Electromag. Waves and Appl.*, **12**, 1337–1347.

Wang, H., J. Pulliainen, and M. Hallikainen (2000a), Extinction behaviour of dry snow at microwave range up to 90 GHz by using strong fluctuation theory, *Progress in Electromag. Res.*, **25**, 39–50.

Wang, H., J. Pulliainen, and M. Hallikainen (2000b), Microwave emission model for dry snow by using radiative transfer and strong fluctuation theory, *Progress in Electromag. Res.*, **29**, 39–55.

Yueh, H. Y., R. T. Shin, and J. A. Kong (1990), Scattering from randomly oriented scatterers with strong permittivity fluctuations, *J. Electromag. Waves and Appl.*, **4**, 983–1004.

Chapter 5

MULTIPLE SCATTERING THEORY FOR DISCRETE SCATTERERS

In this chapter, we study the multiple scattering theory for discrete scatterers. We establish the Foldy-Lax multiple scattering equations rigorously. Configurational averages are taken leading to a hierarchy of equations. However, to solve the equations, approximations need to be made. We study two approximations: the effective field approximation and the quasi-crystalline approximation. The coherent potential is then introduced to impose self-consistent approximations. Energy conservation is expressed in terms of the integrated optical relation. It is shown that if (*i*) the quasi-crystalline approximation with coherent potential applied to the first moment equation and (*ii*) the correlated ladder approximation is applied to the second moment equation, then the integrated optical relation is obeyed. In Section 5, we show the derivation of the radiative transfer equation from the ladder approximation for isotropic point scatterers.

1 Transition Operator

Consider the case of a single discrete scatterer centered at the origin. The permittivity is $\epsilon_p(\overline{r})$. Then the equation that governs the Green's function is

$$\nabla \times \nabla \times \overline{\overline{G}}_s(\overline{r}, \overline{r}') - \omega^2 \mu \epsilon_p(\overline{r}) \overline{\overline{G}}_s(\overline{r}, \overline{r}') = \overline{\overline{I}} \delta(\overline{r} - \overline{r}') \tag{5.1.1}$$

The equation that governs the free-space dyadic Green's function is

$$\nabla \times \nabla \times \overline{\overline{G}}_o(\overline{r}, \overline{r}') - k^2 \overline{\overline{G}}_o(\overline{r}, \overline{r}') = \overline{\overline{I}} \delta(\overline{r} - \overline{r}') \tag{5.1.2}$$

We can rewrite (5.1.1) as

$$\nabla \times \nabla \times \overline{\overline{G}}_s(\overline{r}, \overline{r}') - k^2 \overline{\overline{G}}_s(\overline{r}, \overline{r}') = \overline{\overline{I}} \delta(\overline{r} - \overline{r}') + \left(k_p^2(\overline{r}) - k^2\right) \overline{\overline{G}}_s(\overline{r}, \overline{r}') \tag{5.1.3}$$

where

$$k_p^2(\overline{r}) = \begin{cases} \omega^2 \mu \epsilon_p(\overline{r}) & \text{for } \overline{r} \text{ inside the particle} \\ k^2 & \text{for } \overline{r} \text{ outside the particle} \end{cases} \tag{5.1.4}$$

We can regard the second term on the right-hand side in (5.1.3) as the source term for $\overline{\overline{G}}_s$. Then it follows that the Green's function obeys the integral equation

$$\overline{\overline{G}}_s(\overline{r}, \overline{r}') = \overline{\overline{G}}_o(\overline{r}, \overline{r}') + \int_{V_p} d\overline{r}'' \overline{\overline{G}}_o(\overline{r}, \overline{r}'') \left(k_p^2(\overline{r}'') - k^2\right) \overline{\overline{G}}_s(\overline{r}'', \overline{r}') \tag{5.1.5}$$

where V_p is the volume occupied by the scatterer. The operator notation is next introduced to put the scattering equations in a more compact form. We let $\overline{\overline{G}}_{sop}$ be the dyadic Green's operator and $\overline{\overline{G}}_{oop}$ be the free space dyadic Green's operator. Dirac's notation can be used to represent the operator notation. In coordinate representation we have

$$\overline{\overline{G}}_o(\overline{r},\overline{r}') = \langle \overline{r}|\overline{\overline{G}}_{oop}|\overline{r}'\rangle \tag{5.1.6}$$

$$\overline{\overline{G}}_s(\overline{r},\overline{r}') = \langle \overline{r}|\overline{\overline{G}}_{sop}|\overline{r}'\rangle \tag{5.1.7}$$

Define the potential function as

$$U(\overline{r}) = \begin{cases} 0 & \text{if } \overline{r} \text{ outside } V_p \\ k_p^2(\overline{r}) - k^2 & \text{if } \overline{r} \text{ inside } V_p \end{cases} \tag{5.1.8}$$

The potential operator U_{op} is

$$\langle \overline{r}|U_{op}|\overline{r}'\rangle = \langle \overline{r}|U(\overline{r}_{op})|\overline{r}'\rangle = U(\overline{r})\langle \overline{r}|\overline{r}'\rangle \tag{5.1.9}$$

where $\overline{r}_{op}$ is the position operator such that

$$\overline{r}_{op}|\overline{r}\rangle = \overline{r}|\overline{r}\rangle \tag{5.1.10}$$

We also have the orthonormal relation of a bra and a ket

$$\langle \overline{r}|\overline{r}'\rangle = \delta(\overline{r} - \overline{r}') \tag{5.1.11}$$

and the unit operator

$$\overline{\overline{I}}_{op} = \int d\overline{r}|\overline{r}\rangle\overline{\overline{I}}\langle \overline{r}| \tag{5.1.12}$$

Equation (5.1.12) is also a completeness relation for the coordinate representation. Using the operator notations, the volume integral equation in (5.1.5) can be put in the following form

$$\overline{\overline{G}}_{sop} = \overline{\overline{G}}_{oop} + \overline{\overline{G}}_{oop}U_{op}\overline{\overline{G}}_{sop} \tag{5.1.13}$$

To show that, we take (5.1.13) in the coordinate representation by applying the bra $\langle \overline{r}|$ from the left and the ket $|\overline{r}'\rangle$ from the right. Thus

$$\langle \overline{r}|\overline{\overline{G}}_{sop}|\overline{r}'\rangle = \langle \overline{r}|G_{oop}|\overline{r}'\rangle + \langle \overline{r}|\overline{\overline{G}}_{oop}U_{op}\overline{\overline{G}}_{sop}|\overline{r}'\rangle \tag{5.1.14}$$

We next apply the unit operator of (5.1.12) to the second term in (5.1.14).

$$\begin{aligned}\overline{\overline{G}}_s(\overline{r},\overline{r}') &= \overline{\overline{G}}_o(\overline{r},\overline{r}') + \int d\overline{r}'' \int d\overline{r}''' \langle \overline{r}|\overline{\overline{G}}_{oop}|\overline{r}''\rangle\langle \overline{r}''|U_{op}|\overline{r}'''\rangle\langle \overline{r}'''|\overline{\overline{G}}_{sop}|\overline{r}'\rangle \\ &= \overline{\overline{G}}_o(\overline{r},\overline{r}') + \int d\overline{r}''\overline{\overline{G}}_o(\overline{r},\overline{r}'')U(\overline{r}'')\overline{\overline{G}}_s(\overline{r}'',\overline{r}')\end{aligned} \tag{5.1.15}$$

A momentum representation $|\overline{p}\rangle$ can also be used with the inner product

$$\langle \overline{r}|\overline{p}\rangle = e^{i\overline{p}\cdot\overline{r}} \tag{5.1.16}$$

A complex conjugate is obtained if we exchange the bra and ket.

$$\langle \overline{p}|\overline{r}\rangle = e^{-i\overline{p}\cdot\overline{r}} \tag{5.1.17}$$

The orthogonality relation for momentum states is

$$\langle \overline{p} | \overline{p}' \rangle = (2\pi)^3 \delta(\overline{p} - \overline{p}') \tag{5.1.18}$$

The unit operator can also be represented in the momentum representation as

$$\overline{\overline{I}}_{op} = \int \frac{d\overline{p}}{(2\pi)^3} |\overline{p}\rangle \overline{\overline{I}} \langle \overline{p}| \tag{5.1.19}$$

For the momentum operator $\overline{p}_{op}$ we have

$$\overline{p}_{op} |\overline{p}\rangle = \overline{p} |\overline{p}\rangle \tag{5.1.20}$$

Consider the $\overline{p}_{op}$ acting on an operator $\overline{\overline{A}}$ in the coordinate representation

$$\begin{aligned}
\langle \overline{r} | \overline{p}_{op} \cdot \overline{\overline{A}} | \overline{r}' \rangle &= \int \frac{d\overline{p}}{(2\pi)^3} \int d\overline{r}'' \langle \overline{r} | \overline{p}_{op} | \overline{p} \rangle \langle \overline{p} | \overline{r}'' \rangle \langle \overline{r}'' | \overline{\overline{A}} | \overline{r}' \rangle \\
&= \int \frac{d\overline{p}}{(2\pi)^3} \int d\overline{r}'' e^{i\overline{p}\cdot\overline{r}} e^{-i\overline{p}\cdot\overline{r}''} \overline{p} \cdot \langle \overline{r}'' | \overline{\overline{A}} | \overline{r}' \rangle \\
&= -i\nabla \cdot \int \frac{d\overline{p}}{(2\pi)^3} \int d\overline{r}'' e^{i\overline{p}\cdot(\overline{r}-\overline{r}'')} \langle \overline{r}'' | \overline{\overline{A}} | \overline{r}' \rangle = -i\nabla \cdot \overline{\overline{A}}(\overline{r}, \overline{r}')
\end{aligned}$$

Thus in coordinate representation we have

$$\overline{p}_{op} = -i\nabla \tag{5.1.21}$$

The equation for $\overline{\overline{G}}_o(\overline{r}, \overline{r}')$ as represented by (5.1.2) can be put in the form

$$\nabla \nabla \cdot \overline{\overline{G}}_o(\overline{r}, \overline{r}') - \nabla^2 \overline{\overline{G}}_o(\overline{r}, \overline{r}') - k^2 \overline{\overline{G}}_o(\overline{r}, \overline{r}') = \overline{\overline{I}} \delta(\overline{r} - \overline{r}') \tag{5.1.22}$$

so that

$$\left(-\overline{p}_{op}\overline{p}_{op} + p_{op}^2 - k^2\right) \overline{\overline{G}}_o(\overline{r}, \overline{r}') = \overline{\overline{I}} \delta(\overline{r} - \overline{r}') \tag{5.1.23}$$

In operator notation

$$\left(-\overline{p}_{op}\overline{p}_{op} + p_{op}^2 - k^2\right) \overline{\overline{G}}_{oop} = \overline{\overline{I}}_{op} \tag{5.1.24}$$

giving

$$\overline{\overline{G}}_{oop} = -\left(\overline{p}_{op}\overline{p}_{op} - p_{op}^2 \overline{\overline{I}}_{op} + k^2 \overline{\overline{I}}_{op}\right)^{-1} \tag{5.1.25}$$

We shall, from now on, unless necessary, suppress the *op* subscript since it can be inferred from the context whether a quantity is a function or an operator. The operator equation of (5.1.5) is

$$\overline{\overline{G}}_s = \overline{\overline{G}}_o + \overline{\overline{G}}_o \overline{\overline{U}}\, \overline{\overline{G}}_s \tag{5.1.26}$$

We use the following diagrammatic notations

$$\overline{\overline{G}}_o = \text{——} \tag{5.1.27}$$

$$\overline{\overline{U}} = \bullet \tag{5.1.28}$$

Equation (5.1.26) assumes the following diagrammatic form

$$\overline{\overline{G}}_s = \text{——} + \text{——}\!\bullet \overline{\overline{G}}_s \tag{5.1.29}$$

The formal perturbation expansion of (5.1.29) is

$$\overline{\overline{G}}_s = \text{——} + \text{——}\!\bullet\!\text{——} + \text{——}\!\bullet\!\text{——}\!\bullet\!\text{——} + \text{——}\!\bullet\!\text{——}\!\bullet\!\text{——}\!\bullet\!\text{——} + \cdots \tag{5.1.30}$$

The dyadic transition operator $\overline{\overline{T}}$ is defined to be

$$\begin{aligned}\overline{\overline{T}} &= \bullet + \bullet\!\text{——}\!\bullet + \bullet\!\text{——}\!\bullet\!\text{——}\!\bullet + \cdots \\ &= \bullet + \bullet\!\text{——}(\bullet + \bullet\!\text{——}\!\bullet + \bullet\!\text{——}\!\bullet\!\text{——}\!\bullet + \cdots)\end{aligned} \tag{5.1.31}$$

Using (5.1.31) in (5.1.30),

$$\overline{\overline{G}}_s = \overline{\overline{G}}_o + \overline{\overline{G}}_o \overline{\overline{T}}\, \overline{\overline{G}}_o \tag{5.1.32}$$

From (5.1.31), it also follows that the transition operator $\overline{\overline{T}}$ satisfies the equation

$$\overline{\overline{T}} = \overline{\overline{U}} + \overline{\overline{U}}\, \overline{\overline{G}}_o \overline{\overline{T}} \tag{5.1.33}$$

Equation (5.1.33) is known as the Lippmann-Schwinger equation. From (5.1.33), the operator $\overline{\overline{T}}$ can be formally written as

$$\overline{\overline{T}} = \left(\overline{\overline{I}} - \overline{\overline{U}}\, \overline{\overline{G}}_o\right)^{-1} \overline{\overline{U}} \tag{5.1.34}$$

We can apply the source state ket $|\overline{J}\rangle$ to the right of (5.1.32) to have

$$\overline{\overline{G}}_s|\overline{J}\rangle = \overline{\overline{G}}_o|\overline{J}\rangle + \overline{\overline{G}}_o \overline{\overline{T}}\, \overline{\overline{G}}_o|\overline{J}\rangle \tag{5.1.35}$$

Identifying

$$\overline{\overline{G}}_s|\overline{J}\rangle = |\overline{E}\rangle \tag{5.1.36}$$

as the electric field and identifying

$$\overline{\overline{G}}_o|\overline{J}\rangle = |\overline{E}_{inc}\rangle \tag{5.1.37}$$

as the incident field, we have

$$|\overline{E}\rangle = |\overline{E}_{inc}\rangle + \overline{\overline{G}}_o \overline{\overline{T}} |\overline{E}_{inc}\rangle \tag{5.1.38}$$

We note that

$$|\overline{E}_s\rangle = \overline{\overline{G}}_o\overline{\overline{T}}|\overline{E}_{inc}\rangle \tag{5.1.39}$$

is the scattered field of a single scatterer. To relate the scattering dyad $\overline{\overline{F}}$ to the transition operator, we follow a procedure similar to that of Chapter 3, Section 2 and let the incident field state be given by a momentum eigenstate

$$\langle\overline{r}|\overline{E}_{inc}\rangle = \hat{e}_i e^{i\overline{k}_i\cdot\overline{r}} \tag{5.1.40}$$

The momentum representation of $\overline{\overline{T}}$ is

$$\langle\overline{p}|\overline{\overline{T}}|\overline{p}'\rangle = \overline{\overline{T}}_p(\overline{p},\overline{p}') \tag{5.1.41}$$

Then from (5.1.39), on inserting unit operators in coordinate and momentum representations, we have

$$\overline{E}_s(\overline{r}) = \langle\overline{r}|\overline{E}_s\rangle = \int d\overline{r}'\overline{\overline{G}}_o(\overline{r},\overline{r}')\int\frac{d\overline{p}'}{(2\pi)^3}e^{i\overline{p}'\cdot\overline{r}'}\overline{\overline{T}}_p(\overline{p}',\overline{k}_i)\cdot\hat{e}_i \tag{5.1.42}$$

as in (3.2.6). The far-field scattering dyad is

$$\overline{E}_s(\overline{r}) = \frac{e^{ikr}}{r}\overline{\overline{F}}(\hat{k}_s,\hat{k}_i)\cdot\hat{e}_i \tag{5.1.43}$$

with

$$\overline{\overline{F}}(\hat{k}_s,\hat{k}_i)\cdot\hat{e}_i = \frac{1}{4\pi}(\hat{\theta}_s\hat{\theta}_s + \hat{\phi}_s\hat{\phi}_s)\cdot\overline{\overline{T}}_p(k\hat{k}_s,k\hat{k}_i)\cdot\hat{e}_i \tag{5.1.44}$$

The free-space dyadic Green's operator in momentum representation is, from (5.1.25),

$$\langle\overline{p}|\overline{\overline{G}}_o|\overline{p}'\rangle = \overline{\overline{G}}_o(\overline{p})\langle\overline{p}|\overline{p}'\rangle \tag{5.1.45}$$

where

$$\overline{\overline{G}}_o(\overline{p}) = -\left(\overline{p}\,\overline{p} - p^2\overline{\overline{I}} + k^2\overline{\overline{I}}\right)^{-1} \tag{5.1.46}$$

Since

$$\overline{\overline{I}} = \hat{p}\hat{p} + \hat{\theta}_p\hat{\theta}_p + \hat{\phi}_p\hat{\phi}_p \tag{5.1.47}$$

we have

$$\overline{\overline{G}}_o(\overline{p}) = -\frac{1}{k^2}\hat{p}\hat{p} + \frac{1}{p^2-k^2}(\hat{\theta}_p\hat{\theta}_p + \hat{\phi}_p\hat{\phi}_p) \tag{5.1.48}$$

with $p = |\overline{p}|$. The orthonormal unit vectors $(\hat{p},\hat{\theta}_p,\hat{\phi}_p)$ of the spherical coordinate system for direction (θ_p,ϕ_p) have been employed.

By comparing (5.1.26) and (5.1.32), we have

$$\overline{\overline{T}}\,\overline{\overline{G}}_o = \overline{\overline{U}}\,\overline{\overline{G}}_s \tag{5.1.49}$$

which gives

$$\overline{\overline{T}} = \overline{\overline{U}}\,\overline{\overline{G}}_s \overline{\overline{G}}_o^{-1} \tag{5.1.50}$$

Hence, in momentum representation,

$$\overline{\overline{T}}_p(\overline{p}_1, \overline{p}_2) = \int d\overline{r}_2 \int d\overline{r}_1\, e^{-i\overline{p}_1\cdot\overline{r}_1} \overline{\overline{U}}(\overline{r}_1)\overline{\overline{G}}_s(\overline{r}_1, \overline{r}_2)\, e^{i\overline{p}_2\cdot\overline{r}_2}\, \overline{\overline{G}}_o^{-1}(\overline{p}_2) \tag{5.1.51}$$

Hence if the dyadic Green's function $\overline{\overline{G}}_s$ for a single particle is known, then the $\overline{\overline{T}}$ operator in momentum representation can be calculated via (5.1.51). This has been performed for the case of a spherical dielectric particle [Tsang and Kong, 1980].

2 Multiple Scattering Equations

Consider N particles occupying regions $V_1, V_2, \ldots, V_N$. The jth particle has permittivity ϵ_j, permeability μ, and wavenumber k_j. Let the particle j be centered at $\overline{r}_j$. The integral equation is

$$\overline{\overline{G}}(\overline{r}, \overline{r}') = \overline{\overline{G}}_o(\overline{r}, \overline{r}') + \sum_{l=1}^{N} \int_{V_l} d\overline{r}'' \overline{\overline{G}}_o(\overline{r}, \overline{r}'')(k_l^2 - k^2)\overline{\overline{G}}(\overline{r}'', \overline{r}') \tag{5.2.1}$$

Define the potential function for particle j as

$$U_j(\overline{r} - \overline{r}_j) = \begin{cases} 0 & \text{for } \overline{r} \notin V_j \\ k_j^2 - k^2 & \text{for } \overline{r} \in V_j \end{cases} \tag{5.2.2}$$

Note that $U_j(\overline{r})$ is the jth particle potential translated to the origin with center at origin.

In the operator notations we have

$$\overline{\overline{G}} = \overline{\overline{G}}_o + \sum_{l=1}^{N} \overline{\overline{G}}_o \overline{\overline{U}}_l \overline{\overline{G}} \tag{5.2.3}$$

If we define the jth particle exciting field operator as

$$\overline{\overline{G}}_j = \overline{\overline{G}}_o + \overline{\overline{G}}_o \sum_{\substack{l=1 \\ l\neq j}}^{N} \overline{\overline{U}}_j \overline{\overline{G}} \tag{5.2.4}$$

then from (5.2.3) and (5.2.4) we have

$$\overline{\overline{G}} = \overline{\overline{G}}_j + \overline{\overline{G}}_o \overline{\overline{U}}_j \overline{\overline{G}} \tag{5.2.5}$$

The transition operator for particle j is

$$\overline{\overline{T}}_j = \left(\overline{\overline{I}} - \overline{\overline{U}}_j \overline{\overline{G}}_o\right)^{-1} \overline{\overline{U}}_j \tag{5.2.6}$$

and is the scattering operator for particle j in the absence of other particles. It is assumed that the single-particle scattering properties and its transition operator are known. Premultiplying (5.2.5) by $\overline{\overline{U}}_j$ gives

$$\overline{\overline{U}}_j\overline{\overline{G}} = \overline{\overline{U}}_j\overline{\overline{G}}_j + \overline{\overline{U}}_j\overline{\overline{G}}_o\overline{\overline{U}}_j\overline{\overline{G}} \tag{5.2.7}$$

Thus

$$\overline{\overline{U}}_j\overline{\overline{G}} = \left(\overline{\overline{I}} - \overline{\overline{U}}_j\overline{\overline{G}}_o\right)^{-1}\overline{\overline{U}}_j\overline{\overline{G}}_j \tag{5.2.8}$$

Using (5.2.6) in (5.2.8) gives

$$\overline{\overline{U}}_j\overline{\overline{G}} = \overline{\overline{T}}_j\overline{\overline{G}}_j \tag{5.2.9}$$

Putting (5.2.9) in (5.2.3) and (5.2.4) gives the Foldy-Lax multiple scattering equations in operator form

$$\overline{\overline{G}} = \overline{\overline{G}}_o + \sum_{l=1}^{N} \overline{\overline{G}}_o\overline{\overline{T}}_l\overline{\overline{G}}_l \tag{5.2.10}$$

$$\overline{\overline{G}}_j = \overline{\overline{G}}_o + \overline{\overline{G}}_o \sum_{\substack{l=1 \\ l\neq j}}^{N} \overline{\overline{T}}_l\overline{\overline{G}}_l \tag{5.2.11}$$

We can apply the source state to the right-hand side of (5.2.10). Let

$$|\overline{E}\rangle = \overline{\overline{G}}|\overline{J}\rangle \tag{5.2.12}$$

be the electric field, let

$$|\overline{E}_{inc}\rangle = \overline{\overline{G}}_o|\overline{J}\rangle \tag{5.2.13}$$

be the incident field, and let

$$|\overline{E}_j^{ex}\rangle = \overline{\overline{G}}_j|\overline{J}\rangle \tag{5.2.14}$$

be the exciting field of the jth particle. We then get

$$|\overline{E}\rangle = |\overline{E}_{inc}\rangle + \sum_{l=1}^{N} \overline{\overline{G}}_o\overline{\overline{T}}_l|\overline{E}_l^{ex}\rangle \tag{5.2.15}$$

$$|\overline{E}_j^{ex}\rangle = |\overline{E}_{inc}\rangle + \overline{\overline{G}}_o \sum_{\substack{l=1 \\ l\neq j}}^{N} \overline{\overline{T}}_l|\overline{E}_l^{ex}\rangle \tag{5.2.16}$$

Equations (5.2.15) and (5.2.16) are the Foldy-Lax equations for the electric field that have been rigorously derived here from the Maxwell equations. These equations are used in Volume II for numerical simulations.

3 Approximations of Multiple Scattering Equations

3.1 Configurational Average of Multiple Scattering Equations

To calculate the coherent and incoherent fields, ensemble averages are to be taken of the multiple scattering equations. The concept of configurational averaging is used. The probability density function of finding the first particle at $\overline{r}_1$, the second particle at $\overline{r}_2, \ldots,$ and the Nth particle at $\overline{r}_N$ is designated by $p(\overline{r}_1, \overline{r}_2, \ldots, \overline{r}_N)$. Using Bayes' rule it can be expressed in terms of conditional probabilities

$$\begin{aligned} p(\overline{r}_1, \ldots, \overline{r}_N) &= p(\overline{r}_i)\, p(\overline{r}_1, \overline{r}_2, \ldots, \acute{\overline{r}}_i, \ldots, \overline{r}_N | \overline{r}_i) \\ &= p(\overline{r}_i)\, p(\overline{r}_j | \overline{r}_i)\, p(\overline{r}_1, \overline{r}_2, \ldots, \acute{\overline{r}}_i, \ldots, \acute{\overline{r}}_j, \ldots, \overline{r}_N | \overline{r}_i, \overline{r}_j) \end{aligned} \quad (5.3.1)$$

where the $\prime$ indicates that the term is absent. In (5.3.1), $p(\overline{r}_i)$ is the probability density function of finding particle i at $\overline{r}_i$ and $p(\overline{r}_j|\overline{r}_i)$ is the conditional probability of finding the jth particle at $\overline{r}_j$ given the ith particle at $\overline{r}_i$.

The states of the particles $s_1, s_2, \ldots, s_N$ can also be introduced. The state of the particle can indicate its size and, in the case of the nonspherical particle, its size and orientation. The probability of the set of N particles located in the volume element $d\overline{r}_1 d\overline{r}_2 \cdots d\overline{r}_N$ with their states in the region $ds_1 ds_2 \cdots ds_N$ is given by $p(\overline{r}_1, \overline{r}_2, \ldots, \overline{r}_N; s_1, s_2, \ldots, s_N) d\overline{r}_1 d\overline{r}_2 \cdots d\overline{r}_N ds_1 ds_2 \cdots ds_N$, a quantity whose integral is normalized to unity.

If the particles are uniformly distributed, the single particle probability density function is $p(\overline{r}_i) = 1/V$. We note that while the Green's function and the exciting field Green's functions $\overline{\overline{G}}$ and $\overline{\overline{G}}_l$, $l = 1, 2, \ldots, N$, are functions of all the particles, $\overline{\overline{T}}_l$ is only a function of the lth particle.

Taking a configurational average of (5.2.10) gives

$$E(\overline{\overline{G}}) = \overline{\overline{G}}_o + \overline{\overline{G}}_o \sum_{j=1}^{N} E(\overline{\overline{T}}_j \overline{\overline{G}}_j) \quad (5.3.2)$$

where E is the expectation value. By using Bayes' rule

$$\begin{aligned} E(\overline{\overline{T}}_j \overline{\overline{G}}_j) &= \int d\overline{r}_1 \cdots d\overline{r}_N\, ds_1 \cdots ds_N\, p(\overline{r}_1, \ldots, \overline{r}_N; s_1, \ldots, s_N) \overline{\overline{T}}_j \overline{\overline{G}}_j \\ &= \int d\overline{r}_j\, ds_j\, \overline{\overline{T}}_j\, p(\overline{r}_j; s_j) \int d\overline{r}_1 \cdots d\acute{\overline{r}}_j \cdots d\overline{r}_N\, ds_1 \cdots d\acute{s}_j \cdots ds_N \\ &\quad \times p(\overline{r}_1, \ldots, \acute{\overline{r}}_j, \ldots, \overline{r}_N; s_1, \ldots, \acute{s}_j, \ldots, s_N | \overline{r}_j, s_j) \overline{\overline{G}}_j \end{aligned} \quad (5.3.3a)$$

Let

$$E_j(\overline{\overline{G}}_j) = \int d\overline{r}_1 \cdots d\acute{\overline{r}}_j \cdots d\overline{r}_N \, ds_1 \cdots d\acute{s}_j \cdots ds_N \\ \times p(\overline{r}_1, \ldots, \acute{\overline{r}}_j, \ldots, \overline{r}_N; s_1, \ldots, \acute{s}_j, \ldots, s_N | \overline{r}_j, s_j) \overline{\overline{G}}_j \qquad (5.3.3b)$$

be the conditional average of $\overline{\overline{G}}_j$ while holding $\overline{r}_j$ and s_j fixed. Also note that $E_j(\overline{\overline{G}}_j)$ is only a function of $\overline{r}_j$ and s_j. Thus

$$E(\overline{\overline{T}}_j \overline{\overline{G}}_j) = \int d\overline{r}_j ds_j \overline{\overline{T}}_j p(\overline{r}_j; s_j) E_j(\overline{\overline{G}}_j) = E\left(\overline{\overline{T}}_j E_j(\overline{\overline{G}}_j)\right) \qquad (5.3.3c)$$

On averaging, $E(\overline{\overline{T}}_j \overline{\overline{G}}_j)$ is independent of the positions and states of all the particles. After the average as indicated in (5.3.3) is taken, every term in the summation of (5.3.2) will be identical. Hence,

$$E(\overline{\overline{G}}) = \overline{\overline{G}}_o + N \overline{\overline{G}}_o \, E\left(\overline{\overline{T}}_j E_j(\overline{\overline{G}}_j)\right) \qquad (5.3.4)$$

Similarly, taking the conditional configurational average of (5.2.11) by fixing particle l, we obtain

$$E_l(\overline{\overline{G}}_l) = \overline{\overline{G}}_o + \overline{\overline{G}}_o \sum_{\substack{j=1 \\ j \neq l}}^{N} E_l(\overline{\overline{T}}_j \overline{\overline{G}}_j) \qquad (5.3.5a)$$

We have

$$E_l(\overline{\overline{T}}_j \overline{\overline{G}}_j) = E_l\left(E_{jl}(\overline{\overline{T}}_j \overline{\overline{G}}_j)\right) = E_l\left(\overline{\overline{T}}_j E_{jl}(\overline{\overline{G}}_j)\right) \qquad (5.3.5b)$$

where E_{jl} is the conditional average given the properties and positions of particles j and l. Hence

$$E_l\left(\overline{\overline{T}}_j E_{jl}(\overline{\overline{G}}_j)\right) = \int d\overline{r}_j \, ds_j \overline{\overline{T}}_j \, p(\overline{r}_j; s_j | \overline{r}_l; s_l) E_{jl}(\overline{\overline{G}}_j) \qquad (5.3.6a)$$

with $j \neq l$. On averaging, the summation in (5.3.5a) can be replaced by $(N-1)$. Thus

$$E_l(\overline{\overline{G}}_l) = \overline{\overline{G}}_o + \overline{\overline{G}}_o (N-1) \, E_l\left(\overline{\overline{T}}_j E_{jl}(\overline{\overline{G}}_j)\right) \qquad (5.3.6b)$$

with $j \neq l$.

The equations (5.3.4) to (5.3.6) indicate that the total average is given in terms of the conditional average with one particle fixed, and the conditional average with one particle fixed is given in terms of the conditional average with two particles fixed. In a similar manner, we can express the conditional average with n particles fixed in terms of $n+1$ particles fixed. A hierarchy of equations is generated. To solve the equations, truncations of the hierarchy will be made in later sections.

By the translational theorem of operators, $\overline{\overline{U}}_j$ is related to $\overline{\overline{U}}$ centered at the origin by the operator equation [Merzbacher, 1970]

$$\overline{\overline{U}}_j = e^{-i\overline{p}\cdot\overline{r}_j}\,\overline{\overline{U}}\,e^{i\overline{p}\cdot\overline{r}_j} \tag{5.3.7}$$

Since

$$\overline{\overline{T}}_j = \overline{\overline{U}}_j + \overline{\overline{U}}_j\overline{\overline{G}}_o\overline{\overline{T}}_j \tag{5.3.8}$$

and $\overline{\overline{G}}_o$ commutes with $\overline{p}$, we have

$$\overline{\overline{T}}_j = e^{-i\overline{p}\cdot\overline{r}_j}\,\overline{\overline{T}}\,e^{i\overline{p}\cdot\overline{r}_j} \tag{5.3.9}$$

3.2 Effective Field Approximation (EFA, Foldy's Approximation)

In the effective field approximation (EFA), truncation is carried out at the first equation of the hierarchy. It is assumed that

$$E_j(\overline{\overline{G}}_j) \simeq E(\overline{\overline{G}}) \tag{5.3.10}$$

The approximation is valid for a sparse concentration of particles. Hence, on using the approximation in (5.3.10), (5.3.4) becomes,

$$E(\overline{\overline{G}}) = \overline{\overline{G}}_o + N\overline{\overline{G}}_o E\left(\overline{\overline{T}}_j E(\overline{\overline{G}})\right) = \overline{\overline{G}}_o + N\overline{\overline{G}}_o E(\overline{\overline{T}}_j)E(\overline{\overline{G}}) \tag{5.3.11}$$

The averaged $\overline{\overline{T}}_j$ in momentum representation is

$$\begin{aligned}\langle\overline{p}|E(\overline{\overline{T}}_j)|\overline{p}'\rangle &= \int d\overline{r}_j ds_j p(\overline{r}_j; s_j)\langle\overline{p}|e^{-i\overline{p}_{op}\cdot\overline{r}_j}\overline{\overline{T}}e^{i\overline{p}_{op}\cdot\overline{r}_j}|\overline{p}'\rangle \\ &= \int d\overline{r}_j ds_j p(\overline{r}_j; s_j)e^{-i\overline{p}\cdot\overline{r}_j + i\overline{p}'\cdot\overline{r}_j}\overline{\overline{T}}_p(\overline{p},\overline{p}')\end{aligned} \tag{5.3.12}$$

where

$$\overline{\overline{T}}_p(\overline{p},\overline{p}') = \langle\overline{p}|\overline{\overline{T}}|\overline{p}'\rangle \tag{5.3.13}$$

In deriving (5.3.12), Eq. (5.3.9) has been used. Let the position and state of a particle be independent of each other:

$$p(\overline{r}_j; s_j) = p(\overline{r}_j)p(s_j) = \frac{p(s_j)}{V} \tag{5.3.14}$$

Thus

$$\begin{aligned}\langle\overline{p}|E(\overline{\overline{T}}_j)|\overline{p}'\rangle &= \int d\overline{r}_j \int ds_j \frac{1}{V}e^{-i\overline{p}\cdot\overline{r}_j + i\overline{p}'\cdot\overline{r}_j}p(s_j)\overline{\overline{T}}_p(\overline{p},\overline{p}') \\ &= \frac{1}{V}\langle\overline{p}|\overline{p}'\rangle \int ds\ p(s)\overline{\overline{T}}_p(\overline{p},\overline{p}')\end{aligned} \tag{5.3.15}$$

We then have

$$\langle \overline{p} | E(\overline{\overline{T}}_j) | \overline{p}' \rangle = \frac{1}{V} \langle T_p(\overline{p}, \overline{p}) \rangle_{st} \langle \overline{p} | \overline{p}' \rangle \tag{5.3.16}$$

where

$$\langle T_p(\overline{p}, \overline{p}) \rangle_{st} = \int ds \, T_p(\overline{p}, \overline{p}) \, p(s) \tag{5.3.17}$$

All the terms in (5.3.11), viz. $E(\overline{\overline{G}})$, $\overline{\overline{G}}_o$, and $\overline{\overline{T}}_j$, have diagonal momentum representation. Let

$$\langle \overline{p} | E(\overline{\overline{G}}) | \overline{p}' \rangle = \langle \overline{p} | \overline{p}' \rangle \overline{\overline{G}}(\overline{p}) \tag{5.3.18}$$

Hence using (5.3.16), (5.3.18), and (5.3.11), we have

$$\overline{\overline{G}}(\overline{p}) = \overline{\overline{G}}_o(\overline{p}) + n_o \overline{\overline{G}}_o(\overline{p}) \langle \overline{\overline{T}}_p(\overline{p}, \overline{p}) \rangle_{st} \overline{\overline{G}}(\overline{p}) \tag{5.3.19}$$

so that

$$\overline{\overline{G}}(\overline{p}) = \left(\overline{\overline{I}} - n_o \overline{\overline{G}}_o(\overline{p}) \langle \overline{\overline{T}}_p(\overline{p}, \overline{p}) \rangle_{st} \right)^{-1} \overline{\overline{G}}_o(\overline{p}) = \left(\overline{\overline{G}}_o^{-1} - n_o \langle \overline{\overline{T}}_p(\overline{p}, \overline{p}) \rangle_{st} \right)^{-1} \tag{5.3.20}$$

By setting the inverse of $\overline{\overline{G}}(\overline{p})$ equal to zero, a dispersion relation is obtained.

$$\det \left[\overline{\overline{G}}_o^{-1}(\overline{p}) - n_o \langle \overline{\overline{T}}_p(\overline{p}, \overline{p}) \rangle_{st} \right] = 0 \tag{5.3.21}$$

In (5.3.21), "det" stands for determinant. The value of $\overline{p}$ that satisfies (5.3.21) will be the value of the effective propagation vector $\overline{K}$. Using the expression (5.1.48) for $\overline{\overline{G}}_o(\overline{p})$, we have

$$\overline{\overline{G}}_o^{-1}(\overline{p}) = -k^2 \hat{p}\hat{p} + (p^2 - k^2)(\hat{\theta}_p \hat{\theta}_p + \hat{\phi}_p \hat{\phi}_p) \tag{5.3.22}$$

Thus

$$\det \left[-k^2 \hat{p}\hat{p} + (p^2 - k^2)(\hat{\theta}_p \hat{\theta}_p + \hat{\phi}_p \hat{\phi}_p) - n_o \langle \overline{\overline{T}}_p(\overline{p}, \overline{p}) \rangle_{st} \right] = 0 \tag{5.3.23}$$

We note that the effective field approximation is valid for a sparse concentration of particles so that the coherent effective propagation constant is close to k. We then look for the solution of p that is close to k in (5.3.23). The arguments of $\overline{\overline{T}}_p$ in (5.3.23) can be replaced by $k\hat{p}$. The first term in (5.3.23) is much larger than the terms that have the $(p^2 - k^2)$ factor. Thus, the determinant of (5.3.23) is approximately equal to $-k^2$ times the determinant of the 2×2 matrix formed by $\hat{\theta}_p$ and $\hat{\phi}_p$ components. Hence, the dispersion relation is

$$0 \simeq -k^2 \cdot \left\{ \begin{vmatrix} p^2 - k^2 - n_o \hat{\theta}_p \cdot \langle \overline{\overline{T}}_p(k\hat{p}, k\hat{p}) \rangle_{st} \cdot \hat{\theta}_p & -n_o \hat{\theta}_p \cdot \langle \overline{\overline{T}}_p(k\hat{p}, k\hat{p}) \rangle_{st} \cdot \hat{\phi}_p \\ -n_o \hat{\phi}_p \cdot \langle \overline{\overline{T}}_p(k\hat{p}, k\hat{p}) \rangle_{st} \cdot \hat{\theta}_p & p^2 - k^2 - n_o \hat{\phi}_p \cdot \langle \overline{\overline{T}}_p(k\hat{p}, k\hat{p}) \rangle_{st} \cdot \hat{\phi}_p \end{vmatrix} \right\} \tag{5.3.24}$$

Using the relation between $\overline{\overline{F}}$ and $\overline{\overline{T}}$ in (5.1.44), the dispersion relation can be expressed in terms of the $\overline{\overline{F}}$ dyad for propagation in the direction $\hat{k}_i$. Note that $\hat{\theta}$ corresponds to $\hat{v}$ and that $\hat{\phi}$ corresponds to $\hat{h}$.

$$\begin{vmatrix} K^2 - k^2 - 4\pi n_o \langle f_{vv}(\theta_i, \phi_i; \theta_i, \phi_i) \rangle_{st} & -4\pi n_o \langle f_{vh}(\theta_i, \phi_i; \theta_i, \phi_i) \rangle_{st} \\ -4\pi n_o \langle f_{hv}(\theta_i, \phi_i; \theta_i, \phi_i) \rangle_{st} & K^2 - k^2 - 4\pi n_o \langle f_{hh}(\theta_i, \phi_i; \theta_i, \phi_i) \rangle_{st} \end{vmatrix} = 0 \tag{5.3.25}$$

with $\hat{v}_i$ and $\hat{h}_i$ (vertical and horizontal polarizations) as the basis for (5.3.25). In (5.3.25), $\langle f_{\alpha\beta} \rangle_{st}$ denotes averaging over the state (e.g., orientation and size). The eigenvalues for K in (5.3.25) are the effective propagation constants, and the corresponding eigenvectors give the characteristic wave polarizations associated with the propagation constants. Let

$$M_{\alpha\beta}(\theta_i, \phi_i) = \langle f_{\alpha\beta}(\theta_i, \phi_i; \theta_i, \phi_i) \rangle_{st} \frac{i 2\pi n_o}{k} \tag{5.3.26}$$

Solving (5.3.25) gives values of K^2 and since K is close to k, the two effective coherent propagation constants are K_1 and K_2 with

$$K_{\substack{1\\2}} = k - \frac{i}{2} \left[M_{vv} + M_{hh} \pm \left\{ (M_{vv} - M_{hh})^2 + 4 M_{vh} M_{hv} \right\}^{1/2} \right] \tag{5.3.27}$$

and the associated eigenvectors are, for $M_{vh}, M_{hv} \neq 0$,

$$\frac{\langle E_v \rangle}{\langle E_h \rangle} = \frac{2 M_{vh}}{-M_{vv} + M_{hh} \pm \left\{ (M_{vv} - M_{hh})^2 + 4 M_{vh} M_{hv} \right\}^{1/2}} \tag{5.3.28}$$

where $\langle E_v \rangle / \langle E_h \rangle$ is the ratio of vertical to horizontal polarization in the characteristic wave. Note that there are only two propagation constants and characteristic waves for a specified direction of propagation. For the case $\langle f_{vh} \rangle_{st} = \langle f_{hv} \rangle_{st} = 0$, when there is no coupling between the two polarizations, the eigensolutions are

$$K_1 = k + \frac{2\pi n_o}{k} \langle f_{vv} \rangle_{st} \tag{5.3.29}$$

with the characteristic wave vertically polarized, while

$$K_2 = k + \frac{2\pi n_o}{k} \langle f_{hh} \rangle_{st} \tag{5.3.30}$$

with the characteristic wave horizontally polarized.

Equation (5.3.25) can be expressed in the form of coupled differential equations for the coherent field along the propagation direction $(\theta_i, \phi_i) = \hat{s}$,

$$\frac{d\langle E_v\rangle}{ds} = (ik + M_{vv})\langle E_v\rangle + M_{vh}\langle E_h\rangle \tag{5.3.31}$$

$$\frac{d\langle E_h\rangle}{ds} = M_{hv}\langle E_v\rangle + (\,ik + M_{hh})\langle E_h\rangle \tag{5.3.32}$$

with s as the distance along the direction of propagation. The coherent field equations of (5.3.21) and (5.3.22) were also derived in Chapter 3, Section 3 using the cascade of layers.

3.3 Quasi-crystalline Approximation (QCA)

The quasi-crystalline approximation (QCA) is a higher-order approximation than the effective field approximation. Truncation is made at the second stage of the hierarchy of equations. To impose the quasi-crystalline approximation, it is convenient to use the $\overline{\overline{Q}}$ operator defined by

$$\overline{\overline{Q}}_j = \overline{\overline{T}}_j\overline{\overline{G}}_j\overline{\overline{G}}_o^{-1} \tag{5.3.33}$$

Then the Foldy-Lax multiple scattering equations of (5.2.10) and (5.2.11) assume the following form

$$\overline{\overline{G}} = \overline{\overline{G}}_o + \overline{\overline{G}}_o\sum_{j=1}^{N}\overline{\overline{Q}}_j\overline{\overline{G}}_o \tag{5.3.34}$$

$$\overline{\overline{Q}}_l = \overline{\overline{T}}_l + \overline{\overline{T}}_l\overline{\overline{G}}_o\sum_{\substack{j=1\\ j\neq l}}^{N}\overline{\overline{Q}}_j \tag{5.3.35}$$

The configurational average of the equations becomes

$$E(\overline{\overline{G}}) = \overline{\overline{G}}_o + N\overline{\overline{G}}_o\, E\Big(E_j(\overline{\overline{Q}}_j)\Big)\overline{\overline{G}}_o \tag{5.3.36}$$

$$E_l(\overline{\overline{Q}}_l) = \overline{\overline{T}}_l + (N-1)\overline{\overline{T}}_l\overline{\overline{G}}_o\, E_l\Big(E_{lj}(\overline{\overline{Q}}_j)\Big) \tag{5.3.37}$$

with $j \neq l$. Under the quasi-crystalline approximation, it is assumed that

$$E_{lj}(\overline{\overline{Q}}_j) \simeq E_j(\overline{\overline{Q}}_j) \tag{5.3.38}$$

so that (5.3.37) becomes

$$E_l(\overline{\overline{Q}}_l) = \overline{\overline{T}}_l + (N-1)\overline{\overline{T}}_l\overline{\overline{G}}_o\, E_l\Big(E_j(\overline{\overline{Q}}_j)\Big) \tag{5.3.39}$$

Since $\overline{\overline{T}}_l = (\overline{\overline{I}} - \overline{\overline{U}}_l\overline{\overline{G}}_o)^{-1}\overline{\overline{U}}_l$, we have

$$E_l(\overline{\overline{Q}}_l) = \overline{\overline{U}}_l + \overline{\overline{U}}_l\overline{\overline{G}}_o E_l(\overline{\overline{Q}}_l) + (N-1)\overline{\overline{U}}_l\overline{\overline{G}}_o E_l\left(E_j(\overline{\overline{Q}}_j)\right) \tag{5.3.40}$$

After the conditional average is taken, $E_l(\overline{\overline{Q}}_l)$ is only a function of the lth particle. We shall illustrate for the case of identical particles. Taking (5.3.40) in momentum representation and using the translational property of $\overline{\overline{U}}_j$ give

$$\begin{aligned}\langle\overline{p}|E_l(\overline{\overline{Q}}_l)|\overline{p}'\rangle =& e^{-i(\overline{p}-\overline{p}')\cdot\overline{r}_l}\,\overline{\overline{U}}_p(\overline{p},\overline{p}') + \int\frac{d\overline{p}''}{(2\pi)^3}\,\overline{\overline{U}}_p(\overline{p},\overline{p}'')\,e^{-i(\overline{p}-\overline{p}'')\cdot\overline{r}_l}\\ &\cdot\overline{\overline{G}}_o(\overline{p}'')\langle\overline{p}''|E_l(\overline{\overline{Q}}_l)|\overline{p}'\rangle\\ &+ N\int d\overline{r}_j\,p(\overline{r}_j|\overline{r}_l)\int\frac{d\overline{p}''}{(2\pi)^3}\,e^{-i(\overline{p}-\overline{p}'')\cdot\overline{r}_l}\,\overline{\overline{U}}_p(\overline{p},\overline{p}'')\\ &\cdot\overline{\overline{G}}_o(\overline{p}'')\langle\overline{p}''|E_j(\overline{\overline{Q}}_j)|\overline{p}'\rangle\end{aligned} \tag{5.3.41}$$

The conditional probability $p(\overline{r}_j|\overline{r}_l)$ can be expressed in terms of the pair distribution function as $p(\overline{r}_j|\overline{r}_l) = g(\overline{r}_j - \overline{r}_l)/V$. Since the pair distribution function asymptotically approaches 1, we define the Fourier transform as follows. Let

$$q(\overline{r}) = n_o[g(\overline{r}) - 1] + \delta(\overline{r}) \tag{5.3.42}$$

and

$$q_p(\overline{p}) = \int d\overline{r}\,e^{-i\overline{p}\cdot\overline{r}}\,q(\overline{r}) \tag{5.3.43}$$

Next assume the following trial solution in (5.3.41)

$$\langle\overline{p}|E_l(\overline{\overline{Q}}_l)|\overline{p}'\rangle = e^{-i(\overline{p}-\overline{p}')\cdot\overline{r}_l}\,\overline{\overline{Q}}_p(\overline{p},\overline{p}') \tag{5.3.44}$$

Then, (5.3.41) becomes

$$\begin{aligned}\overline{\overline{Q}}_p(\overline{p},\overline{p}') =& \overline{\overline{U}}_p(\overline{p},\overline{p}') + \int\frac{d\overline{p}''}{(2\pi)^3}q_p(\overline{p}''-\overline{p}')\overline{\overline{U}}_p(\overline{p},\overline{p}'')\overline{\overline{G}}_o(\overline{p}'')\overline{\overline{Q}}_p(\overline{p}'',\overline{p}')\\ &+ n_o\overline{\overline{U}}_p(\overline{p},\overline{p}')\overline{\overline{G}}_o(\overline{p}')\overline{\overline{Q}}_p(\overline{p}',\overline{p}')\end{aligned} \tag{5.3.45}$$

Let

$$\overline{\overline{C}}_p(\overline{p},\overline{p}_1) = \overline{\overline{Q}}_p(\overline{p},\overline{p}_1)\left[\overline{\overline{I}} + n_o\overline{\overline{G}}_o(\overline{p}_1)\overline{\overline{Q}}_p(\overline{p}_1,\overline{p}_1)\right]^{-1} \tag{5.3.46}$$

Postmultiply (5.3.45) by $\left[\overline{\overline{I}} + n_o\overline{\overline{G}}_o(\overline{p}')\overline{\overline{Q}}_p(\overline{p}',\overline{p}')\right]^{-1}$, we get

$$\overline{\overline{C}}_p(\overline{p}_1,\overline{p}_2) = \overline{\overline{U}}_p(\overline{p}_1,\overline{p}_2) + \int\frac{d\overline{p}_3}{(2\pi)^3}\overline{\overline{U}}_p(\overline{p}_1,\overline{p}_3)\overline{\overline{G}}_o(\overline{p}_3)q_p(\overline{p}_3-\overline{p}_2)\overline{\overline{C}}_p(\overline{p}_3,\overline{p}_2) \tag{5.3.47}$$

From (5.3.36), we have, taking the momentum representation,

$$\overline{\overline{G}}(\overline{p}) = \overline{\overline{G}}_o(\overline{p}) + n_o\overline{\overline{G}}_o(\overline{p})\overline{\overline{Q}}_p(\overline{p},\overline{p})\overline{\overline{G}}_o(\overline{p}) \tag{5.3.48}$$

Since

$$\overline{\overline{Q}}_p(\overline{p},\overline{p}) = \left(\overline{\overline{I}} - n_o\overline{\overline{C}}_p(\overline{p},\overline{p})\overline{\overline{G}}_o(\overline{p})\right)^{-1}\overline{\overline{C}}_p(\overline{p},\overline{p}) \tag{5.3.49}$$

substituting (5.3.49) into (5.3.48) gives

$$\overline{\overline{G}}(\overline{p}) = \left\{\overline{\overline{G}}_o^{-1}(\overline{p}) - n_o\overline{\overline{C}}_p(\overline{p},\overline{p})\right\}^{-1} \tag{5.3.50a}$$

which can be expressed as

$$\overline{\overline{G}}(\overline{p}) = \overline{\overline{G}}_o(\overline{p}) + n_o\overline{\overline{G}}_o(\overline{p})\overline{\overline{C}}_p(\overline{p},\overline{p})\overline{\overline{G}}(\overline{p}) \tag{5.3.50b}$$

To calculate the dispersion relation, the unknown $\overline{\overline{C}}_p(\overline{p},\overline{p}')$ is solved through (5.3.47) and then the effective propagation constant is calculated by solving the equation

$$\det\left[\overline{\overline{G}}_o^{-1}(\overline{p}) - n_o\overline{\overline{C}}_p(\overline{p},\overline{p})\right] = 0 \tag{5.3.51}$$

From (5.3.42) and (5.3.43)

$$q_p(\overline{p}) = n_o H_p(\overline{p}) + 1 \tag{5.3.52}$$

where

$$H(\overline{p}) = \frac{1}{(2\pi)^3}\int_{-\infty}^{\infty} d\overline{r}\, h(\overline{r})\, e^{-i\overline{p}\cdot\overline{r}} \tag{5.3.53}$$

and $h(\overline{r}) = g(\overline{r}) - 1$. Since $\overline{\overline{T}} = \left(\overline{\overline{I}} - \overline{\overline{U}}\,\overline{\overline{G}}_o\right)^{-1}\overline{\overline{U}}$, we have from (5.3.47)

$$\overline{\overline{C}}_p(\overline{p}_1,\overline{p}_2) = \overline{\overline{T}}_p(\overline{p}_1,\overline{p}_2) + n_o\int d\overline{p}_3\overline{\overline{T}}_p(\overline{p}_1,\overline{p}_3)\overline{\overline{G}}_o(\overline{p}_3)H(\overline{p}_3 - \overline{p}_2)\overline{\overline{C}}_p(\overline{p}_3,\overline{p}_2) \tag{5.3.54}$$

If we define the operator $\overline{\overline{C}}_j$ that is a function of $\overline{r}_j$ such that

$$\langle\overline{p}_1|\overline{\overline{C}}_j|\overline{p}_2\rangle = e^{-i(\overline{p}_1-\overline{p}_2)\cdot\overline{r}_j}\,\overline{\overline{C}}_p(\overline{p}_1,\overline{p}_2) \tag{5.3.55}$$

then (5.3.54) can be put in operator form

$$\overline{\overline{C}}_j = \overline{\overline{T}}_j + n_o\int d\overline{r}_l\, h(\overline{r}_l - \overline{r}_j)\,\overline{\overline{T}}_j\,\overline{\overline{G}}_o\,\overline{\overline{C}}_l \tag{5.3.56a}$$

Then

$$(\overline{\overline{I}} - \overline{\overline{U}}_j\overline{\overline{G}}_o)\overline{\overline{C}}_j = \overline{\overline{U}}_j + n_o\int d\overline{r}_l h(\overline{r}_l - \overline{r}_j)\overline{\overline{U}}_j\overline{\overline{G}}_o\overline{\overline{C}}_l \tag{5.3.56b}$$

This gives

$$\overline{\overline{C}}_j = \overline{\overline{U}}_j + \int d\overline{r}_l q(\overline{r}_l - \overline{r}_j)\overline{\overline{U}}_j\overline{\overline{G}}_o\overline{\overline{C}}_l \tag{5.3.56c}$$

The operator equation that is equivalent to (5.3.56*b*) is then

$$E(\overline{\overline{G}}) = \overline{\overline{G}}_o + n_o\overline{\overline{G}}_o \int d\overline{r}_j\, \overline{\overline{C}}_j E(\overline{\overline{G}}) \tag{5.3.57}$$

From (5.3.57) it can be concluded that $n_o \int d\overline{r}_j \overline{\overline{C}}_j$ is the mass operator for discrete scatterers under the quasi-crystalline approximation.

For a sparse concentration of scatterers, it is reasonable to assume that the particle positions are independent of each other. The conditional probability $p(\overline{r}_j|\overline{r}_l)$ can be approximated by $p(\overline{r}_j|\overline{r}_l) = p(\overline{r}_j) = 1/V$ so that the pair correlation function becomes $g(\overline{r}) = 1$, and $h(\overline{r}) = 0$ and $\overline{\overline{H}}(\overline{p}) = 0$. It follows then that (5.3.56) gives

$$\overline{\overline{C}}_j = \overline{\overline{T}}_j \tag{5.3.58a}$$

$$\overline{\overline{C}}_p(\overline{p}_1, \overline{p}_2) = \overline{\overline{T}}_p(\overline{p}_1, \overline{p}_2) \tag{5.3.58b}$$

in this limit. Hence the dispersion relation of QCA of (5.3.51) reduces to that of EFA if we set $g(\overline{r}) = 1$.

3.4 Coherent Potential (CP)

The idea of coherent potential (CP) originates from the observation that in the derivation of the multiple scattering equations, the background medium dyadic Green's operator is used. The potential operator $\overline{\overline{U}}$ is proportional to $k_p^2 - k^2$ and is a measure of the difference in permittivity from the background medium. However, as the concentration of particles increases, the coherent wave will be propagating in an effective medium K, and the scattering potential is a result of the difference in wavenumber from K rather than k. The idea of coherent potential is to introduce the Green's operator with wavenumber K and is analogous to the nonlinear approximation in random medium theory of Chapter 4.

Consider the N-particle scattering equation,

$$\left(\overline{\overline{G}}_o^{-1} - \sum_{j=1}^{N} \overline{\overline{U}}_j\right)\overline{\overline{G}} = \overline{\overline{I}} \tag{5.3.59}$$

Suppose that the basic equation (5.3.59) is rewritten by adding and subtracting an operator $n_o\overline{\overline{w}}(\overline{p})$ on the left-hand side of the equation. Then,

$$\left[\overline{\overline{G}}_o^{-1} - n_o\overline{\overline{w}}(\overline{p}) - \sum_{j=1}^{N}\left(\overline{\overline{U}}_j - \frac{\overline{\overline{w}}(\overline{p})}{V}\right)\right]\overline{\overline{G}} = \overline{\overline{I}} \tag{5.3.60}$$

where $\overline{\overline{w}}(\overline{p})$ is the coherent potential operator which is constant in space but may be a function of the momentum operator $\overline{p}$. From (5.3.60) we note that the background Green's operator and the potential operator are modified. Let

$$\overline{\overline{G}}_c^{-1} = \overline{\overline{G}}_o^{-1} - n_o\,\overline{\overline{w}}(\overline{p}) \tag{5.3.61a}$$

be the coherent Green's operator and let

$$\overline{\overline{\overline{U}}}_j = \overline{\overline{U}}_j - \frac{\overline{\overline{w}}(\overline{p})}{V} \tag{5.3.61b}$$

be the modified potential operator for each scatterer. Then (5.3.60) becomes

$$\left[\overline{\overline{G}}_c^{-1} - \sum_{j=1}^{N}\overline{\overline{\overline{U}}}_j\right]\overline{\overline{G}} = \overline{\overline{I}} \tag{5.3.62}$$

and

$$\overline{\overline{G}} = \overline{\overline{G}}_c + \overline{\overline{G}}_c\sum_{j=1}^{N}\overline{\overline{\overline{U}}}_j\,\overline{\overline{G}} \tag{5.3.63}$$

is the N-particle scattering equation. Equation (5.3.63) is analogous to the original N-particle scattering equation, with $\overline{\overline{G}}_o$ replaced by $\overline{\overline{G}}_c$ and $\overline{\overline{U}}_j$ replaced by $\overline{\overline{\overline{U}}}_j$. The new transition operator is

$$\overline{\overline{\overline{T}}} = \overline{\overline{\overline{U}}} + \overline{\overline{\overline{U}}}\,\overline{\overline{G}}_c\overline{\overline{\overline{T}}} \tag{5.3.64}$$

The process of taking configurational averages and truncating the hierarchy of equations can be repeated, giving new dispersion relations. However, these new dispersion relations depend on the choice of the coherent potential operator $\overline{\overline{w}}(\overline{p})$, which has not yet been determined. The *consistent* choice for $\overline{\overline{w}}(\overline{p})$ or the coherent potential choice is choosing $\overline{\overline{w}}(\overline{p})$ such that

$$\langle\overline{\overline{G}}\rangle = \overline{\overline{G}}_c \tag{5.3.65}$$

Hence, the final result of the averaged Green's operator is equal to the original coherent Green's operator. The coherent potential shall be introduced into the effective field approximation and the quasi-crystalline approximation.

Effective Field Approximation with Coherent Potential (EFA–CP)

When the effective field approximation is applied to the N-particle scattering equation (5.3.63) in a manner completely analogous to Section 3.2, with $\overline{\overline{G}}_c$ playing the role of $\overline{\overline{G}}_o$ and $\overline{\overline{\widetilde{U}}}$ playing the role of $\overline{\overline{U}}$, the result for the averaged Green's operator in momentum representation is

$$\overline{\overline{G}}(\overline{p}) = \left[\overline{\overline{G}}_c^{-1}(\overline{p}) - n_o \overline{\overline{\widetilde{T}}}_p(\overline{p},\overline{p})\right]^{-1} \tag{5.3.66}$$

The coherent potential choice of $\overline{\overline{w}}(\overline{p})$ is such that (5.3.65) is satisfied, which, in view of (5.3.66), imposes the condition

$$\overline{\overline{\widetilde{T}}}_p(\overline{p},\overline{p}) = 0 \tag{5.3.67}$$

That is, the momentum representation of $\overline{\overline{\widetilde{T}}}$ with both arguments equal to $\overline{p}$ is zero.

By using (5.3.61b) and $\overline{\overline{\widetilde{T}}} = \overline{\overline{\widetilde{U}}} + \overline{\overline{\widetilde{T}}}\,\overline{\overline{G}}_c \overline{\overline{\widetilde{U}}}$ it follows that $\overline{\overline{\widetilde{T}}} = \overline{\overline{U}} - \frac{\overline{\overline{w}}(\overline{p})}{V} + \overline{\overline{\widetilde{T}}}\,\overline{\overline{G}}_c\left(\overline{\overline{U}} - \frac{\overline{\overline{w}}(\overline{p})}{V}\right)$ so that

$$\overline{\overline{\widetilde{T}}}\,\overline{\overline{G}}_c \overline{\overline{\widehat{G}}}^{-1} + \frac{\overline{\overline{w}}_p}{V} = \left(\overline{\overline{I}} + \overline{\overline{\widetilde{T}}}\,\overline{\overline{G}}_c\right)\overline{\overline{U}} \tag{5.3.68}$$

where

$$\overline{\overline{\widehat{G}}} = \left(\overline{\overline{G}}_c^{-1} + \frac{\overline{\overline{w}}_p}{V}\right)^{-1} \tag{5.3.69}$$

Define $\overline{\overline{\hat{t}}}$ that satisfies the Lippmann-Schwinger equation

$$\overline{\overline{\hat{t}}} = \overline{\overline{U}} + \overline{\overline{U}}\,\overline{\overline{\widehat{G}}}\,\overline{\overline{\hat{t}}} \tag{5.3.70}$$

Eliminating $\overline{\overline{U}}$ from (5.3.68) and (5.3.70) and using (5.3.69), we have

$$\left(\overline{\overline{\widetilde{T}}}\,\overline{\overline{G}}_c\overline{\overline{\widehat{G}}}^{-1} + \frac{\overline{\overline{w}}_p}{V}\right)\left(\overline{\overline{I}} + \overline{\overline{\widehat{G}}}\,\overline{\overline{\hat{t}}}\right) = \overline{\overline{\hat{t}}} + \overline{\overline{\widetilde{T}}}\,\overline{\overline{G}}_c\overline{\overline{\hat{t}}}$$

Simplifying we get

$$\overline{\overline{\widetilde{T}}} = \left(\overline{\overline{I}} - \frac{\overline{\overline{w}}_p(\overline{p})}{V}\overline{\overline{\widehat{G}}}\right)\overline{\overline{\hat{t}}}\,\overline{\overline{\widehat{G}}}\,\overline{\overline{G}}_c^{-1} - \frac{\overline{\overline{w}}_p}{V}\overline{\overline{\widehat{G}}}\,\overline{\overline{G}}_c^{-1}$$

and then

$$\overline{\overline{\widetilde{T}}} = \overline{\overline{G}}_c^{-1}\overline{\overline{\widehat{G}}}\,\overline{\overline{\hat{t}}}\,\overline{\overline{\widehat{G}}}\,\overline{\overline{G}}_c^{-1} - \frac{\overline{\overline{w}}_p}{V}\overline{\overline{\widehat{G}}}\,\overline{\overline{G}}_c^{-1} \tag{5.3.71a}$$

We take in momentum representation and define $\langle \overline{p}|\overline{\overline{G}}_c|\overline{p}_1\rangle = \overline{\overline{G}}_c(\overline{p})\langle \overline{p}|\overline{p}_1\rangle$ and $\langle \overline{p}|\overline{\overline{\hat{G}}}|\overline{p}_1\rangle = \overline{\overline{\hat{G}}}(\overline{p})\langle \overline{p}|\overline{p}_1\rangle$. To normalize the resulting equation, we multiply by $(1/V)\,e^{-i(\overline{p}-\overline{p}_1)\cdot\overline{r}_j}$ and integrate over $d\overline{r}_j$, noting that $\int d\overline{r}_j(1/V)$ $e^{-i(\overline{p}-\overline{p}_1)\cdot\overline{r}_j} = \langle \overline{p}|\overline{p}_1\rangle/V$ and $\int d\overline{r}_j = V$. Thus

$$\overline{\overline{\hat{T}}}_p(\overline{p},\overline{p}) = \left[\overline{\overline{G}}_c^{-1}(\overline{p})\,\overline{\overline{\hat{G}}}(\overline{p})\,\overline{\overline{\hat{t}}}_p(\overline{p},\overline{p}) - \overline{\overline{w}}(\overline{p})\right]\overline{\overline{\hat{G}}}(\overline{p})\,\overline{\overline{G}}_c^{-1}(\overline{p}) \tag{5.3.71b}$$

Hence, the coherent potential condition of $\overline{\overline{\hat{T}}}_p(\overline{p},\overline{p}) = 0$ implies that $\overline{\overline{w}}(\overline{p}) = \overline{\overline{G}}_c^{-1}(\overline{p})\,\overline{\overline{\hat{G}}}(\overline{p})\,\overline{\overline{\hat{t}}}_p(\overline{p},\overline{p})$. Furthermore, by taking $V \to \infty$ in (5.3.69), and noting that $\overline{\overline{w}}_p\overline{\overline{G}}_c(\overline{p})$ remains finite, we have $\overline{\overline{\hat{G}}}(\overline{p}) \to \overline{\overline{G}}_c(\overline{p})$. Thus the choice of coherent potential is

$$\overline{\overline{w}}(\overline{p}) = \overline{\overline{\hat{t}}}_p(\overline{p},\overline{p}) \tag{5.3.72}$$

Hence, the final results for the averaged Green's operator are (5.3.61a) with $\overline{\overline{w}}(\overline{p})$ given by (5.3.72). The quantity $\overline{\overline{\hat{t}}}$ satisfies (5.3.70), the momentum representation of which becomes (on letting $V \to \infty$)

$$\overline{\overline{\hat{t}}}_p(\overline{p}_1,\overline{p}_2) = \overline{\overline{U}}_p(\overline{p}_1,\overline{p}_2) + \int \frac{d\overline{p}_3}{(2\pi)^3}\,\overline{\overline{U}}_p(\overline{p}_1,\overline{p}_3)\,\overline{\overline{G}}_c(\overline{p}_3)\,\overline{\overline{\hat{t}}}_p(\overline{p}_3,\overline{p}_2) \tag{5.3.73}$$

The dispersion relation for EFA–CP is

$$\det\left[\overline{\overline{G}}_o^{-1}(\overline{p}) - n_o\overline{\overline{\hat{t}}}_p(\overline{p},\overline{p})\right] = 0 \tag{5.3.74}$$

Since $\overline{\overline{G}}_c$ depends on $\overline{\overline{\hat{t}}}_p$ via (5.3.61a) and (5.3.72), the integral equation (5.3.73) is nonlinear.

3.5 Quasi-crystalline Approximation with Coherent Potential (QCA–CP)

The quasi-crystalline approximation can be applied to the N-particle scattering equation (5.3.63). Manipulations are performed as in Section 3.3 with $\overline{\overline{G}}_c$ playing the role of $\overline{\overline{G}}_o$ and $\overline{\overline{\hat{U}}}$ playing the role of $\overline{\overline{U}}$. The result for the averaged Green's operator is

$$\overline{\overline{G}}(\overline{p}) = \left[\overline{\overline{G}}_c^{-1}(\overline{p}) - n_o\overline{\overline{\hat{C}}}_p(\overline{p},\overline{p})\right]^{-1} \tag{5.3.75}$$

where

$$\overline{\overline{\hat{C}}}_j = \overline{\overline{\hat{U}}}_j + \int d\overline{r}_l q(\overline{r}_l - \overline{r}_j)\overline{\overline{\hat{U}}}_j\overline{\overline{G}}_c\overline{\overline{\hat{C}}}_l \tag{5.3.76}$$

Putting $\overline{\overline{\tilde{U}}}_j$ in terms of $\overline{\overline{U}}_j$, we have

$$\overline{\overline{\tilde{C}}}_j = \overline{\overline{U}}_j - \frac{\overline{\overline{w}}_p}{V} + \int d\bar{r}_l q(\bar{r}_l - \bar{r}_j)\overline{\overline{U}}_j \overline{\overline{G}}_c \overline{\overline{\tilde{C}}}_l - \int d\bar{r}_l q(\bar{r}_l - \bar{r}_j)\frac{\overline{\overline{w}}_p}{V}\overline{\overline{G}}_c \overline{\overline{\tilde{C}}}_l \tag{5.3.77}$$

Taking (5.3.77) in momentum representation and noting that $\overline{\overline{\tilde{C}}}_j = e^{-i\bar{p}\cdot\bar{r}_j}$ $\overline{\overline{\tilde{C}}} e^{i\bar{p}\cdot\bar{r}_j}$ gives

$$\overline{\overline{\hat{C}}}_p(\bar{p}_1, \bar{p}_2) = \overline{\overline{U}}_p(\bar{p}_1, \bar{p}_2) + \int \frac{d\bar{p}_3}{(2\pi)^3} \overline{\overline{U}}_p(\bar{p}_1, \bar{p}_3)\overline{\overline{G}}_c(\bar{p}_3) q_p(\bar{p}_3 - \bar{p}_2)\overline{\overline{\tilde{C}}}_p(\bar{p}_3, \bar{p}_2) \tag{5.3.78}$$

where

$$\overline{\overline{\hat{C}}}_p(\bar{p}, \bar{p}_1) = \left[\overline{\overline{I}} + \frac{\overline{\overline{w}}_p(\bar{p})}{V}\overline{\overline{G}}_c(\bar{p}) q_p(\bar{p} - \bar{p}_1)\right] \cdot \overline{\overline{\tilde{C}}}_p(\bar{p}, \bar{p}_1) + \frac{\overline{\overline{w}}_p(\bar{p})}{V}\langle \bar{p} | \bar{p}_1 \rangle \tag{5.3.79}$$

Next, use (5.3.79) to express $\overline{\overline{\tilde{C}}}_p$ in terms of $\overline{\overline{\hat{C}}}_p$. We obtain

$$\begin{aligned} \overline{\overline{\tilde{C}}}_p(\bar{p}, \bar{p}_1) = {} & \left[\overline{\overline{I}} + \frac{\overline{\overline{w}}_p(\bar{p})}{V}\overline{\overline{G}}_c(\bar{p}) q_p(\bar{p} - \bar{p}_1)\right]^{-1} \overline{\overline{\hat{C}}}_p(\bar{p}, \bar{p}_1) \\ & - \left[\overline{\overline{I}} + \frac{\overline{\overline{w}}_p(\bar{p})}{V}\overline{\overline{G}}_c(\bar{p}) q_p(0)\right]^{-1} \frac{\overline{\overline{w}}_p(\bar{p})}{V}\langle \bar{p} \mid \bar{p}_1 \rangle \end{aligned} \tag{5.3.80}$$

Putting (5.3.80) in (5.3.78), we then have

$$\begin{aligned} \overline{\overline{\hat{C}}}_p(\bar{p}_1, \bar{p}_2) = {} & \overline{\overline{U}}_p(\bar{p}_1, \bar{p}_2)\overline{\overline{G}}_c(\bar{p}_2)\left[\overline{\overline{I}} + \frac{\overline{\overline{w}}_p(\bar{p}_2)}{V}\overline{\overline{G}}_c(\bar{p}_2) q_p(0)\right]^{-1} \overline{\overline{G}}_c^{-1}(\bar{p}_2) \\ & + \int \frac{d\bar{p}_3}{(2\pi)^3} \overline{\overline{U}}_p(\bar{p}_1, \bar{p}_3)\overline{\overline{G}}_c(\bar{p}_3) q_p(\bar{p}_3 - \bar{p}_2) \\ & \cdot \left[\overline{\overline{I}} + \frac{\overline{\overline{w}}_p(\bar{p}_3)}{V}\overline{\overline{G}}_c(\bar{p}_3) q_p(\bar{p}_3 - \bar{p}_2)\right]^{-1} \overline{\overline{\hat{C}}}_p(\bar{p}_3, \bar{p}_2) \end{aligned} \tag{5.3.81}$$

The QCA approximation gives

$$\overline{\overline{G}}^{-1}(\bar{p}) = \overline{\overline{G}}_c^{-1}(\bar{p}) - n_o \overline{\overline{\tilde{C}}}_p(\bar{p}, \bar{p}) \tag{5.3.82}$$

For coherent potential condition to apply, we choose $\overline{\overline{w}}_p$ such that

$$\overline{\overline{\tilde{C}}}_p(\bar{p}, \bar{p}) = 0 \tag{5.3.83}$$

which we will impose on (5.3.79).

To normalize properly (5.3.79), we use the technique as in EFA–CP by multiplying both sides of the equation with $(1/V)e^{-i(\bar{p}-\bar{p}_1)\cdot\bar{r}_j}$ and integrating over $d\bar{r}_j$. Hence, (5.3.79) becomes

$$\overline{\overline{\hat{C}}}_p(\bar{p}, \bar{p}) = \left[\overline{\overline{I}} + \frac{\overline{\overline{w}}_p(\bar{p})}{V}\,\overline{\overline{G}}_c(\bar{p})\, q_p(0)\right] \overline{\overline{\tilde{C}}}_p(\bar{p}, \bar{p}) + \overline{\overline{w}}_p(\bar{p}) \tag{5.3.84}$$

Taking the limit of $V \to \infty$, we note that $q_p(0)/V \to 0$. Thus

$$\overline{\overline{w}}_p(\overline{p}) = \overline{\overline{\hat{C}}}_p(\overline{p}, \overline{p}) \tag{5.3.85}$$

Taking the limit of $V \to \infty$ in (5.3.81), we have

$$\overline{\overline{\hat{C}}}_p(\overline{p}_1, \overline{p}_2) = \overline{\overline{U}}_p(\overline{p}_1, \overline{p}_2) + \int \frac{d\overline{p}_3}{(2\pi)^3} \overline{\overline{U}}_p(\overline{p}_1, \overline{p}_3) \overline{\overline{G}}_c(\overline{p}_3) q_p(\overline{p}_3 - \overline{p}_2) \overline{\overline{\hat{C}}}_p(\overline{p}_3, \overline{p}_2) \tag{5.3.86}$$

Then we have

$$\overline{\overline{G}}_c^{-1}(\overline{p}) = \overline{\overline{G}}_o^{-1}(\overline{p}) - n_o \overline{\overline{w}}(\overline{p}) = \overline{\overline{G}}_o^{-1}(\overline{p}) - n_o \overline{\overline{\hat{C}}}_p(\overline{p}, \overline{p}) \tag{5.3.87}$$

The dispersion relation is

$$\det \left[\overline{\overline{G}}_o^{-1}(\overline{p}) - n_o \overline{\overline{\hat{C}}}_p(\overline{p}, \overline{p}) \right] = 0 \tag{5.3.88}$$

If we define operator

$$\overline{\overline{\hat{C}}}_j = e^{-i\overline{p}\cdot\overline{r}_j} \overline{\overline{\hat{C}}} e^{i\overline{p}\cdot\overline{r}_j} \tag{5.3.89}$$

then (5.3.86) becomes

$$\overline{\overline{\hat{C}}}_j = \overline{\overline{U}}_j + \int d\overline{r}_l \overline{\overline{U}}_j \overline{\overline{G}}_c q(\overline{r}_l - \overline{r}_j) \overline{\overline{\hat{C}}}_l \tag{5.3.90}$$

Let

$$\overline{\overline{\hat{t}}}_j = \overline{\overline{U}}_j + \overline{\overline{U}}_j \overline{\overline{G}}_c \overline{\overline{\hat{t}}}_j \tag{5.3.91}$$

It follows that $\overline{\overline{\hat{t}}}_j = e^{-i\overline{p}\cdot\overline{r}_j} \overline{\overline{\hat{t}}} e^{i\overline{p}\cdot\overline{r}_j}$. Then putting $q(\overline{r}) = n_o h(\overline{r}) + \delta(\overline{r})$, we have

$$\overline{\overline{\hat{C}}}_j = \overline{\overline{U}}_j + \overline{\overline{U}}_j \overline{\overline{G}}_c \overline{\overline{\hat{C}}}_j + n_o \int d\overline{r}_l \overline{\overline{U}}_j \overline{\overline{G}}_c h(\overline{r}_l - \overline{r}_j) \overline{\overline{\hat{C}}}_l \tag{5.3.92}$$

Using (5.3.91) in (5.3.92) then gives

$$\overline{\overline{\hat{C}}}_j = \overline{\overline{\hat{t}}}_j + n_o \int d\overline{r}_l \overline{\overline{\hat{t}}}_j \overline{\overline{G}}_c h(\overline{r}_l - \overline{r}_j) \overline{\overline{\hat{C}}}_l \tag{5.3.93}$$

The momentum representations are

$$\overline{\overline{\hat{t}}}_p(\overline{p}_1, \overline{p}_2) = \overline{\overline{U}}_p(\overline{p}_1, \overline{p}_2) + \int \frac{d\overline{p}_3}{(2\pi)^3} \overline{\overline{U}}_p(\overline{p}_1, \overline{p}_3) \overline{\overline{G}}_c(\overline{p}_3) \overline{\overline{\hat{t}}}_p(\overline{p}_3, \overline{p}_2) \tag{5.3.94}$$

$$\overline{\overline{\hat{C}}}_p(\overline{p}_1, \overline{p}_2) = \overline{\overline{\hat{t}}}_p(\overline{p}_1, \overline{p}_2) + n_o \int d\overline{p}_3 \overline{\overline{\hat{t}}}_p(\overline{p}_1, \overline{p}_3) \overline{\overline{G}}_c(\overline{p}_3) H(\overline{p}_3 - \overline{p}_2) \overline{\overline{\hat{C}}}_p(\overline{p}_3, \overline{p}_2) \tag{5.3.95}$$

The operator equation for the mean Green's operator is

$$E(\overline{\overline{G}}) = \overline{\overline{G}}_c = \overline{\overline{G}}_o + n_o \overline{\overline{G}}_o \int d\overline{r}_j \overline{\overline{\hat{C}}}_j E(\overline{\overline{G}}) \tag{5.3.96}$$

Hence the mass operator under QCA–CP is $n_o \int d\overline{r}_j \overline{\overline{\widehat{C}}}_j$. A system of nonlinear equations were developed for the case of dielectric spheres under QCA–CP [Tsang and Kong, 1980].

3.6 Low-Frequency Solutions

In this section the effective propagation constants under EFA, QCA, EFA–CP and QCA–CP will be calculated for spherical particles in the low-frequency limit. Both the real part of the effective propagation constant K_r and the imaginary part of the effective propagation constant K_i will be studied. We consider cases where $K_i \ll K_r$ with $K = K_r + iK_i$. In solving the integral equations of the various approximations, the leading term of the real part and the leading term of the imaginary part will be kept while the other terms can be safely discarded.

The introduction of the coherent potential into the effective field approximation and the quasi-crystalline approximation results in a nonlinear equation for the mean Green's function. This is analogous to the relation between nonlinear approximation and the bilocal approximation of random medium, where it was found that the difference between the results of the two approximations is insignificant if the real part of the effective propagation constant K_r is approximately equal to k_m. For the case of random discrete scatterers, the permittivity of the particles can be appreciably different from that of the background medium. When the particles are densely distributed, the real part of the effective wavenumber K_r is appreciably different from k. Because of the difference between K_r and k, the introduction of any nonlinear approximation can significantly change the effective propagation constant. Thus the results of EFA–CP differ significantly from EFA, and the results of QCA–CP differ significantly from QCA.

Effective Field Approximation (EFA)

The Lippmann-Schwinger equation for the transition operator is

$$\overline{\overline{T}} = \overline{\overline{U}} + \overline{\overline{U}}\,\overline{\overline{G}}_o \overline{\overline{T}} \tag{5.3.97}$$

In space representation, $\langle \overline{r} | \overline{\overline{U}} | \overline{r}' \rangle = U(\overline{r})\, \overline{\overline{I}} \langle \overline{r} | \overline{r}' \rangle$ where

$$U(\overline{r}) = \begin{cases} k_s^2 - k^2 & r \le a \\ 0 & r > a \end{cases} \tag{5.3.98}$$

and a is the radius of the particle of permittivity ϵ_s. To solve (5.3.97), a mixed representation $\langle \overline{r} | \overline{\overline{T}} | \overline{p} \rangle \equiv \overline{\overline{T}}_m(\overline{r}, \overline{p})$ can be used. Applying to (5.3.97) gives

$$\overline{\overline{T}}_m(\overline{r},\overline{p}) = U(\overline{r})\,\overline{\overline{I}}\,e^{i\overline{p}\cdot\overline{r}} + U(\overline{r})\,\overline{\overline{I}}\int d\overline{r}'\,\overline{\overline{G}}_o(\overline{r},\overline{r}')\,\overline{\overline{T}}_m(\overline{r}',\overline{p}) \tag{5.3.99}$$

Since the particles are spherical, the singularity of the dyadic Green's function will be separated out in the following manner

$$\overline{\overline{G}}_o(\overline{r},\overline{r}') = PS\,\overline{\overline{G}}_o(\overline{r},\overline{r}') - \frac{\overline{\overline{I}}}{3k^2}\,\delta(\overline{r}-\overline{r}') \tag{5.3.100}$$

where PS stands for principal value with a spherical exclusion volume at $\overline{r}=\overline{r}'$. At low frequencies, $\overline{p}$ can be set equal to zero in (5.3.99). We further let $\overline{\overline{T}}_m(\overline{r}) = \overline{\overline{T}}_m(\overline{r},\overline{p}=0)$. Hence

$$\overline{\overline{T}}_m(\overline{r}) = U(\overline{r})\,\overline{\overline{I}} - \frac{U(\overline{r})}{3k^2}\,\overline{\overline{T}}_m(\overline{r}) + U(\overline{r})\int d\overline{r}'\;PS\,\overline{\overline{G}}_o(\overline{r},\overline{r}')\,\overline{\overline{T}}_m(\overline{r}') \tag{5.3.101}$$

To solve (5.3.101), let $\overline{\overline{T}}_m(\overline{r}) = T_m(\overline{r})\overline{\overline{I}}$ with

$$T_m(\overline{r}) = \begin{cases} T_m & r \le a \\ 0 & r > a \end{cases} \tag{5.3.102}$$

Substitute (5.3.102) into (5.3.101) and set $\overline{r} = 0$. The integral in (5.3.101) is of the same form as that (4.3.32a) and can be evaluated readily. The integral equation in (5.3.101) then reduces to the form

$$T_m = (k_s^2 - k^2) - \frac{(k_s^2-k^2)}{3k^2}\,T_m + T_m(k_s^2-k^2)\frac{2}{3}\int_0^a dr\,r\,e^{ikr} \tag{5.3.103}$$

For small particles, ka is much less than 1, and the real part of the third term in (5.3.103) is much smaller than the real part of the first term and the second term and can be neglected. However, the imaginary part contributes to the leading term of the imaginary part of T_m and cannot be neglected. Hence, keeping the leading term in the real part and the leading term in the imaginary part we obtain

$$T_m = (k_s^2 - k^2) - \frac{(k_s^2-k^2)}{3k^2}\,T_m + T_m(k_s^2-k^2)\frac{i2ka^3}{9} \tag{5.3.104}$$

Solving (5.3.104) gives

$$T_m = 3k^2 y\left[1 + \frac{2}{3}ik^3a^3y\right] \tag{5.3.105}$$

where

$$y = \frac{\epsilon_s - \epsilon}{\epsilon_s + 2\epsilon} \tag{5.3.106}$$

Using the results of (5.3.102) and (5.3.105), the momentum representation of $\overline{\overline{T}}$ can be calculated from the mixed representation. In the low frequency limit we have

$$\overline{\overline{T}}_p(\overline{p}_1, \overline{p}_2) = \langle \overline{p}_1 | \overline{\overline{T}} | \overline{p}_2 \rangle = \int d\overline{r} e^{-i\overline{p}_1 \cdot \overline{r}} \langle \overline{r} | \overline{\overline{T}} | \overline{p}_2 \rangle = \int_{r \le a} d\overline{r}\, T_m \overline{\overline{I}} = T_m v_o \overline{\overline{I}} \tag{5.3.107a}$$

where $v_o = 4\pi a^3/3$ is the volume of the particle. The dispersion relation is given by $\det \left[\overline{\overline{G}}_o^{-1}(\overline{p}) - n_o \overline{\overline{T}}_p(\overline{p}, \overline{p}) \right] = 0$. That is

$$\det \left[\hat{p}\hat{p}(-k^2 + fT_m) + (p^2 - k^2 - fT_m)(\hat{\theta}_p \hat{\theta}_p + \hat{\phi}_p \hat{\phi}_p) \right] = 0 \tag{5.3.107b}$$

with $f = n_o v_o$ as the volume fraction occupied by the particles. The solution for the effective propagation K is

$$K^2 = k^2 + fT_m \tag{5.3.108a}$$

so that

$$K^2 = k^2 + 3k^2 fy \left[1 + i\frac{2}{3} k^3 a^3 y \right] \tag{5.3.108b}$$

Quasi-crystalline Approximation (QCA)

We need to solve for $\overline{\overline{C}}_p(\overline{p}_1, \overline{p}_2)$ governed by (5.3.54). Let $\overline{\overline{T}}_p(\overline{p}_1, \overline{p}_2) = T_m v_o \overline{\overline{I}}$, we have

$$\overline{\overline{C}}_p(\overline{p}_1, \overline{p}_2) = T_m v_o \overline{\overline{I}} + fT_m \int_{-\infty}^{\infty} d\overline{p}_3\, \overline{\overline{G}}_o(\overline{p}_3) H(\overline{p}_3 - \overline{p}_2) \overline{\overline{C}}_p(\overline{p}_3, \overline{p}_2) \tag{5.3.109}$$

In the low frequency limit, $\overline{p}_2$ can be set to zero in (5.3.109). Furthermore, let

$$\overline{\overline{C}}_p(\overline{p}_1, \overline{p}_2) = c\overline{\overline{I}} \tag{5.3.110}$$

Substituting (5.3.110) into (5.3.109), the integral in (5.3.109) can be put in the space domain. We then have

$$c = T_m v_o + \frac{fT_m}{3k^2} c + cfT_m \int_{-\infty}^{\infty} d\overline{r}\, PS\, \overline{\overline{G}}_o(\overline{r}) \left[g(\overline{r}) - 1 \right] \tag{5.3.111}$$

Keeping the leading term of the real and imaginary parts of (5.3.111), we have

$$c = T_m v_o + fT_m c \left\{ \frac{1}{3k^2} + \frac{2}{3} ik \int_0^{\infty} dr\, r^2 \left[g(r) - 1 \right] \right\} \tag{5.3.112}$$

The integral in (5.3.112) is of the order $\mathcal{O}(a^3)$. Solving for c in (5.3.112), using the expression for T_m as given in (5.3.105), and noting that $\text{Re}(T_m) \gg \text{Im}(T_m)$ and $\text{Re}(c) \gg \text{Im}(c)$, gives

$$c = \frac{3v_o k^2 y}{1 - fy}\left\{1 + i\frac{2}{3}\frac{(ka)^3 y}{1 - fy}\left[1 + 4\pi n_o \int_0^\infty dr\, r^2\left[g(r) - 1\right]\right]\right\} \tag{5.3.113a}$$

The dispersion relation is, under QCA,

$$\det\left[\overline{\overline{G}}_o^{-1}(\overline{p}) - n_o \overline{\overline{C}}_p(\overline{p}, \overline{p})\right] = 0 \tag{5.3.113b}$$

so that

$$\det\left[\hat{p}\hat{p}(-k^2 + n_o c) + (p^2 - k^2 - n_o c)(\hat{\theta}_p\hat{\theta}_p + \hat{\phi}_p\hat{\phi}_p)\right] = 0 \tag{5.3.113c}$$

The effective propagation constant K is such that

$$K^2 = k^2 + n_o c \tag{5.3.113d}$$

giving

$$K^2 = \frac{3fk^2 y}{1 - fy}\left\{1 + i\frac{2}{3}\frac{(ka)^3 y}{1 - fy}\left[1 + 4\pi n_o \int_0^\infty dr\; r^2(g(r) - 1)\right]\right\} \tag{5.3.114}$$

Effective Field Approximation with Coherent Potential (EFA–CP)

Under EFA–CP, the equation for the modified transition operator $\hat{\overline{\overline{t}}}$ is

$$\hat{\overline{\overline{t}}} = \overline{\overline{U}} + \overline{\overline{U}}\,\overline{\overline{G}}_c\,\hat{\overline{\overline{t}}} \tag{5.3.115}$$

To solve (5.3.115), we assume that in the low-frequency limit, $\overline{\overline{G}}_c$ assumes the same expression as $\overline{\overline{G}}_o$ with k replaced by K. Hence,

$$\overline{\overline{G}}_c(\overline{r}, \overline{r}') = PS\,\overline{\overline{G}}_c(\overline{r}, \overline{r}') - \frac{\overline{\overline{I}}}{3K^2}\,\delta(\overline{r} - \overline{r}') \tag{5.3.116}$$

The solution of the integral equation in (5.3.115) can be obtained in a manner analogous to that of (5.3.98) through (5.3.104). Thus,

$$\hat{\overline{\overline{t}}}_p(\overline{p}_1, \overline{p}_2) = \hat{t}_m v_o \overline{\overline{I}} \tag{5.3.117}$$

where $\hat{t}_m$ satisfies the following equation analogous to (5.3.104)

$$\hat{t}_m = (k_s^2 - k^2) - \frac{(k_s^2 - k^2)\hat{t}_m}{3K^2} + \hat{t}_m(k_s^2 - k^2)\frac{i2Ka^3}{9} \tag{5.3.118}$$

The $(k_s^2 - k^2)$ factor in (5.3.118) arises from the potential $\overline{\overline{U}}$, which, as indicated in (5.3.115), is unchanged. Other k's in (5.3.104) are replaced by

K to give (5.3.118). Solving (5.3.118) gives

$$\hat{t}_m = \frac{z}{1 + z/(3K^2)}\left[1 + i\frac{2}{9}\frac{Ka^3 z}{1 + z/(3K^2)}\right] \tag{5.3.119}$$

where $z = k_s^2 - k^2$. The dispersion relation is, under EFA–CP,

$$K^2 = k^2 + f\,\hat{t}_m = k^2 + \frac{fz}{1 + \dfrac{z}{3K^2}}\left[1 + i\frac{2}{9}\frac{Ka^3 z}{1 + \dfrac{z}{3K^2}}\right] \tag{5.3.120}$$

which is a nonlinear equation for K^2.

Quasi-crystalline Approximation with Coherent Potential (QCA–CP)

The integral equation is

$$\overline{\overline{\widehat{C}}}_p(\overline{p}_1, \overline{p}_2) = \overline{\overline{\hat{t}}}_p(\overline{p}_1, \overline{p}_2) + n_o\int d\overline{p}_3 \overline{\overline{\hat{t}}}_p(\overline{p}_1, \overline{p}_3)\overline{\overline{G}}_c(\overline{p}_3)H(\overline{p}_3 - \overline{p}_2)\overline{\overline{\widehat{C}}}_p(\overline{p}_3, \overline{p}_2)$$

The procedure of solution is similar to the case of QCA. Hence, in the low-frequency limit we have

$$\overline{\overline{\widehat{C}}}_p(\overline{p}_1, \overline{p}_2) = \hat{c}\,\overline{\overline{I}} \tag{5.3.121}$$

and $\hat{c}$ obeys the following equation that is analogous to (5.3.112)

$$\hat{c} = v_o\,\hat{t}_m + f\hat{t}_m\hat{c}\left[\frac{1}{3K^2} + \frac{2}{3}iK\int_0^\infty dr\, r^2\,[g(r) - 1]\right] \tag{5.3.122}$$

Use the expression of $\hat{t}_m$ as given by (5.3.119), solve $\hat{c}$ from (5.3.122), and retain only the leading term of the real part and the leading term of the imaginary part of $\hat{c}$ to get

$$\hat{c} = \frac{v_o z}{1 + z(1-f)/(3K^2)}\left\{1 + i\frac{2}{9}Ka^3\frac{z}{1 + z(1-f)/(3K^2)}\right.$$
$$\left.\cdot\left[1 + 4\pi n_o\int_0^\infty dr\, r^2\,[g(r) - 1]\right]\right\} \tag{5.3.123}$$

The dispersion relation under QCA–CP is

$$K^2 = k^2 + n_o\hat{c} \tag{5.3.124}$$

which is a nonlinear equation for K^2.

In the very low frequency limit, the scattering attenuation term that is dependent on particle size in (5.3.123) can be neglected, and the mixture formula for $\epsilon_{\text{eff}} = K^2/\omega\mu_o$ is, on using the first term in (5.3.123),

$$\epsilon_{\text{eff}} = \epsilon + \frac{3f(\epsilon_s - \epsilon)\epsilon_{\text{eff}}}{3\epsilon_{\text{eff}} + (\epsilon_s - \epsilon)(1 - f)} \tag{5.3.125}$$

With simple algebra, it can be shown that (5.3.125) can be expressed in the form

$$\epsilon_{\text{eff}} = \frac{\epsilon(1-f) + f\epsilon_s + f\dfrac{(\epsilon - \epsilon_s)\epsilon_s}{3\epsilon_{\text{eff}} - \epsilon + \epsilon_s}}{1 + f\dfrac{\epsilon - \epsilon_s}{3\epsilon_{\text{eff}} + \epsilon_s - \epsilon}} \tag{5.3.126}$$

3.7 QCA–CP for Multiple Species of Particles

In this section, the QCA–CP result is extended to the case where there are multiple species of particles. The different species refer to the fact that particles can be of different shapes, sizes, and permittivities. The different species are denoted by $s_j = 1, 2, \ldots, L$. Each species can have a distinct size, shape, and permittivity ϵ_{s_j}.

The multiple scattering equations become

$$\overline{\overline{G}} = \overline{\overline{G}}_o + \overline{\overline{G}}_o \sum_{j=1}^{N} \overline{\overline{U}}_j^{s_j} \overline{\overline{G}} \tag{5.3.127}$$

where

$$U^{s_j}(\overline{r} - \overline{r}_j) = \begin{cases} 0 & \text{for } \overline{r} \text{ outside particle } j \\ k_{s_j}^2 - k^2 & \text{for } \overline{r} \text{ inside particle } j \end{cases} \tag{5.3.128}$$

with $k_{s_j} = w\sqrt{\mu\epsilon_{s_j}}$ being the wavenumber in the s_j-type scatterers. The QCA–CP equation can be developed in a similar manner. The pair distribution functions for multiple species of particles are described in Chapter 8 of Volume II.

The average Green's operator must be equal to the coherent potential Green's operator.

$$E(\overline{\overline{G}}) = \overline{\overline{G}}_c \tag{5.3.129}$$

The modified transition operator is then, in analogy to Section 3.5,

$$\overline{\overline{\hat{t}}}_j^{s_j} = \overline{\overline{U}}_j^{s_j} + \overline{\overline{U}}_j^{s_j} \overline{\overline{G}}_c \overline{\overline{\hat{t}}}_j^{s_j} \tag{5.3.130}$$

The result for the average Green's operator under QCA–CP becomes

$$\overline{\overline{G}}(\overline{p}) = \overline{\overline{G}}_c(\overline{p}) = \left[\overline{\overline{G}}_o^{-1}(\overline{p}) - \sum_{s_j=1}^{L} n_{s_j} \overline{\overline{\hat{C}}}_p^{s_j}(\overline{p}, \overline{p}) \right]^{-1} \tag{5.3.131}$$

where $\overline{\overline{\widehat{C}}}_p^{s_j}(\overline{p},\overline{p}')$ satisfies the integral equation

$$\overline{\overline{\widehat{C}}}_p^{s_l}(\overline{p},\overline{p}') = \overline{\overline{\hat{t}}}_p^{s_l}(\overline{p},\overline{p}') + \sum_{s_j=1}^{L} n_{s_j} \int d\overline{p}'' \overline{\overline{\hat{t}}}_p^{s_l}(\overline{p},\overline{p}'')\,\overline{\overline{G}}_c(\overline{p}'')\,H_{s_l s_j}(\overline{p}''-\overline{p}') \cdot \overline{\overline{\widehat{C}}}_p^{s_j}(\overline{p}'',\overline{p}') \tag{5.3.132}$$

and $\overline{\overline{\hat{t}}}_p^{s_j}(\overline{p},\overline{p}') = \left\langle \overline{p} \mid \overline{\overline{\hat{t}}}^{s_j} \mid \overline{p}' \right\rangle$ is the momentum representation of $\overline{\overline{\hat{t}}}^{s_j}$. The dispersion relation for QCA–CP is

$$\det\left[\overline{\overline{G}}_o^{-1}(\overline{p}) - \sum_{s_j=1}^{L} n_{s_j} \overline{\overline{\widehat{C}}}_p^{s_j}(\overline{p},\overline{p})\right] = 0 \tag{5.3.133}$$

The dispersion relations of (5.3.130)–(5.3.133) can be expressed in terms of mass operator. Let the mass operator $\overline{\overline{\widehat{M}}}$ under QCA–CP be expressed as

$$\overline{\overline{\widehat{M}}} = \sum_{s_j=1}^{L} n_{s_j} \int d\overline{r}_j \overline{\overline{\widehat{C}}}_j^{s_j} \tag{5.3.134}$$

where the relation analogous to (5.3.93) is

$$\overline{\overline{\widehat{C}}}_l^{s_l} = \overline{\overline{\hat{t}}}_l^{s_l} + \sum_{s_j=1}^{L} n_{s_j} \int d\overline{r}_j h_{s_l s_j}(\overline{r}_j - \overline{r}_l)\,\overline{\overline{\hat{t}}}_l^{s_l}\,\overline{\overline{G}}_c \overline{\overline{\widehat{C}}}_j^{s_j} \tag{5.3.135}$$

The average Green's operator under QCA–CP can be expressed as

$$E(\overline{\overline{G}}) = \overline{\overline{G}}_c = \overline{\overline{G}}_o + \overline{\overline{G}}_o \overline{\overline{\widehat{M}}}\,\overline{\overline{G}}_c \tag{5.3.136}$$

In this case,

$$\langle \overline{p} \mid \overline{\overline{\widehat{M}}} \mid \overline{p}' \rangle = \langle \overline{p} \mid \overline{p}' \rangle \sum_{s_j=1}^{L} n_{s_j} \overline{\overline{\widehat{C}}}_p^{s_j}(\overline{p},\overline{p}) \tag{5.3.137}$$

The mean Green's function then becomes

$$\overline{\overline{G}}(\overline{p}) = \left[\overline{\overline{G}}_o^{-1}(\overline{p}) - \overline{\overline{\widehat{M}}}(\overline{p},\overline{p})\right]^{-1} \tag{5.3.138}$$

and the dispersion relation is

$$\det\left[\overline{\overline{G}}_o^{-1}(\overline{p}) - \overline{\overline{\widehat{M}}}(\overline{p},\overline{p})\right] = 0 \tag{5.3.139}$$

At low frequency, we have

$$\overline{\overline{\hat{t}}}_p^{s_j}(\overline{p},\overline{p}') = \hat{t}_m^{s_j} v_{s_j} \overline{\overline{I}} \tag{5.3.140}$$

where $v_{s_j} = 4\pi a_{s_j}^3/3$, and

$$\hat{t}_m^{s_j} = 3K^2\hat{y}_{s_j}\left[1 + i\frac{2}{3}K^3 a_{s_j}^3 \hat{y}_{s_j}\right] \tag{5.3.141}$$

$$\hat{y}_{s_j} = \frac{k_{s_j}^2 - k^2}{3K^2 + \left(k_{s_j}^2 - k^2\right)} \tag{5.3.142}$$

In (5.3.141) and (5.3.142), K is the effective propagation constant. We also have

$$\overline{\overline{\widehat{C}}}_p^{s_j}(\overline{p}, \overline{p}') = \widehat{C}_{s_j}\overline{\overline{I}} \tag{5.3.143}$$

$$\widehat{C}_{s_j} = \hat{t}_m^{s_j} v_{s_j} + \hat{t}_m^{s_j} v_{s_j}\sum_{s_l=1}^{L} n_{s_l}\widehat{C}_{s_l}\left[\frac{1}{3K^2} + i\frac{4\pi^2 K}{3}H_{s_j s_l}(\overline{p}=0)\right] \tag{5.3.144a}$$

where

$$H_{s_j s_l}(\overline{p}) = \frac{1}{(2\pi)^3}\int_{-\infty}^{\infty} d\overline{r}\left[g_{s_j s_l}(\overline{r}) - 1\right]e^{-i\overline{p}\cdot\overline{r}} \tag{5.3.144b}$$

The effective propagation constant can be calculated from

$$K^2 = k^2 + \sum_{s_j=1}^{L} n_{s_j}\widehat{C}_{s_j} = k^2 + \frac{3K^2}{\widehat{D}}\sum_{s_l=1}^{L} f_{s_l}\hat{y}_{s_l}$$
$$\times\left\{1 + i\frac{2K^3}{3\widehat{D}}\left[a_{s_l}^3\hat{y}_{s_l} + \sum_{s_j=1}^{L} a_{s_j}^3 8\pi^3 n_{s_j}\hat{y}_{s_j}H_{s_j s_l}(\overline{p}=0)\right]\right\} \tag{5.3.145}$$

$$\widehat{D} = 1 - \sum_{s_l=1}^{L} f_{s_l}\hat{y}_{s_l} \tag{5.3.146}$$

Note that $\hat{y}_{s_j}$ depends on K^2 so that (5.3.145) is a nonlinear equation of K^2.

4 Ward's Identity and Energy Conservation

In Chapter 4, it was shown that in the case of random medium, energy conservation is satisfied exactly if the nonlinear approximation of the mean field is used in conjunction with the ladder approximation of the second moment of the field. It was also mentioned that the difference between the results of the nonlinear approximation and the bilocal approximation is small when the real part of the effective propagation constant is approximately equal to that of the background medium or the mean wavenumber. However, in this chapter, there is an appreciable difference between K_r and k so that the results of QCA and QCA–CP can be significantly different. Hence it is important to examine the issue of energy conservation for discrete scatterers. The relation between the first moment and second moment is also known as Ward's identity.

Consider a discrete random medium consisting of discrete scatterers of L different species embedded in a background medium with permittivity ϵ (Fig. 5.4.1). Let $\overline{S}$ be the time-averaged Poynting's vector such that

$$\overline{S} = \frac{1}{4i\omega\mu}\left\{-\overline{E} \times (\nabla \times \overline{E}^*) + \overline{E}^* \times (\nabla \times \overline{E})\right\} \tag{5.4.1}$$

Energy conservation requires that for the case of nonabsorptive background and scatterers when ϵ and ϵ_{s_j} $(j = 1, 2, \ldots, L)$ are real, the integrated optical relation must be satisfied. The integrated optical relation is

$$\int d\overline{r}\langle\nabla \cdot \overline{S}\rangle = 0 \tag{5.4.2}$$

The vector wave equation governing the electric field is

$$\overline{E}\,(\overline{r}) = \overline{E}^{inc}\,(\overline{r}) + \sum_{l=1}^{N} \int d\overline{r}'\overline{\overline{G}}_o\,(\overline{r}, \overline{r}')\, U_l^{s_l}\,(\overline{r}')\,\overline{E}\,(\overline{r}') \tag{5.4.3}$$

Let the operator $\overline{\overline{W}}$ be defined by

$$\overline{\overline{W}}\,(\overline{r}) \cdot \overline{E}\,(\overline{r}) = \nabla \times \nabla \times \overline{E}\,(\overline{r}) - k^2\overline{E}\,(\overline{r}) \tag{5.4.4}$$

so that

$$\overline{\overline{W}}\,(\overline{r}) \cdot \overline{E}\,(\overline{r}) = \sum_{l=1}^{N} U_l^{s_l}\,(\overline{r})\,\overline{E}\,(\overline{r}) \tag{5.4.5}$$

The integrated optical relation of (5.4.2), by using (5.4.1), (5.4.5), and the fact that $U_l^{s_l}$ is real for lossless scatterers, assumes the form

$$\begin{aligned} 4i\omega\mu \int d\overline{r}\,\langle\nabla \cdot \overline{S}\,(\overline{r})\rangle = \int d\overline{r} \sum_{\alpha,\beta} \lim_{\overline{r}' \to \overline{r}} \Big\{ & W_{\alpha\beta}\,(\overline{r}')\,\langle E_\alpha\,(\overline{r})\,E_\beta^*\,(\overline{r}')\rangle \\ & - W_{\alpha\beta}\,(\overline{r})\,\langle E_\beta\,(\overline{r})\,E_\alpha^*\,(\overline{r}')\rangle \Big\} \end{aligned} \tag{5.4.6}$$

In (5.4.6) the Greek index subscripts $\alpha, \beta = 1, 2, 3$ are used to denote the Cartesian field components.

The coherent field $\langle\overline{E}\,(\overline{r})\rangle$ and the coherent Green's function both obey the Dyson's equation

$$\nabla \times \nabla \times \langle\overline{E}\,(\overline{r})\rangle - k^2\,\langle\overline{E}\,(\overline{r})\rangle = \int d\overline{r}'\overline{\overline{M}}\,(\overline{r}, \overline{r}') \cdot \langle\overline{E}\,(\overline{r}')\rangle \tag{5.4.7}$$

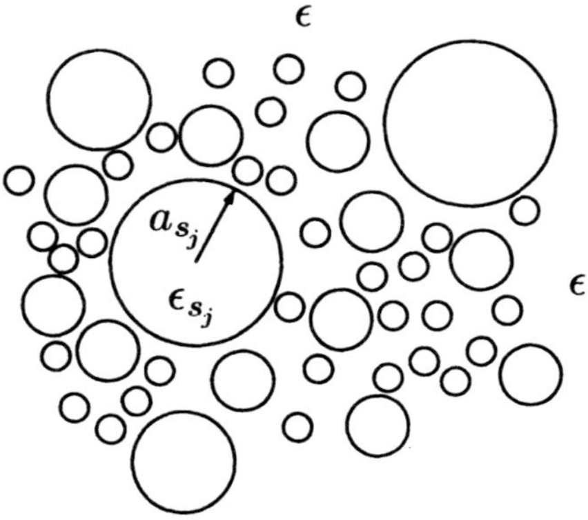

Figure 5.4.1 Discrete random medium consisting of particles of L different species, $s_j = 1, 2, \ldots, L$ with radius a_{s_j} and permittivity ϵ_{s_j} embedded in a background medium of permittivity ϵ.

$$\nabla \times \nabla \times \left\langle \overline{\overline{G}}\left(\overline{r}, \overline{r}'\right)\right\rangle - k^2 \left\langle \overline{\overline{G}}\left(\overline{r}, \overline{r}'\right)\right\rangle$$
$$= \overline{\overline{I}}\delta\left(\overline{r} - \overline{r}'\right) + \int d\overline{r}'' \overline{\overline{M}}\left(\overline{r}, \overline{r}''\right) \cdot \left\langle \overline{\overline{G}}\left(\overline{r}'', \overline{r}'\right)\right\rangle \qquad (5.4.8)$$

where $\overline{\overline{M}}$ is the mass operator. Equations (5.4.7) and (5.4.8) are exact, and the exact mass operator contains an infinite number of terms.

The second moment of the field obeys the Bethe-Salpeter equation which is exact.

$$\left\langle E_\alpha\left(\overline{r}\right) E_\beta^*\left(\overline{r}'\right)\right\rangle = \left\langle E_\alpha\left(\overline{r}\right)\right\rangle \left\langle E_\beta^*\left(\overline{r}'\right)\right\rangle$$
$$+ \sum_{\alpha', \beta'} \sum_{\alpha'', \beta''} \int d\overline{r}_1 \int d\overline{r}_2 \int d\overline{r}_1' \int d\overline{r}_2' \left\langle G_{\alpha\alpha'}\left(\overline{r}, \overline{r}_1\right)\right\rangle \left\langle G_{\beta\beta'}^*\left(\overline{r}', \overline{r}_1'\right)\right\rangle$$
$$I_{\alpha'\alpha'';\beta'\beta''}\left(\overline{r}_1, \overline{r}_2; \overline{r}_1', \overline{r}_2'\right) \left\langle E_{\alpha''}\left(\overline{r}_2\right) E_{\beta''}^*\left(\overline{r}_2'\right)\right\rangle \qquad (5.4.9)$$

where I is the intensity operator and is exact. From (5.4.7)–(5.4.9), it follows that

$$\sum_\beta W_{\alpha\beta}\left(\overline{r}'\right) \left\langle E_\alpha\left(\overline{r}\right) E_\beta^*\left(\overline{r}'\right)\right\rangle$$
$$= \left\langle E_\alpha\left(\overline{r}\right)\right\rangle \sum_\beta \int d\overline{r}_2 M_{\alpha\beta}^*\left(\overline{r}', \overline{r}_2\right) \left\langle E_\beta^*\left(\overline{r}_2\right)\right\rangle$$
$$+ \sum_{\alpha'} \sum_{\alpha'', \beta''} \int d\overline{r}_1 \int d\overline{r}_2 \int d\overline{r}_2' \left\langle G_{\alpha\alpha'}\left(\overline{r}, \overline{r}_1\right)\right\rangle$$
$$\cdot I_{\alpha'\alpha'';\alpha\beta''}\left(\overline{r}_1, \overline{r}_2; \overline{r}', \overline{r}_2'\right) \left\langle E_{\alpha''}\left(\overline{r}_2\right) E_{\beta''}^*\left(\overline{r}_2'\right)\right\rangle$$

$$+\sum_{\beta}\sum_{\alpha',\beta'}\sum_{\alpha'',\beta''}\int d\overline{r}_1\int d\overline{r}_2\int d\overline{r}_1'\int d\overline{r}_2'\left\langle G_{\alpha\alpha'}\left(\overline{r},\overline{r}_1\right)\right\rangle\int d\overline{r}_2'' M^*_{\alpha\beta}\left(\overline{r}',\overline{r}_2''\right)$$
$$\cdot\left\langle G^*_{\beta\beta'}\left(\overline{r}_2'',\overline{r}_1'\right)\right\rangle I_{\alpha'\alpha'';\beta'\beta''}\left(\overline{r}_1,\overline{r}_2;\overline{r}_1',\overline{r}_2'\right)\left\langle E_{\alpha''}\left(\overline{r}_2\right)E^*_{\beta''}\left(\overline{r}_2'\right)\right\rangle \qquad (5.4.10)$$

Next apply the Bethe-Salpeter equation of (5.4.9) to the second term on the right-hand side of (5.4.10). This gives

$$\sum_{\beta}W_{\alpha\beta}\left(\overline{r}'\right)\left\langle E_{\alpha}\left(\overline{r}\right)E^*_{\beta}\left(\overline{r}'\right)\right\rangle$$
$$=\left\langle E_{\alpha}\left(\overline{r}\right)\right\rangle\sum_{\beta}\int d\overline{r}_2 M^*_{\alpha\beta}\left(\overline{r}',\overline{r}_2\right)\left\langle E^*_{\beta}\left(\overline{r}_2\right)\right\rangle$$
$$+\sum_{\alpha'}\sum_{\alpha'',\beta''}\int d\overline{r}_1\int d\overline{r}_2\int d\overline{r}_2'\left\langle G_{\alpha\alpha'}\left(\overline{r},\overline{r}_1\right)\right\rangle$$
$$\cdot I_{\alpha'\alpha'';\alpha\beta''}\left(\overline{r}_1,\overline{r}_2;\overline{r}',\overline{r}_2'\right)\left\langle E_{\alpha''}\left(\overline{r}_2\right)\right\rangle\left\langle E^*_{\beta''}\left(\overline{r}_2'\right)\right\rangle$$
$$+\sum_{\alpha'}\sum_{\alpha'',\beta''}\int d\overline{r}_1\int d\overline{r}_2\int d\overline{r}_2'\left\langle G_{\alpha\alpha'}\left(\overline{r},\overline{r}_1\right)\right\rangle I_{\alpha'\alpha'';\alpha\beta''}\left(\overline{r}_1,\overline{r}_2;\overline{r}',\overline{r}_2'\right)$$
$$\cdot\sum_{\alpha_1,\beta_1}\sum_{\alpha_2,\beta_2}\int d\overline{r}_3\int d\overline{r}_4\int d\overline{r}_3'\int d\overline{r}_4'\left\langle G_{\alpha''\alpha_1}\left(\overline{r}_2,\overline{r}_3\right)\right\rangle$$
$$\cdot\left\langle G^*_{\beta''\beta_1}\left(\overline{r}_2',\overline{r}_3'\right)\right\rangle I_{\alpha_1\alpha_2;\beta_1\beta_2}\left(\overline{r}_3,\overline{r}_4;\overline{r}_3',\overline{r}_4'\right)\left\langle E_{\alpha_2}\left(\overline{r}_4\right)E^*_{\beta_2}\left(\overline{r}_4'\right)\right\rangle$$
$$+\sum_{\beta}\sum_{\alpha',\beta'}\sum_{\alpha'',\beta''}\int d\overline{r}_1\int d\overline{r}_2\int d\overline{r}_1'\int d\overline{r}_2'\left\langle G_{\alpha\alpha'}\left(\overline{r},\overline{r}_1\right)\right\rangle\int d\overline{r}_2'' M^*_{\alpha\beta}\left(\overline{r}',\overline{r}_2''\right)$$
$$\cdot\left\langle G^*_{\beta\beta'}\left(\overline{r}_2'',\overline{r}_1'\right)\right\rangle I_{\alpha'\alpha'';\beta'\beta''}\left(\overline{r}_1,\overline{r}_2;\overline{r}_1',\overline{r}_2'\right)\left\langle E_{\alpha''}\left(\overline{r}_2\right)E^*_{\beta''}\left(\overline{r}_2'\right)\right\rangle \qquad (5.4.11)$$

Similar manipulation can be applied to $\sum_{\beta}W_{\alpha\beta}\left(\overline{r}\right)\left\langle E^*_{\alpha}\left(\overline{r}'\right)E_{\beta}\left(\overline{r}\right)\right\rangle$. The resultant expression and (5.4.11) are then substituted into the right-hand side of (5.4.6). The limit $\overline{r}'\to\overline{r}$ is taken and the integration is carried out over $d\overline{r}$. There are eight terms. By interchanging dummy indices and arguments, the eight terms can be rearranged, factorized, and cast into the following form

$$4i\omega\mu\int d\overline{r}\left\langle\nabla\cdot\overline{S}\left(\overline{r}\right)\right\rangle=\int d\overline{r}_2\int d\overline{r}_2'\sum_{\alpha'',\beta''}\left\{M^*_{\alpha''\beta''}\left(\overline{r}_2,\overline{r}_2'\right)-M_{\beta''\alpha''}\left(\overline{r}_2',\overline{r}_2\right)\right.$$
$$\left.+\int d\overline{r}\int d\overline{r}_1\sum_{\alpha,\alpha''}\left[\left\langle G_{\alpha\alpha'}\left(\overline{r},\overline{r}_1\right)\right\rangle-\left\langle G^*_{\alpha'\alpha}\left(\overline{r}_1,\overline{r}\right)\right\rangle\right]I_{\alpha'\alpha'';\alpha\beta''}\left(\overline{r}_1,\overline{r}_2;\overline{r},\overline{r}_2'\right)\right\}$$
$$\cdot\left\langle E_{\alpha''}\left(\overline{r}_2\right)\right\rangle\left\langle E^*_{\beta''}\left(\overline{r}_2'\right)\right\rangle$$

$$+ \sum_{\alpha_1,\beta_1} \sum_{\alpha_2,\beta_2} \sum_{\alpha'',\beta''} \int d\overline{r}_2 \int d\overline{r}_3 \int d\overline{r}_4 \int d\overline{r}'_3 \int d\overline{r}'_4 \int d\overline{r}'_2$$
$$\cdot \left\langle G_{\alpha''\alpha_1}\left(\overline{r}_2,\overline{r}_3\right)\right\rangle \left\langle G^*_{\beta''\beta_1}\left(\overline{r}'_2,\overline{r}'_3\right)\right\rangle I_{\alpha_1\alpha_2;\beta_1\beta_2}\left(\overline{r}_3,\overline{r}_4;\overline{r}'_3,\overline{r}'_4\right) \left\langle E_{\alpha_2}\left(\overline{r}_4\right) E^*_{\beta_2}\left(\overline{r}'_4\right)\right\rangle$$
$$\cdot \left\{ M^*_{\alpha''\beta''}\left(\overline{r}_2,\overline{r}'_2\right) - M_{\beta''\alpha''}\left(\overline{r}'_2,\overline{r}_2\right) \right.$$
$$\left. + \int d\overline{r}_1 \sum_{\alpha,\alpha'} \int d\overline{r}\left[\left\langle G_{\alpha\alpha'}\left(\overline{r},\overline{r}_1\right)\right\rangle - \left\langle G^*_{\alpha'\alpha}\left(\overline{r}_1,\overline{r}\right)\right\rangle\right] I_{\alpha'\alpha'';\alpha\beta''}\left(\overline{r}_1,\overline{r}_2;\overline{r},\overline{r}'_2\right) \right\} \tag{5.4.12}$$

From (5.4.12), it follows that the integrated optical relation is satisfied exactly if

$$M^*_{\alpha\beta}\left(\overline{r},\overline{r}'\right) - M_{\beta\alpha}\left(\overline{r}',\overline{r}\right) + \sum_{\alpha',\beta'} \int d\overline{r}_1 \int d\overline{r}_2$$
$$\cdot \left[\left\langle G_{\alpha'\beta'}\left(\overline{r}_1,\overline{r}_2\right)\right\rangle - \left\langle G^*_{\beta'\alpha'}\left(\overline{r}_2,\overline{r}_1\right)\right\rangle\right] I_{\beta'\alpha;\alpha'\beta}\left(\overline{r}_2,\overline{r};\overline{r}_1,\overline{r}'\right) = 0 \tag{5.4.13}$$

In the following we shall show that the condition is obeyed by QCA–CP and the correlated ladder approximation for dense media. The correlated ladder approximation is the ladder approximation with particle pair distribution functions included. From (5.4.9), the mass operator under QCA–CP is $\overline{\overline{M}}$:

$$\overline{\overline{M}} = \sum_{s_j=1}^{L} n_{s_j} \int d\overline{r}_j \overline{\overline{\widehat{C}}}_j^{s_j} \tag{5.4.14}$$

Let the superscript † denote Hermitian adjoint such that

$$(\overline{\overline{A}}^\dagger)_{\alpha''\beta''}\left(\overline{r}_2,\overline{r}'_2\right) = \overline{\overline{A}}^*_{\beta''\alpha''}\left(\overline{r}'_2,\overline{r}_2\right) \tag{5.4.15}$$

It follows from (5.3.130) and (5.3.135) that

$$\overline{\overline{\widehat{C}}}_j^{s_j} = \overline{\overline{U}}_j^{s_j} + \sum_{s_l=1}^{L} \int d\overline{r}_l q_{s_j s_l}\left(\overline{r}_j - \overline{r}_l\right) \overline{\overline{U}}_j^{s_j} \overline{\overline{G}}_c \overline{\overline{\widehat{C}}}_l^{s_l} \tag{5.4.16}$$

where

$$q_{s_j s_l}\left(\overline{r}_j - \overline{r}_l\right) = \delta_{s_j s_l} \delta\left(\overline{r}_j - \overline{r}_l\right) + n_{s_l} h_{s_j s_l}\left(\overline{r}_j - \overline{r}_l\right) \tag{5.4.17}$$

From (5.4.14), (5.4.16), and

$$\overline{\overline{U}}_j^{s_j \dagger} = \overline{\overline{U}}_j^{s_j} \tag{5.4.18}$$

for lossless scatterers, it follows that

$$\overline{\overline{M}}^{\dagger}-\overline{\overline{M}}=\sum_{s_j=1}^{L}\sum_{s_l=1}^{L}n_{s_j}\int d\overline{r}_j\int d\overline{r}_l\,q_{s_js_l}(\overline{r}_j-\overline{r}_l)\left[\overline{\overline{\widehat{C}}}_l^{s_l\dagger}\overline{\overline{G}}_c^{\dagger}\overline{\overline{U}}_j^{s_j}-\overline{\overline{U}}_j^{s_j}\overline{\overline{G}}_c\overline{\overline{\widehat{C}}}_l^{s_l}\right] \tag{5.4.19}$$

Next, use $\overline{\overline{U}}_j^{s_j}$ from (5.4.16) to substitute in the first $\overline{\overline{U}}_j^{s_j}$ in (5.4.19). For the second $\overline{\overline{U}}_j^{s_j}$ in (5.4.19), use

$$\overline{\overline{U}}_j^{s_j}=\overline{\overline{\widehat{C}}}_j^{s_j\dagger}-\sum_{s_m=1}^{L}\int d\overline{r}_m q_{s_js_m}(\overline{r}_j-\overline{r}_m)\overline{\overline{\widehat{C}}}_m^{s_m\dagger}\overline{\overline{G}}_c^{\dagger}\overline{\overline{U}}_j^{s_j} \tag{5.4.20}$$

which is a consequence of (5.4.16) and (5.4.18). Thus, we get four terms on the right-hand side of (5.4.19). Two of the terms cancel on exchange of dummy indices and arguments. The result is

$$\overline{\overline{M}}^{\dagger}-\overline{\overline{M}}=\sum_{s_j=1}^{L}\sum_{s_l=1}^{L}n_{s_j}\int d\overline{r}_j\int d\overline{r}_l\,q_{s_js_l}(\overline{r}_j-\overline{r}_l)\left[\overline{\overline{\widehat{C}}}_l^{s_l\dagger}\overline{\overline{G}}_c^{\dagger}\overline{\overline{\widehat{C}}}_j^{s_j}-\overline{\overline{\widehat{C}}}_j^{s_j\dagger}\overline{\overline{G}}_c\overline{\overline{\widehat{C}}}_l^{s_l}\right] \tag{5.4.21}$$

Since $h_{s_js_l}(\overline{r}_j-\overline{r}_l)$ is symmetric with respect to s_j and s_l as well as $\overline{r}_j$ and $\overline{r}_l$, it follows that $n_{s_j}q_{s_js_l}(\overline{r}_j-\overline{r}_l)=n_{s_l}q_{s_ls_j}(\overline{r}_j-\overline{r}_l)$. Thus (5.4.21) is of the same form as the requirement of energy conservation of (5.4.13), provided that the intensity operator is chosen as

$$\begin{aligned}I_{\alpha\beta;\alpha'\beta'}&(\overline{r},\overline{r}_1;\overline{r}',\overline{r}_1')\\&=\sum_{s_j=1}^{L}\sum_{s_l=1}^{L}\int d\overline{r}_j\int d\overline{r}_l\left[n_{s_j}\delta_{s_js_l}\delta(\overline{r}_j-\overline{r}_l)+n_{s_j}n_{s_l}h_{s_js_l}(\overline{r}_j-\overline{r}_l)\right]\\&\quad\cdot\widehat{C}^{s_l}_{l\alpha\beta}(\overline{r},\overline{r}_1)\,\widehat{C}^{s_j*}_{j\alpha'\beta'}(\overline{r}',\overline{r}_1')\end{aligned} \tag{5.4.22}$$

The intensity operator represents a sum of contribution from the same scatterer as represented by the first term in (5.4.22) and from different scatterers weighted by $n_{s_j}n_{s_l}h_{s_js_l}(\overline{r}_j-\overline{r}_l)$ as represented by the second term in (5.4.22). As noted, $\overline{\overline{\widehat{C}}}_j^{s_j}$ represents transition scattering matrix of the jth scatterer in the presence of other scatterers.

It is interesting to compare (5.4.22) with the usually ladder approximation. In the usual ladder approximation, the second term in (5.4.22) is absent and secondly $\overline{\overline{\widehat{C}}}_l$ in (5.4.22) is replaced by $\overline{\overline{T}}_l$, which is the transition matrix of the lth particle in the absence of other scatterers. Obviously, the two cases reduce to each other for sparse concentration of particles when $n_{s_j}n_{s_l}h_{s_js_l}$ can be neglected and $\overline{\overline{\widehat{C}}}_l$ is approximately equal to $\overline{\overline{T}}_l$.

5 Derivation of Radiative Transfer Equation from Ladder Approximation

In this section, we illustrate the derivation of the radiative transfer equation from the ladder approximation. To simplify the derivation, we use isotropic point scatterers. We assume that the extinction coefficient of the coherent wave is much smaller than the wavenumber.

Consider a wave $\psi_{inc}(\overline{r})$ incident on a single point scatterer located at $\overline{r}_\alpha$. Then the total field ψ is given by

$$\psi(\overline{r}) = \psi_{inc}(\overline{r}) + f\,\frac{e^{ik|\overline{r}-\overline{r}_\alpha|}}{|\overline{r}-\overline{r}_\alpha|}\,\psi_{inc}(\overline{r}_\alpha) \tag{5.5.1}$$

where f is the scattering amplitude and k is the wavenumber of the background medium. Since $G^{(0)}(\overline{r}) = \exp(ikr)/4\pi r$ is the Green's function for the unperturbed problem, (5.5.1) can be written as

$$\psi(\overline{r}) = \psi_{inc}(\overline{r}) + \int d\overline{r}' \int d\overline{r}''\, G^{(0)}(\overline{r},\overline{r}')T_\alpha(\overline{r}',\overline{r}'')\psi_{inc}(\overline{r}'') \tag{5.5.2}$$

where

$$T_\alpha(\overline{r}',\overline{r}'') = 4\pi f\,\delta(\overline{r}'-\overline{r}_\alpha)\,\delta(\overline{r}''-\overline{r}_\alpha) \tag{5.5.3}$$

is the transition operator for the point scatterer at $\overline{r}_\alpha$. We also use

$$G^{(0)}(\overline{r},\overline{r}') = \frac{e^{ik|\overline{r}-\overline{r}'|}}{4\pi|\overline{r}-\overline{r}'|} \tag{5.5.4}$$

In diagrammatic notation, let

$$T_\alpha = \overset{\alpha}{\otimes} \tag{5.5.5}$$

Then (5.5.2) becomes

$$\psi = \psi_{inc} + \text{———}\overset{\alpha}{\otimes}\ \psi_{inc} \tag{5.5.6}$$

where ——— stands for $G^{(0)}$.

For the case of isotropic point scatterers, we assume that the particle positions are independent. For the first and second moment equations, retaining the first terms in the mass operator and the intensity operator leads, respectively, to

$$\text{═══} = \text{———} + \text{———}\otimes\text{═══} \tag{5.5.7}$$

for the Foldy-approximated Dyson's equation, and

$$\langle GG^*\rangle = \begin{matrix}\text{═══}\\ \text{═══}\end{matrix} + \begin{matrix}\text{═══}\otimes\\ \text{═══}\otimes\end{matrix}\ \langle GG^*\rangle \tag{5.5.8}$$

for the ladder-approximated Bethe-Salpeter equation, where the diagrammatic notations are used. In (5.5.7) and (5.5.8), $\otimes$ stands for transition operator and a solid line joining two $\otimes$ denotes that the two $\otimes$ belong to the same scatterer. Equation (5.5.7) in analytic form is

$$\langle G(\overline{r},\overline{r}_o)\rangle = G^{(0)}(\overline{r},\overline{r}_o)+n_o\int d\overline{r}_i \int d\overline{r}_1 \int d\overline{r}_2\, G^{(0)}(\overline{r},\overline{r}_1)T_i(\overline{r}_1,\overline{r}_2)\langle G(\overline{r}_2,\overline{r}_o)\rangle \tag{5.5.9}$$

and (5.5.8) in analytic form is

$$\begin{aligned}\langle G(\overline{r},\overline{r}_o)G^*(\overline{r}',\overline{r}_o')\rangle &= \langle G(\overline{r},\overline{r}_o)\rangle\langle G^*(\overline{r}',\overline{r}_o')\rangle + n_o\int d\overline{r}_i\int d\overline{r}_1\int d\overline{r}_2\int d\overline{r}_3\int d\overline{r}_4\\ &\langle G(\overline{r},\overline{r}_1)\rangle\langle G^*(\overline{r}',\overline{r}_2)\rangle T_i(\overline{r}_1,\overline{r}_3)T_i^*(\overline{r}_2,\overline{r}_4)\langle G(\overline{r}_3,\overline{r}_o)G^*(\overline{r}_4,\overline{r}_o')\rangle\end{aligned} \tag{5.5.10}$$

Using (5.5.3) in (5.5.9) and (5.5.10), we have

$$\langle G(\overline{r},\overline{r}_o)\rangle = G^{(0)}(\overline{r},\overline{r}_o) + 4\pi n_o f\int d\overline{r}_i\, G^{(0)}(\overline{r},\overline{r}_i)\langle G(\overline{r}_i,\overline{r}_o)\rangle \tag{5.5.11}$$

and

$$\begin{aligned}\langle G(\overline{r},\overline{r}_o)G^*(\overline{r}',\overline{r}_o')\rangle &= \langle G(\overline{r},\overline{r}_o)\rangle\langle G^*(\overline{r}',\overline{r}_o')\rangle\\ &+ n_o(4\pi)^2\,|f|^2\int d\overline{r}_i\langle G(\overline{r},\overline{r}_i)\rangle\langle G^*(\overline{r}',\overline{r}_i)\rangle\langle G(\overline{r}_i,\overline{r}_o)G^*(\overline{r}_i,\overline{r}_o')\rangle\end{aligned} \tag{5.5.12}$$

The Foldy-approximated Dyson's equation in (5.5.11) can be solved readily. Operating on (5.5.11) by $(\nabla^2 + k^2)$, we have

$$(\nabla^2 + k^2 + 4\pi n_o f)\langle G(\overline{r},\overline{r}_o)\rangle = -\delta(\overline{r}-\overline{r}_o) \tag{5.5.13}$$

Hence, the effective propagation constant K obeys the equation

$$K^2 = k^2 + 4\pi n_o f \tag{5.5.14}$$

Since $|4\pi n_o f| \ll k^2$,

$$K = k + \frac{2\pi n_o f}{k} \tag{5.5.15}$$

and the attenuation rate κ_e for the coherent wave is

$$\kappa_e = 2K'' = \frac{4\pi n_o}{k}\mathrm{Im}\, f \tag{5.5.16}$$

Generally the extinction rate κ_e is much smaller than the wavenumber K'

$$2K'' \ll K' \tag{5.5.17}$$

Next consider a plane wave incident upon a slab of point scatterers of thickness d in the direction (θ_i, ϕ_i) (Fig. 5.5.1). The slab of scatterers lies between $z = 0$ and $z = -d$. Let k be the wavenumber of the upper half-space (region 0) and the lower half-space (region 2) and also the wavenumber

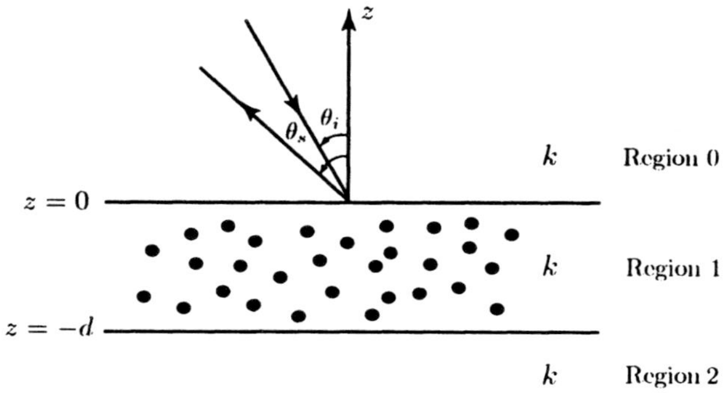

Figure 5.5.1 Scattering of waves by a slab of isotropic point scatterers.

of the background medium of the slab (region 1). It is assumed that the concentration of particles is low so that K' is approximately equal to k. The mean field in region 1 is

$$\psi_m(\overline{r}) = e^{i\overline{K}_i \cdot \overline{r}} \tag{5.5.18}$$

where

$$\overline{K}_i = k \sin\theta_i \cos\phi_i \hat{x} + k \sin\theta_i \sin\phi_i \hat{y} - K_{iz}\hat{z} \tag{5.5.19}$$

with

$$K_{iz} = (K^2 - k^2 \sin^2\theta_i)^{1/2} \simeq k\cos\theta_i + i\frac{K''}{\mu_i} \tag{5.5.20}$$

and $\mu_i = \cos\theta_i$. The approximate relation in (5.5.20) is a result of (5.5.17) that attenuation rate is much less than the wavenumber.

Let $I_L(\overline{r})$ denote intensity in region 0, which is a summation of all the ladder terms. Diagrammatically,

$I_L(\overline{r}) =$ 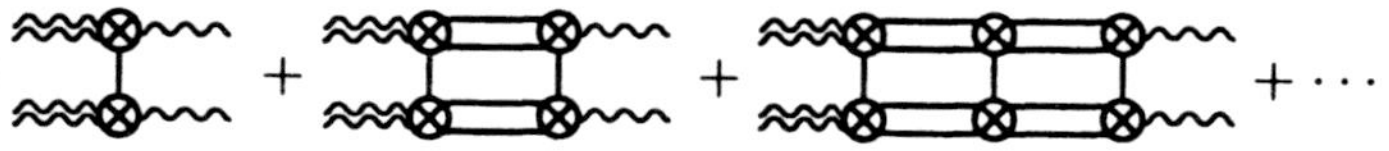 (5.5.21)

where

 = mean Green's function $\langle G_{11}(\overline{r}, \overline{r}')\rangle$ with both observation and source points in region 1

 = mean field ψ_m

 = mean Green's function $\langle G_{01}(\overline{r}, \overline{r}')\rangle$ with observation point in region 0 and source in region 1

Note that (5.5.21) is consistent with (5.5.8). The mean Green's function $\langle G_{11} \rangle$ is

$$\langle G_{11}(\overline{r}_1, \overline{r}_2) \rangle = \frac{e^{iK|\overline{r}_1 - \overline{r}_2|}}{4\pi|\overline{r}_1 - \overline{r}_2|} = \frac{i}{8\pi^2} \int_{-\infty}^{\infty} d\overline{k}_\perp \frac{e^{i\overline{k}_\perp \cdot (\overline{r}_{1\perp} - \overline{r}_{2\perp}) + iK_z|z_1 - z_2|}}{K_z} \tag{5.5.22}$$

with $\overline{k}_\perp = k_x\hat{x} + k_y\hat{y}$ and $K_z = (K^2 - k_x^2 - k_y^2)^{1/2}$. Because the observation point is in far field in region 0, the mean Green's function $\langle G_{01}(\overline{r}, \overline{r}') \rangle$ is

$$\langle G_{01}(\overline{r}, \overline{r}') \rangle = \frac{e^{ikr}}{4\pi r} e^{-i\overline{K}_s \cdot \overline{r}'} \tag{5.5.23}$$

where $\overline{K}_s$ denotes scattered direction and is

$$\overline{K}_s = k \sin\theta_s \cos\phi_s \hat{x} + k \sin\theta_s \sin\phi_s \hat{y} + K_{zs}\hat{z} \tag{5.5.24}$$

where

$$K_{zs} = \sqrt{K^2 - k^2 \sin^2\theta_s} \simeq k\cos\theta_s + i\frac{K''}{\mu_s} \tag{5.5.25}$$

with $\mu_s = \cos\theta_s$.

Next we define $\boxed{L}$ as I_L with ends stripped. Hence diagrammatically,

$$I_L(\overline{r}) = \tag{5.5.26}$$

By summing the ladder series, the operator equation for $\boxed{L}$ is

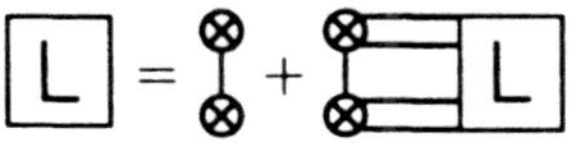

(5.5.27a)

We also let

$$L(\overline{r}_1, \overline{r}_2; \overline{r}_3, \overline{r}_4) = \tag{5.5.27b}$$

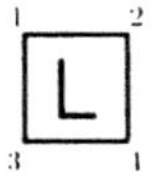

To solve the ladder intensity, let

$$L(\overline{r}_1, \overline{r}_2; \overline{r}_3, \overline{r}_4) = F(\overline{r}_1, \overline{r}_2)\, \delta(\overline{r}_1 - \overline{r}_3)\, \delta(\overline{r}_2 - \overline{r}_4) + n_o(4\pi)^2|f|^2\, \delta(\overline{r}_1 - \overline{r}_2)\, \delta(\overline{r}_1 - \overline{r}_3)\, \delta(\overline{r}_2 - \overline{r}_4) \tag{5.5.28}$$

On substituting (5.5.28) into (5.5.27) and using the transition operator for point scatterers as given by (5.5.3a), the F function in (5.5.28) obeys the equation

$$F(\overline{r}_1, \overline{r}_2) = n_o|f|^2 \int_{V_1} d\overline{r}' \frac{e^{-\kappa_e|\overline{r}_1 - \overline{r}'|}}{|\overline{r}_1 - \overline{r}'|^2} F(\overline{r}', \overline{r}_2) + n_o|f|^4(4\pi)^2 \frac{e^{-\kappa_e|\overline{r}_1 - \overline{r}_2|}}{|\overline{r}_1 - \overline{r}_2|^2} \tag{5.5.29}$$

where V_1 is the volume in the region between $z = 0$ and $z = -d$.

Next define the two-dimensional Fourier transform by making use of the translational invariance property in the horizontal direction:

$$F(\overline{r}_1, \overline{r}_2) = \int_{-\infty}^{\infty} d\overline{k}_\perp \, e^{i\overline{k}_\perp \cdot (\overline{r}_{1\perp} - \overline{r}_{2\perp})} \, \widetilde{F}(\overline{k}_\perp; z_1, z_2) \tag{5.5.30}$$

The bistatic scattering coefficient for ladder intensity is defined as

$$\gamma_L(\mu_s, \mu_i) = \frac{4\pi r^2}{A \cos\theta_i} I_L(\overline{r}) \tag{5.5.31}$$

where A is the area of the target under observation. From (5.5.26) we have

$$\begin{aligned}
& I_L(\overline{r}) \\
&= \iiiint d\overline{r}_1 d\overline{r}_2 d\overline{r}_3 d\overline{r}_4 \langle G_{01}(\overline{r}, \overline{r}_1)\rangle \langle G_{01}^*(\overline{r}, \overline{r}_3)\rangle L(\overline{r}_1, \overline{r}_2, \overline{r}_3, \overline{r}_4) \psi_m(\overline{r}_2) \psi_m^*(\overline{r}_4) \\
&= \frac{1}{16\pi^2 r^2} \iint d\overline{r}_1 d\overline{r}_2 e^{2K''_{sz} z_1} F(\overline{r}_1, \overline{r}_2) e^{2K''_{iz} z_2} + \frac{1}{r^2} n_o |f|^2 \int d\overline{r}_1 e^{2K''_{sz} z_1 + 2K''_{iz} z_1} \\
&= \frac{n_o |f|^2}{r^2} A \frac{1 - e^{-(2K''_{sz} + 2K''_{iz})d}}{2K''_{sz} + 2K''_{iz}} \\
&\quad + \frac{1}{16\pi^2 r^2} \iint d\overline{r}_1 d\overline{r}_2 e^{2K''_{sz} z_1 + 2K''_{iz} z_2} \int_{-\infty}^{\infty} d\overline{k}_\perp e^{i\overline{k}_\perp \cdot (\overline{r}_{1\perp} - \overline{r}_{2\perp})} \widetilde{F}(\overline{k}_\perp; z_1, z_2) \\
&= \frac{n_o |f|^2 A}{r^2} \frac{1 - e^{-2(K''_{sz} + K''_{iz})d}}{2K''_{sz} + 2K''_{iz}} + \frac{A}{4r^2} \int_{-d}^{0} dz_1 \int_{-d}^{0} dz_2 e^{2K''_{sz} z_1 + 2K''_{iz} z_2} \widetilde{F}(0; z_1, z_2)
\end{aligned} \tag{5.5.32}$$

The scattering rate is

$$\kappa_s = n_o 4\pi |f|^2 \tag{5.5.33}$$

and the albedo is

$$\tilde{\omega} = \frac{\kappa_s}{\kappa_e} = \frac{2\pi n_o |f|^2}{K''} \tag{5.5.34}$$

Let

$$\widetilde{F}_o(z', z'') = \widetilde{F}(\overline{k}_\perp = 0; z', z'') \tag{5.5.35}$$

It follows that

$$\begin{aligned}
\gamma_L(\mu_s, \mu_i) &= \frac{\tilde{\omega}}{\mu_i} \frac{1}{1/\mu_s + 1/\mu_i} \left(1 - e^{-\tau_d (1/\mu_s + 1/\mu_i)}\right) \\
&\quad + \frac{\pi}{\mu_i} \int_{-d}^{0} dz' \int_{-d}^{0} dz'' \, \widetilde{F}_o(z', z'') \, e^{\kappa_e (z'/\mu_s + z''/\mu_i)}
\end{aligned} \tag{5.5.36}$$

where $\tau_d = \kappa_e d$ is the optical thickness of the slab. Thus the bistatic scattering coefficient depends only on $\widetilde{F}(\overline{k}_\perp; z', z'')$ at $\overline{k}_\perp = 0$. Next, we make use of the plane wave representation of Green's function to calculate the absolute value of the Green's function squared. This will be in terms of a double integral of $\int d\overline{k}_{1\perp} \int d\overline{k}_{2\perp}$. We then make the transformations of $\overline{k}_{A\perp} = (\overline{k}_{1\perp} + \overline{k}_{2\perp})/2$ and $\overline{k}_{D\perp} = \overline{k}_{1\perp} - \overline{k}_{2\perp}$. Hence

$$\frac{e^{-\kappa_e|\overline{r}_1-\overline{r}_2|}}{|\overline{r}_1-\overline{r}_2|^2} = \frac{1}{4\pi^2}\int_{-\infty}^{\infty} d\overline{k}_{A\perp}\int_{-\infty}^{\infty} d\overline{k}_{D\perp}\frac{e^{i\overline{k}_{D\perp}\cdot(\overline{r}_{1\perp}-\overline{r}_{2\perp})}}{K_{z(A+D)}K^*_{z(A-D)}} \cdot e^{i(K_{z(A+D)}-K^*_{z(A-D)})|z_1-z_2|} \tag{5.5.37}$$

where

$$K_{z(A+D)} = \left[K^2 - \left|\overline{k}_{A\perp} + \frac{\overline{k}_{D\perp}}{2}\right|^2\right]^{1/2} \tag{5.5.38}$$

$$K_{z(A-D)} = \left[K^2 - \left|\overline{k}_{A\perp} - \frac{\overline{k}_{D\perp}}{2}\right|^2\right]^{1/2} \tag{5.5.39}$$

Substitute (5.5.30) and (5.5.37) into (5.5.29), balance the terms on both sides of the resulting equation, and finally set $\overline{k}_{D\perp}$ to zero to obtain an equation for the quantity $\widetilde{F}_o(z_1, z_2)$. We have

$$\widetilde{F}_o(z_1,z_2) = n_o|f|^2\int_{-d}^{0} dz'\int_{-\infty}^{\infty} d\overline{k}_{A\perp}\frac{e^{-2K''_{zA}|z_1-z'|}}{|K_{zA}|^2}\widetilde{F}_o(z',z_2) + 4n_o^2|f|^4\int_{-\infty}^{\infty} d\overline{k}_{A\perp}\frac{e^{-2K''_{zA}|z_1-z_2|}}{|K_{zA}|^2} \tag{5.5.40}$$

where $K_{zA} = (K^2 - k_{A\perp}^2)^{1/2}$ since $\overline{k}_{D\perp}$ has been set to zero to obtain equation for $\widetilde{F}_o$. Let

$$\alpha = \frac{|\overline{k}_{D\perp}|}{\kappa_e} \tag{5.5.41}$$

We consider the kernel represented by

$$W(\tau,\alpha) = \frac{1}{2\pi}\int_{-\infty}^{\infty} d\overline{k}_{A\perp}\frac{e^{i(K_{z(A+D)}-K^*_{z(A-D)})\tau/\kappa_e}}{K_{z(A+D)}K^*_{z(A-D)}} \tag{5.5.42}$$

which is the $d\overline{k}_{A\perp}$ part of Eq. (5.5.37). To evaluate the kernel $W(\tau,\alpha)$, we take the inverse Fourier transform of (5.5.37) with respect to $d\overline{r}_\perp$. It follows that

$$W(\tau,\alpha) = \frac{1}{2\pi}\int_{-\infty}^{\infty} d\overline{r}_\perp\, e^{-i\overline{k}_{D\perp}\cdot\overline{r}_\perp}\frac{e^{-\kappa_e(r_\perp^2+z^2)^{1/2}}}{r_\perp^2+z^2} \tag{5.5.43}$$

Converting to polar coordinates and carrying out the angular integration gives

$$W(\tau,\alpha)=\int_0^\infty ds\,s\,J_o(\alpha s)\,\frac{\exp[-(s^2+\tau^2)^{1/2}]}{s^2+\tau^2} \tag{5.5.44}$$

We next make use of the Sommerfeld integral identity, [Kong, 2000] which can be cast in the form

$$\frac{\exp[-\tau(\alpha^2+t^2)^{1/2}]}{(\alpha^2+t^2)^{1/2}}=\int_0^\infty ds\,s\,\frac{J_o(\alpha s)}{(s^2+\tau^2)^{1/2}}\exp[-(s^2+\tau^2)^{1/2}t] \tag{5.5.45}$$

Integration of the right hand side of (5.5.45) over t from 1 to ∞ is $W(\tau,\alpha)$. Thus

$$W(\tau,\alpha)=\int_1^\infty dt\,\frac{\exp[-\tau(\alpha^2+t^2)^{1/2}]}{(\alpha^2+t^2)^{1/2}} \tag{5.5.46}$$

Note that setting $\overline{k}_{D\perp}=0$ in (5.5.42) and (5.5.46) gives

$$W(\tau,\alpha=0)=E_1(\tau) \tag{5.5.47}$$

where $E_1(\tau)$ is exponential integral. Another useful formula for W is obtained by making an integration variable transformation of $t=1/\mu$

$$W(\tau,\alpha)=\int_0^1\frac{d\mu}{\mu}\,\frac{\exp[-(\tau/\mu)(1+\alpha^2\mu^2)^{1/2}]}{(1+\alpha^2\mu^2)^{1/2}} \tag{5.5.48}$$

The integration variable μ physically corresponds to $\cos\theta$ with θ being the angle of the direction of propagation.

For $\alpha=0$ we have

$$W(\tau,0)=E_1(\tau)=\frac{1}{2\pi}\int_{-\infty}^\infty d\overline{k}_{A\perp}\,\frac{e^{-2K''_{zA}\tau/\kappa_e}}{|K_{zA}|^2} \tag{5.5.49}$$

Using (5.5.49) in (5.5.40) gives

$$\widetilde{F}_o(z_1,z_2)=n_o|f|^2 2\pi\int_{-d}^0 dz' E_1(\kappa_e|z_1-z'|)\widetilde{F}_o(z',z_2)+8\pi n_o^2|f|^4E_1(\kappa_e|z_1-z_2|) \tag{5.5.50}$$

Let $\tau=\kappa_e z$ be the optical distance. Also let

$$J_s(\tau)=\frac{1}{4n_o|f|^2}\int_{-d}^0 dz_2\,\widetilde{F}_o(z,z_2)\,e^{\kappa_e z_2/\mu_i} \tag{5.5.51}$$

Multiply (5.5.50) by $\exp(\kappa_e z_2/\mu_i)$ and integrate, and using (5.5.51) we get

$$J_s(\tau)=\frac{\tilde{\omega}}{2}\int_{-\tau_d}^0 d\tau' E_1(|\tau-\tau'|)J_s(\tau')+\frac{\tilde{\omega}}{2}\int_{-\tau_d}^0 d\tau' E_1(|\tau-\tau'|)\,e^{\tau'/\mu_i} \tag{5.5.52}$$

The ladder bistatic scattering coefficient is obtained by using (5.5.51) in (5.5.36).

$$\gamma_L(\mu_s, \mu_i) = \frac{\tilde{\omega}}{\mu_i} \frac{1}{1/\mu_s + 1/\mu_i} \left(1 - e^{-\tau_d(1/\mu_s + 1/\mu_i)}\right) + \frac{\tilde{\omega}}{\mu_i} \int_{-\tau_d}^{0} d\tau' \, e^{\tau'/\mu_s} \, J_s(\tau') \tag{5.5.53}$$

Equation (5.5.52) is of the form of the Schwarzschild-Milne problem [Chandrasekhar, 1960]. Its equivalence to radiative transfer equations can be demonstrated as follows. Define upward-going higher-order diffuse specific intensity as

$$I(\tau, \mu) = \frac{\tilde{\omega}}{4\pi\mu} \int_{-\tau_d}^{\tau} d\tau' \, J_s(\tau') \, e^{(\tau' - \tau)/\mu} \tag{5.5.54}$$

for $\mu > 0$, and define higher-order downward-going diffuse specific intensity as

$$I(\tau, -\mu) = \frac{\tilde{\omega}}{4\pi\mu} \int_{\tau}^{0} d\tau' \, J_s(\tau') \, e^{(\tau - \tau')/\mu} \tag{5.5.55}$$

for $\mu > 0$. Differentiating (5.5.54) and (5.5.55) with respect to τ gives

$$\frac{dI(\tau, \mu)}{d\tau} = -\frac{I(\tau, \mu)}{\mu} + \frac{\tilde{\omega}}{4\pi\mu} J_s(\tau) \tag{5.5.56a}$$

$$\frac{dI(\tau, -\mu)}{d\tau} = -\frac{I(\tau, -\mu)}{\mu} - \frac{\tilde{\omega}}{4\pi\mu} J_s(\tau) \tag{5.5.56b}$$

Adding (5.5.54) and (5.5.55) and integrating our μ from 0 to 1 give

$$\int_0^1 d\mu \, (I(\tau, \mu) + I(\tau - \mu)) = \frac{\tilde{\omega}}{4\pi} \int_{-\tau_d}^{0} d\tau' J_s(\tau') \int_0^1 \frac{d\mu}{\mu} e^{|\tau' - \tau|\mu} \tag{5.5.57}$$

Since

$$E_1(z) = \int_0^1 \frac{d\mu}{\mu} e^{-z/\mu} \tag{5.5.58}$$

for $z > 0$, we have

$$\int_0^1 d\mu \, (I(\tau, \mu) + I(\tau, -\mu)) = \frac{\tilde{\omega}}{4\pi} \int_{-\tau_d}^{0} d\tau' J_s(\tau') E_1(|\tau' - \tau|) \tag{5.5.59}$$

Using (5.5.52) and (5.5.59), we obtain

$$\int_0^1 d\mu \, (I(\tau, \mu) + I(\tau, -\mu)) = \frac{J_s(\tau)}{2\pi} - \frac{\tilde{\omega}}{4\pi} \int_{-\tau_d}^{0} d\tau' E_1(|\tau - \tau'|) e^{\tau'/\mu_i} \tag{5.5.60}$$

Substituting (5.5.60) into (5.5.56) gives

$$\frac{dI(\tau,\mu)}{d\tau} = -\frac{I(\tau,\mu)}{\mu} + \frac{\tilde{\omega}}{2\mu}\int_0^1 d\mu\,(I(\tau,\mu)+I(\tau,-\mu)) + \frac{\tilde{\omega}^2}{8\pi\mu}\int_{-\tau_d}^{0} d\tau' E_1(|\tau-\tau'|)e^{\tau'/\mu_i} \tag{5.5.61a}$$

$$\frac{dI(\tau,-\mu)}{d\tau} = -\frac{I(\tau,-\mu)}{\mu} + \frac{\tilde{\omega}}{2\mu}\int_0^1 d\mu\,(I(\tau,\mu)+I(\tau,-\mu)) + \frac{\tilde{\omega}^2}{8\pi\mu}\int_{-\tau_d}^{0} d\tau' E_1(|\tau-\tau'|)e^{\tau'/\mu_i} \tag{5.5.61b}$$

This is the form of radiative transfer equation for isotropic point scatterers. The radiative transfer theory is treated extensively in Volume I.

REFERENCES AND ADDITIONAL READINGS

Chandrasekhar, S. (1960), *Radiative Transfer*, Dover, New York.

Chow, P. L., W. E. Kohler, and G. C. Papanicolaou, Eds. (1981), *Multiple Scattering and Waves in Random Media*, North-Holland Publishing Company, Amsterdam.

Davis, V. A. and L. Schwartz (1985), Electromagnetic propagation in close-packed disordered suspensions, *Phys. Rev. B*, **31**, 5155–5165.

Devaney, A. J. (1980), Multiple scattering theory for discrete elastic random media, *J. Math. Phys.*, **21**, 2603–2611.

de Vries, P., D. V. van Coevorden, and A. Lagendijk (1998), Point scatterers for classical waves, *Rev. Modern Phys.*, **70**, 447–466.

Ding, K. H. and L. Tsang (1988), Effective propagation constants of dense nontenuous media with multi-species of particles, *J. Electromag. Waves and Appl.*, **2(8)**, 757–777.

Ding, K. H. and L. Tsang (1989), Effective propagation constants in media with densely distributed dielectric particles of multiple sizes and permittivities, *Progress in Electromag. Res.*, J. A. Kong, Ed., **1**, Ch. 3, 241–295, Elsevier Science Publishing Co., New York.

Ding, K. H. and L. Tsang (1991), Effective propagation constants and attenuation rates in media of densely distributed dielectric particles with size distributions, *J. Electromag. Waves and Appl.*, **5**, 117–142.

Feng, S., C. Kane, P. A. Lee, and A. D. Stone (1988), Correlations and fluctuations of coherent wave transmission through disordered media, *Phys. Rev. Lett.*, **61**, 834–837.

Foldy, L. L. (1945), The multiple scattering of waves, *Phys. Rev.*, **67**, 107–119.

Fikioris, J. G. and P. C. Waterman (1964), Multiple scattering of waves, II. Hole corrections in the scalar case, *J. Math. Phys.*, **5**, 1413–1420.

Frisch, V. (1968), Wave propagation in random medium, *Probabilistic Methods in Applied Mathematics*, **1**, A. T. Bharuch-Reid, Ed., Academic Press, New York.

Gyorffy, B. L. (1970), Electronic states in liquid metals: A generalization of the coherent-potential approximation for a system with short-range order, *Phys. Rev. B*, **1**, 3290–3299.

Ishimaru, A. and Y. Kuga (1982), Attenuation constant of a coherent field in a dense distribution of particles, *J. Opt. Soc. Am.*, **72**, 1317–1320.

Kong, J. A. (2000), *Electromagnetic Wave Theory*, EMW Publishing, Cambridge.

Koh, G. (1992), Experimental study of electromagnetic wave propagation in dense random media, *Waves in Random Media*, **2**, 39–48.

Kohler, W. E. and G. C. Papanicolaou (1981), Some applications of the coherent potential approximation, *Multiple Scattering and Waves in Random Media*, P. L. Chow, W. K. Kohler, and G. C. Papanicolaou, Eds., 199–223, North-Holland, New York.

Korringa, J. and R. L. Mills (1972), Coherent-potential approximation for random systems with short-range correlations, *Phys. Rev. B*, **5**, 1654–1655.

Kuga, Y., A. Ishimaru, and D. Rice (1993), Velocity of coherent and incoherent electromagnetic waves in a dense strongly scattering medium, *Phys. Rev. B*, **48**, 13155–13158.

Kuga, Y. and P. Phu (1995), Experimental techniques in random media and rough surface, *Progress in Electromagnetics Research*, **9**, Elsevier Science Publishers, Cambridge, Mass.

Lagendijk, A. and B. A. van Tiggelen (1996), Resonant multiple scattering of light, *Phys. Reports*, **270(3)**, 143-215.

Lax, M. (1951), Multiple scattering of waves, *Rev. Modern Phys.*, **23**, 287–310.

Lax, M. (1952), Multiple scattering of waves II. The effective field in dense systems, *Phys. Rev.*, **85**, 261–269.

Mandt, C. E., Y. Kuga, L. Tsang, and A. Ishimaru (1992), Microwave propagation and scattering in a dense distribution of nontenuous spheres: experiment and theory, *Waves in Random Media*, **2(3)**, 225–234.

Merzbacher, E. (1970), *Quantum Mechanics*, 2nd edition, John Wiley & Sons, New York.

Roth, L. (1974), Effective medium approximation for liquid metals, *Phys. Rev. B*, **4**, 2476–2484.

Schwartz, L. and T. J. Plona (1984), Ultrasonic propagation in close-packed ordered and disordered suspensions, *J. Appl. Phys.*, **55**, 3971–3977.

Sheng, P., Ed. (1990), *Scattering and Localization of Classical Waves in Random Media*, World Scientific, Singapore.

Sheng, P. (1995), *Introduction to Wave Scattering, Localization, and Mesoscopic Phenomena*, Academic, San Diego.

Soven, P. (1967), Coherent-potential model of substitutional disordered alloys, *Phys. Rev.*, **156**, 809–813.

Tsang, L. (1992), Dense media radiative transfer theory for dense discrete random media with spherical particles of multiple sizes and permittivities, *Progress in Electromag. Res.*, **6**, Dielectric Properties of Heterogeneous Materials, A. Priou, ed., Ch. 5, 181–230, Elsevier, New York.

Tsang, L., K. H. Ding, and B. Wen (1990), Polarimetric signatures of random discrete scatterers based on vector radiative transfer theory. Part I. Nonspherical particles and Part II. Dense media, *PIER 3*, Polarimetric Remote Sensing, J. A. Kong, Ed., Chapter 2, 75–142.

Tsang, L. and A. Ishimaru (1985), Radiative wave and cyclical transfer equations for dense nontenuous media, *J. Opt. Soc. Am. A*, **2**, 2187–2194.

Tsang, L. and A. Ishimaru (1992), Radiative wave equations for vector electromagnetic propagation in dense nontenuous media, *J. Electromag. Waves and Appl.*, **1**, 809–813.

Tsang, L. and J. A. Kong (1980), Multiple scattering of electromagnetic waves by random distribution of discrete scatterers with coherent potential and quantum mechanical formulism, *J. Appl. Phys.*, **51**, 3465–3485.

Tsang, L. and J. A. Kong (1982), Effective propagation constants for coherent electromagnetic wave propagation in media embedded with dielectric scatterers, *J. Appl. Phys.*, **53**, 7162–7173.

Tsang, L., J. A. Kong, and R. T. Shin (1985), *Theory of Microwave Remote Sensing*, Wiley-Interscience, New York.

Twersky, V. (1962), On scattering of waves by random distributions I. Free space scatterer formalism, *J. Math. Phys.*, **3**, 700–715.

Twersky, V. (1964), On propagation in random media of scatterers, *Proc. Symp. Appl. Math.*, **16**, 84–116, Am. Math. Soc., Providence, RI.

Twersky, V. (1977), Coherent scalar field in pair-correlated random distributions of aligned scatterers, *J. Math. Phys.*, **18**, 2468–2486.

Twersky, V. (1978), Coherent electromagnetic waves in pair-correlated random distributions of aligned scatterers, *J. Math. Phys.*, **19**, 215–230.

van Albada, M. P., B. A. van Tiggelen, A. Lagendijk, and A. Tip (1991), Speed of propagation of classical waves in strongly scattering media, *Phys. Rev. Lett.*, **66(24)**, 3132–3135.

Varadan, V. K., V. N. Bringi, V. A. Varadan, and A. Ishimaru (1983), Multiple scattering theory for waves in discrete random media and comparison with experiments, *Radio Sci.*, **18**, 321–327.

Varadan, V. V., V. K. Varadan, Y. Ma, and W. A. Steele (1987), Effects of nonspherical statistics on EM wave propagation in discrete random media, *Radio Sci.*, **22**, 491–498.

Wen, B., L. Tsang, D. P. Winebrenner, and A. Ishimaru (1990), Dense medium radiative transfer theory: Comparison with experiment and applications to microwave remote sensing and polarimetry, *IEEE Trans. Geosci. Remote Sens.*, **28**, 46–59.

West, R., D. Gibbs, L. Tsang, and A. K. Fung (1994), Comparison of optical scattering experiments and the quasi-crystalline approximation for dense media, *J. Opt. Soc. Am.*, **11(6)**, 1854–1858.

Zhu, P. A., A. K. Fung, and K. W. Wong (1987), Effective propagation constants in dense random media under effective medium approximation, *Radio Sci.*, **22**, 234–250.

Chapter 6

QUASI-CRYSTALLINE APPROXIMATION IN DENSE MEDIA SCATTERING

In this chapter, we study the quasi-crystalline approximation (QCA) in dense media scattering using coordinate representation. Using QCA, the higher order multipoles can be included in the model. Thus the results are applicable to particles of moderate sizes. The results are also applicable to dense medium consisting of sticky (adhesive) particles where particles adhere to form aggregates. Even even when the individual particles are small, they can adhere together to become aggregates of moderate sizes. The physical parameters of the dense medium model are the sizes of the particles and the sticky parameter τ.

We study the effective propagation constants. Generalized Lorentz-Lorenz law and generalized Ewald-Oseen extinction theorem are derived. To study bistatic scattering, we also applied the distorted Born approximation. In Section 4, the distorted Born approximation is applied in combination with QCA to develop a QCA based dense media radiative transfer theory (DMRT).

1 Scattering of Electromagnetic Waves from a Half-Space of Dielectric Scatterers—Normal Incidence

In this section we consider a plane electromagnetic wave incident onto a half-space of identical dielectric spherical scatterers. The simpler case of normal incidence is first treated (Fig. 6.1.1). The quasi-crystalline approximation and T-matrix formalism will be used. The case of point dipoles were treated in Born and Wolf [1975]. It was shown that the original incident wave was extinguished giving the Ewald-Oseen extinction theorem. A transmitted wave is obtained that has an effective propagation constant governed by Lorentz-Lorenz law. For our case here, finite size spheres are used. Two sets of equations are obtained, namely, the generalized Ewald-Oseen extinction theorem and the generalized Lorentz-Lorenz law. The generalized Lorentz-Lorenz law is a set of homogeneous equations. Setting the determinant of the coefficient matrix equal to zero gives the dispersion relation for the coherent effective propagation constant K. For this new case, unlike the point dipoles treated in Born and Wolf [1975], K has an imaginary part that accounts for the decay due to scattering.

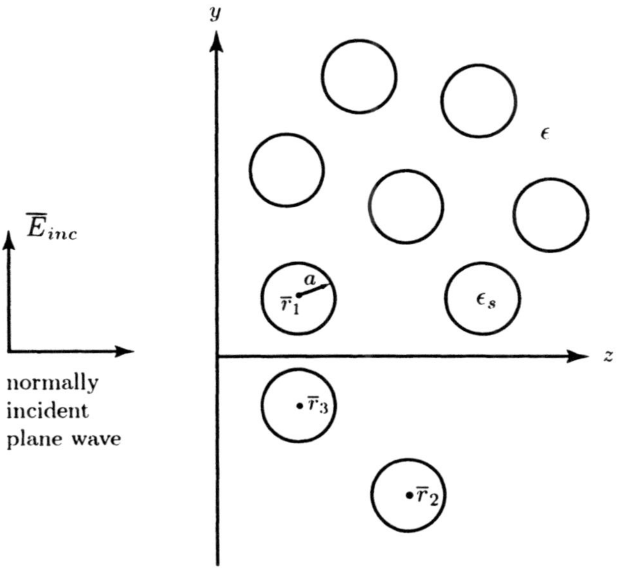

Figure 6.1.1 Plane electromagnetic wave normally incident on a half-space of dielectric scatterers.

1.1 Coherent Wave Propagation

Consider a plane wave impinging upon a half-space of N identical dielectric spherical scatterers. The incident field is

$$\overline{E}_{inc}(\overline{r}) = \overline{A}e^{i\overline{k}_i \cdot \overline{r}} \tag{6.1.1}$$

where

$$\overline{k}_i = k(\sin\theta_i \cos\phi_i \hat{x} + \sin\theta_i \sin\phi_i \hat{y} + \cos\theta_i \hat{z}) \tag{6.1.2}$$

Equation (6.1.1) can be written in terms of spherical waves

$$\overline{E}_{inc}(\overline{r}) = \sum_{mn} \left[a_{mn}^{(M)} \; Rg\, \overline{M}_{mn}(k\overline{r}) + a_{mn}^{(N)} \; Rg\, \overline{N}_{mn}(k\overline{r}) \right] \tag{6.1.3}$$

In terms of matrix notation, we use

$$\overline{a}_{inc} = \begin{bmatrix} a_{mn}^{(M)} \\ a_{mn}^{(N)} \end{bmatrix} \tag{6.1.4}$$

where m, n go through the vector spherical wave coefficients $n = 1, 2, \ldots, N_{\max}$, $m = 0, \pm 1, \pm 2, \ldots, \pm n$. Then the field exciting the lth scatterer is

$$\overline{E}_{\text{ex}}^{l}(\overline{r}) = \sum_{mn} \left[w_{mn}^{(M)(l)} \; Rg\, \overline{M}_{mn}(k\overline{r}\overline{r}_l) + w_{mn}^{(N)(l)} \; Rg\, \overline{N}_{mn}(k\overline{r}\overline{r}_l) \right] \tag{6.1.5}$$

We use the matrix notation

$$\overline{w}^{(l)} = \begin{bmatrix} w_{mn}^{(M)(l)} \\ w_{mn}^{(N)(l)} \end{bmatrix} \tag{6.1.6}$$

as the column matrix containing the exciting field coefficients of particle l.

In Chapter 10 of Volume II, the Foldy-Lax multiple scattering equations in matrix form were derived. Using the same notations,

$$\overline{w}^{(l)} = \sum_{\substack{j=1 \\ j\neq l}}^{N} \overline{\overline{\sigma}}(k\overline{r_l r_j})\overline{\overline{T}}^{(j)}\overline{w}^{(j)} + e^{i\overline{k}_i\cdot\overline{r}_l}\overline{a}_{inc} \tag{6.1.7}$$

Equation (6.1.7) can be interpreted by saying that the field exciting the lth scatterer is the sum of incident wave and scattered wave from all scatterers j except $j = l$. The matrix $\overline{\overline{\sigma}}(k\overline{r_l r_j})$ represents vector translation formula translating the center at $\overline{r}_j$ to spherical waves centered at $\overline{r}_l$. The matrix elements of $\overline{\overline{\sigma}}$ are given in Chapter 10 of Volume II. In index notation, we have

$$\begin{aligned}
&\sum_{mn}\left[w_{mn}^{(M)(l)}\, Rg\,\overline{M}_{mn}(k\overline{rr_l}) + w_{mn}^{(N)(l)}\, Rg\,\overline{N}_{mn}(k\overline{rr_l})\right] = \sum_{\substack{j=1 \\ j\neq l}}^{N}\sum_{\substack{mn \\ \mu\nu}} \\
&\Big\{Rg\,\overline{M}_{mn}(k\overline{rr_l})\left[A_{mn\mu\nu}(k\overline{r_l r_j})T_\nu^{(M)}w_{\mu\nu}^{(M)(j)} + B_{mn\mu\nu}(k\overline{r_l r_j})T_\nu^{(N)}w_{\mu\nu}^{(N)(j)}\right] \\
&+\ Rg\,\overline{N}_{mn}(k\overline{rr_l})\left[B_{mn\mu\nu}(k\overline{r_l r_j})T_\nu^{(M)}w_{\mu\nu}^{(M)(j)} + A_{mn\mu\nu}(k\overline{r_l r_j})T_\nu^{(N)}w_{\mu\nu}^{(N)(j)}\right]\Big\} \\
&+ e^{i\overline{k}_i\cdot\overline{r}_l}\sum_{mn}\left[a_{mn}^{(M)}\, Rg\,\overline{M}_{mn}(k\overline{rr_l}) + a_{mn}^{(N)}\, Rg\,\overline{N}_{mn}(k\overline{rr_l})\right]
\end{aligned} \tag{6.1.8}$$

Equating the spherical wave coefficients, we have

$$\begin{aligned}
w_{mn}^{(M)(l)} = &\sum_{\substack{j=1 \\ j\neq l}}^{N}\sum_{\mu\nu}\left\{A_{mn\mu\nu}(k\overline{r_l r_j})T_\nu^{(M)}w_{\mu\nu}^{(M)(j)} + B_{mn\mu\nu}(k\overline{r_l r_j})T_\nu^{(N)}w_{\mu\nu}^{(N)(j)}\right\} \\
&+ e^{i\overline{k}_i\cdot\overline{r}_l}a_{mn}^{(M)}
\end{aligned} \tag{6.1.9a}$$

$$\begin{aligned}
w_{mn}^{(N)(l)} = &\sum_{\substack{j=1 \\ j\neq l}}^{N}\sum_{\mu\nu}\left\{B_{mn\mu\nu}(k\overline{r_l r_j})T_\nu^{(M)}w_{\mu\nu}^{(M)(j)} + A_{mn\mu\nu}(k\overline{r_l r_j})T_\nu^{(N)}w_{\mu\nu}^{(N)(j)}\right\} \\
&+ e^{i\overline{k}_i\cdot\overline{r}_l}a_{mn}^{(N)}
\end{aligned} \tag{6.1.9b}$$

We note that in general the coefficients depend on the positions of the N

particles, $\overline{r}_1, \overline{r}_2, \ldots, \overline{r}_N$. In Volume II, we carried out Monte Carlo simulations and for every realization, $\overline{r}_1, \overline{r}_2, \ldots, \overline{r}_N$ are fixed. Here, we perform statistical configurational averaging using conditional averaging. Let $p(\overline{r}_1, \overline{r}_2, \ldots, \overline{r}_l', \ldots, \overline{r}_N \mid \overline{r}_l)$ (where $'$ denotes that term is absent) denote the conditional probability density function. The equation

$$E_l(\overline{w}^{(l)}) = \int d\overline{r}_1 d\overline{r}_2 \ldots d\overline{r}_l' \ldots d\overline{r}_N \, \overline{w}^{(l)} p(\overline{r}_1, \overline{r}_2, \ldots, \overline{r}_l', \ldots, \overline{r}_N | \overline{r}_l) = \overline{w}(\overline{r}_l) \tag{6.1.10}$$

is the conditional average based on holding position $\overline{r}_l$ fixed. Note that on averaging in (6.1.10), $\overline{r}_1, \overline{r}_2, \ldots, \overline{r}_N$ become dummy variables except $\overline{r}_l$. Equation (6.1.7) becomes

$$\overline{w}(\overline{r}_l) = \sum_{\substack{j=1 \\ j \neq l}}^{N} \int d\overline{r}_1 \ldots d\overline{r}_l' \ldots d\overline{r}_N \overline{\overline{\sigma}}(k\overline{r_l r_j}) \overline{\overline{T}} \, \overline{w}^{(j)} p(\overline{r}_1 \ldots \overline{r}_l' \ldots \overline{r}_N | \overline{r}_l) + e^{i\overline{k}_i \cdot \overline{r}_l} \overline{a}_{inc} \tag{6.1.11}$$

Since $\overline{\overline{\sigma}}(k\overline{r_l r_j})$ depends on both $\overline{r}_l$ and $\overline{r}_j$, we use Bayes' rule

$$p(\overline{r}_1 \ldots \overline{r}_l' \ldots \overline{r}_N \mid \overline{r}_l) = p(\overline{r}_1 \ldots \overline{r}_j' \ldots \overline{r}_l' \ldots \overline{r}_N \mid \overline{r}_j, \overline{r}_l) p(\overline{r}_j | \overline{r}_l) \tag{6.1.12}$$

Thus (6.1.11) becomes

$$\overline{w}(\overline{r}_l) = \sum_{\substack{j=1 \\ j \neq l}}^{N} \int d\overline{r}_j \overline{\overline{\sigma}}(k\overline{r_l r_j}) p(\overline{r}_j | \overline{r}_l) \overline{\overline{T}} \, E_{jl}(\overline{w}^{(j)}) + e^{i\overline{k}_i \cdot \overline{r}_l} \overline{a}_{inc} \tag{6.1.13}$$

where $E_{jl}(\overline{w}^{(j)})$ is the conditional average holding both $\overline{r}_j$ and $\overline{r}_l$ fixed. Based on the quasi-crystalline approximation,

$$E_{jl}(\overline{w}^{(j)}) \simeq E_j(\overline{w}^{(j)}) = \overline{w}(\overline{r}_j) \tag{6.1.14}$$

we have

$$\overline{w}(\overline{r}_l) = \sum_{\substack{j=1 \\ j \neq l}}^{N} \int d\overline{r}_j \overline{\overline{\sigma}}(k\overline{r_l r_j}) p(\overline{r}_j | \overline{r}_l) \overline{\overline{T}} \, \overline{w}(\overline{r}_j) + e^{i\overline{k}_i \cdot \overline{r}_l} \overline{a}_{inc} \tag{6.1.15}$$

Since $\overline{r}_j$ has become a dummy variable, we can replace $\sum_{\substack{j=1 \\ j \neq l}}^{N}$ by $(N-1)$. Also,

$$(N-1) p(\overline{r}_j | \overline{r}_l) = n_o g(\overline{r}_j - \overline{r}_l) \tag{6.1.16}$$

where n_o is the number of spheres per unit volume and $g(\overline{r}_j - \overline{r}_l)$ is the pair distribution function.

$$\overline{w}(\overline{r}_l) = n_o \int_{z_j>0} d\overline{r}_j g(\overline{r}_j - \overline{r}_l)\overline{\overline{\sigma}}(k\overline{r_l r_j})\overline{\overline{T}}\,\overline{w}(\overline{r}_j) + e^{i\overline{k}_i\cdot\overline{r}_l}\overline{a}_{inc} \tag{6.1.17}$$

The domain of $z_j > 0$ is imposed because the particles are in the region of $z > 0$. In index notation, we have

$$w_{mn}^{(M)}(\overline{r}_1) = \sum_{\mu\nu} n_o \int_{z_2>0} d\overline{r}_2 \left[A_{mn\mu\nu}(k\overline{r_1 r_2})T_\nu^{(M)} w_{\mu\nu}^{(M)}(\overline{r}_2) + B_{mn\mu\nu}(k\overline{r_1 r_2})T_\nu^{(N)} w_{\mu\nu}^{(N)}(\overline{r}_2)\right] g(\overline{r}_2 - \overline{r}_1) + e^{i\overline{k}_i\cdot\overline{r}_1} a_{mn}^{(M)} \tag{6.1.18a}$$

$$w_{mn}^{(N)}(\overline{r}_1) = \sum_{\mu\nu} n_o \int_{z_2>0} d\overline{r}_2 \left[B_{mn\mu\nu}(k\overline{r_1 r_2})T_\nu^{(M)} w_{\mu\nu}^{(M)}(\overline{r}_2) + A_{mn\mu\nu}(k\overline{r_1 r_2})T_\nu^{(N)} w_{\mu\nu}^{(N)}(\overline{r}_2)\right] g(\overline{r}_2 - \overline{r}_1) + e^{i\overline{k}_i\cdot\overline{r}_1} a_{mn}^{(N)} \tag{6.1.18b}$$

For the case of normal incidence, $\overline{k}_i = k\hat{z}$. Then

$$\overline{w}(\overline{r}_1) = n_o \int_{z_2>0} d\overline{r}_2\, g(\overline{r}_2 - \overline{r}_1)\,\overline{\overline{\sigma}}(k\overline{r_1 r_2})\,\overline{\overline{T}}\,\overline{w}(\overline{r}_2) + e^{ikz_1}\,\overline{a}_{inc} \tag{6.1.19}$$

The integration volume in (6.1.19) is the half-space $z_2 > 0$, excluding a spherical volume of radius b centered about $\overline{r}_1$. This is due to the fact that particles do not interpenetrate each other so that $p(\overline{r}_2|\overline{r}_1) = 0$ for $|\overline{r}_2 - \overline{r}_1| < b$. For $z_1 \geq b$, the excluded volume is a complete sphere of radius b centered at $\overline{r}_1$.

To solve (6.1.19), assume the following approximate solution

$$\overline{w}(\overline{r}_1) = \overline{a}^E\, e^{+iKz_1} \tag{6.1.20}$$

The representation in (6.1.20) is assumed to be valid for all $z_1 \geq 0$, even in the boundary layer region of $b_1 > z_1 \geq 0$ where a more complicated dependence on z_1 should be expected. The approximation solution of (6.1.20) ignores this boundary layer effect.

Let the incident field be $\overline{E}_{inc} = \hat{y}e^{ikz}$. The $\hat{y}$ polarization gives the azimuthal harmonic dependence of $m = \pm 1$ in the vector spherical waves.

$$\overline{E}_{inc} = \sum_{n=1}^{\infty} \frac{1}{2} i^n [4\pi(2n+1)]^{1/2} \Big\{ Rg\overline{M}_{1n}(kr,\theta,\phi) - Rg\overline{M}_{-1n}(kr,\theta,\phi) + Rg\overline{N}_{1n}(kr,\theta,\phi) + Rg\overline{N}_{-1n}(kr,\theta,\phi) \Big\} \tag{6.1.21}$$

Thus

$$a_{1n}^{(M)} = \frac{1}{2} i^n \left[4\pi(2n+1)\right]^{1/2} \tag{6.1.21a}$$

$$a_{1n}^{(N)} = \frac{1}{2} i^n \left[4\pi(2n+1)\right]^{1/2} \tag{6.1.21b}$$

$$a_{-1n}^{(M)} = -a_{1n}^{(M)} \tag{6.1.21c}$$

$$a_{-1n}^{(N)} = a_{1n}^{(N)} \tag{6.1.21d}$$

To solve (6.1.21), we let

$$w_{mn}^{(M)}(\overline{r}_1) = a_{mn}^{E(M)} e^{iKz_1} \tag{6.1.22a}$$

$$w_{mn}^{(N)}(\overline{r}_1) = a_{mn}^{E(N)} e^{iKz_1} \tag{6.1.22b}$$

and $a_{mn}^{E(M)}$ and $a_{mn}^{E(N)}$ are equal to 0 except when $m = \pm 1$. In view of (6.1.21), let

$$a_{-1n}^{E(M)} = -a_{1n}^{E(M)} \tag{6.1.23a}$$

$$a_{-1n}^{E(N)} = a_{1n}^{E(N)} \tag{6.1.23b}$$

Thus (6.1.18a) becomes

$$a_{mn}^{E(M)} e^{iKz_1} = \sum_{\mu\nu} n_o \int_{z_2>0} d\overline{r}_2 \Big[A_{mn\mu\nu}(k\overline{r_1 r_2}) T_\nu^{(M)} a_{\mu\nu}^{E(M)} e^{iKz_2} + B_{mn\mu\nu}(k\overline{r_1 r_2}) T_\nu^{(N)} a_{\mu\nu}^{E(N)} e^{iKz_2} \Big] g(\overline{r}_2 - \overline{r}_1) + e^{ikz_1} a_{mn}^{(M)} \tag{6.1.24}$$

Because of statistical translational invariance in the x-y plane, let $\overline{r}_1 = (0,0,z_1)$ without loss of generality. Then, we need to integrate terms like, for example,

$$\begin{aligned}
&\int_{z_2>0} d\overline{r}_2 A_{mn\mu\nu}(k\overline{r_1 r_2}) e^{iK(z_2-z_1)} g(\overline{r}_2 - \overline{r}_1) \\
&= \frac{\gamma_{\mu\nu}}{\gamma_{mn}} (-1)^m \sum_p a(\mu,\nu|-m,n|p) a(\nu,n,p) \\
&\quad \times \int_{z_2>0} d\overline{r}_2 h_p(k|\overline{r_1 r_2}|) Y_p^{\mu-m}(\theta_{\overline{r_1 r_2}}, \phi_{\overline{r_1 r_2}}) e^{iK(z_2-z_1)} g(|\overline{r}_2 - \overline{r}_1|) \\
&= \delta_{\mu m} \frac{\gamma_{\mu\nu}}{\gamma_{mn}} (-1)^m \sum_p a(\mu,\nu|-m,n|p) a(\nu,n,p) \\
&\quad \times \int_{z_2>0} d\overline{r}_2 h_p(k|\overline{r_1 r_2}|) P_p(\cos\theta_{\overline{r_1 r_2}}) e^{iK(z_2-z_1)} g(|\overline{r}_2 - \overline{r}_1|)
\end{aligned} \tag{6.1.25}$$

The Kronecker delta of $\delta_{\mu m}$ is a result of integration over $d\phi_{\overline{r_2 r_1}}$ that gives $\mu = m$. A similar result applies to the $B_{mn\mu\nu}(k\overline{r_1 r_2})$ terms. Thus

$$a_{mn}^{E(M)} e^{iKz_1} = \sum_{\nu} n_o \frac{\gamma_{m\nu}}{\gamma_{mn}} \left[(-1)^m \sum_p a(m,\nu|-m,n|p) a(\nu,n,p) T_\nu^{(M)} a_{m\nu}^{E(M)} \right.$$
$$\left. +(-1)^{m+1} \sum_p a(m,\nu|-m,n|p,p-1) b(\nu,n,p) T_\nu^{(N)} a_{m\nu}^{E(N)} \right] I$$
$$+ e^{ikz_1} a_{mn}^{(M)} \tag{6.1.26}$$

where

$$I = \int_{\substack{|\overline{r}_2 - \overline{r}_1| > b \\ z_2 \geq 0}} d\overline{r}_2 \, g(\overline{r}_2 - \overline{r}_1) \, e^{iKz_2} \, h_p(k \, |\overline{r_1 r_2}|) P_p(\cos\theta_{\overline{r_1 r_2}}) \tag{6.1.27}$$

We can write

$$g(\overline{r}_2 - \overline{r}_1) = 1 + (g(\overline{r}_2 - \overline{r}_1) - 1)$$

for $|\overline{r}_2 - \overline{r}_1| > b$. The term $g(\overline{r}_2 - \overline{r}_1) - 1$ quickly becomes small for $|\overline{r}_2 - \overline{r}_1|$ larger than a few diameters. The expressions of $a(\mu,\nu| - m,n|p)$, $a(m,\nu| - m,n|p,p-1)$, $a(\nu,n,p)$ and $b(\nu,n,p)$ in (6.1.25)–(6.1.26) have been provided in Chapter 10 of Volume II.

Then, the integral in (6.1.27) can be decomposed into two terms

$$I = I_1 + I_2 \tag{6.1.28}$$

where

$$I_1 = e^{iKz_1} \int_{\substack{|\overline{r}_2 - \overline{r}_1| > b \\ z_2 \geq 0}} d\overline{r}_2 \, e^{iK(z_2 - z_1)} \, h_p(k \, |\overline{r_1 r_2}|) P_p(\cos\theta_{\overline{r_1 r_2}}) \tag{6.1.29}$$

and

$$I_2 = e^{iKz_1} \int_{\substack{|\overline{r}_2 - \overline{r}_1| > b \\ z_2 \geq 0}} d\overline{r}_2 \, [g(\overline{r}_2 - \overline{r}_1) - 1] \, e^{iK(z_2 - z_1)} h_p(k \, |\overline{r_1 r_2}|) P_p(\cos\theta_{\overline{r_1 r_2}}) \tag{6.1.30}$$

Let $\overline{r}_2 - \overline{r}_1 = \overline{r}$. Then $h_p(k|\overline{r_1 r_2}|) P_p(\cos\theta_{\overline{r_1 r_2}}) = h_p(kr) P_p(\cos\theta)(-1)^p$. To evaluate integral I_1, note that, with $y_p = h_p(kr) P_p(\cos\theta)$,

$$e^{iKz} y_p = \frac{1}{K^2 - k^2} [e^{iKz} \nabla^2 y_p - y_p \nabla^2 e^{iKz}] \tag{6.1.31}$$

Use (6.1.31) and Green's theorem to transform the volume integral into three surface integrals over S_d, S_e, and S_∞ (Fig. 6.1.2). The integral over S_∞ can be shown to vanish.

$$I_1 = e^{iKz_1} \frac{(-1)^p}{K^2 - k^2} \left\{ - \int_{S_e} dS \left[e^{iKz} \frac{\partial y_p}{\partial r} - y_p \frac{\partial e^{iKz}}{\partial r} \right]_{r=b} \right.$$
$$\left. - \int_{S_d} dS \left[e^{iKz} \frac{\partial y_p}{\partial z} - y_p \frac{\partial e^{iKz}}{\partial z} \right]_{z=-z_1} \right\} \tag{6.1.32}$$

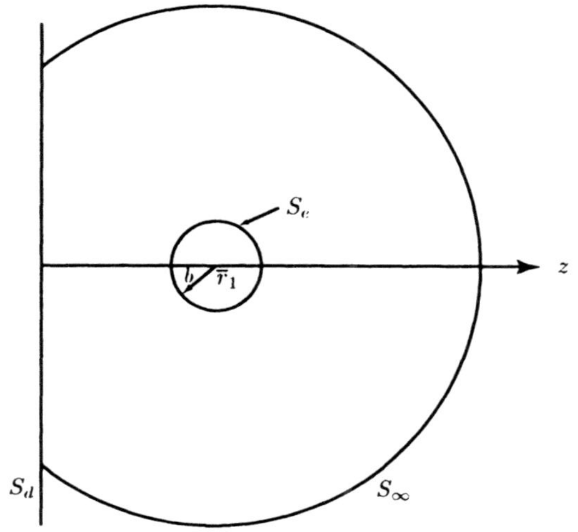

Figure 6.1.2 Volume integration bounded by surface S_d, S_e, and S_∞.

The integral over S_e is

$$\int_{S_e} dS \left[e^{iKz} \frac{\partial y_p}{\partial r} - y_p \frac{\partial e^{iKz}}{\partial r} \right]_{r=b} = -4\pi i^n (K^2 - k^2) L_p(k, K|b) \qquad (6.1.33)$$

where

$$L_p(k, K|b) = -\frac{b^2}{(K^2 - k^2)} \left[k h_p'(kb) j_p(Kb) - K h_p(kb) j_p'(Kb) \right] \qquad (6.1.34)$$

For the surface integral over S_d, the following Kasterin's representation [Waterman and Truell, 1961] can be used.

$$y_p = h_p(kr) P_p(\cos\theta) = (-i)^p P_p \left(\frac{1}{ik} \frac{\partial}{\partial z} \right) h_o(kr) \qquad (6.1.35)$$

Thus

$$\int_{S_d} dS \left[e^{iKz} \frac{\partial y_p}{\partial z} - y_p \frac{\partial e^{iKz}}{\partial z} \right]_{z=-z_1}$$

$$= 2\pi\, e^{-iKz_1} (-i)^{p+1} \int_0^\infty \rho\, d\rho \left\{ \left[\frac{\partial}{\partial z} - iK \right] P_p \left(\frac{1}{ik} \frac{\partial}{\partial z} \right) \frac{e^{ik\sqrt{\rho^2+z^2}}}{\sqrt{\rho^2 + z^2}} \right\}_{z=-z_1}$$

$$= -2\pi\, i^{p+1} \frac{(k + K)}{k^2}\, e^{-iKz_1 + ikz_1} \qquad (6.1.36)$$

Substituting (6.1.33) and (6.1.36) into (6.1.32) gives

$$I_1 = \frac{2\pi i (-i)^p}{(K - k)} \frac{e^{ikz_1}}{k^2} + 4\pi (-i)^p L_p(k, K|b)\, e^{iKz_1} \qquad (6.1.37)$$

To evaluate I_2 of (6.1.30), we note that the volume of integration in (6.1.30) consists of the half-space $z_2 > 0$ minus a sphere of radius b centered at the point $\bar{r}_1$. We have seen that for $0 \leq f \leq 0.4$, $g(\bar{r}_2 - \bar{r}_1) - 1$ is practically zero for $|\bar{r}_2 - \bar{r}_1|$ larger than a few b's. If the point $\bar{r}_1$ is at least several diameters deep in the scattering medium (thus ignoring boundary layer effects), the volume of integration in (6.1.30) can be extended to infinite space. Thus, letting $\bar{r} = \bar{r}_2 - \bar{r}_1$ in (6.1.30) and making the above assumption give

$$I_2 = 4\pi(-i)^p \, e^{iKz_1} M_p(k, K|b) \tag{6.1.38}$$

where

$$M_p(k, K|b) = \int_b^\infty dr \, r^2 [g(r) - 1] h_p(kr) j_p(Kr) \tag{6.1.39}$$

Substituting $I = I_1 + I_2$ in (6.1.26), we have

$$\begin{aligned} a_{mn}^{E(M)} e^{iKz_1} = & \sum_\nu n_o \frac{\gamma_{m\nu}}{\gamma_{mn}} \Bigg[(-1)^m \sum_p a(m,\nu|-m,n|p) a(\nu,n,p) T_\nu^{(M)} a_{m\nu}^{E(M)} \\ & + (-1)^{m+1} \sum_p a(m,\nu|-m,n|p,p-1) b(\nu,n,p) T_\nu^{(N)} a_{m\nu}^{E(N)} \Bigg] \\ & \cdot \left\{ \frac{2\pi i(-i)^p e^{ikz_1}}{(K-k)k^2} + 4\pi(-i)^p e^{iKz_1} \Big[L_p(k,K|b) + M_p(k,K|b) \Big] \right\} \\ & + e^{ikz_1} a_{mn}^{(M)} \end{aligned} \tag{6.1.40}$$

There are two kinds of wave dependences in (6.1.40) as characterized by their respective phase terms.

One type of term has a $\exp(ikz_1)$ dependence corresponding to waves traveling with the propagation constant of the incident wave. The other type of term has a $\exp(iKz_1)$ dependence corresponding to waves travelling with the propagation constant of the effective medium. The terms with propagation constant k should balance each other, giving the generalized Ewald-Oseen extinction theorem. The medium generates a wave that *extinguishes* the original incident wave. Balancing the terms with propagation constant K gives the generalized Lorentz-Lorenz law. These laws were derived for the case of point dipoles in Born and Wolf [1975]. In the treatment here, the Ewald-Oseen extinction theorem and Lorentz-Lorenz law are generalized to the use of finite-size spheres that give scattering attenuation of the coherent wave.

The generalized Lorentz-Lorenz law is, on balancing e^{iKz_1} terms in

(6.1.40) and replacing $(\mu,\nu)\leftrightarrow(m,n)$,

$$a_{\mu\nu}^{E(M)}=\sum_n n_o\frac{\gamma_{\mu n}}{\gamma_{\mu\nu}}\left[(-1)^\mu\sum_p a(\mu,n|-\mu,\nu|p)a(n,\nu,p)T_n^{(M)}a_{\mu n}^{E(M)}\right.$$
$$\left.+(-1)^{\mu+1}\sum_p a(\mu,n|-\mu,\nu|p,p-1)b(n,\nu,p)T_n^{(N)}a_{\mu n}^{E(N)}\right]$$
$$\cdot 4\pi(-i)^p\left[L_p(k,K|b)+M_p(k,K|b)\right] \tag{6.1.41a}$$

for $\mu=\pm1$. Equation (6.1.41a) is the result of starting with (6.1.18a). If we start with (6.1.18b) and use similar derivation, we get

$$a_{\mu\nu}^{E(N)}=\sum_n n_o\frac{\gamma_{\mu n}}{\gamma_{\mu\nu}}\left[(-1)^{\mu+1}\sum_p a(\mu,n|-\mu,\nu|p,p-1)b(n,\nu,p)T_n^{(M)}a_{\mu n}^{E(M)}\right.$$
$$\left.+(-1)^{\mu}\sum_p a(\mu,n|-\mu,\nu|p)a(n,\nu,p)T_n^{(N)}a_{\mu n}^{E(N)}\right]$$
$$\cdot 4\pi(-i)^p\left[L_p(k,K|b)+M_p(k,K|b)\right] \tag{6.1.41b}$$

The generalized Ewald-Oseen extinction theorem is obtained by balancing the $\exp(ikz_1)$ terms in (6.1.20).

$$\sqrt{4\pi(2\nu+1)}\frac{1}{2}i^\nu\delta_{\mu1}+n_o\sum_{n,p}(-1)^{p+\mu}\frac{\gamma_{1n}}{\gamma_{1\nu}}\frac{(2\pi i)\,i^p}{(K-k)k^2}$$
$$\times\left\{T_n^{(M)}a_{1n}^{E(M)}a(\mu,n|-\mu,\nu|p)a(n,\nu,p)\right.$$
$$\left.-T_n^{(N)}a_{1n}^{E(N)}a(\mu,n|-\mu,\nu|p,p-1)b(n,\nu,p)\right\}=0 \tag{6.1.42a}$$

If we start with (6.1.18b) and use similar derivation, we get

$$\sqrt{4\pi(2\nu+1)}\frac{1}{2}i^\nu\delta_{\mu1}+n_o\sum_{n,p}(-1)^{p+\mu}\frac{\gamma_{1n}}{\gamma_{1\nu}}\frac{(2\pi i)\,i^p}{(K-k)k^2}$$
$$\times\left\{-T_n^{(M)}a_{1n}^{E(M)}a(\mu,n|-\mu,\nu|p,p-1)b(n,\nu,p)\right.$$
$$\left.+T_n^{(N)}a_{1n}^{E(N)}a(\mu,n|-\mu,\nu|p)a(n,\nu,p)\right\}=0 \tag{6.1.42b}$$

Let

$$a(n,\nu,p)=\frac{(2\nu+1)n(n+1)}{2\nu(\nu+1)}i^{\nu-n+p}A(n,\nu,p) \tag{6.1.43a}$$

$$b(n,\nu,p)=-\frac{(2\nu+1)n(n+1)}{2\nu(\nu+1)}i^{\nu-n+p}B(n,\nu,p) \tag{6.1.43b}$$

where

$$A(n,\nu,p) = \frac{1}{n(n+1)(2\nu+1)}\Big[2\nu(\nu+1)(2\nu+1) + (\nu+1)(n+\nu-p)$$
$$\cdot(n+p-\nu+1) - \nu(n+\nu+p+2)(\nu+p-n+1)\Big]$$
$$= \frac{\nu(\nu+1)+n(n+1)-p(p+1)}{n(n+1)} \tag{6.1.44a}$$

$$B(n,\nu,p) =$$
$$\frac{1}{n(n+1)}\Big[(n+\nu+p+1)(\nu+p-n)(n+p-\nu)(n+\nu-p+1)\Big]^{1/2} \tag{6.1.44b}$$

Also let

$$a_{1n}^{E(M)} = \frac{i^n\sqrt{4\pi(2n+1)}}{2}X_n^{(M)} \tag{6.1.45a}$$

$$a_{1n}^{E(N)} = \frac{i^n\sqrt{4\pi(2n+1)}}{2}X_n^{(N)} \tag{6.1.45b}$$

Then putting into (6.1.41a) and (6.1.41b) of L-L law, we then have

$$X_\nu^{(M)} = -2\pi n_o\sum_{n,p}(2n+1)\Big[L_p(k,K|b)+M_p(k,K|b)\Big]$$
$$\times\Big\{T_n^{(M)}X_n^{(M)}a(1,n|-1,\nu|p)A(n,\nu,p)$$
$$+T_n^{(N)}X_n^{(N)}a(1,n|-1,\nu|p,p-1)B(n,\nu,p)\Big\} \tag{6.1.46a}$$

$$X_\nu^{(N)} = -2\pi n_o\sum_{n,p}(2n+1)\Big[L_p(k,K|b)+M_p(k,K|b)\Big]$$
$$\times\Big\{T_n^{(M)}X_n^{(M)}a(1,n|-1,\nu|p,p-1)B(n,\nu,p)$$
$$+T_n^{(N)}X_n^{(N)}a(1,n|-1,\nu|p)A(n,\nu,p)\Big\} \tag{6.1.46b}$$

The Ewald-Oseen extinction theorem can be simplified. We make use of the following summation identities

$$-\sum_p(-i)^p a(1,n|-1,\nu|p,p-1)\,b(n,\nu,p)$$
$$= \sum_p(-i)^p a(1,n|-1,\nu|p)\,a(n,\nu,p)$$
$$= -i^{\nu-n}\frac{(2\nu+1)\,n(n+1)}{2\nu(\nu+1)} \tag{6.1.47}$$

Then (6.1.42a) and (6.1.42b) become the same equation of

$$\sqrt{4\pi(2\nu+1)}\frac{i^\nu}{2} = -n_o\frac{2\pi i}{(K-k)k^2}\sum_n i^{\nu-n}\frac{(2\nu+1)n(n+1)}{2\nu(\nu+1)}\frac{\gamma_{1n}}{\gamma_{1\nu}} \cdot\left(T_n^{(M)}a_{1n}^{E(M)} + T_n^{(N)}a_{1n}^{E(N)}\right) \tag{6.1.48}$$

Substituting (6.1.45a) and (6.1.45b) into (6.1.48), we obtain

$$K - k = -\frac{\pi i n_o}{k^2}\sum_n \left(T_n^{(M)}X_n^{(M)} + T_n^{(N)}X_n^{(N)}\right)(2n+1) \tag{6.1.49}$$

To solve the L-L law and E-O theorem, we note that (6.1.46a) and (6.1.46b) form a system of simultaneous homogeneous equations for the unknown amplitudes $X_\nu^{(M)}$ and $X_\nu^{(N)}$. For a nontrivial solution, the determinant of the coefficients must vanish, yielding an equation for the effective wavenumber K in terms of k, the scattering coefficients $T_n^{(M)}$ and $T_n^{(N)}$, and the fractional volume of the scatterers. This is the dispersion relation for the composite medium. Hence, the procedure is to first find the solution K of the generalized Lorentz-Lorenz law, leaving an arbitrary constant to be next determined by the single scalar equation (6.1.49) of the generalized Ewald-Oseen extinction theorem.

1.2 Effective Phase Velocity and Attenuation Rate in the Low-Frequency Limit

A closed-form solution can be obtained for the effective propagation constant K in the low-frequency Rayleigh limit. In such a case, only the electric dipole term $T_1^{(N)}$ contributes in (6.1.46a) and (6.1.46b). Thus including $n = 1$ term only in (6.1.46b) gives an equation involving the unknown amplitude $X_1^{(N)}$. Setting the coefficient equal to zero gives the dispersion relation

$$1 + 6\pi n_o T_1^{(N)}\sum_p a(1,1|-1,1|p)A(1,1,p)\Big(L_p(k,K|b) + M_p(k,K|b)\Big) = 0 \tag{6.1.50}$$

As discussed in Chapter 2, Section 8 of Volume I, in the low-frequency limit, $T_1^{(N)} = T_{1r}^{(N)} + iT_{1i}^{(N)}$ and $T_{1r}^{(N)} = -T_{1i}^{(N)2}$. In calculating the effective propagation constant K, it is to be noted that $K_i = \mathrm{Im}(K)$ is a combination of absorption and scattering effects. We note that $K_r = \mathrm{Re}(K)$ is of order $\mathcal{O}(1)$ and K_i is of order $\mathcal{O}(f)$ for absorption and is of order $\mathcal{O}(fk^3a^3)$ for scattering. Since $|T_{1i}^{(N)}| \ll 1$, we have $|T_{1r}^{(N)}| \ll |T_{1i}^{(N)}|$. However, $T_{1r}^{(N)}$ cannot be neglected because it contributes to the attenuation rate K_i. Thus, in performing the low-frequency approximation of terms in (6.1.50), we shall (*i*) keep only the leading term of the real part and the leading term of the imaginary part, (*ii*) ignore the real part of a term if it is less than $\mathcal{O}(k^3a^3)$ of the corresponding imaginary part, and (*iii*) ignore the imaginary part of a term if it is less than $\mathcal{O}(k^3a^3)$ of the corresponding real part.

Using these rules, the low-frequency approximations for $T_1^{(N)}$ is

$$T_1^{(N)} \simeq i\frac{2}{3}(ka)^3 y \left[1 + i\frac{2}{3}(ka)^3 y\right] \tag{6.1.51}$$

where

$$y = \frac{\epsilon_s - \epsilon}{\epsilon_s + 2\epsilon} \tag{6.1.52}$$

The low-frequency approximation of L_p is

$$L_p(k, K|b) = -\frac{i}{k(K^2 - k^2)}\left(\frac{K}{k}\right)^p - \delta_{p0}\frac{b^3}{3} \tag{6.1.53}$$

Only the $p = 0$ term in L_p has a nonzero second term in (6.1.53) because for $p \geq 1$, the real part is of an order smaller than $\mathcal{O}(k^3a^3)$ of the imaginary part. Similarly, $M_p(k, K|b) = \delta_{p0}M_0$ with

$$M_0 = \int_b^\infty dr\, r^2\,[g(r) - 1] = \int_0^\infty dr\, r^2\,[g(r) - 1] + \frac{b^3}{3} \tag{6.1.54}$$

Only the real part of $M_p(k, K|b)$ for $p = 0$ contributes. Also note that $g(r) = 0$ for $r < b$. Other real and imaginary terms are smaller than the corresponding real and imaginary terms of $L_p(k, K|b)$. Thus, in the Rayleigh limit, only the knowledge of the integral of $g(r) - 1$ is required rather than a detailed behavior of the pair-distribution function $g(r)$ as a function of r.

Substituting the low-frequency approximation in (6.1.50) gives

$$1 + \frac{2\pi n_o T_1^{(N)}}{k}\frac{i}{(K^2 - k^2)}\left[2 + \left(\frac{K}{k}\right)^2\right] + 4\pi n_o T_1^{(N)}\left(\frac{b^3}{3} - M_0\right) = 0 \tag{6.1.55}$$

Note that

$$\frac{i2\pi n_o T_1^{(N)}}{k^3} = -fy\left[1 + i\frac{2}{3}(ka)^3 y\right] \tag{6.1.56}$$

Solving (6.1.55) gives

$$\frac{K^2}{k^2} = 1 + \frac{3fy\left[1 + i\frac{2}{3}(ka)^3 y\right]}{1 - fy\left[1 + i\frac{2}{3}(ka)^3 y\right] + 4\pi n_o T_1^{(N)}\left(\frac{b^3}{3} - M_o\right)} \tag{6.1.57}$$

We also have

$$4\pi n_o T_1^{(N)} \approx 2ifyk^3 \int_0^\infty dr\, r^2 \left[g(r) - 1\right] \tag{6.1.58}$$

Then

$$\frac{K^2}{k^2} = 1 + \frac{3fy\left[1 + i\frac{2}{3}(ka)^3 y\right]}{(1 - fy)\left[1 - i\frac{2}{3}\frac{(ka)^3 y}{1-fy}\left(fy + 4\pi n_o \int_0^\infty dr\, r^2\left[g(r) - 1\right]\right)\right]} \tag{6.1.59}$$

The solution can be further simplified by making use of the property $K_i \ll K_r$. The final result is

$$K^2 = k^2 + \frac{3fk^2 y}{1 - fy}\left[1 + i\frac{2}{3}k^3 a^3 \frac{y}{1 - fy}\left\{1 + 4\pi n_o \int_0^\infty dr\, r^2 [g(r) - 1]\right\}\right] \tag{6.1.60}$$

where $f = n_o 4\pi a^3/3$ is the fractional volume of scatterers. The result in (6.1.60) is the same as the result obtained through the simple model in Chapter 10 of Volume II.

Using the Percus-Yevick pair-distribution function from Chapter 8 of Volume II for $g(r)$, the result for the effective propagation constant is

$$K^2 = k^2 + \frac{3fk^2 y}{1 - fy}\left[1 + i\frac{2}{3}k^3 a^3 y \frac{(1 - f)^4}{(1 - fy)(1 + 2f)^2}\right] \tag{6.1.61}$$

In Figs. 6.1.3 and 6.1.4, we show, respectively, the normalized phase velocity $v_p = k/K_r$ and the effective loss tangent $LT = 2K_i/K_r$ as a function of fractional volume of scatterers f using the result of (6.1.61) for $ka = 0.1$. We note that the phase velocity decreases with f. The loss tangent first increases with f, then saturates and decreases as f further increases.

1.3 Dispersion Relations at Higher Frequencies

For larger values of ka, (6.1.46*a*) and (6.1.46*b*) can be solved numerically. In this section we shall show the numerical results of the effective propagation constant K for ka ranging from 0.05 to 2.5 and f ranging from 0 to 0.4. For these values of ka and f, the determinant of the coefficient matrix was computed numerically by retaining a maximum of eight simultaneous homogeneous complex equations. The Wigner $3j$ symbols are generated by the computer code, as discussed in Chapter 10 of Volume II, and checked against tabulated values [Rotenberg et al. 1959]. The elements of $M_p(k, K|b)$ for $p = 0, 1, \ldots, 8$ were computed by numerically evaluating the integral in (6.1.39), for r between b and $4b$. For f between 0 and 0.4, the value of $g(r) - 1$ is practically zero for r larger than $4b$.

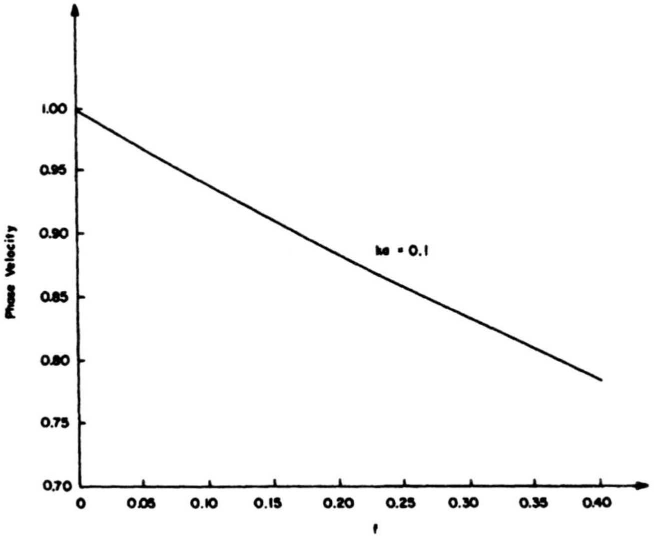

Figure 6.1.3 Normalized phase velocity k/K_r in the Rayleigh limit as a function of fractional volume of scatterers f for $ka = 0.1$, and $\epsilon_s = 3.24\epsilon_o$.

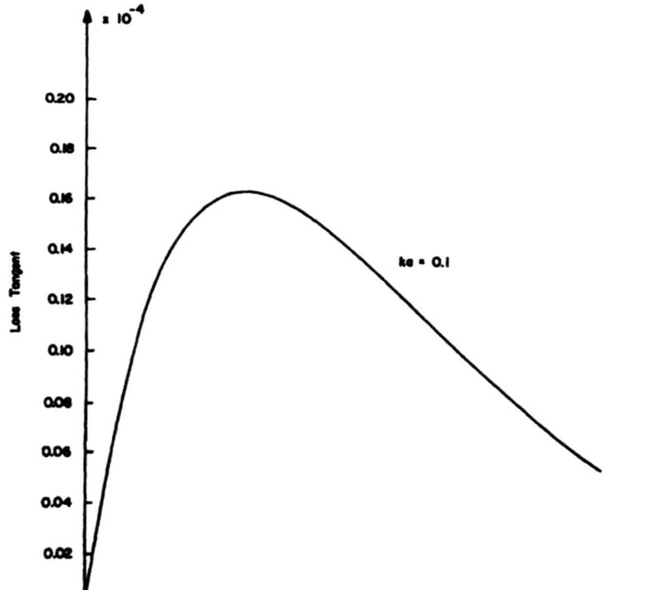

Figure 6.1.4 Effective loss tangent $2K_i/K_r$ as a function of fractional volume of scatterers f for $ka = 0.1$ and $\epsilon_s = 3.24\epsilon_o$.

For a given value of ka and f, the roots of the determinant were searched for in the complex K plane ($K_r + iK_i$) using Muller's method. There are two good initial guesses:

(*i*) We first note that for media with sparse concentration of particles, the exciting field will be approximately the same as the incident field, so that $X_n^{(M)}$ and $X_n^{(N)}$ both are approximately equal to 1. Replacing them by 1 in the generalized Ewald-Oseen extinction theorem of (6.1.49) gives the result of an effective propagation constant $K^{(F)}$ under Foldy's approximation

$$K^{(F)} = k - \frac{\pi i n_o}{k^2} \sum_{n=1}^{\infty} (2n+1) \left(T_n^{(M)} + T_n^{(N)} \right) \tag{6.1.62}$$

(*ii*) The second initial guess is provided by the low-frequency solution in (6.1.60). These two guesses could be used systematically to obtain quick convergence of roots at increasingly higher values of ka. In the following, we shall illustrate results for (a) the normalized phase velocity v_p defined as k/K_r and (b) the effective loss tangent (LT) defined as $2K_i/K_r$. The Percus-Yevick pair-distribution function is used unless otherwise specified.

In Figs. 6.1.5 and 6.1.6, we plot, respectively, $v_p = k/K_r$ and $LT = 2K_i/K_r$ versus ka for two different f values and for real ϵ_s. We note that the phase velocity first decreases with increasing frequency and then oscillates as frequency increases further. The oscillation is a characteristic of resonant scattering at higher frequencies. The loss tangent first increases rapidly with frequency and then saturates at high frequencies.

In Figs. 6.1.7 and 6.1.8, we plot v_p and LT versus ka for two different f values and for complex ϵ_s representing absorptive scatterers. On comparing the results of Figs. 6.1.6 and 6.1.8, we note that at low frequencies, absorption dominates over scattering as the loss tangent in Fig. 6.1.8 is much larger than that in Fig. 6.1.6. At higher frequencies the loss tangents are comparable in magnitude in the two cases.

In Figs. 6.1.9 and 6.1.10 we show, respectively, v_p and LT versus f for two different ka values. We note that the phase velocity decreases as f increases since the phase velocity for the scatterer is lower. As a function of f, LT first rises to a peak and then decreases as f further increases.

In Figs. 6.1.11 and 6.1.12 we compare the results based on the Percus-Yevick (PY) pair function and the hole-correction (HC) pair function for $ka = 0.5$. For the HC pair function we have $g_{HC}(r) = 0$ for $r \leq b$ and $g_{HC}(r) = 1$ for $r > b$. We note that the HC result for v_p agrees very well

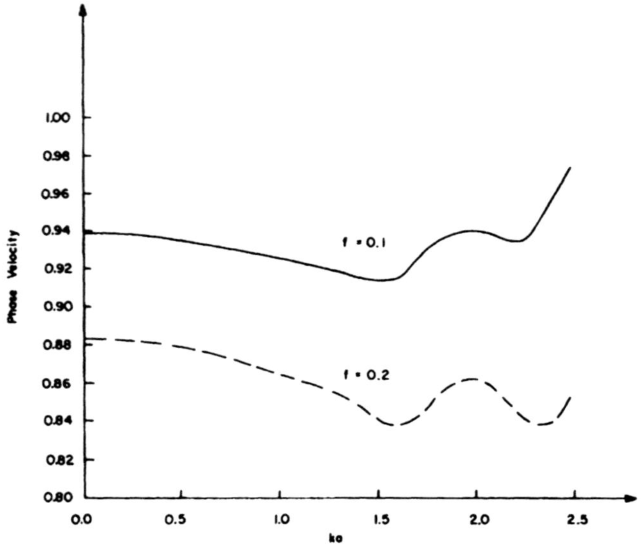

Figure 6.1.5 Normalized phase velocity k/K_r as a function of ka for $\epsilon_s = 3.24\epsilon_o$ and $f = 0.1$ and 0.2.

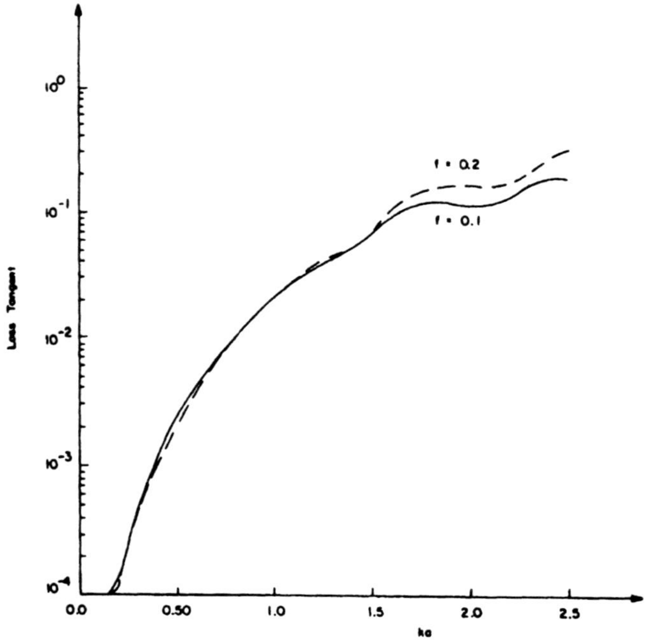

Figure 6.1.6 Effective loss tangent $2K_i/K_r$ as a function of ka for $\epsilon_s = 3.24\epsilon_o$ and $f = 0.1$ and 0.2.

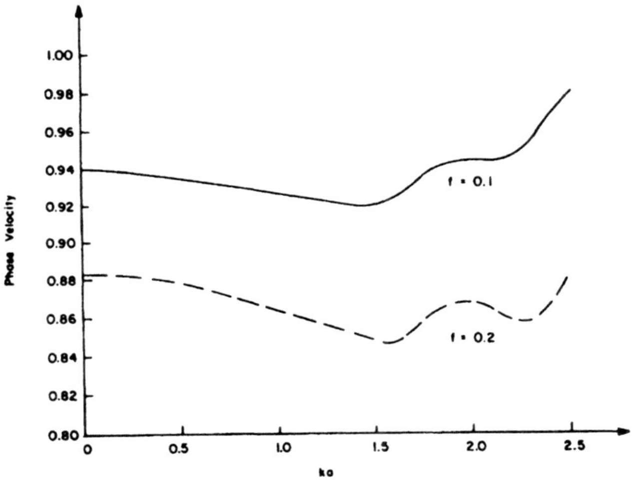

Figure 6.1.7 Normalized phase velocity k/K_r as a function of ka for $\epsilon_s = (3.2375 + i0.18)\epsilon_o$ and $f = 0.1$ and 0.2.

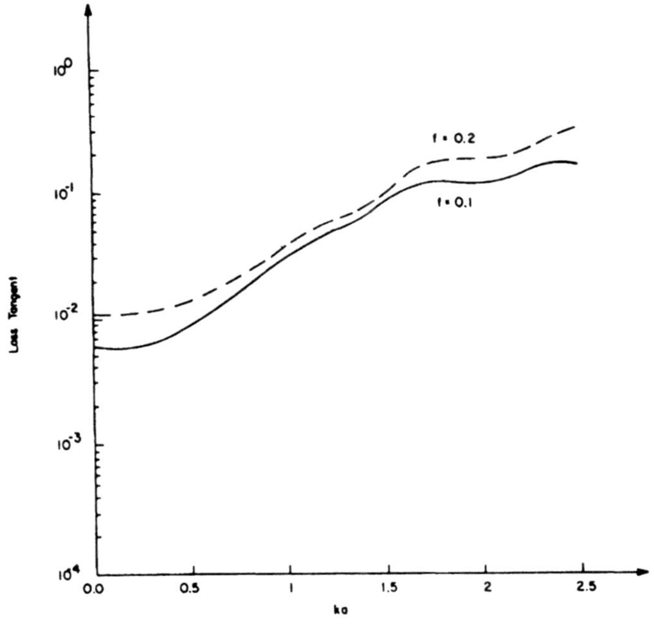

Figure 6.1.8 Effective loss tangent $2K_i/K_r$ as a function of ka for $\epsilon_s = (3.2375 + i0.18)\epsilon_o$ and $f = 0.1$ and 0.2.

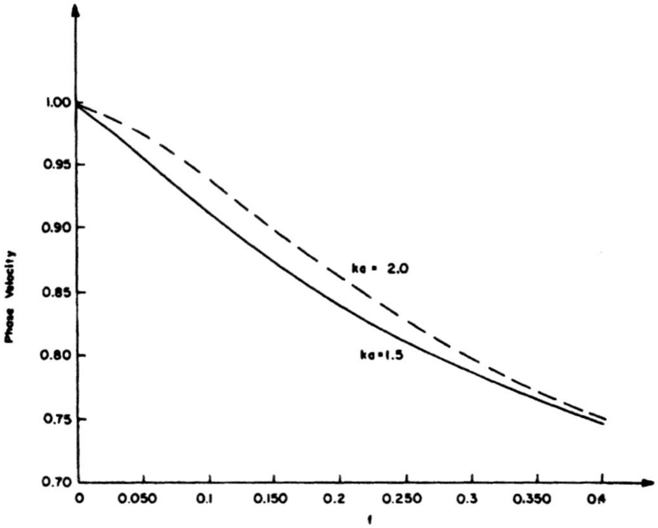

Figure 6.1.9 Normalized phase velocity k/K_r as a function of f for $\epsilon_s = 3.24\epsilon_o$ and $ka = 1.5$ and 2.0.

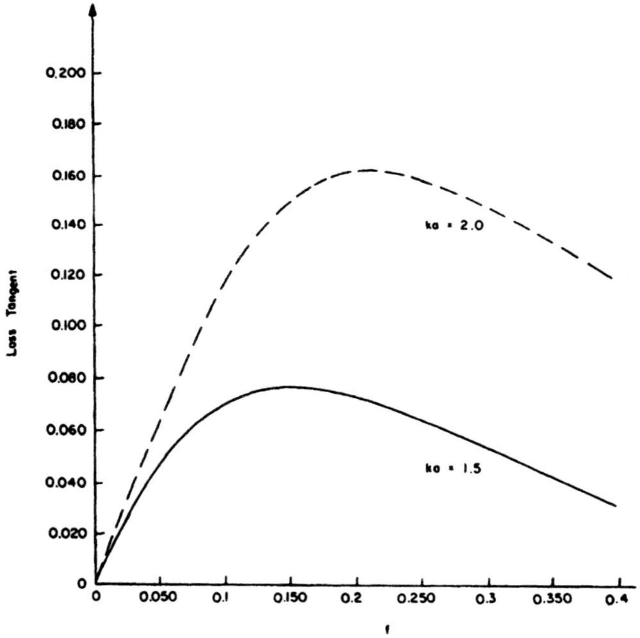

Figure 6.1.10 Effective loss tangent $2K_i/K_r$ as a function of f for $\epsilon_s = 3.24\epsilon_o$ and $ka = 1.5$ and 2.0.

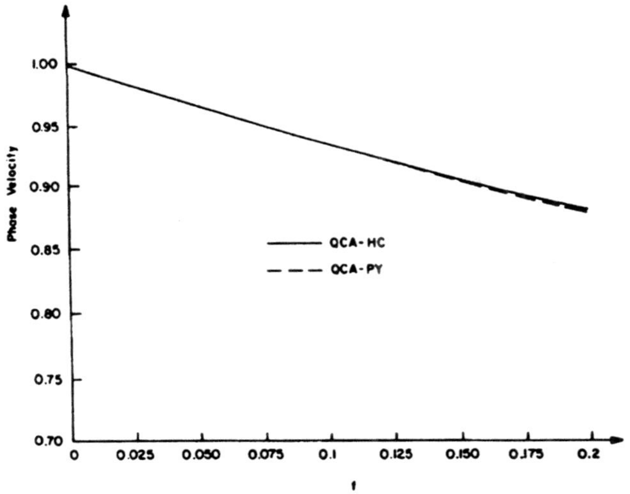

Figure 6.1.11 Normalized phase velocity k/K_r as a function of f for $\epsilon_s = 3.24\epsilon_o$ and $ka = 0.5$. The QCA-PY and QCA-HC results are compared.

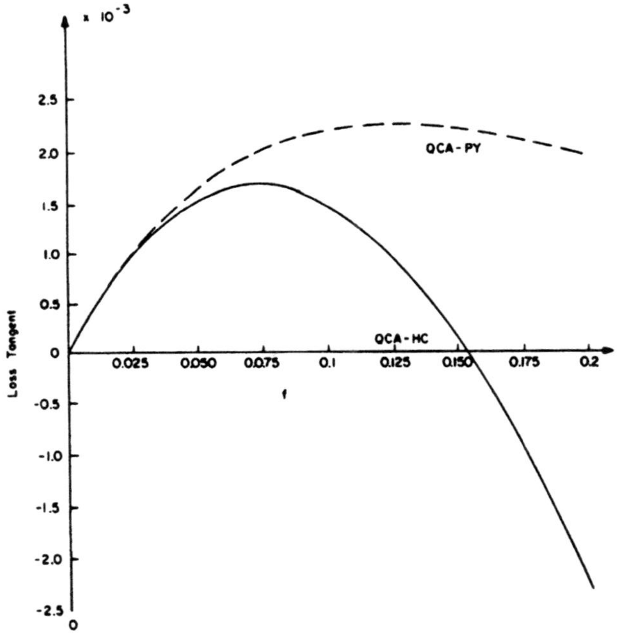

Figure 6.1.12 Effective loss tangent $2K_i/K_r$ as a function of f for $\epsilon_s = 3.24\epsilon_o$ and $ka = 0.5$. The QCA-PY and QCA-HC results are compared.

with the PY result. However, for the loss tangent, the two results agree only for small f (up to 0.04). Furthermore, the HC result, after increasing to a peak, decreases rapidly as f further increases. For f larger than 0.15 (approximately), it decreases to negative values which are physically unacceptable because that indicates negative scattering and growth of the coherent wave. Thus the HC pair function is not acceptable for high concentration of scatterers.

2 Scattering of Electromagnetic Waves from a Half-Space of Dielectric Scatterers—Oblique Incidence

In this section we extend the normal incidence case of the previous section to oblique incidence. Both vertically and horizontally polarized incident waves are considered. The coherent reflectivity will be calculated. The T-matrix, together with the quasi-crystalline approximation, is used to generate a set of homogeneous equations (the generalized Lorentz-Lorenz law) and a set of inhomogeneous equations (the generalized Ewald-Oseen extinction theorem). The coherent reflected wave is calculated by solving these two sets of equations. For the case of spherical scatterers, it is also shown that the effective propagation constant is independent of direction and polarization. In the low-frequency limit, the scattered wave produced by the scatterers contains a reflected coherent part that obeys Snell's law, gives the Fresnel reflection coefficients for TE and TM waves, and reproduces the Brewster angle effect and the Clausius-Mosotti relation. At higher frequencies, the coherent reflected wave is calculated numerically.

2.1 Dispersion Relation and Coherent Reflected Wave

Consider a plane wave incident on a half-space of N identical dielectric scatterers with radius a and permittivity ϵ_s and centered at $\overline{r}_1, \overline{r}_2, \ldots, \overline{r}_N$ (Fig. 6.2.1). The particles occupy the lower half-space region of $z \leq 0$. The background medium has permittivity ϵ. The same medium also occupies the region $z \geq 0$. The incident wave is propagating downward, and we let $\overline{k}_{id}$ to denote the downward propagating incident wave vector

$$\overline{k}_{id} = k(\sin\theta_i \cos\phi_i \hat{x} + \sin\theta_i \sin\phi_i \hat{y} - \cos\theta_i \hat{z}) = k_{ix}\hat{x} + k_{iy}\hat{y} - k_{iz}\hat{z} \quad (6.2.1)$$

Thus the incident direction is $(\theta, \phi) = (\pi - \theta_i, \phi_i)$.

For the case of vertically polarized incident plane wave, the coefficients

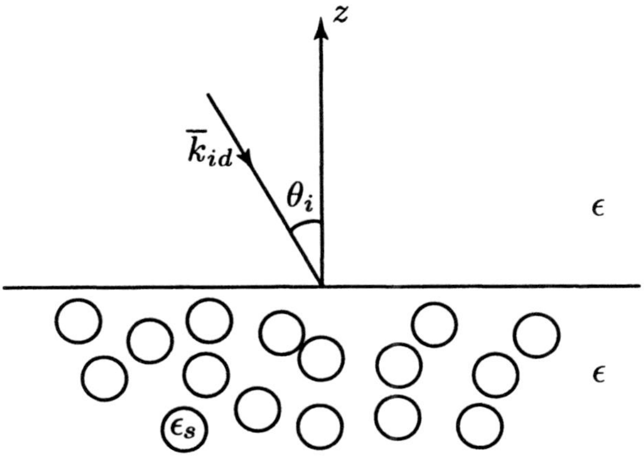

Figure 6.2.1 Plane electromagnetic wave obliquely incident on a half-space of densely packed dielectric scatterers.

$\overline{a}^{inc}$ are, using the results of Chapter 1 of Volume I,

$$a_{mn}^{inc(M)} = (-1)^m \frac{1}{\gamma_{mn}} \frac{(2n+1)}{n(n+1)} i^n \,\hat{\theta}_{id} \cdot \overline{C}_{-mn}(\pi - \theta_i, \phi_i) \tag{6.2.2}$$

$$a_{mn}^{inc(N)} = (-1)^m \frac{1}{\gamma_{mn}} \frac{(2n+1)}{n(n+1)} i^{n-1} \,\hat{\theta}_{id} \cdot \overline{B}_{-mn}(\pi - \theta_i, \phi_i) \tag{6.2.3}$$

where $\hat{\theta}_{id}$ is the vertically polarized vector for the downward-propagating wave $\overline{k}_{id}$:

$$\hat{\theta}_{id} = -\hat{x}\cos\theta_i \cos\phi_i - \hat{y}\cos\theta_i \sin\phi_i - \hat{z}\sin\theta_i \tag{6.2.4}$$

For an incident horizontally polarized wave, the coefficients $a_{mn}^{inc(M)}$ and $a_{mn}^{inc(N)}$ are obtained by taking the dot product in (6.2.2) and (6.2.3) with $\hat{\phi}_i$ instead of $\hat{\theta}_{id}$. Using quasi-crystalline approximation, the equation for the conditional-averaged exciting field coefficients is

$$\overline{w}_E(\overline{r}_1) = n_o \int_{z_2<0} d\overline{r}_2 \, g(\overline{r}_2 - \overline{r}_1) \overline{\overline{\sigma}}(k\overline{r_1 r_2}) \overline{\overline{T}} \, \overline{w}_E(\overline{r}_2) + e^{i\overline{k}_{id}\cdot\overline{r}_1} \, \overline{a}^{inc} \tag{6.2.5}$$

To solve (6.2.5), let

$$\overline{w}_E(\overline{r}_1) = \overline{a}^E \, e^{i\overline{K}_d\cdot\overline{r}_1} \tag{6.2.6}$$

where $\overline{K}_d$ denotes the downward-going effective wavevector

$$\overline{K}_d = K_x\hat{x} + K_y\hat{y} - K_z\hat{z} \tag{6.2.7}$$

and the effective wavenumber is $K = (K_x^2 + K_y^2 + K_z^2)^{1/2}$. After substituting (6.2.6) into the integral in (6.2.5), manipulations can be performed in a manner similar to those of the case of normal incidence in Section 1.

In index notations, we have

$$w_{mn}^{(M)}(\overline{r}_1) = \sum_{\mu\nu} n_o \int_{z_2<0} d\overline{r}_2 \left[A_{mn\mu\nu}(k\overline{r_1 r_2}) T_\nu^{(M)} w_{\mu\nu}^{(M)}(\overline{r}_2) + B_{mn\mu\nu}(k\overline{r_1 r_2}) T_\nu^{(N)} w_{\mu\nu}^{(N)}(\overline{r}_2)\right] g(\overline{r}_2 - \overline{r}_1) + a_{mn}^{inc(M)} e^{i\overline{k}_{id}\cdot\overline{r}_1} \tag{6.2.8a}$$

$$w_{mn}^{(N)}(\overline{r}_1) = \sum_{\mu\nu} n_o \int_{z_2<0} d\overline{r}_2 \left[B_{mn\mu\nu}(k\overline{r_1 r_2}) T_\nu^{(M)} w_{\mu\nu}^{(M)}(\overline{r}_2) + A_{mn\mu\nu}(k\overline{r_1 r_2}) T_\nu^{(N)} w_{\mu\nu}^{(N)}(\overline{r}_2)\right] g(\overline{r}_2 - \overline{r}_1) + a_{mn}^{inc(N)} e^{i\overline{k}_{id}\cdot\overline{r}_1} \tag{6.2.8b}$$

Putting (6.2.6) in (6.2.8a), we have

$$a_{mn}^{E(M)} e^{i\overline{K}_d\cdot\overline{r}_1} = \sum_{\mu\nu} n_o e^{i\overline{K}_d\cdot\overline{r}_1} \int_{z_2<0} d\overline{r}_2 \left[A_{mn\mu\nu}(k\overline{r_1 r_2}) T_\nu^{(M)} a_{\mu\nu}^{E(M)} + B_{mn\mu\nu}(k\overline{r_1 r_2}) T_\nu^{(N)} a_{\mu\nu}^{E(N)}\right] g(\overline{r}_2 - \overline{r}_1) e^{i\overline{K}_d\cdot(\overline{r}_2 - \overline{r}_1)} + a_{mn}^{inc(M)} e^{i\overline{k}_{id}\cdot\overline{r}_1} \tag{6.2.9}$$

The integral to be calculated is

$$I = \int_{z_2<0} d\overline{r}_2 h_p(k|\overline{r_1 r_2}|) Y_p^{\mu-m}(\theta_{\overline{r_1 r_2}}, \phi_{\overline{r_1 r_2}}) g(|\overline{r}_2 - \overline{r}_1|) e^{i\overline{K}_d\cdot(\overline{r}_2 - \overline{r}_1)} \tag{6.2.10}$$

Using the property that

$$P_n^m(-x) = (-1)^{n+m} P_n^m(x) \tag{6.2.11}$$

$\theta_{\overline{r_1 r_2}} = \pi - \theta_{\overline{r_2 r_1}}$, $\phi_{\overline{r_1 r_2}} = \pi + \phi_{\overline{r_2 r_1}}$, and

$$Y_p^m(\theta_{\overline{r_1 r_2}}, \phi_{\overline{r_1 r_2}}) = (-1)^p Y_p^m(\theta_{\overline{r_2 r_1}}, \phi_{\overline{r_2 r_1}}) \tag{6.2.12}$$

The integral in (6.2.10) becomes

$$I = (-1)^p \int_{\substack{z_2<0 \\ |\overline{r}_2 - \overline{r}_1| > b}} d\overline{r}_2 h_p(k|\overline{r_2 r_1}|) Y_p^{\mu-m}(\theta_{\overline{r_2 r_1}}, \phi_{\overline{r_2 r_1}}) g(|\overline{r}_2 - \overline{r}_1|) e^{i\overline{K}_d\cdot(\overline{r}_2 - \overline{r}_1)} \tag{6.2.13}$$

$$= I_1 + I_2 \tag{6.2.14}$$

where

$$I_1 = (-1)^p \int_{\substack{z_2<0 \\ |\overline{r}_2 - \overline{r}_1| > b}} d\overline{r}_2 h_p(k|\overline{r_2 r_1}|) Y_p^{\mu-m}(\theta_{\overline{r_2 r_1}}, \phi_{\overline{r_2 r_1}}) e^{i\overline{K}_d\cdot(\overline{r}_2 - \overline{r}_1)} \tag{6.2.15a}$$

$$I_2 = (-1)^p \int_{\substack{z_2<0 \\ |\overline{r}_2-\overline{r}_1|>b}} d\overline{r}_2 h_p(k|\overline{r_2 r_1}|) Y_p^{\mu-m}(\theta_{\overline{r_2 r_1}}, \phi_{\overline{r_2 r_1}})(g(|\overline{r}_2-\overline{r}_1|)-1) e^{i\overline{K}_d\cdot(\overline{r}_2-\overline{r}_1)} \tag{6.2.15b}$$

To evaluate the integral I_1, we let $\overline{r}_2 - \overline{r}_1 = \overline{r}$, and

$$y_p = h_p(kr) Y_p^{\mu-m}(\theta, \phi) \tag{6.2.16}$$

Then

$$e^{i\overline{K}_d\cdot\overline{r}} y_p = \frac{1}{K^2 - k^2}\left[e^{i\overline{K}_d\cdot\overline{r}}\nabla^2 y_p - y_p \nabla^2 e^{i\overline{K}_d\cdot\overline{r}}\right] \tag{6.2.17}$$

Using Green's theorem, we obtain

$$\begin{aligned} I_1 = \frac{(-1)^p}{K^2-k^2}\Bigg\{ &- \int_{S_e} dS\left[e^{i\overline{K}_d\cdot\overline{r}}\frac{\partial y_p}{\partial r} - y_p\frac{\partial}{\partial r}e^{i\overline{K}_d\cdot\overline{r}}\right]_{r=b} \\ &+ \int_{S_d} dS\left[e^{i\overline{K}_d\cdot\overline{r}}\frac{\partial y_p}{\partial z} - y_p\frac{\partial}{\partial z}e^{i\overline{K}_d\cdot\overline{r}}\right]_{z=-z_1}\Bigg\} \end{aligned} \tag{6.2.18}$$

In performing the integration over S_e, we note that if Θ is the angle between $\overline{K}_d$ and $\overline{r}$, then

$$\begin{aligned} e^{i\overline{K}_d\cdot\overline{r}} &= \sum_{n=0}^{\infty}(2n+1)i^n P_n(\cos\Theta) j_n(Kr) \\ &= \sum_{n=0}^{\infty}(2n+1)i^n j_n(Kr)\sum_{m=-n}^{n}(-1)^m Y_n^m(\pi-\theta_t,\phi_t) Y_n^{-m}(\theta,\phi) \end{aligned} \tag{6.2.19}$$

Note that $\overline{K}_d$ is described by the angular variables $(\pi-\theta_t, \phi_t)$ for its direction.

Thus,

$$\begin{aligned} &-\frac{(-1)^p}{K^2-k^2}\int_{S_e} dS\left[e^{i\overline{K}_d\cdot\overline{r}}\frac{\partial y_p}{\partial r} - y_p\frac{\partial}{\partial r}e^{i\overline{K}_d\cdot\overline{r}}\right]_{r=b} \\ &= -\frac{(-1)^p}{K^2-k^2}\int_0^{\pi} d\theta\sin\theta\int_0^{2\pi} d\phi b^2\left[e^{i\overline{K}_d\cdot\overline{r}}\frac{\partial y_p}{\partial r} - y_p\frac{\partial}{\partial r}e^{i\overline{K}_d\cdot\overline{r}}\right] \\ &= -\frac{(-1)^p}{K^2-k^2}\int_0^{\pi} d\theta\sin\theta\int_0^{2\pi} d\phi b^2\sum_{p'}(2p'+1)i^{p'}\sum_{m'=-p'}^{p'}(-1)^{m'} \\ &\quad\cdot Y_{p'}^{m'}(\pi-\theta_t,\phi_t)Y_{p'}^{m'}(\theta,\phi)Y_p^{\mu-m}(\theta,\phi)\left[j_{p'}(Kb)kh_p'(kb) - h_p(kb)Kj_{p'}(Kb)\right] \\ &= -\frac{(-1)^p}{K^2-k^2}b^2 i^p Y_p^{\mu-m}(\pi-\theta_t,\phi_t)4\pi\left(j_p(Kb)kh_p'(kb) - h_p(kb)Kj_p'(Kb)\right) \\ &= (-i)^p 4\pi Y_p^{\mu-m}(\pi-\theta_t,\phi_t)L_p(k,K|b) \end{aligned} \tag{6.2.20}$$

For I_2, we can extend to an infinite domain because g is equal to unity for large $|\overline{r}_2 - \overline{r}_1|$

$$I_2 = (-1)^p \int_{|\overline{r}|>b} d\overline{r} y_p(g(r) - 1) e^{i\overline{K}_d \cdot \overline{r}} = (-i)^p 4\pi M_p(k, K|b) Y_p^{\mu-m}(\pi - \theta_t, \phi_t) \tag{6.2.21}$$

To calculate the S_d integral in (6.2.18), we note that

$$\frac{e^{ik|\overline{r}_1 - \overline{r}_2|}}{4\pi|\overline{r}_1 - \overline{r}_2|} = \frac{ik}{4\pi} \sum_{n=0}^{\infty} (2n+1) \sum_m (-1)^m Y_n^{-m}(\theta_2, \phi_2) Y_n^m(\theta_1, \phi_1) j_n(kr_2) h_n(kr_1) \quad \text{for } r_1 > r_2 \tag{6.2.22}$$

so that

$$ikj_n(kr_2) h_n(kr) Y_n^m(\theta, \phi) = \int_0^{\pi} d\theta_2 \sin\theta_2 \int_0^{2\pi} d\phi_2 \frac{e^{ik|\overline{r} - \overline{r}_2|}}{4\pi|\overline{r} - \overline{r}_2|} Y_n^m(\theta_2, \phi_2) \quad \text{for } r > r_2 \tag{6.2.23}$$

A plane wave representation can be used for $\exp(ik|\overline{r} - \overline{r}_2|)/4\pi|\overline{r} - \overline{r}_2|$ in (6.2.23), so that

$$h_p(kr) Y_p^{\mu-m}(\theta, \phi) = \frac{i^p}{k\pi} \int_{-\infty}^{\infty} dk_x \int_{-\infty}^{\infty} dk_y \frac{e^{-ik_x x - ik_y y} e^{ik_z z}}{2k_z} Y_p^{\mu-m}(\pi - \theta_k, \phi_k) \quad \text{for } z > 0 \tag{6.2.24}$$

The $\int_{S_d}$ integral will produce a phase dependence $e^{-i\sqrt{k^2 - K_x^2 - K_y^2} z_1} \cdot e^{iK_z z_1}$. This phase term when combined with the $\exp(i\overline{K}_d \cdot \overline{r}_1)$ in (6.2.9), will give a term with overall phase dependence

$$e^{i\overline{K}_d \cdot \overline{r}_1 + iK_z z_1 - i\sqrt{k^2 - K_x^2 - K_y^2} z_1} = e^{iK_x x_1 + iK_y y_1 - i\sqrt{k^2 - K_x^2 - K_y^2} z_1} \tag{6.2.25}$$

This is to be balanced with the incident wave $e^{i\overline{k}_{id} \cdot \overline{r}_1}$ to give the generalized Ewald-Oseen extinction theorem. Thus we have

$$K_x = k_{ix} \tag{6.2.26a}$$

$$K_y = k_{iy} \tag{6.2.26b}$$

giving the phase matching condition as a consequence of the Ewald-Oseen extinction theorem. Also

$$K_z = K_{iz} = \sqrt{K^2 - k_{ix}^2 - k_{iy}^2} \tag{6.2.26c}$$

$$\phi_t = \phi_i \tag{6.2.26d}$$

Based on (6.2.26a)–(6.2.26d), we have Snell's law of $k\sin\theta_i = K\sin\theta_t$. The integral over S_d is

$$\frac{(-1)^p}{K^2-k^2}\int_{S_d} dS\left[e^{i\overline{K}_d\cdot\overline{r}}\frac{\partial y_p}{\partial z} - y_p\frac{\partial}{\partial z}e^{i\overline{K}_d\cdot\overline{r}}\right]_{z=-z_1}$$

$$= \frac{(-1)^p}{K^2-k^2}\iint dxdy\Big\{e^{ik_{ix}x+ik_{iy}y-iK_{iz}z}$$

$$\frac{i^p}{k\pi}\iint dk_xdk_y\frac{e^{-ik_xx-ik_yy}e^{ik_zz}}{2k_z}Y_p^{\mu-m}(\pi-\theta_k,\phi_k)\Big\}_{z=-z_1}(ik_z+iK_z)$$

$$= \frac{4\pi(-1)^pi^p}{k(K^2-k^2)}\frac{i(k_{iz}+K_{iz})}{2k_{iz}}e^{iK_{iz}z_1}e^{-ik_{iz}z_1}Y_p^{\mu-m}(\pi-\theta_i,\phi_i) \qquad (6.2.27)$$

Thus

$$I = (-i)^p4\pi(L_p(k,K|b)+M_p(k,K|b))Y_p^{\mu-m}(\pi-\theta_t,\phi_t)$$
$$+\frac{4\pi(-1)^pi^{p+1}}{k(K_{iz}-k_{iz})2k_{iz}}e^{iK_{iz}z_1-ik_{iz}z_1}Y_p^{\mu-m}(\pi-\theta_i,\phi_i) \qquad (6.2.28)$$

Putting the solution of the integrals into (6.2.9) gives

$$a_{mn}^{E(M)}e^{i\overline{K}_d\cdot\overline{r}_1} = \sum_{\mu\nu}n_o\frac{\gamma_{\mu\nu}}{\gamma_{mn}}(-1)^m$$

$$\sum_p\Big[(-1)^p4\pi i^p(L_p(k,K|b)+M_p(k,K|b))e^{i\overline{K}_d\cdot\overline{r}_1}Y_p^{\mu-m}(\pi-\theta_t,\phi_t)$$

$$+\frac{4\pi(-1)^pi^{p+1}}{k(K_{iz}-k_{iz})2k_{iz}}e^{i\overline{k}_{id}\cdot\overline{r}_1}Y_p^{\mu-m}(\pi-\theta_i,\phi_i)\Big]$$

$$\Big[a(\mu,\nu|-m,n|p)a(\nu,n,p)T_\nu^{(M)}a_{\mu\nu}^{E(M)}$$

$$-a(\mu,\nu|-m,n|p,p-1)b(\nu,n,p)T_\nu^{(N)}a_{\mu\nu}^{E(N)}\Big]+a_{mn}^{(M)}e^{i\overline{k}_{id}\cdot\overline{r}_1} \qquad (6.2.29a)$$

We manipulate (6.2.8b) in the same manner and get

$$a_{mn}^{E(N)}e^{i\overline{K}_d\cdot\overline{r}_1} = \sum_{\mu\nu}n_o\frac{\gamma_{\mu\nu}}{\gamma_{mn}}(-1)^m$$

$$\sum_p\Big[(-1)^p4\pi i^p(L_p(k,K|b)+M_p(k,K|b))e^{i\overline{K}_d\cdot\overline{r}_1}Y_p^{\mu-m}(\pi-\theta_t,\phi_t)$$

$$+\frac{4\pi(-1)^pi^{p+1}}{k(K_{iz}-k_{iz})2k_{iz}}e^{i\overline{k}_{id}\cdot\overline{r}_1}Y_p^{\mu-m}(\pi-\theta_i,\phi_i)\Big]$$

$$\Big[-a(\mu,\nu|-m,n|p,p-1)b(\nu,n,p)T_\nu^{(M)}a_{\mu\nu}^{E(M)}$$

$$+a(\mu,\nu|-m,n|p)a(\nu,n,p)T_\nu^{(N)}a_{\mu\nu}^{E(N)}\Big]+a_{mn}^{(N)}e^{i\overline{k}_{id}\cdot\overline{r}_1} \qquad (6.2.29b)$$

Next we match the $e^{i\overline{K}_d\cdot\overline{r}_1}$ dependence in (6.2.29a) and (6.2.29b) to get the generalized Lorentz-Lorenz law.

Let

$$X_{mn\mu\nu}^{(K)} = \sum_p \frac{\gamma_{\mu\nu}}{\gamma_{mn}}(-1)^m a(\mu,\nu|-m,n|p)a(\nu,n,p)S_{p(\mu-m)}^{(K)} \tag{6.2.30}$$

$$Z_{mn\mu\nu}^{(K)} = \sum_p \frac{\gamma_{\mu\nu}}{\gamma_{mn}}(-1)^{m+1} a(\mu,\nu|-m,n|p,p-1)b(\nu,n,p)S_{p(\mu-m)}^{(K)} \tag{6.2.31}$$

$$S_{p(\mu-m)}^{(K)} = S_p(-1)^p 4\pi i^p P_p^{\mu-m}\left(\cos(\pi-\theta_t)\right)e^{i(\mu-m)\phi_t} \tag{6.2.32}$$

$$\begin{aligned} S_p &= \int_b^\infty dr\, r^2\,[g(r)-1] j_p(Kr)h_p(kr) \\ &\quad - \frac{b^2}{K^2-k^2}\left[kj_p(Kb)h_p'(kb) - Kj_p'(Kb)h_p(kb)\right] \\ &= L_p(k,K|b) + M_p(k,K|b) \end{aligned} \tag{6.2.33}$$

Then from (6.2.29a) and (6.2.29b), we get the system of equations of the generalized Lorentz-Lorenz law

$$a_{mn}^{E(M)} = \sum_{\mu\nu}\left[n_o X_{mn\mu\nu}^{(K)} T_\nu^{(M)} a_{\mu\nu}^{E(M)} + n_o Z_{mn\mu\nu}^{(K)} T_\nu^{(N)} a_{\mu\nu}^{E(N)}\right] \tag{6.2.34a}$$

$$a_{mn}^{E(N)} = \sum_{\mu\nu}\left[n_o Z_{mn\mu\nu}^{(K)} T_\nu^{(M)} a_{\mu\nu}^{E(M)} + n_o X_{mn\mu\nu}^{(K)} T_\nu^{(N)} a_{\mu\nu}^{E(N)}\right] \tag{6.2.34b}$$

In matrix notation, we write

$$\overline{a}_E = n_o \overline{\overline{U}}^{(K)}\overline{\overline{T}}\overline{a}_E \tag{6.2.35}$$

$$\overline{\overline{U}}^{(K)} = \begin{bmatrix} \overline{\overline{X}}^{(K)} & \overline{\overline{Z}}^{(K)} \\ \overline{\overline{Z}}^{(K)} & \overline{\overline{X}}^{(K)} \end{bmatrix} \tag{6.2.36}$$

Balancing the $e^{i\overline{k}_{id}\cdot\overline{r}_1}$ dependence in (6.2.29a) and (6.2.29b) to get the generalized Ewald-Oseen theorem, we obtain

$$\begin{aligned} &\sum_{\mu\nu} n_o \frac{\gamma_{\mu\nu}}{\gamma_{mn}}(-1)^m \sum_p \left[\frac{4\pi(-1)^p i^{p+1}}{k(K_{iz}-k_{iz})2k_{iz}} Y_p^{\mu-m}(\pi-\theta_i,\phi_i)\right] \\ &\left[a(\mu,\nu|-m,n|p)a(\nu,n,p)T_\nu^{(M)}a_{\mu\nu}^{E(M)}\right. \\ &\left.- a(\mu,\nu|-m,n|p,p-1)b(\nu,n,p)T_\nu^{(N)}a_{\mu\nu}^{E(N)}\right] \\ &= -a_{mn}^{(M)} \end{aligned} \tag{6.2.37a}$$

$$\sum_{\mu\nu} n_o \frac{\gamma_{\mu\nu}}{\gamma_{mn}} (-1)^m \sum_p \left[\frac{4\pi(-1)^p i^{p+1}}{k(K_{iz}-k_{iz})2k_{iz}} Y_p^{\mu-m}(\pi-\theta_i, \phi_i) \right]$$
$$\Big[-a(\mu,\nu|-m,n|p,p-1) b(\nu,n,p) T_\nu^{(M)} a_{\mu\nu}^{E(M)}$$
$$+ a(\mu,\nu|-m,n|p) a(\nu,n,p) T_\nu^{(N)} a_{\mu\nu}^{E(N)} \Big]$$
$$= -\, a_{mn}^{(N)} \tag{6.2.37b}$$

The results of (6.2.37a) and (6.2.37b) can be simplified by making use of the following two relations

$$\sum_p (-i)^p a(\nu,n,p) a(\mu,\nu|-m,n|p) P_p^{\mu-m}(\cos\theta) e^{i(\mu-m)\phi}$$
$$= i^{\nu-n}(-1)^{\nu+n} \frac{(2n+1)}{n(n+1)} \overline{B}_{\mu\nu}(\theta,\phi)\cdot \overline{B}_{-mn}(\theta,\phi) \tag{6.2.38}$$

$$\sum_p (-i)^p b(\nu,n,p) a(\mu,\nu|-m,n|p,p-1) P_p^{\mu-m}(\cos\theta) e^{i(\mu-m)\phi}$$
$$= i^{\nu-n+1}(-1)^{\nu+n+1} \frac{(2n+1)}{n(n+1)} \overline{B}_{\mu\nu}(\theta,\phi)\cdot \overline{C}_{-mn}(\theta,\phi) \tag{6.2.39}$$

They can be derived from the orthogonality property of vector spherical harmonics, the completeness relation of spherical harmonics, and the following relation, which expresses a product of spherical harmonics and vector spherical harmonics in terms of vector spherical harmonics

$$Y_p^{\mu-m}(\theta,\phi)\overline{B}_{-\mu\nu}(\theta,\phi)$$
$$= \sum_l \frac{i^{l-\nu-p}(-1)^{p+\mu}}{2p+1} a(-\mu,\nu|m,l|p) a(\nu,l,p) \overline{B}_{-ml}(\theta,\phi)$$
$$+ \sum_l \frac{i^{l-\nu-p-1}(-1)^{p+\mu}}{2p+1} a(-\mu,\nu|m,l|p,p-1) b(\nu,l,p) \overline{C}_{-ml}(\theta,\phi) \tag{6.2.40}$$

A relation for $Y_p^{\mu-m}(\theta,\phi)\overline{C}_{-\mu\nu}(\theta,\phi)$ similar to (6.2.40) can be obtained by taking the cross product of (6.2.40) with $\hat{r}$. Using the results of (6.2.38) and (6.2.39), the generalized Ewald-Oseen extinction theorem (6.2.37a) assumes the following form

$$\sum_{\mu\nu} n_o \frac{\gamma_{\mu\nu}}{\gamma_{mn}} (-1)^m \frac{4\pi i}{k(K_{iz}-k_{iz})2k_{iz}}$$
$$\cdot\, i^{\nu-n}(-1)^{\nu+n} \frac{2n+1}{n(n+1)} \overline{B}_{\mu\nu}(\pi-\theta_i,\phi_i)\cdot \overline{B}_{-mn}(\pi-\theta_i,\phi_i) T_\nu^{(M)} a_{\mu\nu}^{E(M)}$$

$$+ \sum_{\mu\nu} n_o \frac{\gamma_{\mu\nu}}{\gamma_{mn}} (-1)^m \frac{4\pi i}{k(K_{iz} - k_{iz}) 2k_{iz}}$$

$$\cdot i^{\nu-n+1} (-1)^{\nu+n} \frac{2n+1}{n(n+1)} \overline{B}_{\mu\nu}(\pi - \theta_i, \phi_i)$$

$$\cdot \overline{C}_{-mn}(\pi - \theta_i, \phi_i) T_\nu^{(N)} a_{\mu\nu}^{E(N)} = -a_{mn}^{(M)} \tag{6.2.41}$$

A similar equation is obtained for (6.2.37*b*). These equations can be put in the compact form of

$$n_o \overline{\overline{U}}^{(k)} \overline{\overline{T}} \overline{a}_E = -\overline{a}^{inc} \tag{6.2.42}$$

where

$$\overline{\overline{U}}^{(k)} = \begin{bmatrix} \overline{\overline{X}}^{(k)} & \overline{\overline{Z}}^{(k)} \\ \overline{\overline{Z}}^{(k)} & \overline{\overline{X}}^{(k)} \end{bmatrix} \tag{6.2.43}$$

with $\overline{\overline{X}}^{(k)}$ and $\overline{\overline{Z}}^{(k)}$ being $L_{max} \times L_{max}$ matrices, and

$$X_{mn\mu\nu}^{(k)} = \frac{\gamma_{\mu\nu}}{\gamma_{mn}} (-1)^{m+n+\nu} 2\pi i^{\nu-n+1} \cdot \frac{2n+1}{n(n+1)} \frac{1}{(K_{iz} - k_{iz}) k_{iz} k}$$

$$\cdot \overline{B}_{\mu\nu}(\pi - \theta_i, \phi_i) \cdot \overline{B}_{-mn}(\pi - \theta_i, \phi_i) \tag{6.2.44a}$$

$$Z_{mn\mu\nu}^{(k)} = \frac{\gamma_{\mu\nu}}{\gamma_{mn}} (-1)^{m+n+\nu+1} i^{\nu-n} \frac{2n+1}{n(n+1)} \frac{2\pi}{(K_{iz} - k_{iz}) k_{iz} k}$$

$$\cdot \overline{B}_{\mu\nu}(\pi - \theta_i, \phi_i) \cdot \overline{C}_{-mn}(\pi - \theta_i, \phi_i) \tag{6.2.44b}$$

The generalized Lorentz-Lorenz law of (6.2.35) is a system of homogeneous equations, whereas the generalized Ewald-Oseen extinction theorem (6.2.42) is a system of inhomogeneous equations. In the following sections we shall show that for the case of spherical scatterers, the set of inhomogeneous equations in the Ewald-Oseen theorem (6.2.42) can be reduced to a single scalar equation that is polarization dependent. Hence, the procedure is first to find the solution of the generalized Lorentz-Lorenz law together with the dispersion relation for K, leaving an arbitrary constant to be next determined by that single scalar equation that is equivalent to (6.2.42).

After the conditional average of the exciting field has been evaluated in the manner outlined above, the coherent reflected wave can be calculated by taking an ensemble average of the scattered field. The ensemble average of the coherent reflected field is

$$\langle \overline{E}^s(\overline{r}) \rangle = n_o \sum_{mn} \int d\overline{r}_j \left[\overline{M}_{mn}(k\overline{r}\overline{r}_j) T_n^{(M)} a_{mn}^{E(M)} e^{i\overline{K}_d \cdot \overline{r}_j} \right.$$

$$\left. + \overline{N}_{mn}(k\overline{r}\overline{r}_j) T_n^{(N)} a_{mn}^{E(N)} e^{i\overline{K}_d \cdot \overline{r}_j} \right] \tag{6.2.45}$$

2.2 Vertically and Horizontally Polarized Incidence

We calculate the coherent reflected wave by solving the generalized Lorentz-Lorenz law and generalized Ewald-Oseen extinction theorem for the cases of vertically and horizontally polarized incidence. It will be shown that the generalized Lorentz-Lorenz law is independent of polarization and direction while the generalized Ewald-Oseen extinction theorem is dependent on both.

(A) Vertically Polarized Incidence

For the case of vertically polarized incidence, the incident field coefficients are given in (6.2.2) and (6.2.3). To solve the generalized Lorentz-Lorenz law, assume the following solution

$$\overline{a}_E = \begin{bmatrix} \overline{a}^{E(M)} \\ \overline{a}^{E(N)} \end{bmatrix} \tag{6.2.46}$$

where $\overline{a}^{E(M)}$ and $\overline{a}^{E(N)}$ are column matrices of dimension L_{max} and

$$\overline{a}^{E(M)}_{mn} = (-1)^m \frac{1}{\gamma_{mn}} \frac{(2n+1)}{n(n+1)} i^n Y_n^{(M)} \left[\hat{\theta}_{td} \cdot \overline{C}_{-mn}(\pi - \theta_t, \phi_i) \right] \tag{6.2.47}$$

$$\overline{a}^{E(N)}_{mn} = (-1)^m \frac{1}{\gamma_{mn}} \frac{(2n+1)}{n(n+1)} i^{n-1} Y_n^{(N)} \left[\hat{\theta}_{td} \cdot \overline{B}_{-mn}(\pi - \theta_t, \phi_i) \right] \tag{6.2.48}$$

Note that in (6.2.47) and (6.2.48), we have assumed that the coefficients $Y_n^{(M)}$ and $Y_n^{(N)}$ are independent of the m index. This is applicable for the case of spherical scatterers. Substitute (6.2.47) and (6.2.48) in the Lorentz-Lorenz law of (6.2.34a) and (6.2.34b), and noting that the T-matrix elements of spheres are diagonal and independent of index m, the summation over index μ can be carried out by using the following relations:

$$\begin{aligned} a(\nu, n, p) \sum_{\mu} & a(\mu, \nu | -m, n | p) a(-\mu, \nu | m, l | p) \\ &= -\delta_{ln} (2p+1) i^{n-\nu+p} a(1, \nu | -1, n | p) \end{aligned} \tag{6.2.49}$$

$$\sum_{\mu} a(\mu, \nu | -m, n | p) a(-\mu, \nu | m, l | p, p-1) = 0 \tag{6.2.50}$$

$$\begin{aligned} b(\nu, n, p) \sum_{\mu} & a(\mu, \nu | -m, n | p, p-1) a(-\mu, \nu | m, l | p, p-1) \\ &= -\delta_{ln} i^{n+p-\nu} a(1, \nu | -1, n | p, p-1)(2p+1) \end{aligned} \tag{6.2.51}$$

which can be derived by using the orthogonality relation of Wigner $3j$ symbols. We arrive at the following equations for the Lorentz-Lorenz law.

$$\begin{aligned} Y_n^{(M)} &= 4\pi n_o \sum_{\nu p} \frac{(2\nu+1)}{\nu(\nu+1)} \frac{n(n+1)}{(2n+1)} i^{\nu-n-p} S_p \\ &\quad \cdot \Big\{ -T_\nu^{(M)} Y_\nu^{(M)} a(1,\nu|-1,n|p) a(\nu,n,p) \\ &\quad + T_\nu^{(N)} Y_\nu^{(N)} a(1,\nu|-1,n|p,p-1) b(\nu,n,p) \Big\} \end{aligned} \tag{6.2.52}$$

$$\begin{aligned} Y_n^{(N)} &= 4\pi n_o \sum_{\nu p} \frac{(2\nu+1)}{\nu(\nu+1)} \frac{n(n+1)}{(2n+1)} i^{\nu-n-p} S_p \\ &\quad \cdot \Big\{ T_\nu^{(M)} Y_\nu^{(M)} a(1,\nu|-1,n|p,p-1) b(\nu,n,p) \\ &\quad - T_\nu^{(N)} Y_\nu^{(N)} a(1,\nu|-1,n|p) a(\nu,n,p) \Big\} \end{aligned} \tag{6.2.53}$$

Making the substitution in (6.2.41), we have

$$\begin{aligned} & \frac{2\pi i}{k k_{iz}(K_{iz} - k_{iz})} \times \left\{ \sum_{\mu\nu} n_o (-1)^\mu \frac{(2\nu+1)}{\nu(\nu+1)} \overline{B}_{\mu\nu}(\pi-\theta_i,\phi_i) \right. \\ & \cdot \overline{B}_{-mn}(\pi-\theta_i,\phi_i) Y_\nu^{(M)} T_\nu^{(M)} \left(\hat{\theta}_{td} \cdot \overline{C}_{-\mu\nu}(\pi-\theta_t,\phi_i) \right) \\ & + \sum_{\mu\nu} n_o (-1)^\mu \frac{(2\nu+1)}{\nu(\nu+1)} \overline{B}_{\mu\nu}(\pi-\theta_i,\phi_i) \\ & \left. \cdot \overline{C}_{-mn}(\pi-\theta_i,\phi_i) Y_\nu^{(N)} T_\nu^{(N)} \left(\hat{\theta}_{td} \cdot \overline{B}_{-\mu\nu}(\pi-\theta_t,\phi_i) \right) \right\} \\ & = -\hat{\theta}_{id} \cdot \overline{C}_{-mn}(\pi-\theta_i,\phi_i) \end{aligned} \tag{6.2.54}$$

The homogeneous system of equations in (6.2.52) and (6.2.53) is identical to that of the case of normal incidence. Thus the dispersion relation is directionally independent. Also note that θ_t and θ_i are related by Snell's law of $k \sin\theta_i = K \sin\theta_t$.

The generalized Ewald-Oseen extinction theorem can also be simplified for spherical scatterers. On substituting (6.2.46) through (6.2.48) into the Ewald-Oseen theorem of (6.2.42), we can make use of the addition theorem for vector spherical harmonics,

$$\begin{aligned} & \hat{\theta}\hat{\theta}' \frac{\partial^2 P_\nu}{\partial\theta\partial\theta'} + \hat{\phi}\hat{\theta}' \frac{1}{\sin\theta} \frac{\partial^2 P_\nu}{\partial\phi\partial\theta'} + \hat{\theta}\hat{\phi}' \frac{1}{\sin\theta'} \frac{\partial^2 P_\nu}{\partial\theta\partial\phi'} + \hat{\phi}\hat{\phi}' \frac{1}{\sin\theta\sin\theta'} \frac{\partial^2 P_\nu}{\partial\phi\partial\phi'} \\ & = \sum_\mu (-1)^\mu \overline{B}_{\mu\nu}(\theta,\phi) \overline{B}_{-\mu\nu}(\theta',\phi') \end{aligned} \tag{6.2.55a}$$

where the argument of the Legendre polynomial P_ν is $(\hat{r}\cdot\hat{r}') = \cos\theta\cos\theta' + \sin\theta\sin\theta'\cos(\phi-\phi')$. Other relations with $\sum_\mu(-1)^\mu \overline{B}_{\mu\nu}(\theta,\phi)\overline{C}_{-\mu\nu}(\theta',\phi')$, and so on, can be obtained easily by noting that $\overline{C}_{mn}\times\hat{r} = \overline{B}_{mn}$. For example,

$$\hat{\phi}\hat{\phi}'\frac{\partial^2 P_\nu}{\partial\theta\partial\theta'} - \hat{\theta}\hat{\phi}'\frac{1}{\sin\theta}\frac{\partial^2 P_\nu}{\partial\phi\partial\theta'} - \hat{\phi}\hat{\theta}'\frac{1}{\sin\theta'}\frac{\partial^2 P_\nu}{\partial\theta\partial\phi'} + \hat{\theta}\hat{\theta}'\frac{1}{\sin\theta\sin\theta'}\frac{\partial^2 P_\nu}{\partial\phi\partial\phi'}$$
$$= \sum_\mu(-1)^\mu\overline{C}_{\mu\nu}(\theta,\phi)\overline{C}_{-\mu\nu}(\theta',\phi') \quad (6.2.55b)$$

Then (6.2.54) gives

$$in_o\sum_\nu\frac{(2\nu+1)}{\nu(\nu+1)}\Bigg\{Y_\nu^{(M)}T_\nu^{(M)}$$
$$\cdot\left[\hat{\phi}\frac{1}{\sin\theta\sin\theta'}\frac{\partial^2 P_\nu}{\partial\phi\partial\phi'} + \hat{\theta}\frac{1}{\sin\theta'}\frac{\partial^2 P_\nu}{\partial\theta\partial\phi'}\right]_{\substack{\theta=\pi-\theta_i,\,\theta'=\pi-\theta_t\\ \phi=\phi'=\phi_i}}\overline{B}_{-mn}(\pi-\theta_i,\phi_i)$$
$$+\left[\hat{\phi}\frac{1}{\sin\theta}\frac{\partial^2 P_\nu}{\partial\phi\partial\theta'} + \hat{\theta}\frac{\partial^2 P_\nu}{\partial\theta\partial\theta'}\right]_{\substack{\theta=\pi-\theta_i,\,\theta'=\pi-\theta_t\\ \phi=\phi'=\phi_i}} Y_\nu^{(N)}T_\nu^{(N)}\cdot\overline{C}_{-mn}(\pi-\theta_i,\phi_i)\Bigg\}$$
$$= -\hat{\theta}_{id}\cdot\overline{C}_{-mn}(\pi-\theta_i,\phi_i)\frac{(K_{iz}-k_{iz})k_{iz}k}{2\pi} \quad (6.2.56)$$

Since

$$\left[\hat{\phi}\frac{1}{\sin\theta\sin\theta'}\frac{\partial^2 P_\nu}{\partial\phi\partial\phi'} + \hat{\theta}\frac{1}{\sin\theta'}\frac{\partial^2 P_\nu}{\partial\theta\partial\phi'}\right]_{\substack{\theta=\pi-\theta_i,\,\theta'=\pi-\theta_t\\ \phi=\phi'=\phi_i}} = -\frac{P_\nu^1(\cos(\theta_i-\theta_t))}{\sin(\theta_i-\theta_t)}\hat{\phi}_i \quad (6.2.57a)$$

$$\left[\hat{\phi}\frac{1}{\sin\theta}\frac{\partial^2 P_\nu}{\partial\phi\partial\theta'} + \hat{\theta}\frac{\partial^2 P_\nu}{\partial\theta\partial\theta'}\right]_{\substack{\theta=\pi-\theta_i,\,\theta'=\pi-\theta_t\\ \phi=\phi'=\phi_i}}$$
$$= \hat{\theta}_{id}\left\{\cot(\theta_i-\theta_t)P_\nu^1(\cos(\theta_i-\theta_t)) + \nu(\nu+1)P_\nu(\cos(\theta_i-\theta_t))\right\} \quad (6.2.57b)$$

Substituting (6.2.57a) and (6.2.57b) into (6.2.56), Ewald-Oseen theorem reduces to a single inhomogeneous scalar equation

$$-\frac{(K_z-k_{iz})k_{iz}k}{2\pi} = -in_o\sum_\nu T_\nu^{(M)}\frac{2\nu+1}{\nu(\nu+1)}Y_\nu^{(M)}\frac{P_\nu^1(\cos(\theta_i-\theta_t))}{|\sin(\theta_i-\theta_t)|}$$
$$+ in_o\sum_\nu T_\nu^{(N)}\frac{2\nu+1}{\nu(\nu+1)}Y_\nu^{(N)}\left\{\frac{\cos(\theta_i-\theta_t)}{|\sin(\theta_i-\theta_t)|}\right.$$
$$\left.\cdot P_\nu^1\Big(\cos(\theta_i-\theta_t)\Big) + \nu(\nu+1)P_\nu\Big(\cos(\theta_i-\theta_t)\Big)\right\} \quad (6.2.58)$$

By using (6.2.52) through (6.2.58), the conditional average of the exciting field can be solved. Equations (6.2.52) and (6.2.53) are homogeneous linear equations for the coefficients $Y_n^{(M)}$ and $Y_n^{(N)}$. For a nontrivial solution to exist, the determinant must vanish giving the dispersion equation from which K is determined. The solution for K is the same as that for normal incidence as in Section 1. There is also an eigenvector associated with the solution of K with an arbitrary constant multiplying the eigenvector. That arbitrary constant is then determined by using the single inhomogeneous equation of (6.2.58). The values of $Y_n^{(M)}$ and $Y_n^{(N)}$ are then uniquely determined. The conditional-averaged exciting field is then given by (6.2.47) and (6.2.48).

To calculate the coherent reflected field, substitute (6.2.47) and (6.2.48) in (6.2.45). The $d\overline{r}_j$ integration can be carried out readily. It follows that

$$\int_{z_j<0} d\overline{r}_j \overline{M}_{mn}(k\overline{rr_j})e^{i\overline{K}_d\cdot\overline{r}_j} = \frac{2\pi\gamma_{mn}i^{-n+1}}{kk_{iz}(K_z+k_{iz})}e^{i\overline{k}_i\cdot\overline{r}}\overline{C}_{mn}(\theta_i,\phi_i) \tag{6.2.59a}$$

$$\int_{z_j<0} d\overline{r}_j \overline{N}_{mn}(k\overline{rr_j})e^{i\overline{K}_d\cdot\overline{r}_j} = -\frac{2\pi\gamma_{mn}i^{-n}}{kk_{iz}(K_z+k_{iz})}e^{i\overline{k}_i\cdot\overline{r}}\overline{B}_{mn}(\theta_i,\phi_i) \tag{6.2.59b}$$

Note that because of statistical translational invariance in the horizontal plane, the integration of $d\overline{r}_j$ over the lower half-space in (6.2.59a) and (6.2.59b) gives a specularly reflected wave for the coherent field. Substituting (6.2.59a) and (6.2.59b) into (6.2.45), we have the coherent reflected field as

$$\langle\overline{E}_s(\overline{r})\rangle = \frac{2\pi n_o i}{kk_{iz}(K_z+k_{iz})}e^{i\overline{k}_i\cdot\overline{r}}\sum_{mn}(-1)^m\frac{(2n+1)}{n(n+1)}\Big\{T_n^{(M)}Y_n^{(M)}\overline{C}_{mn}(\theta_i,\phi_i)\hat{\theta}_{td}$$
$$\cdot\,\overline{C}_{-mn}(\pi-\theta_t,\phi_i) + T_n^{(N)}Y_n^{(N)}\overline{B}_{mn}(\theta_i,\phi_i)\hat{\theta}_{td}\cdot\overline{B}_{-mn}(\pi-\theta_t,\phi_i)\Big\} \tag{6.2.60}$$

Next, we make use of the vector addition theorem of (6.2.55a) and (6.2.55b). For this case $\theta=\theta_i$, $\phi=\phi_i$, $\theta'=\pi-\theta_t$, and $\phi'=\phi_i$, so that $(\hat{r}\cdot\hat{r}') = -\cos\theta_i\cos\theta_t+\sin\theta_i\sin\theta_t = -\cos(\theta_i+\theta_t)$. The final result for the coherent reflected field is

$$\langle\overline{E}^s(\overline{r})\rangle = \hat{\theta}_i\,2\pi n_o\,e^{i\overline{k}_i\cdot\overline{r}}\frac{i}{kk_{iz}}\frac{1}{(k_{iz}+K_z)}\sum_n(-1)^n\frac{(2n+1)}{n(n+1)}\Big\{Y_n^{(M)}T_n^{(M)}$$
$$\cdot\frac{P_n^1(\cos(\theta_i+\theta_t))}{\sin(\theta_i+\theta_t)} + Y_n^{(N)}T_n^{(N)}\Big[\cot(\theta_i+\theta_t)P_n^1(\cos(\theta_i+\theta_t))$$
$$+\,n(n+1)P_n(\cos(\theta_i+\theta_t))\Big]\Big\} \tag{6.2.61}$$

where $\overline{k}_i = k(\sin\theta_i\cos\phi_i\hat{x}+\sin\theta_i\sin\phi_i\hat{y}+\cos\theta_i\hat{z})$ denotes the upward propagating direction that is specularly related to the downward incident wavevector $\overline{k}_{id}$. Thus, the coherent reflected wave is in the specular direction and contains no depolarization. We let R_V to be the reflection coefficient so that from (6.2.61),

$$R_V = \frac{2\pi n_o i}{k k_{iz}(k_{iz}+K_z)} \sum_n (-1)^n \frac{(2n+1)}{n(n+1)} \left\{ Y_n^{(M)} T_n^{(M)} \frac{P_n^1(\cos(\theta_i+\theta_t))}{\sin(\theta_i+\theta_t)} \right.$$
$$\left. + Y_n^{(N)} T_n^{(N)} \left[\cot(\theta_i+\theta_t) P_n^1(\cos(\theta_i+\theta_t)) + n(n+1) P_n(\cos(\theta_i+\theta_t)) \right] \right\} \tag{6.2.62}$$

and

$$\langle \overline{E}^s(\overline{r}) \rangle = \hat{\theta}_i R_V e^{i\overline{k}_i \cdot \overline{r}} \tag{6.2.63}$$

The incident wave has $\overline{E}_{inc}(\overline{r}) = \hat{\theta}_{id} e^{i\overline{k}_{id}\cdot\overline{r}}$ and the coherent transmitted wave is $\langle \overline{E}(\overline{r}) \rangle = \hat{\theta}_{td} T_V e^{i\overline{K}_d \cdot \overline{r}}$. Hence by matching the continuity of the tangential electric field for the coherent field, we have

$$T_V = (1 - R_V) \frac{\cos\theta_i}{\cos\theta_t} \tag{6.2.64}$$

Horizontally Polarized Incidence

A similar derivation can be made for the case of a horizontally polarized incident wave. To solve the generalized Lorentz-Lorenz law, we let $\overline{a}^E$ assume the expressions as in (6.2.47) and (6.2.48) with $\hat{\theta}_{td}$ replaced by $\hat{\phi}_i$. Substituting into the generalized Lorentz-Lorenz law, we obtain equations that are identical to (6.2.52) and (6.2.53). This means that the generalized Lorentz-Lorenz law and dispersion relation for the horizontally polarized wave is the same as that for the case of the vertically polarized incident wave. Thus, the dispersion relation is polarization-independent besides being angular-independent for the case of spherical scatterers. Substituting into the Ewald-Oseen extinction theorem, and performing algebraic simplification as in the vertically polarized case, the Ewald-Oseen theorem reduces to the following single scalar equation

$$-\frac{(K_z - k_{iz}) k_{iz} k}{2\pi} = i n_o \sum_\nu \frac{2\nu+1}{\nu(\nu+1)} Y_\nu^{(M)} T_\nu^{(M)} \left\{ \frac{\cos(\theta_i-\theta_t)}{|\sin(\theta_i-\theta_t)|} \right.$$
$$\left. \cdot P_\nu^1(\cos(\theta_i-\theta_t)) + \nu(\nu+1) P_\nu(\cos(\theta_i-\theta_t)) \right\}$$
$$- i n_o \sum_\nu \frac{2\nu+1}{\nu(\nu+1)} Y_\nu^{(N)} T_\nu^{(N)} \frac{P_\nu^1(\cos(\theta_i-\theta_t))}{|\sin(\theta_i-\theta_t)|} \tag{6.2.65}$$

Since (6.2.65) is different from (6.2.58), the generalized Ewald-Oseen extinction theorem is dependent on both polarization and directions. The procedure to calculate the coherent exciting field is to first solve for the propagation constant K through (6.2.52) and (6.2.53) with its associated eigenvector, leaving the arbitrary constant to be calculated by the single inhomogeneous equation of (6.2.65). The quantities $Y_\nu^{(M)}$ and $Y_\nu^{(N)}$ will be uniquely determined by these two steps. The coherent reflected field is calculated similarly. We have the coherent reflected field as

$$\langle \overline{E}^s(\overline{r})\rangle = \hat{\phi}_i R_H e^{i\overline{k}_i \cdot \overline{r}} \tag{6.2.66}$$

where

$$\begin{aligned} R_H = & \, i\frac{n_o 2\pi}{(k_{iz}+K_z)kk_{iz}} \sum_n \frac{(2n+1)}{n(n+1)}(-1)^n \Big\{ T_n^{(M)} Y_n^{(M)} \Big[\cot(\theta_i+\theta_t) \\ & \cdot P_n^1(\cos(\theta_i+\theta_t)) + n(n+1)P_n(\cos(\theta_i+\theta_t))\Big] \\ & + T_n^{(N)} Y_n^{(N)} \frac{P_n^1(\cos(\theta_i+\theta_t))}{\sin(\theta_i+\theta_t)} \Big\} \end{aligned} \tag{6.2.67}$$

The coherent field is in the specular direction and is horizontally polarized. There is no depolarization in the coherent reflected field. However, the Ewald-Oseen extinction theorem is different between TE and TM. This means that the coefficients Y_n's will be different and the coherent reflected field is different. The coherent transmitted field for the horizontally polarized incidence is

$$\langle E(\overline{r})\rangle = \hat{\phi}_i T_H e^{i\overline{K}_d \cdot \overline{r}} \tag{6.2.68}$$

where

$$T_H = 1 + R_H. \tag{6.2.69}$$

3 Cases with Size Distributions

In this section we study the scattering of a plane electromagnetic wave obliquely incident on a layer of dense medium consisting of dielectric spherical particles with size distributions using the quasi-crystalline approximation.

We use the Percus-Yevick cross-pair distribution functions of multiple sizes. The incoherent scattered wave is calculated with the distorted Born approximation with the result expressed in terms of a product of the T-matrices of particles of different sizes and permittivities and the Fourier transform of the cross-pair distribution functions. Numerical examples are chosen to illustrate microwave and millimeter wave scattering from snow cover in the frequency range of 5 GHz to 95 GHz. The Rayleigh size distribution is used to illustrate the scattering. Salient features of the numerical results for scattering from snow with Rayleigh size distribution are as follows: (1) Correlated dense medium scattering is less than independent scattering at low frequency. (2) Scattering from dense medium of a Rayleigh size distribution with an average radius can be much larger than the case of monodisperse particles of the sizes equal to that average radius. Particles with larger size can contribute to scattering even though they are fewer in number. (3) The scattering attenuation rate increases rapidly with frequency at a low-frequency regime and begins to level off at a high-frequency regime. This is to be contrasted with Rayleigh scattering where scattering attenuation just increases rapidly with frequency. (4) The coherent wave scattering attenuation rate can be large at frequencies above 15 GHz. Comparisons are made with extinction measurements of dry snow at 18 GHz, 35 GHz, 60 GHz, and 90 GHz.

3.1 Coherent Field

Consider a plane electromagnetic plane wave impinging in direction $(\pi - \theta_i, \phi_i)$ upon a half-space of spherical discrete scatterers of multiple sizes and multiple permittivities (Fig. 6.3.1). Let s_j be the species index $j = 1, 2, \ldots, L$. Each species is described by the permittivity ϵ_{sj}, radius a_{sj} and number density n_{sj}. The number density n_{sj} is related to the volume fraction of species f_{sj}, $f_{sj} = 4\pi n_{sj} a_{sj}^3/3$. The scatterers are embedded in the lower half-space (region 1) with background permittivity equal to ϵ, which is the same permittivity as that of the region 0 (Fig. 6.3.1).

The electric field of the incident wave is given by

$$\overline{E}_i(\overline{r}) = (E_{vi}\hat{\theta}_{id} + E_{hi}\hat{\phi}_i)\, e^{i\overline{k}_{id}\cdot\overline{r}} \tag{6.3.1}$$

where subscripts v and h denote vertical and horizontal polarizations, respectively, $\overline{k}_{id}$ denotes the downward propagating incident wave vector, and $\hat{\theta}_{id}$ and $\hat{\phi}_i$ are the associated polarization vectors

$$\overline{k}_{id} = k(\sin\theta_i \cos\phi_i\, \hat{x} + \sin\theta_i \sin\phi_i\, \hat{y} - \cos\theta_i \hat{z}) = k_{ix}\, \hat{x} + k_{iy}\, \hat{y} - k_{iz}\, \hat{z} \tag{6.3.2}$$

$$\hat{\theta}_{id} = -\hat{x}\cos\theta_i \cos\phi_i - \hat{y}\cos\theta_i \sin\phi_i - \hat{z}\sin\theta_i \tag{6.3.3}$$

$$\hat{\phi}_i = -\hat{x}\, \sin\phi_i + \hat{y}\, \cos\phi_i \tag{6.3.4}$$

Under the quasi-crystalline approximation for multiple species, the coherent

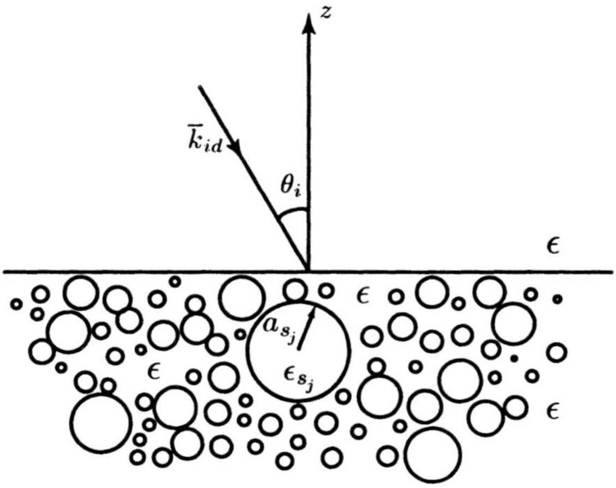

Figure 6.3.1 Geometry of the problem: an incident electromagnetic plane wave impinging upon a dense medium consisting of a spherical particles with a size distribution.

field obeys the equation

$$\overline{w}^{(s_l)}(\overline{r}_l) = \sum_{s_j=1}^{L} n_{s_j} \int_{V_j} d\overline{r}_j g_{s_j s_l}(\overline{r}_j - \overline{r}_l)\,\overline{\overline{\sigma}}(k\overline{r_l r_j})\,\overline{\overline{T}}^{(s_j)}\,\overline{w}^{(s_j)}(\overline{r}_j) + e^{i\overline{k}_{id}\cdot\overline{r}_l}\overline{a}_{inc} \tag{6.3.5}$$

where $\overline{w}^{(s_l)}(\overline{r}_l)$ is the conditional average of the exciting field of particles at $\overline{r}_l$ of species s_l conditioning on the particle at $\overline{r}_l$, $\overline{\overline{T}}^{(s_j)}$ is the matrix representation of the T matrix of species s_j, and $\overline{\overline{\sigma}}(k\overline{r_l r_j})$ is a transformation matrix transforming from vector spherical waves with center at $\overline{r}_j$ to vector spherical waves with center at $\overline{r}_l$. The function $g_{s_j s_l}(\overline{r}_j - \overline{r}_l)$ is the cross-pair distribution function of two species s_j and s_l. These can be computed readily from the Percus-Yevick approximation as discussed in Chapter 8 of Volume II. In (6.3.5) the volume of integration V_j for $d\overline{r}_j$ is the lower half-space excluding a "hole" of radius $(a_{s_j} + a_{s_l})$ around vector $\overline{r}_l$. This is because

$$g_{s_j s_l}(\overline{r}_j - \overline{r}_l) = 0 \qquad \text{for} \quad |\overline{r}_j - \overline{r}_l| < a_{s_j} + a_{s_l} \tag{6.3.6}$$

In (6.3.5) $\overline{a}_{inc}$ is the column matrix representing the coefficients of the incident wave when expanded into vector spherical waves. To solve (6.3.5), we assume the solution as in the Section 2.

$$\overline{w}^{(s_l)}(\overline{r}_l) = \overline{a}_E^{(s_l)} e^{i\overline{K}_d\cdot\overline{r}_l} \tag{6.3.7}$$

where $\overline{a}_E^{(s_l)}$ is a column matrix representing the coefficients of the exciting field and $\overline{K}_d$ is the downward-going effective wave vector

$$\overline{K}_d = K_x\,\hat{x} + K_y\,\hat{y} - K_z\,\hat{z} \tag{6.3.8}$$

and the effective wavenumber is

$$K = \sqrt{K_x^2 + K_y^2 + K_z^2} \tag{6.3.9}$$

Substitution of (6.3.7) into the integrand of the integral (6.3.5), integrations can be performed as in Section 2. The integral yields two terms that correspond to two waves with dependencies

$$e^{i\overline{K}_d\cdot\overline{r}_l} \tag{6.3.10}$$

and

$$e^{iK_x x_l + iK_y y_l - i(k^2 - K_x^2 - K_y^2)^{1/2} z_l} \tag{6.3.11}$$

The first wave given by (6.3.10) propagates with $\overline{K}_d$ and is to be balanced with the term $\overline{w}^{s_l}(\overline{r}_l)$ on the left-hand side of equation since they both have the same propagation constant. Balancing these two terms give rise to the generalized Lorentz-Lorenz law. The second wave of (6.3.11) with incident wave wavenumber k in the square root will cancel the second term on the right-hand side of (6.3.5) which is the incident wave term. This cancellation or extinction is the generalized Ewald-Oseen extinction theorem. This gives rise to the phase-matching condition

$$K_x = k_{ix} \tag{6.3.12}$$

$$K_y = k_{iy} \tag{6.3.13}$$

so that from (6.3.8)–(6.3.10), we obtain Snell's law for the effective propagation constant,

$$K\sin\theta_t = k\sin\theta_i \tag{6.3.14}$$

$$K_z = (K^2 - k^2\sin^2\theta_i)^{1/2} \tag{6.3.15}$$

Generalized Lorentz-Lorenz Law

The $\overline{a}_E^{(s_l)}$ column matrix for the exciting field can be expressed in terms of $Y_n^{(s_j)(M)}$ and $Y_n^{(s_j)(N)}$ coefficients. The $Y_n^{(s_j)(M)}$ and $Y_n^{(s_j)(N)}$ coefficients satisfy a homogeneous set of equations. The homogeneous set of equations is, analogous to the single size case,

$$Y_\nu^{(s_j)(M)} = -2\pi \sum_{s_l=1}^{L} n_{s_l} \sum_{n,p} (2n+1)\, S_p(k, K|R_{s_j s_l})$$
$$\Big\{ a\,(1, n| -1, \nu|p)\, A(n,\nu,p)\, T_n^{(s_l)(M)}\, Y_n^{(s_l)(M)}$$
$$+ a\,(1, n| -1, \nu|p, p-1)\, B(n,\nu,p)\, T_n^{(s_l)(N)}\, Y_n^{(s_l)(N)} \Big\} \quad (6.3.16)$$

$$Y_\nu^{(s_j)(N)} = -2\pi \sum_{s_l=1}^{L} n_{s_l} \sum_{n,p} (2n+1)\, S_p(k, K|R_{s_j s_l})$$
$$\times \Big\{ a\,(1, n| -1, \nu|p, p-1)\, B(n,\nu,p)\, T_n^{(s_l)(M)}\, Y_n^{(s_l)(M)}$$
$$+ a\,(1, n| -1, \nu|p) A(n,\nu,p)\, T_n^{(s_l)(N)}\, Y_n^{(s_l)(N)} \Big\} \quad (6.3.17)$$

In (6.3.16) and (6.3.17) we have

$$A\,(n,\nu,p) = \frac{\nu\,(\nu+1) + n\,(n+1) - p\,(p+1)}{n\,(n+1)}$$
$$= \frac{2\nu(\nu+1)}{(2\nu+1)} \frac{i^{n-\nu-p}}{n(n+1)} a(n,\nu,p) \quad (6.3.18)$$

$$B(n,\nu,p) = \frac{\sqrt{(n+\nu+p+1)(\nu+p-n)(n+p-\nu)(n+\nu-p+1)}}{n\,(n+1)}$$
$$= -\frac{b(n,\nu,p)}{n(n+1)} \frac{2\nu(\nu+1)}{(2\nu+1)}\, i^{n-\nu-p} \quad (6.3.19)$$

$$S_p(k, K|R_{s_j s_l}) = -\frac{R_{s_j s_l}^2}{K^2 - k^2}$$
$$\cdot \left[k h_p'(kR_{s_j s_l})\, j_p(KR_{s_j s_l}) - K h_p(kR_{s_j s_l})\, j_p'(KR_{s_j s_l}) \right]$$
$$+ \int_{R_{s_j s_l}}^{\infty} dr\, r^2 \left[g_{s_j s_l}(r) - 1 \right] h_p(kr) j_p(Kr) \quad (6.3.20)$$

$$R_{s_j s_l} = a_{s_j} + a_{s_l} \quad (6.3.21)$$

and

$$T_n^{(s_l)(M)} = -\frac{[\rho_{s_l} j_n(\rho_{s_l})]' j_n(\zeta_{s_l}) - [\zeta_{s_l} j_n(\zeta_{s_l})]' j_n(\rho_{s_l})}{[\rho_{s_l} h_n(\rho_{s_l})]' j_n(\zeta_{s_l}) - [\zeta_{s_l} j_n(\zeta_{s_l})]' h_n(\rho_{s_l})} \quad (6.3.22)$$

$$T_n^{(s_l)(N)} = -\frac{[\rho_{s_l} j_n(\rho_{s_l})]' \zeta_{s_l}^2 j_n(\zeta_{s_l}) - [\zeta_{s_l} j_n(\zeta_{s_l})]' \rho_{s_l}^2 j_n(\rho_{s_l})}{[\rho_{s_l} h_n(\rho_{s_l})]' \zeta_{s_l}^2 j_n(\zeta_{s_l}) - [\zeta_{s_l} j_n(\zeta_{s_l})]' \rho_{s_l}^2 h_n(\rho_{s_l})} \quad (6.3.23)$$

with $\rho_{s_l} = ka_{s_l}$, $\zeta_{s_l} = k_{s_l}a_{s_l}$, and $k_{s_l} = \omega\sqrt{\mu\epsilon_{s_l}}$. $T_n^{(s_l)((M)}$ and $T_n^{(s_l)((N)}$ are the Mie scattering T-matrix coefficients for $\overline{M}$ and $\overline{N}$ vector spherical waves for species s_l.

For each species s_l, we terminate at a multipole N_{s_l} depending on the radius a_{s_l} of that species. Thus (6.3.16) and (6.3.17) form N_e homogeneous equations for the N_e unknowns $Y_n^{(s_l)(M)}$ and $Y_n^{(s_l)(N)}$, $n = 1, 2, \ldots, N_{s_l}$ and $s_l = 1, 2, \ldots, L$. Here $N_e = \sum_{l=1}^{L} 2N_{s_l}$.

The generalized Lorentz-Lorenz law determines the effective propagation K. Setting the determinant of (6.3.16)–(6.3.17) equal to zero gives the nonlinear equation for the effective propagation constant K. After K is calculated, the complex transmitted angle θ_t and K_z can be determined from (6.3.12)–(6.3.15).

The N_e coefficients $Y_n^{(s_l)(M)}$ and $Y_n^{(s_l)(N)}$ are then reduced to only one arbitrary constant that is to be determined by the incident wave. That one equation to determine the arbitrary constant is provided by the generalized Ewald-Oseen extinction theorem, which gives a single inhomogeneous equation.

Generalized Ewald-Oseen Extinction Theorem

The generalized Ewald-Oseen extinction theorem is obtained by balancing the incident wave term of the second term of (6.3.5) and the term of the same phase dependence that is a result of the integral in (6.3.5). The result matrix equation can be reduced to one single equation for the set of coefficients $Y_n^{(s_l)(M)}$ and $Y_n^{(s_l)(N)}$.

Vertically Polarized Incidence

In this case $E_{vi} \neq 0$ and $E_{hi} = 0$. The coherent transmitted wave is also vertically polarized. The generalized Ewald-Oseen extinction theorem gives rise to the following inhomogeneous equation

$$\sum_{s_l=1}^{L}\sum_{n} n_{s_l} i \frac{(2n+1)}{n(n+1)} \Bigg[-T_n^{(s_l)(M)} Y_n^{(s_l)(M)(V)} \frac{P_n^1(\cos(\theta_i - \theta_t))}{|\sin(\theta_i - \theta_t)|}$$
$$+ Y_n^{(s_l)(N)(V)} T_n^{(s_l)(N)}$$
$$\cdot \left\{ \cos(\theta_i - \theta_t) \frac{P_n^1(\cos(\theta_i - \theta_t))}{|\sin(\theta_i - \theta_t)|} + n(n+1) P_n(\cos(\theta_i - \theta_t)) \right\} \Bigg]$$
$$= -\frac{(K_z - k_{iz})\, k_{iz} k}{2\pi} E_{vi} \tag{6.3.24}$$

where the additional superscript (V) denotes the vertically polarized case. Note that the set of coefficients $Y_n^{(s_l)(M)(V)}$ and $Y_n^{(s_l)(N)(V)}$, $n = 1, \ldots, N_{s_l}$, $s_l = 1, \ldots, L$, have been determined to within one arbitrary constant by the generalized Lorentz-Lorenz law of (6.3.16) and (6.3.17). Thus (6.3.24) provides the last equation that determines all those coefficients uniquely. The coherent reflected field is

$$\langle \overline{E}_s(\overline{r}) \rangle = \hat{\theta}_i R_V e^{i\overline{k}_i \cdot \overline{r}} \tag{6.3.25a}$$

where

$$\begin{aligned} R_V = & \sum_{s_j=1}^{L} \frac{2\pi n_{s_j} i}{k k_{iz}(K_z + k_{iz})} \sum_n (-1)^n \frac{(2n+1)}{n(n+1)} \\ & \cdot \left[T_n^{(M)(s_j)} Y_n^{(s_j)(M)(V)} \frac{P_n^1(\cos(\theta_i + \theta_t))}{\sin(\theta_i + \theta_t)} \right. \\ & + T_n^{(N)(s_j)} Y_n^{(s_j)(N)(V)} \Big(n(n+1) P_n(\cos(\theta_i + \theta_t)) \\ & \left. + \cot(\theta_i + \theta_t) P_n^1(\cos(\theta_i + \theta_t)) \Big) \right] \end{aligned} \tag{6.3.25b}$$

The Snell's law is $K \sin\theta_t = k \sin\theta_i$. The coherent transmitted field is the same as in Section 2 of (6.2.63) and (6.2.64).

Horizontal Polarized Incidence

For this case $E_{vi} = 0$ and $E_{hi} \neq 0$. The coherent transmitted wave is horizontally polarized. The inhomogeneous equation for the generalized Ewald-Oseen extinction theorem is

$$\begin{aligned} & \sum_{s_l=1}^{L} \sum_n n_{s_l} i \frac{(2n+1)}{n(n+1)} \left[T_n^{(s_l)(M)} Y_n^{(s_l)(M)(H)} \right. \\ & \left\{ \cos(\theta_i - \theta_t) \frac{P_n^1(\cos(\theta_i - \theta_t))}{|\sin(\theta_i - \theta_t)|} + n(n+1) P_n(\cos(\theta_i - \theta_t)) \right\} \\ & \left. + T_n^{(s_l)(N)} Y_n^{(s_l)(N)(H)} \left(- \frac{P_n^1(\cos(\theta_i - \theta_t))}{|\sin(\theta_i - \theta_t)|} \right) \right] \\ & = -\frac{(K_z - k_{iz}) k_{iz} k}{2\pi} E_{hi} \end{aligned} \tag{6.3.26}$$

where the additional superscript (H) denotes horizontal polarization. Thus the homogeneous system of equations of (6.3.16) and (6.3.17) combined with the inhomogeneous equation (6.3.26) determines the coefficients $Y_n^{(s_l)(M)(H)}$

and $Y_n^{(s_l)(N)(H)}$ uniquely for the use of horizontally polarized incidence. The coherent reflected field is

$$\langle \overline{E}_s(\overline{r}) \rangle = \hat{\phi}_i R_H e^{i\overline{k}_i \cdot \overline{r}} \tag{6.3.27a}$$

where

$$\begin{aligned} R_H = & \sum_{s_j=1}^{L} \frac{2\pi n_{s_j} i}{k k_{iz}(K_z + k_{iz})} \sum_n (-1)^n \frac{(2n+1)}{n(n+1)} \left\{ T_n^{(s_j)(M)} Y_n^{(s_j)(M)(H)} \right. \\ & \cdot \left[n(n+1) P_n(\cos(\theta_i + \theta_t)) + \cot(\theta_i + \theta_t) P_n^1(\cos(\theta_i + \theta_t)) \right] \\ & \left. + T_n^{(s_j)(N)} Y_n^{(s_j)(N)(H)} \frac{P_n^1(\cos(\theta_i + \theta_t))}{\sin(\theta_i + \theta_t)} \right\} \end{aligned} \tag{6.3.27b}$$

The coherent transmitted field is the same expression as in Section 2 of (6.2.68) and (6.2.69).

3.2 Incoherent Field Using Distorted Born Approximation

The coherent scattered field is in the specular direction. To calculate the backscattering coefficients, we need to study the incoherent intensity. To the first-order approximation, the incoherent intensity can be calculated by the distorted Born approximation, which takes into account single scattering of the coherent field. The incoherent scattered intensity is

$$\langle \overline{\mathcal{E}}^s(\overline{r}) \cdot \overline{\mathcal{E}}^{s*}(\overline{r}) \rangle = \langle \overline{E}^s(\overline{r}) \cdot \overline{E}^{s*}(\overline{r}) \rangle - \langle \overline{E}^s(\overline{r}) \rangle \cdot \langle \overline{E}^{s*}(\overline{r}) \rangle \tag{6.3.28}$$

The scattered field is

$$\begin{aligned} \overline{E}^s(\overline{r}) = & \sum_{s_j=1}^{L} \sum_{j=1}^{N_{s_j}} \sum_{mn} \left[T_n^{(s_j)(M)} w_{mn}^{(M)(j)} \overline{M}_{mn}(k\overline{r r_j}) \right. \\ & \left. + T_n^{(s_j)(N)} w_{mn}^{(N)(j)} \overline{N}_{mn}(k\overline{r r_j}) \right] \\ = & \sum_{s_j=1}^{L} \sum_{j=1}^{N_{s_j}} \sum_{m} T_m^{(s_j)} w_m^{(j)} \overline{\psi}_m(k\overline{r r_j}) \end{aligned} \tag{6.3.29}$$

where $\overline{\psi}_m$ can mean $\overline{M}_{mn}$ or $\overline{N}_{mn}$.

Using the quasi-crystalline approximation,

$$\langle \overline{E}^s(\overline{r}) \rangle = \sum_{s_j=1}^{L} \sum_{m} n_{s_j} \int d\overline{r}_j T_m^{(s_j)} E_j \left(w_m^{(j)} \right) \overline{\psi}_m(k\overline{r r_j}) \tag{6.3.30}$$

Thus the coherent intensity is

$$\langle \overline{E}^s(\overline{r})\rangle \cdot \langle \overline{E}^{s*}(\overline{r})\rangle = \sum_{s_j=1}^{L}\sum_{s_l=1}^{L}\sum_{m}\sum_{m'} n_{s_j} n_{s_l} \int d\overline{r}_j \int d\overline{r}_l T_m^{(s_j)} \overline{\psi}_m(k\overline{r}\overline{r}_j)$$
$$\cdot T_{m'}^{(s_l)*} \overline{\psi}_{m'}(k\overline{r}\overline{r}_l) E_j\left(w_m^{(j)}\right) E_l\left(w_m^{(l)}\right) \tag{6.3.31}$$

When we compute the total intensity, we can decompose the double summation over particles $\sum_j\sum_l = \sum_j + \sum_j\sum_{l\neq j}$ so that

$$\langle \overline{E}^s(\overline{r}) \cdot \overline{E}^{s*}(\overline{r})\rangle =$$
$$\sum_{s_j=1}^{L}\sum_{mm'} n_{s_j} \int d\overline{r}_j T_m^{(s_j)} \overline{\psi}_m(k\overline{r}\overline{r}_j) \cdot T_{m'}^{(s_j)*} \overline{\psi}^*_{m'}(k\overline{r}\overline{r}_j) E_j\left(w_m^{(j)} w_{m'}^{(j)*}\right)$$
$$+ \sum_{s_j=1}^{L}\sum_{s_l=1}^{L}\sum_{mm'} n_{s_j} n_{s_l} \int d\overline{r}_j \int d\overline{r}_l \; g_{s_j s_l}(\overline{r}_j, \overline{r}_l)$$
$$\cdot T_m^{(s_j)} \overline{\psi}_m(k\overline{r}\overline{r}_j) \cdot T_{m'}^{(s_l)*} \overline{\psi}^*_{m'}(k\overline{r}\overline{r}_l) E_{jl}\left(w_m^{(j)} w_{m'}^{(l)*}\right) \tag{6.3.32}$$

Using the quasi-crystalline approximation and the distorted Born approximation, we approximate

$$E_j\left(w_m^{(j)} w_{m'}^{(j)*}\right) \simeq E_j\left(w_m^{(j)}\right) E_j\left(w_{m'}^{(j)*}\right) \tag{6.3.33}$$

$$E_{jl}\left(w_m^{(j)} w_{m'}^{(l)*}\right) \simeq E_j\left(w_m^{(j)}\right) E_l\left(w_{m'}^{(l)*}\right) \tag{6.3.34}$$

Also the incoherent field is

$$\overline{\mathcal{E}}_s = \overline{E}_s - \langle \overline{E}_s \rangle \tag{6.3.35}$$

To get incoherent intensity, we subtract the coherent intensity from the intensity. This is done by subtracting (6.3.31) from (6.3.32).

$$\langle \overline{\mathcal{E}}_s(\overline{r}) \cdot \overline{\mathcal{E}}_s^*(\overline{r})\rangle = \sum_{s_j=1}^{L}\sum_{n,m}\sum_{n',m'} n_{s_j} \int d\overline{r}_j$$
$$\cdot \left[T_n^{(s_j)(M)} E_j\left(w_{mn}^{(j)(M)}\right) \overline{M}_{mn}(k\overline{r}\overline{r}_j) + T_n^{(s_j)(N)} E_j\left(w_{mn}^{(j)(N)}\right) \overline{N}_{mn}(k\overline{r}\overline{r}_j)\right]$$
$$\cdot \left[T_{n'}^{(s_j)(M)*} E_j\left(w_{m'n'}^{(j)(M)*}\right) \overline{M}^*_{m'n'}(k\overline{r}\overline{r}_j) + T_{n'}^{(s_j)(N)*} E_j\left(w_{m'n'}^{(j)(N)*}\right) \overline{N}^*_{m'n'}(k\overline{r}\overline{r}_j)\right]$$
$$+ \sum_{s_j=1}^{L}\sum_{s_l=1}^{L} n_{s_j} n_{s_l} \int d\overline{r}_j \int d\overline{r}_l \left(g_{s_j s_l}(\overline{r}_j, \overline{r}_l) - 1\right)$$

$$\cdot \left[T_n^{(s_j)(M)} E_j\left(w_{mn}^{(j)(M)}\right) \overline{M}_{mn}(k\overline{r r_j}) + T_n^{(s_j)(N)} E_j\left(w_{mn}^{(j)(N)}\right) \overline{N}_{mn}(k\overline{r r_j})\right]$$
$$\cdot \left[T_{n'}^{(s_l)(M)*} E_l\left(w_{m'n'}^{(l)(M)*}\right) \overline{M}_{m'n'}^*(k\overline{r r_l}) + T_{n'}^{(s_l)(N)*} E_l\left(w_{m'n'}^{(l)(N)*}\right) \overline{N}_{m'n'}^*(k\overline{r r_l})\right] \tag{6.3.36}$$

We shall evaluate (6.3.36) by considering a layer of particles with thickness d.

A typical integral in (6.3.36) can be evaluated as follows. The scattered field in the distorted Born approximation is calculated in the far-field to obtain the bistatic scattering coefficient. Substitute in

$$E_j(w_{mn}^{(P)(s)}) = w_{mn}^{(P)(s_j)}(\overline{r}_j) = a_{mn}^{E(P)(s_j)}, \quad (P = M, N) \tag{6.3.37}$$

we have

$$\begin{aligned} I &= \int d\overline{r}_j \int d\overline{r}_l \left(g_{s_j s_l}(\overline{r}_j, \overline{r}_l) - 1\right) E_j\left(w_{mn}^{(j)(M)}\right) \overline{M}_{mn}(k\overline{r r_j}) \\ &\quad \cdot E_l\left(w_{m'n'}^{(l)(N)}\right) \overline{N}_{m'n'}(k\overline{r r_l}) \\ &= \frac{\gamma_{mn}\gamma_{m'n'} i^{-n-1} i^{n'}}{k^2 r^2} a_{mn}^{E(s_j)(M)} a_{m'n'}^{E(s_l)(N)*} \overline{C}_{mn}(\theta_s, \phi_s) \cdot \overline{B}_{m'n'}^*(\theta_s, \phi_s) \\ &\quad \cdot \int d\overline{r}_j \int d\overline{r}_l \left(g_{s_j s_l}(\overline{r}_j - \overline{r}_l) - 1\right) e^{i\overline{K}_d \cdot \overline{r}_j} e^{-i\overline{K}_d^* \cdot \overline{r}_l} e^{-i\overline{k}_s \cdot \overline{r}_j} e^{i\overline{k}_s \cdot \overline{r}_l} \end{aligned} \tag{6.3.38}$$

We note that $g_{s_j s_l}(\overline{r}_j - \overline{r}_l) - 1$ is equal to zero for $|\overline{r}_j - \overline{r}_l|$ larger than a few particle diameters. The particle diameter is much smaller than the mean free path of $1/2K_z''$, where $K_z'' = \mathrm{Im}(K_z)$. Let $\overline{r} = \overline{r}_j - \overline{r}_l$ and $\overline{r}_A = (\overline{r}_j + \overline{r}_l)/2$ in (6.3.38). Then, the term in (6.3.38) becomes

$$\begin{aligned} &\int d\overline{r}_j \int d\overline{r}_l \left(g_{s_j s_l}(\overline{r}_j - \overline{r}_l) - 1\right) e^{i\overline{K}_d \cdot \overline{r}_j} e^{-i\overline{K}_d^* \cdot \overline{r}_l} e^{-i\overline{k}_s \cdot \overline{r}_j + i\overline{k}_s \cdot \overline{r}_l} \\ &= \int d\overline{r}_A \int d\overline{r} \left(g_{s_j s_l}(\overline{r}) - 1\right) e^{i(\mathrm{Re}\overline{K}_d - \overline{k}_s) \cdot \overline{r}} e^{2K_z'' z_A} \\ &= A_o (2\pi)^3 H_{s_j s_l}(\overline{k}_s - \mathrm{Re}\overline{K}_d) \frac{F_d}{2K_z''} \end{aligned} \tag{6.3.39}$$

where A_o is the area of target, and

$$\overline{k}_s = k(\sin\theta_s \cos\phi_s \hat{x} + \sin\theta_s \sin\phi_s \hat{y} + \cos\theta_s \hat{z}) \tag{6.3.40}$$

$$H_{s_j s_l}(\overline{p}) = \frac{1}{(2\pi)^3} \int_{-\infty}^{\infty} d\overline{r} \, [g_{s_j s_l}(\overline{r}) - 1] e^{-i\overline{p} \cdot \overline{r}} \tag{6.3.41}$$

$$F_d = \left(1 - e^{-2K_z'' d}\right) \tag{6.3.42}$$

We calculated the second term of (6.3.36) in this manner. For the first term in (6.3.36), we note that

$$\int d\overline{r}_j e^{i\overline{K}_d\cdot\overline{r}_j - i\overline{K}_d^*\cdot\overline{r}_j} = A_o \frac{F_d}{2K_z''} \tag{6.3.43}$$

Then (6.3.36) becomes

$$\begin{aligned}\langle \overline{\mathcal{E}}_s(\overline{r})\cdot\overline{\mathcal{E}}_s^*(\overline{r})\rangle &= \frac{F_d A_o}{2K_z'' k^2 r^2}\Bigg\{\sum_{s_j=1}^{L} n_{s_j}\sum_{n,m}\sum_{n',m'}\gamma_{mn}\gamma_{m'n'} i^{-n} i^{n'}\\
&\cdot\left[T_n^{(s_j)(M)} a_{mn}^{E(s_j)(M)}\frac{\overline{C}_{mn}(\theta_s,\phi_s)}{i} + T_n^{(s_j)(N)} a_{mn}^{E(s_j)(N)}\overline{B}_{mn}(\theta_s,\phi_s)\right]\\
&\cdot\left[T_{n'}^{(s_j)(M)*} a_{m'n'}^{E*(s_j)(M)}\frac{\overline{C}_{m'n'}^*(\theta_s,\phi_s)}{-i} + T_{n'}^{(s_j)(N)*} a_{m'n'}^{E*(s_j)(N)}\overline{B}_{m'n'}^*(\theta_s,\phi_s)\right]\\
&+\sum_{s_j=1}^{L}\sum_{s_l=1}^{L} n_{s_j} n_{s_l}\sum_{nm}\sum_{m'n'}\gamma_{mn}\gamma_{m'n'} i^{-n} i^{n'}\\
&\cdot\left[T_n^{(s_j)(M)} a_{mn}^{E(s_j)(M)}\frac{\overline{C}_{mn}(\theta_s,\phi_s)}{i} + T_n^{(s_j)(N)} a_{mn}^{E(s_j)(N)}\overline{B}_{mn}(\theta_s,\phi_s)\right]\\
&\cdot\left[T_{n'}^{(s_l)(M)*} a_{m'n'}^{E(s_l)(M)*}\frac{\overline{C}_{m'n'}^*(\theta_s,\phi_s)}{-i} + T_{n'}^{(s_l)(N)*} a_{m'n'}^{E(s_l)(N)*}\overline{B}_{m'n'}^*(\theta_s,\phi_s)\right]\\
&\cdot (2\pi)^3 H_{s_j s_l}(\overline{k}_s - \mathrm{Re}\overline{K}_d)\Bigg\}\end{aligned} \tag{6.3.44}$$

We note that

$$\overline{a}_{mn}^{E(s_j)(M)} = (-1)^m \frac{1}{\gamma_{mn}}\frac{(2n+1)}{n(n+1)} i^n Y_n^{(s_j)(M)}\left[\hat{\alpha}\cdot\overline{C}_{-mn}(\pi-\theta_t,\phi_i)\right] \tag{6.3.45a}$$

$$\overline{a}_{mn}^{E(s_j)(N)} = (-1)^m \frac{1}{\gamma_{mn}}\frac{(2n+1)}{n(n+1)} i^{n-1} Y_n^{(s_j)(N)}\left[\hat{\alpha}\cdot\overline{B}_{-mn}(\pi-\theta_t,\phi_i)\right] \tag{6.3.45b}$$

where $\hat{\alpha} = \hat{\theta}_{td}$ for vertical polarization and $\alpha = \hat{\phi}_i$ for horizontal polarization.

Substituting into (6.3.44), we have

$$\begin{aligned}\langle \overline{\mathcal{E}}_s(\overline{r})\cdot\overline{\mathcal{E}}_s^*(\overline{r})\rangle &= \frac{F_d}{2K_z''}\frac{A_o}{k^2 r^2}\Bigg\{\sum_{s_j=1}^{L} n_{s_j}\sum_{n,m}\sum_{n',m'}(-1)^{m+m'}\frac{(2n+1)}{n(n+1)}\frac{(2n'+1)}{n'(n'+1)}\\
&\cdot\Big[T_n^{(s_j)(M)} Y_n^{(s_j)(M)}\hat{\alpha}\cdot\overline{C}_{-mn}(\pi-\theta_t,\phi_t)\overline{C}_{mn}(\theta_s,\phi_s)\\
&\quad + T_n^{(s_j)(N)} Y_n^{(s_j)(N)}\hat{\alpha}\cdot\overline{B}_{-mn}(\pi-\theta_t,\phi_t)\overline{B}_{mn}(\theta_s,\phi_s)\Big]\\
&\cdot\Big[T_{n'}^{(s_j)(M)*} Y_{n'}^{(s_j)(M)*}\hat{\alpha}\cdot\overline{C}_{-m'n'}^*(\pi-\theta_t,\phi_t)\overline{C}_{mn}^*(\theta_s,\phi_s)\end{aligned}$$

$$+T_{n'}^{(s_j)(N)*}Y_{n'}^{(s_j)(N)*}\hat{\alpha}\cdot\overline{B}^*_{-m'n'}(\pi-\theta_t,\phi_t)\overline{B}^*_{mn}(\theta_s,\phi_s)\Big]$$

$$+\sum_{s_j=1}^{L}\sum_{s_l=1}^{L}n_{s_j}n_{s_l}\sum_{n,m}\sum_{n'm'}(-1)^{m+m'}\frac{(2n+1)}{n(n+1)}\frac{(2n'+1)}{n'(n'+1)}$$

$$\cdot\Big[T_n^{(s_j)(M)}Y_n^{(s_j)(M)}\hat{\alpha}\cdot\overline{C}_{-mn}(\pi-\theta_t,\phi_t)\overline{C}_{mn}(\theta_s,\phi_s)$$

$$+T_n^{(s_j)(N)}Y_n^{(s_j)(N)}\hat{\alpha}\cdot\overline{B}_{-mn}(\pi-\theta_t,\phi_t)\overline{B}_{mn}(\theta_s,\phi_s)\Big]$$

$$\cdot\Big[T_{n'}^{(s_l)(M)*}Y_{n'}^{(s_l)(M)*}\hat{\alpha}\cdot\overline{C}^*_{-m'n'}(\pi-\theta_t,\phi_t)\overline{C}^*_{m'n'}(\theta_s,\phi_s)$$

$$+T_{n'}^{(s_l)(N)*}Y_{n'}^{(s_l)(N)*}\hat{\alpha}\cdot\overline{B}^*_{-m'n'}(\pi-\theta_t,\phi_t)\overline{B}^*_{m'n'}(\theta_s,\phi_s)\Big]$$

$$\cdot(2\pi)^3H_{s_js_l}(\overline{k}_s-\mathrm{Re}\overline{K}_d)\Big\}\tag{6.3.46}$$

Define the bistatic scattering coefficients as

$$\gamma_{\beta\alpha}(\theta_s,\phi_s;\pi-\theta_i,\phi_i)=\frac{4\pi r^2\left|\overline{\mathcal{E}}^s_\beta\right|^2}{A_o\cos\theta_i}\tag{6.3.47}$$

Use the addition theorem for vector spherical harmonics and let

$$W_{VV}^{(s_j)}(\theta_s,\phi_s;\pi-\theta_t,\phi_t)=\sum_n\frac{(2n+1)}{n(n+1)}\left[T_n^{(s_j)(M)}Y_n^{(s_j)(M)}\frac{1}{\sin\theta\sin\theta'}\frac{\partial^2P_n}{\partial\phi\partial\phi'}\right.$$

$$\left.+\;T_n^{(s_j)(N)}Y_n^{(s_j)(N)}\frac{\partial^2P_n}{\partial\theta\partial\theta'}\right]\tag{6.3.48a}$$

$$W_{HV}^{(s_j)}(\theta_s,\phi_s;\pi-\theta_t,\phi_t)=\sum_n\frac{(2n+1)}{n(n+1)}\left[-T_n^{(s_j)(M)}Y_n^{(s_j)(M)}\frac{1}{\sin\theta'}\frac{\partial^2P_n}{\partial\theta\partial\phi'}\right.$$

$$\left.+\;T_n^{(s_j)(N)}Y_n^{(s_j)(N)}\frac{1}{\sin\theta}\frac{\partial^2P_n}{\partial\phi\partial\theta'}\right]\tag{6.3.48b}$$

$$W_{VH}^{(s_j)}(\theta_s,\phi_s;\pi-\theta_t,\phi_t)=\sum_n\frac{(2n+1)}{n(n+1)}\left[-T_n^{(s_j)(M)}Y_n^{(s_j)(M)}\frac{1}{\sin\theta}\frac{\partial^2P_n}{\partial\phi\partial\theta'}\right.$$

$$\left.+\;T_n^{(s_j)(N)}Y_n^{(s_j)(N)}\frac{1}{\sin\theta'}\frac{\partial^2P_n}{\partial\theta\partial\phi'}\right]\tag{6.3.48c}$$

$$W_{HH}^{(s_j)}(\theta_s,\phi_s;\pi-\theta_t,\phi_t)=\sum_n\frac{(2n+1)}{n(n+1)}\left[T_n^{(s_j)(M)}Y_n^{(s_j)(M)}\frac{\partial^2P_n}{\partial\theta\partial\theta'}\right.$$

$$\left.+\;T_n^{(s_j)(N)}Y_n^{(s_j)(N)}\frac{1}{\sin\theta\sin\theta'}\frac{\partial^2P_n}{\partial\phi\partial\phi'}\right]\tag{6.3.48d}$$

where the argument of P_n is $\cos\theta\cos\theta' + \sin\theta\sin\theta'\cos(\phi - \phi')$ and $\theta = \theta_s$, $\phi = \phi_s, \theta' = \pi - \theta_t$, $\phi' = \phi_i = \phi_t$. In (6.3.48a)–(6.3.48d)

$$\frac{1}{\sin\theta\sin\theta'}\frac{\partial^2 P_n}{\partial\phi\partial\phi'} = -P_n''\sin\theta\sin\theta'\sin^2(\phi-\phi') + P_n'\cos(\phi-\phi') \tag{6.3.48e}$$

$$\frac{1}{\sin\theta'}\frac{\partial^2 P_n}{\partial\theta\partial\phi'} = P_n''\sin\theta\sin(\phi-\phi')(-\sin\theta\cos\theta' + \cos\theta\sin\theta'\cos(\phi-\phi')) + P_n'\cos\theta\sin(\phi-\phi') \tag{6.3.48f}$$

$$\frac{1}{\sin\theta}\frac{\partial^2 P_n}{\partial\phi\partial\theta'} = -P_n''\sin\theta'\sin(\phi-\phi')(-\cos\theta\sin\theta' + \sin\theta\cos\theta'\cos(\phi-\phi')) - P_n'\cos\theta'\sin(\phi-\phi') \tag{6.3.48g}$$

$$\frac{\partial^2 P_n}{\partial\theta\partial\theta'} = P_n''(-\cos\theta\sin\theta' + \sin\theta\cos\theta'\cos(\phi-\phi'))(-\sin\theta\cos\theta' + \cos\theta\sin\theta'\cos(\phi-\phi')) + P_n'(\sin\theta\sin\theta' + \cos\theta\cos\theta'\cos(\phi-\phi')) \tag{6.3.48h}$$

where P_n' and P_n'' are the first and second derivatives of P_n and having the same argument as P_n. Then the bistatic scattering coefficients can be written as

$$\gamma_{\beta\alpha}(\theta_s,\phi_s;\pi-\theta_i,\phi_i) = \frac{4\pi}{k^2\cos\theta_i}\frac{F_d}{2K_z''}\left\{\sum_{s_j=1}^{L} n_{s_j}\left|W_{\beta\alpha}^{(s_j)}(\theta_s,\phi_s;\pi-\theta_t,\phi_t)\right|^2 + \sum_{s_j=1}^{L}\sum_{s_l=1}^{L} n_{s_j}n_{s_l}W_{\beta\alpha}^{(s_j)}(\theta_s,\phi_s;\pi-\theta_t,\phi_t)W_{\beta\alpha}^{(s_l)*}(\theta_s,\phi_s;\pi-\theta_t,\phi_t) \cdot (2\pi)^3 H_{s_j s_l}(\overline{k}_s - \mathrm{Re}\overline{K}_d)\right\} \tag{6.3.49}$$

For the special case that $\phi_s = \phi_i + \pi = \phi_t + \pi$, we have

$$W_{VH}^{(s_j)}(\theta_s,\phi_t+\pi;\pi-\theta_t,\phi_t) = W_{HV}^{(s_j)}(\theta_s,\phi_t+\pi;\pi-\theta_t,\phi_t) = 0 \tag{6.3.50}$$

$$W_{VV}^{(s_j)}(\theta_s,\phi_t+\pi;\pi-\theta_t,\phi_t) = \sum_n (-1)^n\frac{(2n+1)}{n(n+1)}\left\{T_n^{(s_j)(M)}Y_n^{(s_j)(M)(V)}\frac{P_n^1(\cos(\theta_s-\theta_t))}{|\sin(\theta_s-\theta_t)|} + T_n^{(s_j)(N)}Y_n^{(s_j)(N)(V)}\left(\cos(\theta_s-\theta_t)\frac{P_n^1(\cos(\theta_s-\theta_t))}{|\sin(\theta_s-\theta_t)|} + n(n+1)P_n(\cos(\theta_s-\theta_t))\right)\right\} \tag{6.3.51a}$$

$$W_{HH}^{(s_j)}(\theta_s, \phi_t + \pi; \pi - \theta_t, \phi_t)$$

$$= \sum_n (-1)^n \frac{(2n+1)}{n(n+1)} \left\{ T_n^{(s_j)(N)} Y_n^{(s_j)(N)(H)} \frac{P_n^1(\cos(\theta_s - \theta_t))}{|\sin(\theta_s - \theta_t)|} \right.$$

$$+ T_n^{(s_j)(M)} Y_n^{(s_j)(M)(H)} \left(\cos(\theta_s - \theta_t) \frac{P_n^1(\cos(\theta_s - \theta_t))}{|\sin(\theta_s - \theta_t)|} \right.$$

$$\left. \left. + n(n+1) P_n(\cos(\theta_s - \theta_t)) \right) \right\} \qquad (6.3.51b)$$

The bistatic scattering coefficient is bistatic cross section per unit area. We can also define a bistatic cross section per unit volume for a small thickness d. Such a definition will be used for defining phase matrix. For small d we have

$$\frac{F_d}{2K_z''} \rightarrow d$$

Then the bistatic cross section per unit volume is

$$\frac{\cos\theta_i \gamma_{\beta\alpha}(\theta_s, \phi_s; \pi - \theta_i, \phi_i)}{d}$$

$$= \frac{4\pi}{k^2} \left\{ \sum_{s_j=1}^{L} n_{s_j} \left| W_{\beta\alpha}^{(s_j)}(\theta_s, \phi_s; \pi - \theta_t, \phi_t) \right|^2 \right.$$

$$+ \sum_{s_j=1}^{L} \sum_{s_l=1}^{L} n_{s_j} n_{s_l} W_{\beta\alpha}^{(s_j)}(\theta_s, \phi_s; \pi - \theta_t, \phi_t) W_{\beta\alpha}^{(s_l)*}(\theta_s, \phi_s; \pi - \theta_t, \phi_t)$$

$$\left. \cdot (2\pi)^3 H_{s_j s_l}(\overline{k}_s - \mathrm{Re}\overline{K}_d) \right\} \qquad (6.3.51c)$$

Numerical Results and Discussion

We shall illustrate the numerical results of the effective propagation constants and backscattering coefficients using parameters of snow. The results are illustrated with the Rayleigh size distribution, which is a special case of the modified gamma size distribution. The size distribution is represented by $n(a)$, which is the number of particles per unit volume with radii between a and $a + da$. The Rayleigh size distribution is given by

$$n(a) = \frac{\pi f a}{16\langle a \rangle^5} \exp\left(-\frac{\pi a^2}{4\langle a \rangle^2} \right) \qquad (6.3.52)$$

where $\langle a \rangle$ is the mean radius. The fractional volume f, the mean radius $\langle a \rangle$, and the mode radius a_m are defined by

$$f = \int_0^\infty da \, \frac{4\pi}{3} a^3 n(a) \tag{6.3.53}$$

$$\langle a \rangle = \frac{\int_0^\infty da \, a n(a)}{\int_0^\infty da \, n(a)} \tag{6.3.54}$$

$$a_m = \sqrt{\frac{2}{\pi}} \langle a \rangle \tag{6.3.55}$$

In (6.3.55), a_m is the mode radius at which the size distribution is at a maximum. Thus, given the fractional volume f, the Rayleigh size distribution contains only one parameter, namely, the mean radius. In the numerical calculations, we discretize the Rayleigh size distribution into 20 different equally spaced sizes (i.e., $L = 20$ species). The pair distribution functions for a medium with 20 sizes are then calculated by the Percus-Yevick approximation as given in Chapter 8 of Volume II. The maximum size for these 20 sizes is chosen such that at least 99.7% of the scattering contributions and the fractional volume of particles are included. That usually occurs around 2.5 to 3 times the mean radius. Comparison will be made between the scattering of a medium with Rayleigh size distribution of a specified value of $\langle a \rangle$ and the case of monodisperse particles with radii equal to that same value. The results of QCA will also be compared with that of independent scattering. For the case of independent scattering, the extinction rate is

$$(\kappa_e)_{\text{independent}} = -\frac{2\pi}{k^2} \sum_{s_j=1}^{L} n_{s_j} \sum_n (2n+1) \operatorname{Re} \left[T_n^{(s_j)(M)} + T_n^{(s_j)(N)} \right] \tag{6.3.56}$$

In Figs. 6.3.2 and 6.3.3, we plot respectively the effective loss tangent $2K_i/K_r$ and the normalized phase velocity k/K_r as a function of fractional volume at frequency = 37 GHz for two Rayleigh size distributions with $\langle a \rangle = 0.05$ cm and $\langle a \rangle = 0.09$ cm. We note that the effective loss tangents first increase with fractional volume, rise to maxima, and then decrease with further increase in fractional volume. The peak of the larger size of $\langle a \rangle = 0.09$ cm occurs at $f = 0.19$, and the peak for $\langle a \rangle = 0.05$ cm occurs at $f = 0.16$. This indicates that scattering by larger particles is "more independent." Figure 6.3.3 shows that the phase velocity decreases with fractional volume, with smaller velocity for larger particles indicating that scattering decreases phase velocity. In Fig. 6.3.4 the backscattering coefficients are plotted as a function of fractional volume for a layer thickness of $d = 3$ cm. The backscattering coefficients are $\sigma_{\beta\beta} = \cos\theta_i \, \gamma_{\beta\beta}$ with $\beta = v, h$.

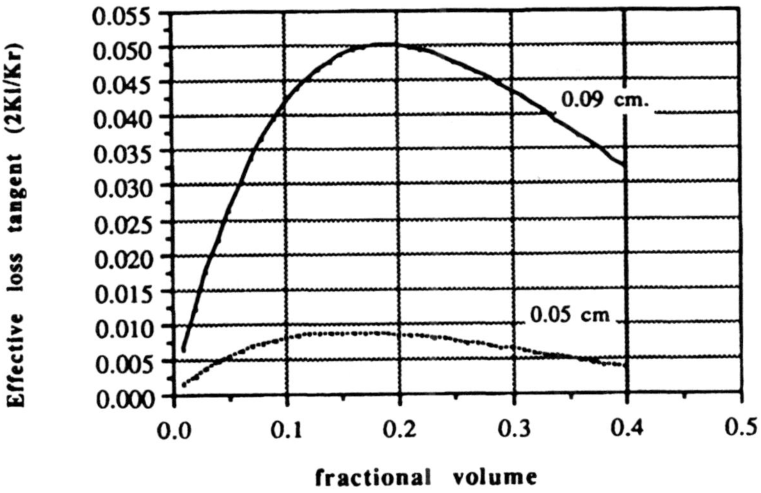

Figure 6.3.2 Effective loss tangent $2K_i/K_r$ of QCA as a function of fractional volume. Frequency is at 37 GHz. The permittivity of the particles is $\epsilon_s = (3.2 + i0.001)\epsilon_0$. Two cases of Rayleigh size distributions are considered with $\langle a \rangle = 0.05$ cm and $\langle a \rangle = 0.09$ cm.

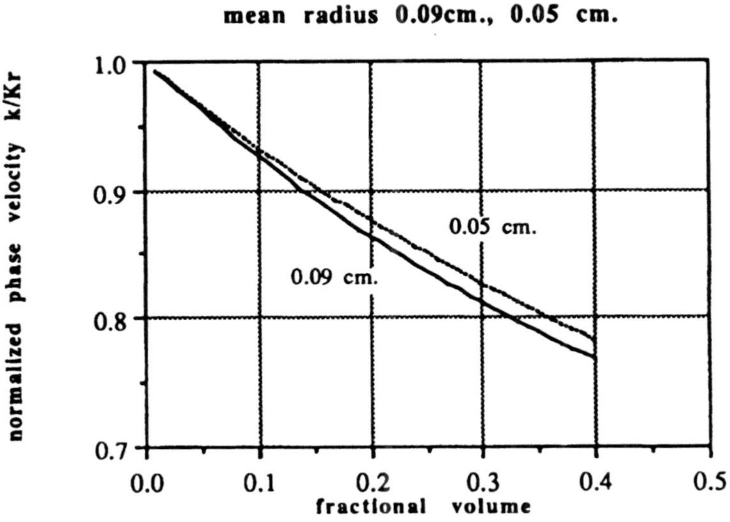

Figure 6.3.3 Normalized phase velocity k/K_r of QCA as a function of fractional volume. Frequency is at 37 GHz. The permittivity of the particles is $\epsilon_s = (3.2 + i0.001)\epsilon_0$. Two cases of Rayleigh size distributions are considered with $\langle a \rangle = 0.05$ cm and $\langle a \rangle = 0.09$ cm.

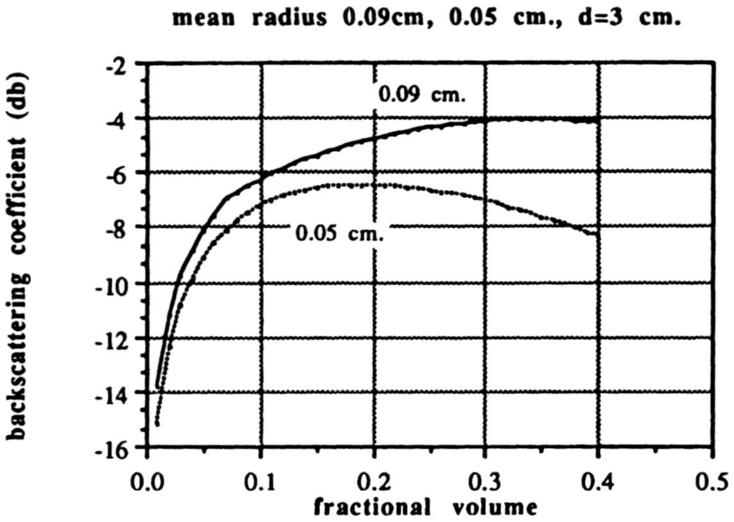

Figure 6.3.4 Backscattering coefficient σ_{hh} (in decibels) of QCA as a function of fractional volume. Frequency is at 37 GHz. The permittivity of the particles is $\epsilon_s = (3.2 + i0.001)\epsilon_0$. Two cases of Rayleigh size distributions are considered with $\langle a \rangle = 0.05$ cm and $\langle a \rangle = 0.09$ cm. The layer thickness is $d = 3$ cm and incident angle is 30°.

In Figs. 6.3.5, 6.3.6, and 6.3.7, we show the coherent wave attenuation rates for a smaller mean grain radius of 0.035 cm. The corresponding mode radius is 0.0279 cm. Figure 6.3.5 compares the extinction rate $2K_i$ between independent scattering and QCA. We note that attenuation rate of independent scattering is larger than QCA. The results show that the extinction rate first increases rapidly with frequency as in Rayleigh scattering. Then it departs from Rayleigh scattering around 20 GHz and starts to gradually level off. This indicates the importance of including Mie scattering effects beyond 20 GHz for $\langle a \rangle = 0.035$ cm. Figure 6.3.6 compares the results of QCA between the extinction rates of monodisperse particles with radii $= 0.035$ cm and the case of Rayleigh size distribution with $\langle a \rangle = 0.035$ cm. We note that the case of Rayleigh size distribution has a considerably larger scattering rate than monodisperse case. This indicates the importance of including size distribution in modeling scattering from realistic geophysical terrain when grain sizes are small. The numerical results indicate that the extinction rate is at 1.056 cm^{-1} based on Rayleigh size distribution at frequency $= 95$ GHz, meaning that there is strong multiple scattering of incoherent waves at such high frequency even for moderate snow layer thicknesses. Figure 6.3.7 shows the results of HH backscattering coefficient for a layer thickness of $d = 10$ cm.

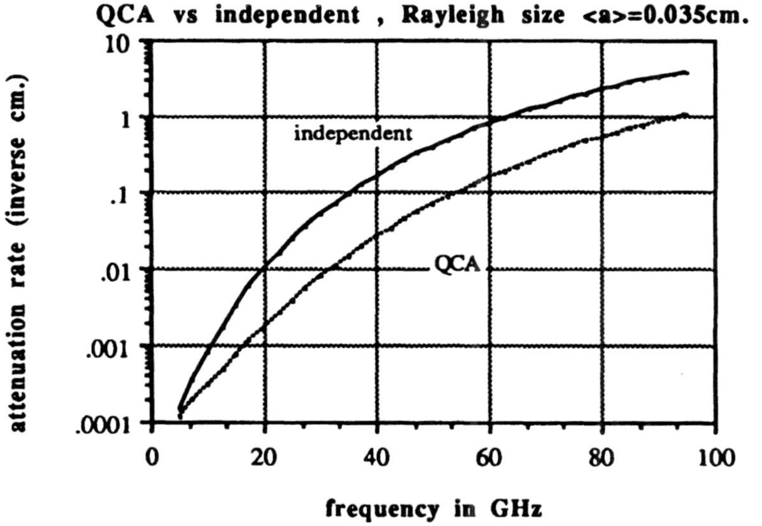

Figure 6.3.5 Coherent wave attenuation rate $2K_i$ in cm^{-1} as a function of frequency. The permittivity of the particles is $\epsilon_s = (3.2 + i0.001)\epsilon_0$. Rayleigh size distribution is considered with $\langle a \rangle = 0.035$ cm and fractional volume $f = 0.3$. The cases of independent scattering and QCA are compared.

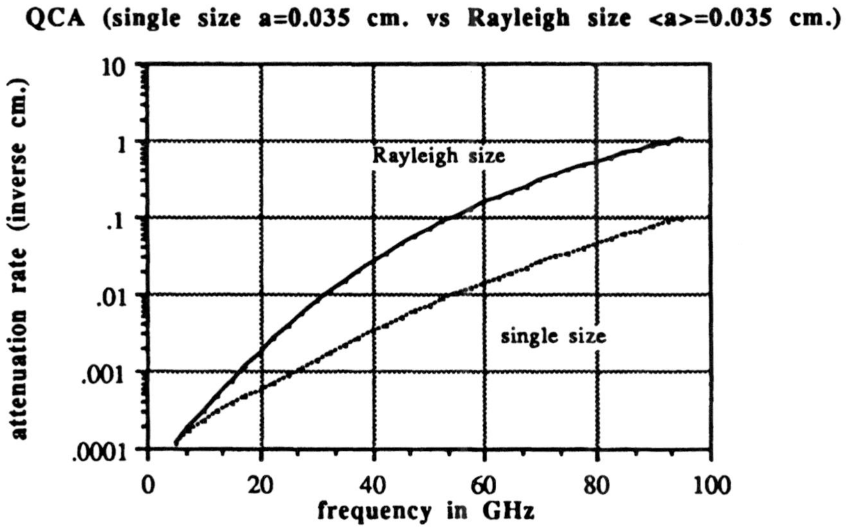

Figure 6.3.6 Coherent wave attenuation rate $2K_i$ in cm^{-1} under QCA as a function of frequency. The permittivity of the particles is $\epsilon_s = (3.2 + i0.001)\epsilon_0$. The fractional volume is $f = 0.3$. The cases of identical sizes of $a = 0.035$ cm and Rayleigh size distribution with $\langle a \rangle = 0.035$ cm are compared.

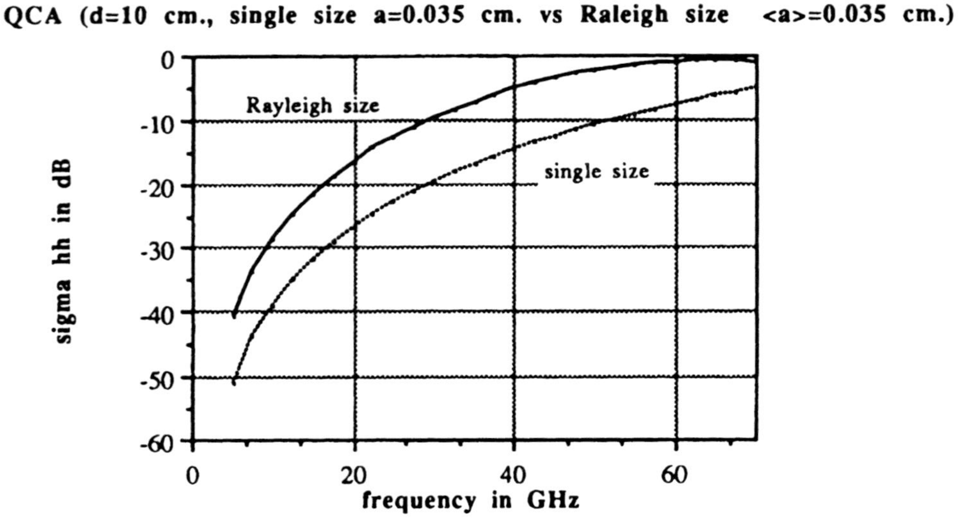

Figure 6.3.7 Backscattering coefficient σ_{hh} in dB of QCA as a function of frequency. The permittivity of the particles is $\epsilon_s = (3.2 + i0.001)\epsilon_0$. The layer thickness is $d = 10$ cm and incident angle is 30°. The cases of identical sizes of $a = 0.035$ cm and Rayleigh size distribution with $\langle a \rangle = 0.035$ cm are compared.

The curves show saturation effects due to the limitations of distorted Born approximation.

Comparison with Extinction Measurements of Dry Snow Between 18 GHz and 90 GHz

We next apply the dense media QCA-PY-Mie theory to compare with extinction measurements of dry snow. Hallikanen et al. [1987] measured the extinction behavior of different types of dry snow at four frequencies: 18 GHz, 35 GHz, 60 GHz, and 90 GHz. Thus, it is important to match the data for a sample over the entire frequency range with one set of physical parameters.

In snow, the standard deviation of grain size radius can be several times larger than the mode radius, showing that snow has a broad grain size distribution. The maximum grain diameter is between 3 mm to 4 mm. The number ratio of large grains to grains with mean size is in the range of 1 to several hundred.

In the following, we shall compare the snow data with ground truth of mean grain diameter = 1 mm and snow density = 0.385 g/cc. This corresponds to a fractional volume of 0.423. The reported measurements of extinc-

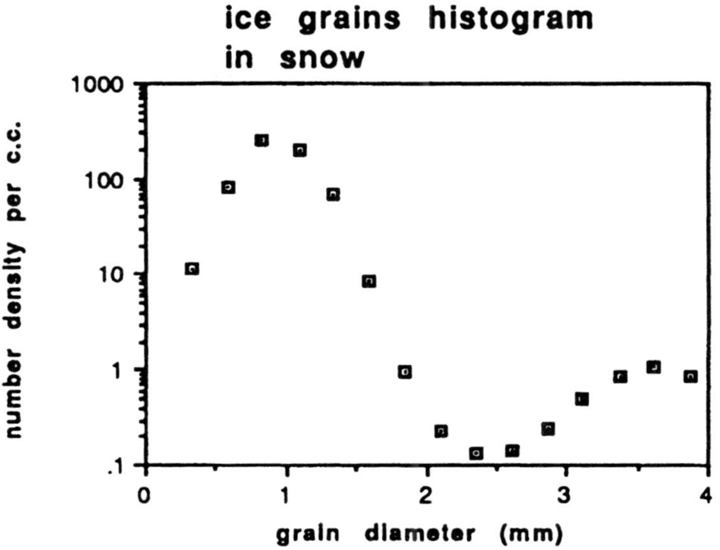

Figure 6.3.8 The ice grain size histogram in snow in number per c.c. as a function of grain diameter. The histogram is used for QCA-PY-Mie theory. There are 15 sizes, with mean grain radius of 0.0479 cm and fractional volume of 0.4134.

tion are 10.2 dB/m, 58.8 dB/m, 247.2 dB/m, and 304.0 dB/m, respectively, at frequencies of 18 GHz, 35 GHz, 60 GHz, and 90 GHz. Comparisons are made with QCA-PY-Mie theory.

The size distribution that we used is in Fig. 6.3.8, which shows the histogram number density (number per cc) for 15 sizes with mean grain radius of 0.0479 cm and fractional volume of 0.4134. Note that there are large grains with diameters between 3 mm to 4 mm. The number ratio of large grains to mean size is about 1 to 300.

In Fig. 6.3.9 we make a comparison with data based on the histogram size distribution of Fig. 6.3.8. The permittivity of ice grains chosen is $\epsilon_s = 3.2\epsilon_0 + i\epsilon_s''$ with $\epsilon_s''/\epsilon_0 = 0.007, 0.009, 0.011$, and 0.014, respectively, at 18 GHz, 35 GHz, 60 GHz, and 90 GHz, which corresponds to a salinity of 0.12 parts per thousand [Mätzler and Wegmuller, 1987]. The agreement is good. It seems that with size distribution, the QCA-PY-Mie theory can match the frequency dependence of the extinction of dry snow between 18 GHz and 90 GHz. The results of independent scattering are also shown for comparison. The independent scattering result predicts a much larger extinction.

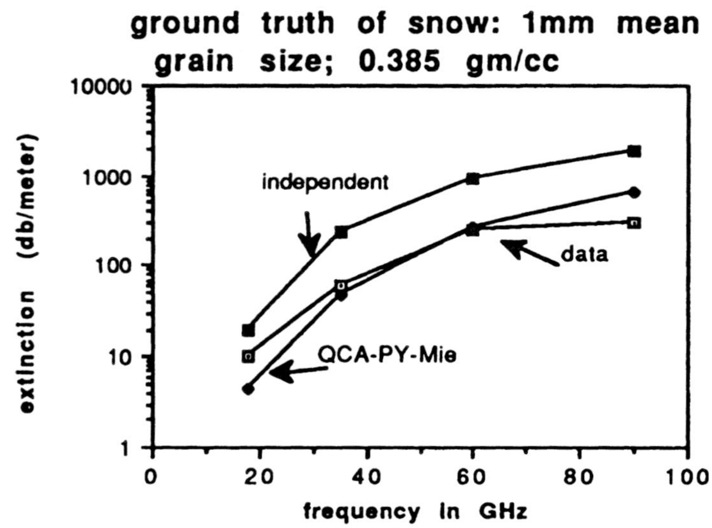

Figure 6.3.9 Comparison between QCA-PY-Mie theory and independent scattering with snow extinction data at 18 GHz, 35 GHz, 60 GHz, and 90 GHz of Hallikanen et al. [1987] with ground truth of mean grain diameter of 1 mm and snow density $= 0.385$ g/cc. The theory is based on the histogram of Fig. 6.3.8 with $\langle a \rangle = 0.0479$ cm and $f = 0.4134$. The permittivity of ice grains chosen is $\epsilon_s = 3.2\epsilon_o + i\epsilon_s''$ with $\epsilon_s''/\epsilon_o = 0.007, 0.009, 0.011$, and 0.014, respectively, at 18 GHz, 35 GHz, 60 GHz, and 90 GHz, which correspond to a salinity of 0.12 parts per thousand [Mätzler and Wegmuller, 1987].

4 Dense Media Radiative Transfer Theory Based on Quasi-crystalline Approximation

In Chapter 5, the radiative transfer equations is rigorously derived from Foldy approximation and ladder approximation for isotropic point scatterers. In Chapter 7, we will use QCA-CP and correlated ladder approximation to rigorously derive the dense medium radiative transfer equation (DMRT) for small particles. Particle correlation effects will be included. However, the QCA theory has the advantage that the higher order multipoles terms can be included readily. The theory is applicable to moderate size particles. Thus it is useful to develop a QCA-based DMRT to include multiple scattering of the incoherent fields. We will adopt such an approach in this section. The phase matrices are derived by using QCA and the distorted Born approximation for a thin layer. These are inserted into radiative transfer equations. In this model the multiple scattering effects will be multiple orders of the distorted Born approximation.

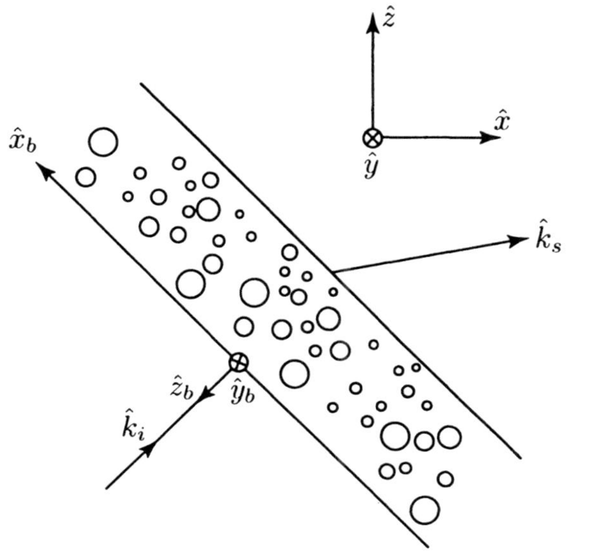

Figure 6.4.1 Illustration of the body frame $(\hat{x}_b, \hat{y}_b, \hat{z}_b)$ and principal frame $(\hat{x}, \hat{y})$ of a thin layer of spherical particles with multiple sizes. Incident wave is imping on the slab in the direction $\hat{k}_i = -\hat{z}_b$.

4.1 Phase Matrix, Extinction, Scattering, and Absorption Coefficients

Consider an incident wave impinging upon a volume of dense media consisting of spheres of moderate sizes and size distribution (Fig. 6.4.1). The incident wave direction is $\hat{k}_i$ and polarization directions are $\hat{v}_i$ and $\hat{h}_i$.

$$\hat{k}_i = \sin\theta_i \hat{x} + \cos\theta_i \hat{z} \tag{6.4.1}$$

$$\hat{v}_i = \cos\theta_i \hat{x} - \sin\theta_i \hat{z} \tag{6.4.2}$$

$$\hat{h}_i = \hat{y} \tag{6.4.3}$$

We want to calculate the extinction and the phase matrix using the results in the previous sections. This requires a transformation of coordinates. To do that we assume a body frame of the medium such that $\hat{z}_b$ points in the negative $\hat{k}_i$ direction (Fig. 6.4.1):

$$\hat{z}_b = -\hat{k}_i \tag{6.4.4}$$

For the body frame we further let,

$$\hat{y}_b = \hat{y} \tag{6.4.5}$$

so that

$$\hat{x}_b = -\cos\theta_i \hat{x} + \sin\theta_i \hat{z} \tag{6.4.6}$$

We also let the vertical polarization in the body frame be

$$\hat{v}_{ib} = -\hat{x}_b = -\hat{v}_i \tag{6.4.7}$$

and let the horizontal polarization in the body frame be

$$\hat{h}_{ib} = \hat{y}_b = \hat{h}_i \tag{6.4.8}$$

In the body frame the wave is "normally" impinging upon the dense medium.

The averaged exciting field coefficients obey the homogeneous system of equations of (6.3.16) and (6.3.17). Setting the determinant equal to zero provides a solution for the effective propagation K with

$$K = K' + iK'' \tag{6.4.9}$$

The determination of the real part K' from (6.3.16) and (6.3.17) presents no numerical difficulty. However, the determination of the imaginary part K'' from (6.3.16) and (6.3.17) can be difficult numerically because

(*i*) The imaginary part K'' is much smaller than the real part so that using the set of equations in (6.3.16) and (6.3.17) to determine both real and imaginary parts together may cause accuracy problem for the imaginary part.

(*ii*) The imaginary part K'' is strongly dependent on the pair distribution function. The function $S_p(k, K|R_{s_j s_l})$ includes an integration of the pair distribution function in the spatial domain. We note that the pair distribution function can have sharp spikes particularly for sticky particles. This further complicates the accuracy of K''.

Thus we use (6.3.16) and (6.3.17) to determine only the real part K'. As for the imaginary part K'', we note that it is related to the extinction coefficient by $\kappa_e = 2K''$. As shown later, we perform angular integrations of the phase functions to obtain the scattering coefficient κ_s. Absorption coefficient κ_a is next determined. Then the extinction coefficient is obtained as $\kappa_e = \kappa_a + \kappa_s$.

After solving the determinantal equation, the exciting field coefficients $Y_n^{(s_j)(M)}$ and $Y_n^{(s_j)(N)}$ are determined to within a single arbitrary constant. That is, the ratios of the $Y_n^{(s_j)(M)}$ and $Y_n^{(s_j)(N)}$ to each other are determined by solving (6.3.16) and (6.3.17). The one single constant is determined by the generalized Ewald-Oseen extinction theorem.

(A) Vertical Polarization Incidence in the Body Frame

$\overline{E}_i = \hat{v}_i = -\hat{v}_{ib}$. The arbitrary constant is determined by the generalized Ewald-Oseen extinction theorem. In the body frame, the incident and transmitted angles are normal. Thus replacing $\theta_i = \theta_t = 0$ in (6.3.24), we have

$$\sum_{s_l=1}^{L} \sum_n n_{s_l} \frac{i(2n+1)}{2} \left[T_n^{(s_l)(M)} Y_n^{(s_l)(M)(V)} + T_n^{(s_l)(N)} Y_n^{(s_l)(N)(V)} \right] = -\frac{(K-k)k^2}{2\pi} \tag{6.4.10}$$

where the additional superscript (V) denotes the vertically polarized incident case for the body frame.

The coherent reflected wave is also expressed by putting $\theta_i = \theta_t = 0$ into (6.3.25), and we express the coherent reflection coefficient as

$$R_V = \sum_{s_j=1}^{L} \sum_n \frac{2\pi i n_{s_j}}{k^2(K+k)} \frac{(-1)^n(2n+1)}{2} \cdot \left[-T_n^{(s_j)(M)} Y_n^{(s_j)(M)(V)} + T_n^{(s_j)(N)} Y_n^{(s_j)(N)(V)} \right] \tag{6.4.11}$$

To ensure energy conservation for QCA-based DMRT, we calculate absorption directly. The absorption coefficient can be calculated by using (2.8.84) of Volume I, (6.2.6), (6.2.47), and (6.2.48).

$$\kappa_a = \frac{2\pi}{k^2|1-R_V|^2} \sum_{s_j=1}^{L} n_{s_j} \sum_n (2n+1) \left[\left| Y_n^{(s_j)(M)(V)} \right|^2 \cdot \left(-\mathrm{Re}\, T_n^{(s_j)(M)} - \left| T_n^{(s_j)(M)} \right|^2 \right) + \left| Y_n^{(s_j)(N)(V)} \right|^2 \left(-\mathrm{Re}\, T_n^{(s_j)(N)} - \left| T_n^{(s_j)(N)} \right|^2 \right) \right] \tag{6.4.12}$$

(B) Horizontal Polarization Incidence in the Body Frame

For horizontally polarized incidence in the body frame, from (6.3.26), the generalized Ewald-Oseen extinction theorem gives, with $\theta_i = \theta_t = 0$ and $E_{h_i} = 1$,

$$\sum_{s_l=1}^{L}\sum_n n_{s_l}\frac{i(2n+1)}{2}\left[T_n^{(s_l)(M)}Y_n^{(s_l)(M)(H)}+T_n^{(s_l)(N)}Y_n^{(s_l)(N)(H)}\right]$$

$$=-\frac{(K-k)k^2}{2\pi} \tag{6.4.13}$$

The coherent reflection coefficient is

$$R_H=\sum_{s_j=1}^{L}\frac{2\pi n_{s_j}i}{k^2(K+k)}\sum_n(-1)^n\frac{(2n+1)}{2}$$

$$\cdot\left\{T_n^{(s_j)(M)}Y_n^{(s_j)(M)(H)}-T_n^{(s_j)(N)}Y_n^{(s_j)(N)(H)}\right\} \tag{6.4.14}$$

The absorption coefficient is

$$\kappa_a=\frac{2\pi}{k^2|1+R_H|^2}\sum_{s_j=1}^{L}n_{s_j}\sum_n(2n+1)$$

$$\cdot\left[\left|Y_n^{(s_j)(M)(H)}\right|^2\left(-\text{Re}\,T_n^{(s_j)(M)}-\left|T_n^{(s_j)(M)}\right|^2\right)\right.$$

$$\left.+\left|Y_n^{(s_j)(N)(H)}\right|^2\left(-\text{Re}\,T_n^{(s_j)(N)}-\left|T_n^{(s_j)(N)}\right|^2\right)\right] \tag{6.4.15}$$

These equations become the same as the vertical polarized incidence for the body frame. For the incoherent scattered wave, we note that in the body frame we have

$$\hat{k}_s=\sin\theta_{bs}\cos\phi_{bs}\hat{x}_b+\sin\theta_{bs}\sin\phi_{bs}\hat{y}_b+\cos\theta_{bs}\hat{z}_b \tag{6.4.16}$$

$$\hat{v}_{bs}=\cos\theta_{bs}\cos\phi_{bs}\hat{x}_b+\cos\theta_{bs}\sin\phi_s\hat{y}_b-\sin\theta_{bs}\hat{z}_b \tag{6.4.17}$$

$$\hat{h}_{bs}=-\sin\phi_{bs}\hat{x}_b+\cos\phi_{bs}\hat{y}_b \tag{6.4.18}$$

Then since $\theta_{bt}=\pi$, the argument of P_n in (6.3.48a–h) is $-\cos\theta_{bs}$. Thus

$$W_{VV}^{(s_j)}(\theta_{bs},\phi_{bs};\pi,0)=\cos\phi_{bs}w_V^{(s_j)}(\theta_{bs}) \tag{6.4.19}$$

$$W_{VH}^{(s_j)}(\theta_{bs},\phi_{bs};\pi,0)=-\sin\phi_{bs}w_V^{(s_j)}(\theta_{bs}) \tag{6.4.20}$$

$$W_{HV}^{(s_j)}(\theta_{bs},\phi_{bs};\pi,0)=\sin\phi_{bs}w_H^{(s_j)}(\theta_{bs}) \tag{6.4.21}$$

$$W_{HH}^{(s_j)}(\theta_{bs},\phi_{bs};\pi,0)=\cos\phi_{bs}w_H^{(s_j)}(\theta_{bs}) \tag{6.4.22}$$

where

$$w_V^{(s_j)}(\theta_{bs})=\sum_n\frac{(2n+1)}{n(n+1)}\left[P_n'(-\cos\theta_{bs})T_n^{(s_j)(M)}Y_n^{(s_j)(M)}+T_n^{(s_j)(N)}Y_n^{(s_j)(N)}\right.$$

$$\left.\cdot\left(\cos\theta_{bs}P_n'(-\cos\theta_{bs})+n(n+1)P_n(-\cos\theta_{bs})\right)\right] \tag{6.4.23}$$

$$w_H^{(s_j)}(\theta_{bs}) = \sum_n \frac{(2n+1)}{n(n+1)} \Big[T_n^{(s_j)(M)} Y_n^{(s_j)(M)} \Big(\cos\theta_{bs} P_n'(-\cos\theta_{bs}) + n(n+1) \cdot P_n(-\cos\theta_{bs}) \Big) + T_n^{(s_j)(N)} Y_n^{(s_j)(N)} P_n'(-\cos\theta_{bs}) \Big] \tag{6.4.24}$$

The bistatic scattering cross section per unit volume (i.e., the phase matrix in body frame) is, from (6.3.51c),

$$P_{v_b v_b}(\theta_{bs}, \phi_{bs}; \pi, 0) = \frac{1}{|1-R_V|^2} \frac{\cos^2\phi_{bs}}{k^2} \Bigg\{ \sum_{s_j=1}^{L} n_{s_j} \left| w_V^{(s_j)}(\theta_{bs}) \right|^2 + \sum_{s_j=1}^{L} \sum_{s_l=1}^{L} n_{s_j} n_{s_l} w_V^{(s_j)}(\theta_{bs}) w_V^{(s_l)*}(\theta_{bs}) (2\pi)^3 H_{s_j s_l} \Bigg\} \tag{6.4.25}$$

where the argument of $H_{s_j s_l}$ is

$$\begin{aligned} \left|\overline{k}_{bs} - \mathrm{Re}\overline{K}_d\right| &= |k\sin\theta_{bs}\cos\phi_{bs}\hat{x} + k\sin\theta_{bs}\sin\phi_{bs}\hat{y} + k\cos\theta_{bs}\hat{z} + K\hat{z}| \\ &= \sqrt{k^2\sin^2\theta_{bs} + (k\cos\theta_{bs} + K')^2} \end{aligned} \tag{6.4.26}$$

The division by $|1-R_V|^2$ is due to the fact that in calculating the bistatic scattering coefficients per unit volume in the previous section of (6.3.51c) based on the distorted Born approximation, there is a transmission across the boundary to the effective medium. The phase function should not carry such boundary effects.

Similarly,

$$P_{h_b v_b}(\theta_{bs}, \phi_{bs}; \pi, 0) = \frac{1}{|1-R_V|^2} \frac{\sin^2\phi_{bs}}{k^2} \Bigg\{ \sum_{s_j=1}^{L} n_{s_j} \left| w_H^{(s_j)}(\theta_{bs}) \right|^2 + \sum_{s_j=1}^{L} \sum_{s_l=1}^{L} n_{s_j} n_{s_l} w_H^{(s_j)}(\theta_{bs}) w_H^{(s_l)*}(\theta_{bs}) (2\pi)^3 H_{s_j s_l} \Bigg\} \tag{6.4.27}$$

$$P_{v_b h_b}(\theta_{bs}, \phi_{bs}; \pi, 0) = \frac{1}{|1+R_H|^2} \frac{\sin^2\phi_{bs}}{k^2} \Bigg\{ \sum_{s_j=1}^{L} n_{s_j} \left| w_V^{(s_j)}(\theta_{bs}) \right|^2 + \sum_{s_j=1}^{L} \sum_{s_l=1}^{L} n_{s_j} n_{s_l} w_V^{(s_j)}(\theta_{bs}) w_V^{(s_l)*}(\theta_{bs}) (2\pi)^3 H_{s_j s_l} \Bigg\} \tag{6.4.28}$$

$$P_{h_b h_b}(\theta_{bs}, \phi_{bs}; \pi, 0) = \frac{1}{|1 + R_H|^2} \frac{\cos^2 \phi_{bs}}{k^2} \left\{ \sum_{s_j=1}^{L} n_{s_j} \left| w_H^{(s_j)}(\theta_{bs}) \right|^2 \right.$$
$$\left. + \sum_{s_j=1}^{L} \sum_{s_l=1}^{L} n_{s_j} n_{s_l} w_H^{(s_j)}(\theta_{bs}) w_H^{(s_l)*}(\theta_{bs}) (2\pi)^3 H_{s_j s_l} \right\} \qquad (6.4.29)$$

For transformation to the principal frame, we also need to calculate the third Stokes parameter. The scattering from vertical polarization to the third Stokes parameter in body frame as

$$P_{3_b v_b}(\theta_{bs}, \phi_{bs}; \pi, 0)$$
$$= \frac{1}{|1 - R_V|^2} \frac{\cos \phi_{bs} \sin \phi_{bs}}{k^2} 2\text{Re} \left\{ \sum_{s_j=1}^{L} n_{s_j} w_V^{(s_j)}(\theta_{bs}) w_H^{(s_j)*}(\theta_{bs}) \right.$$
$$\left. + \sum_{s_j=1}^{L} \sum_{s_l=1}^{L} n_{s_j} n_{s_l} w_V^{(s_j)}(\theta_{bs}) w_H^{(s_l)*}(\theta_{bs}) (2\pi)^3 H_{s_j s_l} \right\} \qquad (6.4.30)$$

Scattering from horizontal polarization to the third Stokes parameter in body frame is

$$P_{3_b h_b}(\theta_{bs}, \phi_{bs}; \pi, 0)$$
$$= -\frac{1}{|1 + R_H|^2} \frac{\sin \phi_{bs} \cos \phi_{bs}}{k^2} 2\text{Re} \left\{ \sum_{s_j=1}^{L} n_{s_j} w_V^{(s_j)}(\theta_{bs}) w_H^{(s_j)*}(\theta_{bs}) \right.$$
$$\left. + \sum_{s_j=1}^{L} \sum_{s_l=1}^{L} n_{s_j} n_{s_l} w_V^{(s_j)}(\theta_{bs}) w_H^{(s_l)*}(\theta_{bs}) (2\pi)^3 H_{s_j s_l} \right\} \qquad (6.4.31)$$

To transform the phase matrix to the principal frame, we note that in the principal frame we use the relations

$$\hat{k}_s = \sin \theta_s \cos \phi_s \hat{x} + \sin \theta_s \sin \phi_s \hat{y} + \cos \theta_s \hat{z} \qquad (6.4.32a)$$
$$\hat{v}_s = \cos \theta_s \cos \phi_s \hat{x} + \cos \theta_s \sin \phi_s \hat{y} - \sin \theta_s \hat{z} \qquad (6.4.32b)$$
$$\hat{h}_s = -\sin \phi_s \hat{x} + \cos \phi_s \hat{y} \qquad (6.4.32c)$$

Given θ_s and ϕ_s, we can equate

$$\sin \theta_s \cos \phi_s \hat{x} + \sin \theta_s \cos \phi_s \hat{y} + \cos \theta_s \hat{z}$$
$$= \sin \theta_{bs} \cos \phi_{bs} \hat{x}_b + \sin \theta_{bs} \sin \phi_{bs} \hat{y}_b + \cos \theta_{bs} \hat{z}_b \qquad (6.4.32d)$$

and calculate θ_{bs} and ϕ_{bs} in terms of θ_s and ϕ_s.

To transform the phase matrix elements to the principal frame, we use the relations

$$\begin{aligned} P_{VV}(\theta_s, \phi_s; \theta_i, \phi_i = 0) &= P_{v_b v_b}(\hat{v}_s \cdot \hat{v}_{sb})^2 + P_{h_b v_b}(\hat{v}_s \cdot \hat{h}_{sb})^2 \\ &+ P_{3_b v_b}(\hat{v}_s \cdot \hat{v}_{sb})(\hat{v}_s \cdot \hat{h}_{sb}) \end{aligned} \quad (6.4.33)$$

$$\begin{aligned} P_{HV}(\theta_s, \phi_s; \theta_i, \phi_i = 0) &= P_{v_b v_b}(\hat{h}_s \cdot \hat{v}_{sb})^2 + P_{h_b v_b}(\hat{h}_s \cdot \hat{h}_{sb})^2 \\ &+ P_{3_b v_b}(\hat{h}_s \cdot \hat{v}_{sb})(\hat{h}_s \cdot \hat{h}_{sb}) \end{aligned} \quad (6.4.34)$$

$$\begin{aligned} P_{VH}(\theta_s, \phi_s; \theta_i, \phi_i = 0) &= P_{v_b h_b}(\hat{v}_s \cdot \hat{v}_{sb})^2 + P_{h_b h_b}(\hat{v}_s \cdot \hat{h}_{sb})^2 \\ &+ P_{3_b h_b}(\hat{v}_s \cdot \hat{v}_{sb})(\hat{v}_s \cdot \hat{h}_{sb}) \end{aligned} \quad (6.4.35)$$

$$\begin{aligned} P_{HH}(\theta_s, \phi_s; \theta_i, \phi_i = 0) &= P_{v_b h_b}(\hat{h}_s \cdot \hat{v}_{sb})^2 + P_{h_b h_b}(\hat{h}_s \cdot \hat{h}_{sb})^2 \\ &+ P_{3_b h_b}(\hat{h}_s \cdot \hat{v}_{sb})(\hat{h}_s \cdot \hat{h}_{sb}) \end{aligned} \quad (6.4.36)$$

4.2 Brightness Temperatures Computed with QCA-Based DMRT

We next illustrate the results of QCA-based DMRT in passive remote sensing. We first calculate the phase functions for passive remote sensing assuming azimuthal symmetry:

$$p_{vv}(\theta, \theta') = \int_0^{2\pi} d\phi P_{VV}(\theta, \phi; \theta', \phi' = 0) \quad (6.4.37)$$

$$p_{vh}(\theta, \theta') = \int_0^{2\pi} d\phi P_{VH}(\theta, \phi; \theta', \phi' = 0) \quad (6.4.38)$$

$$p_{hv}(\theta, \theta') = \int_0^{2\pi} d\phi P_{HV}(\theta, \phi; \theta', \phi' = 0) \quad (6.4.39)$$

$$p_{hh}(\theta, \theta') = \int_0^{2\pi} d\phi P_{HH}(\theta, \phi; \theta', \phi' = 0) \quad (6.4.40)$$

The scattering coefficient is

$$\begin{aligned} \kappa_s &= \int_0^{\pi} d\theta \sin\theta \left(p_{vv}(\theta, \theta') + p_{hv}(\theta, \theta') \right) \\ &= \int_0^{\pi} d\theta \sin\theta \left(p_{vh}(\theta, \theta') + p_{hh}(\theta, \theta') \right) \end{aligned} \quad (6.4.41)$$

which is independent of angle θ' and polarization. The absorption coefficient κ_a is as given in (6.4.15) and is also independent of angle and polarization. The extinction coefficient is

$$\kappa_e = \kappa_s + \kappa_a \quad (6.4.42)$$

The albedo is

$$\tilde{\omega} = \frac{\kappa_s}{\kappa_e} \tag{6.4.43}$$

The effective propagation constant K is

$$K = K' + i\frac{\kappa_e}{2} \tag{6.4.44}$$

where κ_e is as given by (6.4.42) and K' is as calculated by the solutions of (6.3.16) and (6.3.17). We use (6.4.42) and (6.4.44) to calculate the scattering coefficient rather than relying on the generalized Lorentz-Lorenz law of (6.3.16) and (6.3.17). As commented previously, this is done to avoid numerical accuracy problem in the determination of K'' for sticky particle pair functions.

Thus we have

$$\cos\theta \frac{dI_v}{dz}(\theta, z) = -\kappa_e I_v(\theta, z) + \kappa_a CT + \int_0^{\pi} d\theta' \sin\theta' \Big[p_{vv}(\theta, \theta') I_v(\theta', z) + p_{vh}(\theta, \theta') I_h(\theta', z)\Big] \tag{6.4.45}$$

$$\cos\theta \frac{dI_h}{dz}(\theta, z) = -\kappa_e I_h(\theta, z) + \kappa_a CT + \int_0^{\pi} d\theta' \sin\theta' \Big[p_{hv}(\theta, \theta') I_v(\theta', z) + p_{hh}(\theta, \theta') I_h(\theta', z)\Big] \tag{6.4.46}$$

The boundary conditions at $z = 0$ and at $z = -d$ are the Fresnel type using the effective propagation constant K of (6.4.44) for the dense medium.

Consider a two layer medium, the boundary conditions are the Fresnel type using the effective propagation constant for the dense medium. For the upper boundary, the air-medium 1 interface at $z = 0$,

$$R_v(\theta) = r_v(\theta) \tag{6.4.47a}$$

$$R_h(\theta) = r_h(\theta) \tag{6.4.47b}$$

where r_v and r_h are the Fresnel reflectivities with effective propagation constant K.

$$r_v(\theta) = \left| \frac{k^2 \cos\theta - K' \left(k^2 - K'^2 \sin^2\theta\right)^{\frac{1}{2}}}{k^2 \cos\theta + K' \left(k^2 - K'^2 \sin^2\theta\right)^{\frac{1}{2}}} \right|^2 \tag{6.4.48}$$

$$r_h(\theta) = \left| \frac{K' \cos\theta - \left(k^2 - K'^2 \sin^2\theta\right)^{\frac{1}{2}}}{K' \cos\theta + \left(k^2 - K'^2 \sin^2\theta\right)^{\frac{1}{2}}} \right|^2 \tag{6.4.49}$$

In the lower boundary $z = -d$, the interface between medium 1 and medium 2,

$$R_{vg}(\theta) = r_{vg}(\theta) \tag{6.4.50}$$

$$R_{hg}(\theta) = r_{vg}(\theta) \tag{6.4.51}$$

with

$$r_{vg}(\theta) = \left| \frac{\frac{\epsilon_2}{\epsilon_0} k^2 \cos\theta - K' \left(\frac{\epsilon_2}{\epsilon_0} k^2 - K'^2 \sin^2\theta \right)^{\frac{1}{2}}}{k^2 \cos\theta + K' \left(\frac{\epsilon_2}{\epsilon_0} k^2 - K'^2 \sin^2\theta \right)^{\frac{1}{2}}} \right|^2 \tag{6.4.52}$$

$$r_{hg}(\theta) = \left| \frac{K' \cos\theta - \left(\frac{\epsilon_2}{\epsilon_0} k^2 - K'^2 \sin^2\theta \right)^{\frac{1}{2}}}{K' \cos\theta + \left(\frac{\epsilon_2}{\epsilon_0} k^2 - K'^2 \sin^2\theta \right)^{\frac{1}{2}}} \right|^2 \tag{6.4.53}$$

The set of DMRT equations can be solved using the method of Gaussian quadrature that is described in Volume I. The boundary conditions at the air-snow interface $z = 0$ are

$$I_v(z = 0, \pi - \theta) = R_v(\theta) I_v(z = 0, \theta) \tag{6.4.54}$$

$$I_h(z = 0, \pi - \theta) = R_h(\theta) I_h(z = 0, \theta) \tag{6.4.55}$$

and for the snow-ground interface $z = -d$, they are

$$I_v(z = -d, \theta) = R_{vg}(\theta) I_v(z = -d, \pi - \theta) + (1 - R_{vg}(\theta)) C T_g \tag{6.4.56}$$

$$I_h(z = -d, \theta) = R_{hg}(\theta) I_h(z = -d, \pi - \theta) + (1 - R_{hg}(\theta)) C T_g \tag{6.4.57}$$

where T_g is the ground temperature.

After solving the eigenvalue problem and imposing on the boundary conditions, the brightness temperatures in the direction θ_0, where $\theta_0 = \sin^{-1}(K' \sin\theta / k)$ is related to θ by Snell's law, are given by

$$T_{B_v}(\theta_0) = \frac{1}{C}\Big(1 - R_v(\theta)\Big) I_v(z = 0, \theta) \tag{6.4.58}$$

$$T_{B_h}(\theta_0) = \frac{1}{C}\Big(1 - R_h(\theta)\Big) I_h(z = 0, \theta) \tag{6.4.59}$$

4.3 Numerical Results for Sticky and Non-Sticky Particles

In applying QCA, the pair distribution function of particle positions must be specified. The Percus-Yevick approximation is used in calculating the particle distribution functions in the study.

Pair Distribution Functions of Multiple Sizes

For the Percus-Yevick approximation in the case of general multiple species, the solution is obtained by Baxter [1970] and given in Chapter 8 of Volume II. The pair distribution functions are calculated by assuming that the particles are non-penetrable spheres without interparticle force. The total correlation function $h_{ij}(r)$ is defined as

$$h_{ij}(r) = g_{ij}(r) - 1 \tag{6.4.60}$$

where $r = |\overline{r}|$, and the structure factor $H_{ij}(p)$ is the Fourier transform of $h_{ij}(r)$.

$$\widetilde{H}_{ij}(\overline{p}) = (n_i n_j)^{\frac{1}{2}} \int d\overline{r} e^{i\overline{p}\cdot\overline{r}} h_{ij}(\overline{r}) = (2\pi)^3 H_{ij}(p) \tag{6.4.61}$$

where $p = |\overline{p}|$.

Pair Distribution Functions of Sticky Particles

In natural media, the densely packed particles can have adhesive force. They adhere together to form aggregates. Thus the model of sticky particles was developed. The sticky particle has a stickiness parameter τ. The smaller the τ is, the more sticky the particles are. The Percus-Yevick approximation of the pair distribution function $g(r)$ for the sticky spherical particles can be solved analytically using the factorization method of Baxter [1970]. The calculations of $g(r)$ can be found in Chapter 8 of Volume II.

We will study the use of dense media with a single particle size a and with a sticky parameter Sticky parameter τ. In Fig. 6.4.2, the pair functions for several τ's are plotted. The scattering depends on the structure factor $H(p)$. In Fig. 6.4.3, the structure factor $H(p)$ is plotted as a function of pb, where $b = 2a$ is the diameter of the particles, for several τ's. For non-small τ, the interparticle force is small. $H(0)$ is negative indicating that scattering is smaller than independent scattering. For small τ, $H(0)$ is positive indicating that scattering is larger than independent scattering. We note in Fig. 6.4.3 that for sticky particles, there are sharp variations of pair functions. These make it difficult to use the Lorentz-Lorenz law to calculate ImK which is much smaller than ReK. Thus, in implementations, we use Lorentz-Lorenz law to calculate ReK but use (6.4.41) to calculate κ_s and (6.4.15) to calculate κ_a. Then $2\text{Im}K = \kappa_a + \kappa_s$.

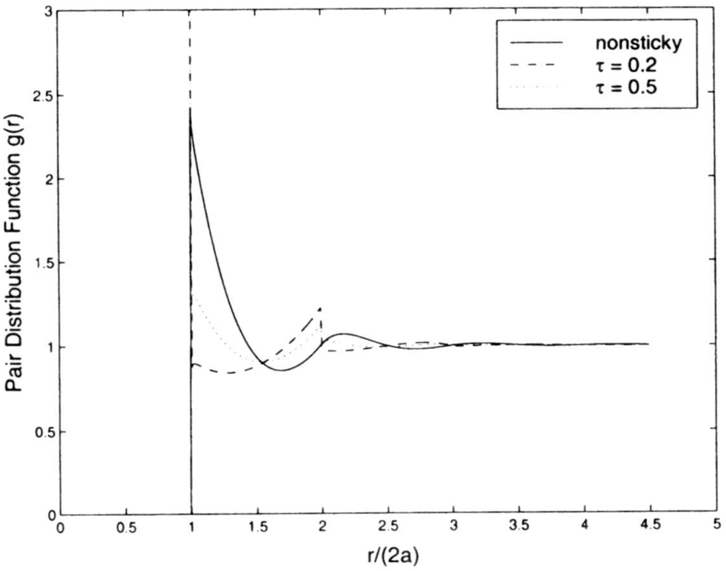

Figure 6.4.2 Pair distribution function of sticky particles.

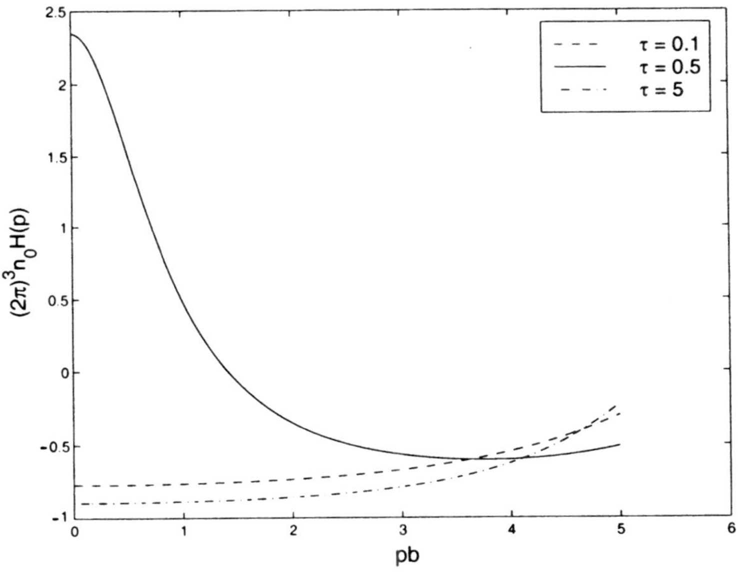

Figure 6.4.3 Structure factor $H(p)$ as a function of pb for sticky particles, where b is the diameter of the particles.

species	1	2	3	4	5
diameter (cm)	0.05	0.1	0.2	0.3	0.4
fractional volume	0.1	0.12	0.05	0.02	0.01

Table 6.4.1 Numerical values for the QCA non-sticky extinction rate calculation.

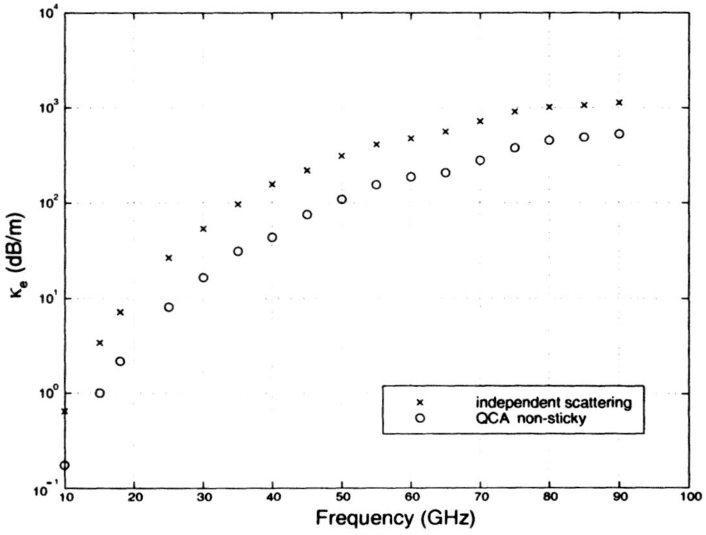

Figure 6.4.4 Extinction rate as a function of frequency for multiple moderate size particles for QCA and independent scattering. Parameters are listed in Table 6.4.1.

Results of Multiple Sizes Non-Sticky Particle Model

To analyze the effect of multiple sizes, we first illustrate the use of a medium with particles of five sizes. The size and the fractional volume of each species are listed in Table 6.4.1. The relative permittivity of particles is assumed to be $\epsilon_{rs} = 3.2$ for all frequencies. In Fig. 6.4.4, the extinction rate is plotted as a function of frequency and compared with the results of independent scattering.

For brightness temperatures calculation, we illustrate the use of a medium with particles of two sizes. The size and the fractional volume of each species are listed in Table 6.4.2. The first species is of the small particles which occupy a higher volume fraction while the second species is of the large particles. Assume that the first species has the size of 1 mm in diameter and occupies a volume of 30% while the second species has the size of 4 mm

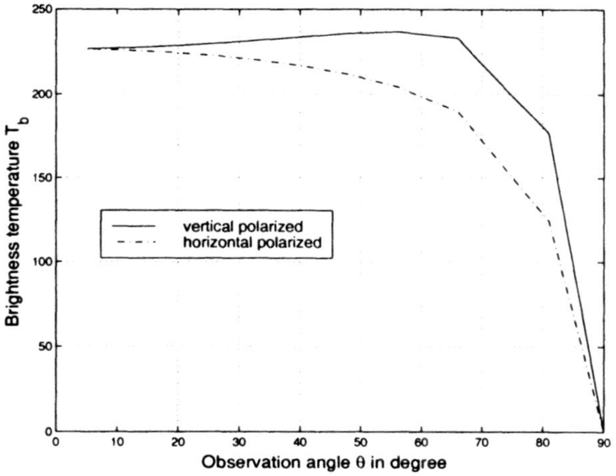

(a) 19 GHz

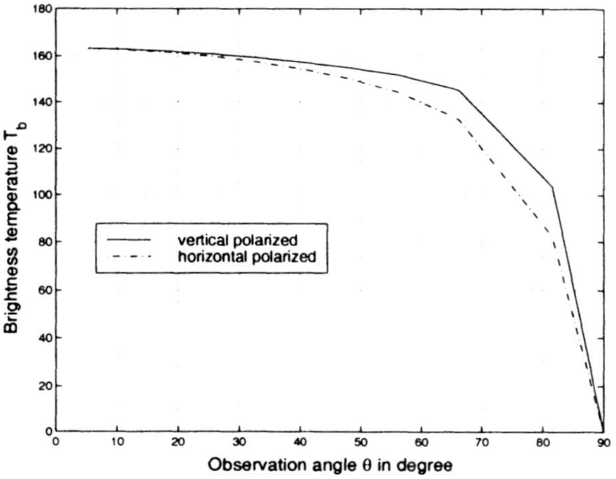

(b) 37 GHz

Figure 6.4.5 Brightness temperatures versus observation angles based on the no interparticle force model. Medium consists of two particles sizes of 0.1 and 0.4 cm in diameter are used with small particles occupying a much larger fractional volume. Parameters are found in Table 6.4.2. (a) 19 GHz, (b) 37 GHz.

species	1	2
radius (cm)	0.05	0.2
fractional volume	0.3	0.03
n_0 (cm^{-3})	572	0.89

Table 6.4.2 Numerical values of the physical parameters for the QCA non-sticky particle simulation.

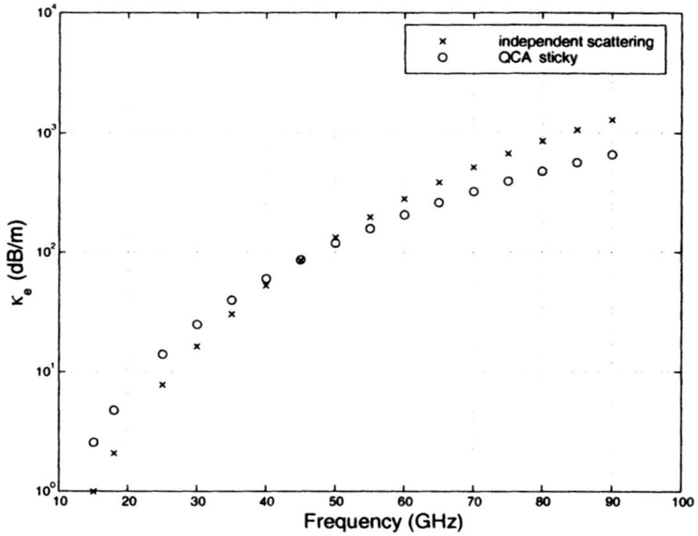

Figure 6.4.6 Comparisons of extinction rate as a function of frequency for sticky moderate size particles of QCA and independent scattering. Fractional volume is 30%, the other parameters are found in Table 6.4.3.

in diameter but only occupies the fractional volume of 3%. The number density ratio of the two sizes is 643. The permittivity of particles is assumed to be $\epsilon_p = 3.2 + i0.003\epsilon_o$ for both frequencies. The brightness temperatures for 19 and 37 GHz are presented in Figs. 6.4.5a and 6.4.5b as a function of observation angle. The snow pack physical temperature and the ground temperature are both assumed to be 270 K. The thickness of the snow layer is 50 cm and the ground permittivity is $\epsilon_2 = (6 + i0.1)\epsilon_o$.

The model of multiple size particles without interparticle force exhibits a strong frequency dependence similar to independent scattering. We next study the sticky particle model which exhibits a weaker frequency dependence.

Results of Sticky Particle Model

In the sticky particle model, the particles are allowed to stick together to form clusters. A single-size sticky particle model is used in this section. The comparison of the frequency dependence for independent scattering and QCA-sticky is shown in Fig. 6.4.6. The particle size is 1.2 mm in diameter and the fractional volume is 30%. The permittivity of the particles is $\epsilon_s = 3.2\epsilon_0$ and the stickiness parameter τ for the sticky particle model is 0.1.

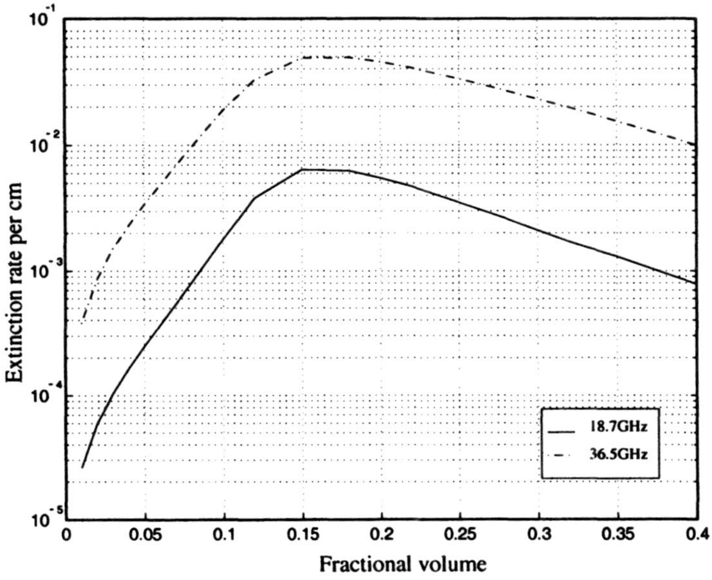

Figure 6.4.7 Extinction rate versus fractional volume for 18.7 and 36.5 GHz. The ratio of extinction rate between 18.7 and 36.5 GHz is in the range of 8 to 12. Parameters are found in Table 6.4.3.

Parameter	Value
diameter (cm)	0.06
stickiness τ	0.1
relative permittivity of particles	
18.7 GHz	$3.2 + i0.002$
36.5 GHz	$3.2 + i0.01$

Table 6.4.3 Numerical values of physical parameters for the QCA-sticky simulation shown in Fig. 6.4.6 and Fig. 6.4.7.

We note that at low frequency, the scattering is larger than independent scattering. This is because the particles adhere to form clusters. At low frequencies, they scatter like a particle with larger effective size. At high frequency, the clustering effect of scattering disappears and the particles scatter like the usual dense media. Scattering is less than independent scattering. Thus, the frequency dependence of scattering is less than independent scattering.

Figure 6.4.7 shows the extinction rate as a function of fractional volume.

Parameter	Value
stickiness τ	0.1
fractional volume	20%
relative permittivity of particles	
18.7 GHz	$3.2 + i0.002$
36.5 GHz	$3.2 + i0.01$
snow temperature	260
ground temperature	260
relative ground permittivity	3.2

Table 6.4.4 Numerical values of physical parameters for Figs. 6.4.8, 6.4.9, and 6.4.10.

The ratio of the extinction rate $(\kappa_e)_{37\text{GHz}}/(\kappa_e)_{19\text{GHz}}$ is about 8–12 times for volume fraction between 0.02 to 0.4. The parameters for the simulation are listed in Table 6.4.3. The extinction rate does not increase monotonically as fractional volume. It saturates at fractional volume of 15% and decreases for volume larger than 15%.

After calculating the extinction rate, scattering rate, and the phase functions by QCA, they are inserted into DMRT equations to calculate the brightness temperatures. Figures 6.4.8 and 6.4.9 are the calculated brightness temperatures from RT equations based on QCA-sticky model and represent typical values in passive remote sensing of snow. In Fig. 6.4.8, we plot the brightness temperatures of horizontal polarization at 19 GHz as a function of thickness of the scattering layer for various particle sizes. The observation angle is 55°. In Fig. 6.4.9, we plot the brightness temperatures for 37 GHz. Figure 6.4.10 shows the brightness temperature difference $T_{B19H} - T_{B37H}$. The parameters are listed in Table 6.4.4. The brightness temperatures for 19 GHz decreases as layer thickness increases. However, at 37 GHz, the brightness temperatures saturate at thickness around 50 cm. This is expected since the penetration depth for 37 GHz is smaller. The differences of the brightness temperatures (ΔT_B) between two frequencies increase linearly with snow depth at small snow depth and small grain size. For large grain size, due to the saturation at 37 GHz, ΔT_B may saturate or even decrease with snow depth as shown in Fig. 6.4.10.

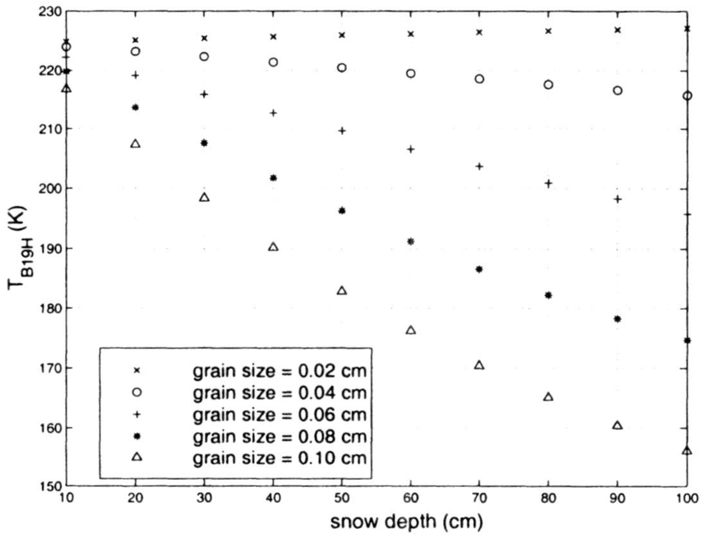

Figure 6.4.8 Brightness temperatures of horizontal polarization at 19 GHz as a function of snow depth with various grain sizes based on the sticky particle model with $\tau = 0.1$ and fractional volume = 20%. The other parameters are listed in Table 6.4.4.

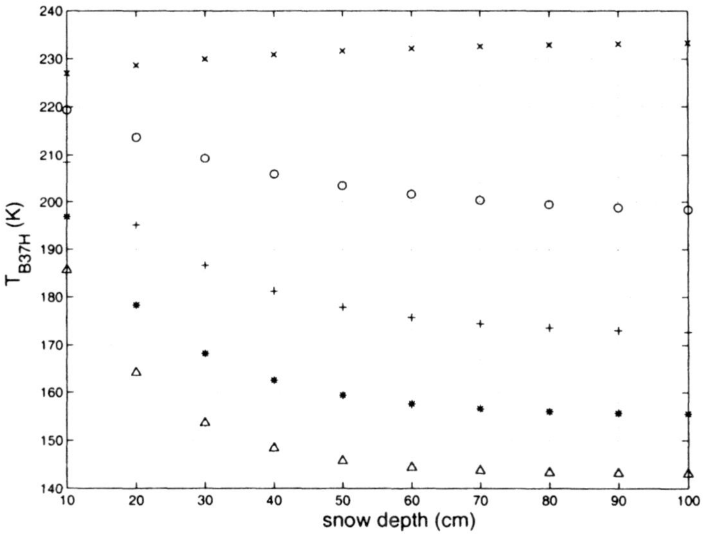

Figure 6.4.9 Brightness temperatures of horizontal polarization at 37 GHz as a function of snow depth with various grain sizes based on the sticky particle model with $\tau = 0.1$ and fractional volume = 20%. The other parameters are listed in Table 6.4.4.

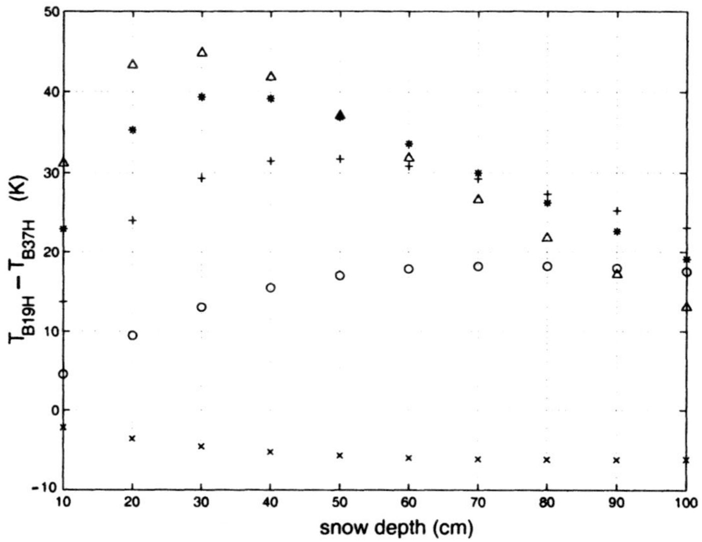

Figure 6.4.10 Brightness temperature differences of horizontal polarization as a function of snow depth with various grain sizes based on the sticky particle model with $\tau = 0.1$ and fractional volume = 20%. The other parameters are listed in Table 6.4.4.

REFERENCES AND ADDITIONAL READINGS

Baxter, R. J. (1968), Percus-Yevick equation for hard spheres with surface adhesion, *J. Chem. Phys.*, **49(6)**, 2770–2773.

Baxter, R. J. (1970), Ornstein-Zernike relation and Percus-Yevick approximation for fluid mixtures, *J. Chem. Phys.*, **52**, 4559–4562.

Born, M. and E. Wolf (1975), *Principles of Optics*, 5th edition, Pergamon Press, New York.

Chow, P. L., W. E. Kohler, and G. C. Papanicolaou, Eds. (1981), *Multiple Scattering and Waves in Random Media*, North-Holland Publishing Company, Amsterdam.

Cruzan, O. R. (1961), Translational addition theorems for spherical vector wave functions, **TR-906**, Diamond Ordinance Fuse Laboratories, Department of the Army, Washington DC.

Cruzan, O. R. (1962), Translational addition theorems for spherical vector wave functions, *Quart. J. Appl. Math.*, **20**, 33–40.

Ding, K. H. (1989), Electromagnetic wave propagation and scattering in dense media, Ph.D. thesis, University of Washington, Seattle.

Ding, K. H., C. Mandt, L. Tsang, and J. A. Kong (1992), Monte Carlo simulations of pair distribution functions of dense discrete random media with multiple sizes of particles, *J. Electromag. Waves and Appl.*, **6(8)**, 1015–1030.

Ding, K. H. and L. Tsang (1988), Effective propagation constants of dense nontenuous media with multi-species of particles, *J. Electromag. Waves and Appl.*, **2(8)**, 757–777.

Ding, K. H. and L. Tsang (1989), *Progress in Electromag. Res.*, J. A. Kong, Ed., **1**, Ch. 3, Elsevier Science Publishing Co., New York.

Ding, K. H. and L. Tsang (1992), Monte Carlo simulations of pair distribution functions of dense discrete random media with multiple sizes of particles, *J. Electromag. Waves and Appl.*, **6(8)**, 1015–1030.

Ding, K. H., L. M. Zurk, and L. Tsang (1994), Pair distribution functions and attenuation rates for sticky particles in dense media, *J. Electromag. Waves and Appl.*, **8(12)**, 1585–1604.

Foldy, L. L. (1945), The multiple scattering of waves, *Phys. Rev.*, **67**, 107–119.

Hallikainen, M. T., F. T. Ulaby, and T. E. V. Deventer (1987), Extinction behavior of dry snow in the 18- to 90-GHz range, *IEEE Trans. Geosci. Remote Sens.*, **25(6)**, 737–750.

Ishimaru, A. and Y. Kuga (1982), Attenuation constant of a coherent field in a dense distribution of particles, *J. Opt. Soc. Am.*, **72**, 1317–1320.

Koh, G. (1992), Experimental study of electromagnetic wave propagation in dense random media, *Waves in Random Media*, **2**, 39–48.

Kuga, Y., A. Ishimaru, and D. Rice (1993), Velocity of coherent and incoherent electromagnetic waves in a dense strongly scattering medium, *Phys. Rev. B*, **48**, 13155–13158.

Kuga, Y. and P. Phu (1995), Experimental techniques in random media and rough surface, *Progress in Electromagnetics Research*, **9**, Elsevier Science Publishers, Cambridge, Mass.

Lam, C. M. and A. Ishimaru (1993), Mueller matrix representation for a slab of random medium with discrete particles and random rough surfaces with moderate surface roughness, *Waves in Random Media*, **3(2)**, 111–125.

Lam, C. M. and A. Ishimaru (1994), Mueller matrix calculation for a slab of random medium with both random rough surfaces and discrete particles, *IEEE Trans. Antennas Propagat.*, **42(2)**, 145–146.

Lax, M. (1951), Multiple scattering of waves, *Rev. Modern Phys.*, **23**, 287–310.

Lax, M. (1952), Multiple scattering of waves II. The effective field in dense systems, *Phys. Rev.*, **85**, 261–269.

Lim, H. H., M. E. Veysoglu, S. H. Yueh, and R. T. Shin (1994), Random medium model approach to scattering from a random collection of discrete scatterers, *J. Electromag. Waves and Appl.*, **8**, 801–817.

Mandt, C. E., Y. Kuga, L. Tsang, and A. Ishimaru (1992), Microwave propagation and scattering in a dense distribution of nontenuous spheres: experiment and theory, *Waves in Random Media*, **2(3)**, 225–234.

Mandt, C. and L. Tsang (1992), Backscattering enhancement from a random distribution of large discrete spherical scatterers with a size distribution, *J. Opt. Soc. Am. A*, **9(12)**, 2246–2251.

Mätzler, C. and U. Wegmuller (1987), Dielectric properties of fresh-water ice at microwave frequencies, *J. Phys. D.*, **20**, 1623–1630.

Rotenberg, M., R. Bivius, N. Metroplis, and J. K. Wooten, Jr. (1959), *The 3-j and 6-j Symbols,* Technology Press, Cambridge, MA.

Siqueira, P. R. and K. Sarabandi (1996), Method of moment evaluation of the 2-D quasi-crystalline approximation, *IEEE Trans. Antennas Propagat.*, **44(8)**, 1067–1077.

Tsang, L. (1984), Scattering of electromagnetic waves from a half space of nonspherical particles, *Radio Sci.*, **19**, 1450–1460.

Tsang, L. (1987), Passive remote sensing of dense nontenuous media, *J. Electromag. Waves and Appl.*, **1(2)**, 159–173.

Tsang, L. (1991), Polarimetric passive microwave remote sensing of random discrete scatterers and rough surfaces, *J. Electromag. Waves and Appl.*, **5(1)**, 41–57.

Tsang, L. (1992), Dense media radiative transfer theory for dense discrete random media with spherical particles of multiple sizes and permittivities, *Progress in Electromag. Res.*, **6**, Dielectric Properties of Heterogeneous Materials, A. Priou, ed., Ch. 5, 181–230, Elsevier Science Publishing Co., New York.

Tsang, L., C. T. Chen, A. T. C. Chang, J. Guo, and K. H. Ding (2000), Dense media radiative transfer theory based on quasi cyotalline approximation with application to passive microwave remote sensing of snow, *Radio Sci.*, **35**, 731–749.

Tsang, L. and A. Ishimaru (1984), Backscattering enhancement of random discrete scatterers, *J. Opt. Soc. Am. A*, **1**, 836–839.

Tsang, L. and A. Ishimaru (1985), Radiative wave and cyclical transfer equations for dense nontenuous media, *J. Opt. Soc. Am. A*, **2**, 2187–2194.

Tsang, L. and A. Ishimaru (1992), Radiative wave equations for vector electromagnetic propagation in dense nontenuous media, *J. Electromag. Waves and Appl.*, **1**, 809–813.

Tsang, L. and J. A. Kong (1980), Multiple scattering of electromagnetic waves by random distribution of discrete scatterers with coherent potential and quantum mechanical formulism, *J. Appl. Phys.*, **51**, 3465–3485.

Tsang, L. and J. A. Kong (1982), Effective propagation constants for coherent electromagnetic wave propagation in media embedded with dielectric scatterers, *J. Appl. Phys.*, **53(11)**, 7162–7173.

Tsang, L. and J. A. Kong (1983), Scattering of electromagnetic waves from a half space of densely distributed dielectric scatterers, *Radio Sci.*, **18**, 1260–1272.

Tsang, L. and J. A. Kong (1992), Scattering of electromagnetic waves from a dense medium consisting of correlated Mie scatterers with size distributions and applications to dry snow, *J. Electromag. Waves and Appl.*, **6(3)**, 265–286.

Tsang, L., J. A. Kong, and R. T. Shin (1985), *Theory of Microwave Remote Sensing*, Wiley-Interscience, New York.

Tsang, L., C. Mandt, and K. H. Ding (1992), Monte Carlo simulations of the extinction rate of dense media with randomly distributed dielectric spheres based on solution of Maxwell's equations, *Optics Lett.*, **17(5)**, 314–316.

Twersky, V. (1962), On scattering of waves by random distributions I. Free space scatterer formalism, *J. Math. Phys.*, **3**, 700–715.

Twersky, V. (1964), On propagation in random media of scatterers, *Proc. Symp. Appl. Math.*, **16**, 84–116, Am. Math. Soc., Providence, RI.

Twersky, V. (1977), Coherent scalar field in pair-correlated random distributions of aligned scatterers, *J. Math. Phys.*, **18**, 2468–2486.

Twersky, V. (1978), Coherent electromagnetic waves in pair-correlated random distributions of aligned scatterers, *J. Math. Phys.*, **19**, 215–230.

van Albada, M. P., B. A. van Tiggelen, A. Langendijk, and A. Tip (1991), Speed of propagation of classical waves in strongly scattering media, *Phys. Rev. Lett.*, **66(24)**, 3132–3135.

Varadan, V. K., V. N. Bringi, V. A. Varadan, and A. Ishimaru (1983), Multiple scattering theory for waves in discrete random media and comparison with experiments, *Radio Sci.*, **18**, 321–327.

Waterman, P. C. (1965), Matrix formulation of electromagnetic scattering, *Proc. IEEE*, **53**, 805–811.

Waterman, P. C., and R. Truell (1961), Multiple scattering of waves, *J. Math. Phys.*, **2(4)**, 512–537.

West, R., D. Gibbs, L. Tsang, and A. K. Fung (1994), Comparison of optical scattering experiments and the quasi-crystalline approximation for dense media, *J. Opt. Soc. Am.*, **11(6)**, 1854–1858.

West, R., L. Tsang, and D. P. Winebrenner (1993), Dense medium radiative transfer theory for two scattering layers with a Kayleigh distribution of particle sizes, *IEEE Trans. Geosci. Remote Sens.*, **31**, 426–437.

Winebrenner, D. P., L. Tsang, B. Wen, and R. West (1989), Sea-ice characterization measurements needed for testing of microwave remote sensing models, *IEEE J. Ocean. Eng.*, **14(2)**, 149–158.

Winebrenner, D. P., L. Tsang, B. Wen, and R. West (1992), Sensitivities for two polarimetric backscattering models for sea ice to geophysical parameters, in *Direct and Inverse Methods in Radar Polarimetry, Part 2*, edited by W. Boerner, 1191–1212, Kluwer Academic Publishers, The Netherlands.

Zurk, L. M. (1995), Electromagnetic wave propagation and scattering in dense, discrete random media with application to remote sensing of snow, Ph.D. thesis, University of Washington, Seattle.

Zurk, L. M., L. Tsang, K. H. Ding, and D. P. Winebrenner (1995), Monte Carlo simulations of the extinction rate of densely packed spheres with clustered and non-clustered geometries, *J. Opt. Soc. Am.*, **12(8)**, 1772–1781.

Zurk, L. M., L. Tsang, and D. P. Winebrenner (1996), Scattering properties of dense media from Monte Carlo simulations with application to active remote sensing of snow, *Radio Sci.*, **31(4)**, 803–819.

Zurk, L. M., L. Tsang, D. P. Winebrenner, J. Shi, and R. E. Davis (1997), Electromagnetic scattering calculated from pair distribution functions retrieved from planar snow sections, *IEEE Trans. Geosci. Remote Sens.*, **35(6)**, 1419–1428.

Chapter 7

DENSE MEDIA SCATTERING

1 Introduction

In Chapter 5, Section 4, we showed that the QCA–CP on the Dyson's equation and the correlated ladder approximation on the Bethe-Salpeter equation obeys energy conservation. In this chapter, we derive the dense media radiative transfer equation (DMRT) rigorously from these two approximations for the case of small dielectric scatterers.

We consider dense discrete random media with multiple species of particles. The multiple species refers to the fact that the medium is a mixture of particles with different sizes and permittivities. The governing equations of Dyson's equation under QCA–CP approximation and the Bethe-Salpeter equation under the correlated ladder approximation for multiple species of particles that are correlated in positions. Expressions for the mass operator and the intensity operator are given. In Section 2, we calculate the effective propagation constants, the mean Green's function, and the mean field for a half-space discrete random medium of multiple species based on QCA–CP. In Section 3, we derive rigorously the dense media radiative transfer equations as well as the boundary conditions from the integral form of the correlated ladder approximation of multiple species. The pair distribution functions of multiple species are calculated by using the Percus-Yevick equation for non-interpenetrable spheres. The final results of dense media radiative transfer equations for active and passive remote sensing are respectively in Sections 4 and 6. All four Stokes parameters are included so that the results are applicable to polarimetric remote sensing of dense media. The numerical results of dense media radiative transfer equations for active and passive remote sensing are illustrated in Section 7 for media with non-sticky particles of multiple sizes and permittivities. The input physical parameters of the dense medium radiative transfer theory are the background medium permittivity, the particles permittivities, and their size distributions.

Dense media scattering is a subject of continual interests because of the importance of the effects from the correlation of scatterers [Tsang and Kong, 1980, 1982; Tsang, 1987, 1991; Ding and Tsang, 1988]. It is also important in the subject of backscattering enhancement and wave localization [Tsang and Ishimaru, 1984; Barabanenkov et al. 1991a,b; Barabanenkov and Kryukov, 1992; Sheng, 1990, 1995].

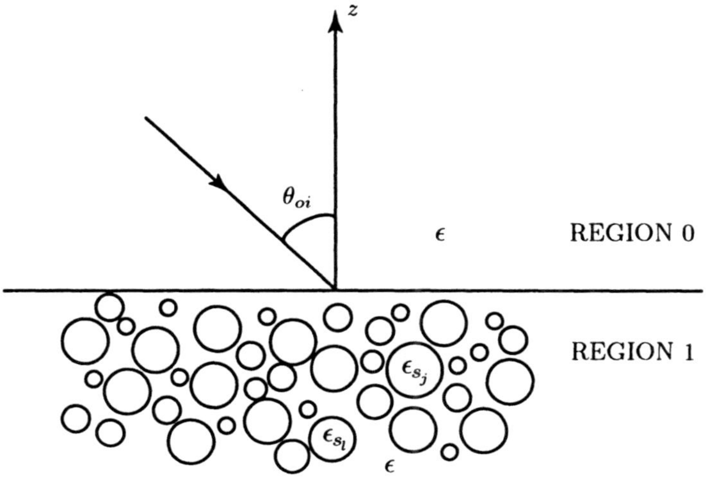

Figure 7.2.1 Active remote sensing with an incident electromagnetic wave impinging upon a half-space discrete random medium of multiple species in the direction $(\pi - \theta_{oi}, \phi_i)$.

2 Effective Propagation Constants, Mean Green's Function, and Mean Field for Half-Space Discrete Random Medium of Multiple Species

Consider a half-space of small spherical particles occupying the region $z < 0$ (Fig. 7.2.1). The upper and lower half-spaces are denoted as regions 0 and 1, respectively. The particles are dense and nontenuous so that the volume fraction $f_{s_j} = 4\pi n_{s_j} a_{s_j}^3/3$ of the s_j species can be appreciable and the wavenumber $k_{s_j} = \omega\sqrt{\mu\epsilon_{s_j}}$ of the particle of species s_j can be substantially different from the background wavenumber k, $s_i = 1, 2, \ldots, L$. The mean Green's function and the mean field contain reflection terms due to this effective boundary. We will first use the QCA–CP equation for multiple species as derived in Section 3.7 of Chapter 5. From (5.3.96), the half-space mean Green's function obeys the equation

$$\langle\overline{\overline{G}}(\overline{r},\overline{r}')\rangle = \overline{\overline{G}}_o\left(\overline{r},\overline{r}'\right) + \sum_{s_j=1}^{L} n_{s_j} \int d\overline{r}'' \int d\overline{r}''' \int_{z_j<0} d\overline{r}_j \overline{\overline{G}}_o\left(\overline{r},\overline{r}''\right) \cdot \overline{\overline{C}}_j^{s_j}\left(\overline{r}'',\overline{r}'''\right) \langle\overline{\overline{G}}(\overline{r}''',\overline{r}')\rangle \tag{7.2.1}$$

To solve (7.2.1) in the small particles limit, we let the mean Green's function assume the following form. For z, $z' < 0$, we have

$$\langle\overline{\overline{G}}(\overline{r},\overline{r}')\rangle = \overline{\overline{G}}_P\left(\overline{r},\overline{r}'\right) + \overline{\overline{G}}_R\left(\overline{r},\overline{r}'\right) \tag{7.2.2}$$

where

$$\overline{\overline{G}}_P(\overline{r},\overline{r}') = \left(\overline{\overline{I}} + \frac{1}{K^2}\nabla\nabla\right)\frac{e^{iK|\overline{r}-\overline{r}'|}}{4\pi|\overline{r}-\overline{r}'|} \tag{7.2.3}$$

is the infinite space Green's function with effective propagation constant K. In (7.2.2) $\overline{\overline{G}}_R$ is the part of the Green's function that accounts for the boundary reflection term.

We next consider Eq. (5.3.91) that governs $\overline{\overline{t}}_j^{s_j}$. To simplify notation, we have dropped the hat (^) symbol used in Section 3 of Chapter 5 in this chapter for QCA–CP operators. We note that the scattering operator $\overline{\overline{U}}_j^{s_j}$ in the coordinate representation is

$$\langle\overline{r}|\overline{\overline{U}}_j^{s_j}|\overline{r}'\rangle = U^{s_j}(\overline{r}-\overline{r}_j)\overline{\overline{I}}\delta(\overline{r}-\overline{r}') \tag{7.2.4}$$

where

$$U^{s_j}(\overline{r}-\overline{r}_j) = \begin{cases} 0 & \text{for } |\overline{r}-\overline{r}_j| > a_{s_j} \\ k_{s_j}^2 - k^2 & \text{for } |\overline{r}-\overline{r}_j| \le a_{s_j} \end{cases} \tag{7.2.5}$$

Because $ka_{s_j} \ll 1$, $\overline{\overline{U}}_j^{s_j}$ is short-ranged. Hence the transition operator $\overline{\overline{t}}_j^{s_j}$ is short-ranged. Thus, if $\overline{r}$ and $\overline{r}'$ are several diameters deep in the lower half-space, the contribution of $\overline{\overline{G}}_R$ in the equation for $\overline{\overline{t}}_j^{s_j}$ will be much smaller than that of $\overline{\overline{G}}_P$. Hence, we have, from (5.3.91),

$$\overline{\overline{t}}_j^{s_j} = \overline{\overline{U}}_j^{s_j} + \overline{\overline{U}}_j^{s_j}\overline{\overline{G}}_P\,\overline{\overline{t}}_j^{s_j} \tag{7.2.6}$$

A similar argument applies Eq. (5.3.93) for $\overline{\overline{C}}_j^{s_j}$, which is also short-ranged, and it is nonzero in the vicinity of the particle located at $\overline{r}_j$ so that the boundary effect at $z=0$ can be ignored. Thus (5.3.93) becomes

$$\overline{\overline{C}}_j^{s_j} = \overline{\overline{t}}_j^{s_j} + \sum_{s_l=1}^{L} n_{s_l}\int d\overline{r}_l\,\overline{\overline{t}}_j^{s_j}\overline{\overline{G}}_P h_{s_j s_l}(\overline{r}_j-\overline{r}_l)\overline{\overline{C}}_l^{s_l} \tag{7.2.7}$$

Thus (7.2.6) and (7.2.7) for $\overline{\overline{t}}_j^{s_j}$ and $\overline{\overline{C}}_j^{s_j}$ become identical to that of the infinite space case which was solved in Chapter 5, Section 3.7. The solution for $\overline{\overline{t}}_j^{s_j}$ is, in momentum representation,

$$\langle\overline{p}|\overline{\overline{t}}^{s_j}|\overline{p}'\rangle = 3v_{s_j}K^2y_{s_j}\left(1+i\frac{2}{3}K^3a_{s_j}^3y_{s_j}\right)\overline{\overline{I}} \tag{7.2.8}$$

where $v_{s_j} = 4\pi a_{s_j}^3/3$ is the volume of the particle of s_j species. In (7.2.8)

$$y_{s_j}(K) = \frac{k_{s_j}^2 - k^2}{3K^2 + (k_{s_j}^2 - k^2)} \tag{7.2.9}$$

The solution for $\overline{\overline{C}}^{s_j}$, in momentum representation, is, in the small particles limit,

$$\langle \overline{p} | \overline{\overline{C}}^{s_j} | \overline{p}' \rangle = C_{s_j} \overline{\overline{I}} \tag{7.2.10}$$

where

$$C_{s_j} = \frac{3K^2 v_{s_j} y_{s_j}}{D(K)} \left\{ 1 + i\frac{2}{3} K^3 a_{s_j}^3 y_{s_j} + \frac{i2K^3}{3D(K)} \sum_{s_l=1}^{L} f_{s_l} y_{s_l} \right.$$
$$\left. \times \left[y_{s_l} a_{s_l}^3 + \sum_{s_m=1}^{L} n_{s_m} y_{s_m} a_{s_m}^3 8\pi^3 H_{s_l s_m}(\overline{p}=0) + 6\pi^2 D(K) H_{s_j s_l}(\overline{p}=0) \right] \right\} \tag{7.2.11}$$

and

$$D(K) = 1 - \sum_{s_l=1}^{L} f_{s_l} y_{s_l} \tag{7.2.12a}$$

$$H_{s_j s_l}(\overline{p}=0) = \frac{1}{(2\pi)^3} \int d\overline{r}\, h_{s_j s_l}(\overline{r}) \tag{7.2.12b}$$

In solving (7.2.6) and (7.2.7), we have retained the leading term in the real part and the leading term in the imaginary part of $\overline{\overline{t}}_j^{s_j}$ and $\overline{\overline{C}}_j^{s_j}$. This enables us to keep track of the real part of the effective propagation constant and the effective attenuation rate. In the coordinate representation, we have

$$\overline{\overline{C}}_j^{s_j}(\overline{r}, \overline{r}') = C_{s_j} \delta(\overline{r} - \overline{r}_j) \delta(\overline{r}' - \overline{r}_j) \overline{\overline{I}} \tag{7.2.13}$$

Substituting (7.2.13) into (7.2.1) gives

$$\langle \overline{\overline{G}}(\overline{r}, \overline{r}') \rangle = \overline{\overline{G}}_o(\overline{r}, \overline{r}') + \sum_{s_j=1}^{L} n_{s_j} C_{s_j} \int_{z_j<0} d\overline{r}_j\, \overline{\overline{G}}_o(\overline{r}, \overline{r}_j) \langle \overline{\overline{G}}(\overline{r}_j, \overline{r}') \rangle \tag{7.2.14}$$

If we take $\nabla \times \nabla \times -k^2$ of (7.2.14), then

$$\nabla \times \nabla \times \langle \overline{\overline{G}}(\overline{r}, \overline{r}') \rangle - k^2 \langle \overline{\overline{G}}(\overline{r}, \overline{r}') \rangle$$
$$= \begin{cases} \overline{\overline{I}} \delta(\overline{r} - \overline{r}'), & \text{for } z > 0 \\ \overline{\overline{I}} \delta(\overline{r} - \overline{r}') + \sum_{s_j=1}^{L} n_{s_j} C_{s_j} \langle \overline{\overline{G}}(\overline{r}, \overline{r}') \rangle, & \text{for } z < 0 \end{cases} \tag{7.2.15}$$

It can be recognized that (7.2.14) is the volume integral equations for a half-space Green's function with upper half-space wavenumber k and lower half-space wavenumber K if we identify in (7.2.15) that

$$\sum_{s_j=1}^{L} n_{s_j} C_{s_j} = K^2 - k^2 \tag{7.2.16}$$

From (7.2.11) and (7.2.16), it follows that the effective propagation constant K is governed by the relation

$$K^2 = k^2 + \frac{3K^2}{D(K)} \sum_{s_l=1}^{L} f_{s_l} y_{s_l} \left\{ 1 + i \frac{2K^3}{3D(K)} \left[a_{s_l}^3 y_{s_l} + \sum_{s_j=1}^{L} y_{s_j} a_{s_j}^3 n_{s_j} 8\pi^3 H_{s_j s_l}(\overline{p} = 0) \right] \right\} \tag{7.2.17}$$

Equation (7.2.15) is the standard volume integral equation for the half-space Green's function. The solution of the volume integral equation for z, $z' < 0$ is given by (7.2.2) with $\overline{\overline{G}}_P(\overline{r}, \overline{r}')$ given by (7.2.3) and

$$\overline{\overline{G}}_R(\overline{r}, \overline{r}') = \frac{i}{8\pi^2} \int_{-\infty}^{\infty} d\overline{k}_\perp \frac{e^{i\overline{k}_\perp \cdot (\overline{r}_\perp - \overline{r}'_\perp) - iK_z(z+z')}}{K_z} \cdot \left\{ R_e(\overline{k}_\perp) \hat{e}(-K_z) \hat{e}(K_z) + R_h(\overline{k}_\perp) \hat{h}(-K_z) \hat{h}(K_z) \right\} \tag{7.2.18}$$

where $\overline{k}_\perp = k_x \hat{x} + k_y \hat{y}$, $k_\perp = (k_x^2 + k_y^2)^{1/2}$, $K_z = (K^2 - k_\perp^2)^{1/2}$, $k_z = (k^2 - k_\perp^2)^{1/2}$, and $\hat{e}$ and $\hat{h}$ are TE and TM unit vectors respectively with $\hat{e}(K_z) = (\hat{x}k_y - \hat{y}k_x)/k_\perp$ and $\hat{h}(K_z) = -K_z(\hat{x}k_x + \hat{y}k_y)/(Kk_\perp) + k_\perp \hat{z}/K$. The quantities R_e and R_h are TE and TM reflection coefficients, respectively, with

$$R_e(\overline{k}_\perp) = \frac{K_z - k_z}{K_z + k_z} \tag{7.2.19}$$

$$R_h(\overline{k}_\perp) = \frac{k^2 K_z - K^2 k_z}{k^2 K_z + K^2 k_z} \tag{7.2.20}$$

Similar expressions hold for $\langle \overline{\overline{G}}(\overline{r}, \overline{r}') \rangle$ for $\overline{r}$ in region 0 and $\overline{r}'$ in region 1 that can be deduced from Chapter 5, Section 2.2 of Volume I and using effective propagation constant K for medium 1.

Next, we calculate the coherent transmitted and reflected fields for an incident plane wave impinging from region 0 in the direction $(\pi - \theta_{oi}, \phi_i)$ upon the lower half-space (Fig. 7.2.1). The incident electric field is

$$\overline{E}^{inc}(\overline{r}) = \left[E_{ih} \hat{h}(-k_{iz}) + E_{ie} \hat{e}(-k_{iz}) \right] e^{ik_{ix}x + ik_{iy}y - ik_{iz}z} \tag{7.2.21}$$

where $k_{ix} = k \sin\theta_{oi} \cos\phi_i$, $k_{iy} = k \sin\theta_{oi} \sin\phi_i$, and $k_{iz} = k \cos\theta_{oi}$. For $\overline{r}$ in region 1, the coherent field obeys the equation, for $z < 0$,

$$\nabla \times \nabla \times \langle \overline{E}(\overline{r}) \rangle = k^2 \langle \overline{E}(\overline{r}) \rangle + \sum_{s_j=1}^{L} n_{s_j} C_{s_j} \langle \overline{E}(\overline{r}) \rangle = K^2 \langle \overline{E}(\overline{r}) \rangle \tag{7.2.22}$$

which is the vector wave equation with wavenumber K. Hence, the coherent transmitted field in region 1 is

$$\langle \overline{E}(\overline{r}) \rangle = \left[E_{ih} T_{ih} \hat{h}(-K_{iz}) + E_{ie} T_{ie} \hat{e}(-K_{iz}) \right] e^{ik_{ix}x + ik_{iy}y - iK_{iz}z} \tag{7.2.23}$$

where $K_{iz} = (K^2 - k_{ix}^2 - k_{iy}^2)^{1/2}$, and

$$T_{ih} = \frac{2k_{iz}}{K\dfrac{k_{iz}}{k} + k\dfrac{K_{iz}}{K}} \tag{7.2.24}$$

$$T_{ie} = \frac{2k_{iz}}{K_{iz} + k_{iz}} \tag{7.2.25}$$

Similarly, for $\overline{r}$ in region 0, the coherent field will consist of the incident field and the field of the coherent reflection wave.

$$\langle \overline{E}(\overline{r}) \rangle = \overline{E}^{inc}(\overline{r}) + e^{ik_{ix}x + ik_{iy}y + ik_{iz}z} \cdot \left\{ \hat{e}(k_{iz}) E_{ie} \frac{k_{iz} - K_{iz}}{k_{iz} + K_{iz}} + \hat{h}(k_{iz}) E_{ih} \frac{K^2 k_{iz} - k^2 K_{iz}}{K^2 k_{iz} + k^2 K_{iz}} \right\} \tag{7.2.26}$$

3 Derivation of Dense Media Radiative Transfer Equation (DMRT)

In this section we derived the dense media radiative transfer equation from correlated ladder approximation taking into account the position correlation of the scatterers. The basic physical idea is that because the particles are randomly distributed, the phases of the scattered fields are random so that the products of scattered fields generally average to zero except those terms in the multiple scattering equations that result in constructive interference. The mathematical approach is to identify and retain only the constructive interference terms in the correlated ladder approximation. Putting these terms together gives the dense medium radiative transfer equation.

The second moment equation under ladder approximation is obtained by using the intensity operator of (5.4.22). We retain the zeroth-order solution of C_{s_j} in the expansion of (5.4.22). Let

$$C_{s_j} \simeq C_{s_j}^{(0)} \tag{7.3.1}$$

where from (7.2.11)

$$C_{s_j}^{(0)} = \frac{3K^2 v_{s_j} y_{s_j}}{D(K)} \tag{7.3.2}$$

The total electrical field can be decomposed into the coherent part $\langle \overline{E}(\overline{r}) \rangle$ and the incoherent part $\overline{\mathcal{E}}(\overline{r})$, so that

$$\overline{E}(\overline{r}) = \langle \overline{E}(\overline{r}) \rangle + \overline{\mathcal{E}}(\overline{r}) \tag{7.3.3}$$

The second moment can also be decomposed into coherent and incoherent parts.

$$\langle E_\alpha(\overline{r}) E_\beta^*(\overline{r}') \rangle = \langle E_\alpha(\overline{r}) \rangle \langle E_\beta^*(\overline{r}') \rangle + \langle \mathcal{E}_\alpha(\overline{r}) \mathcal{E}_\beta^*(\overline{r}') \rangle \tag{7.3.4}$$

From (5.4.9)

$$\begin{aligned} &\big\langle \mathcal{E}_\alpha(\overline{r}) \mathcal{E}_\beta^*(\overline{r}') \big\rangle = \\ &\sum_{\alpha',\beta'} \sum_{\alpha'',\beta''} \int d\overline{r}_1 \int d\overline{r}_2 \int d\overline{r}_1' \int d\overline{r}_2' \langle G_{\alpha\alpha'}(\overline{r}, \overline{r}_1) \rangle \big\langle G_{\beta\beta'}^*(\overline{r}', \overline{r}_1') \big\rangle \\ &I_{\alpha'\alpha'';\beta'\beta''}(\overline{r}_1, \overline{r}_2; \overline{r}_1', \overline{r}_2') \big\langle E_{\alpha''}(\overline{r}_2) E_{\beta''}^*(\overline{r}_2') \big\rangle \end{aligned} \tag{7.3.5}$$

From (5.4.22) and (7.3.2), we then have

$$\begin{aligned} \langle \mathcal{E}_\alpha(\overline{r}) \mathcal{E}_\beta^*(\overline{r}') \rangle = \sum_{\alpha',\beta'} \int_{z_1<0} d\overline{r}_1 \int_{z_1'<0} d\overline{r}_1' \langle G_{\alpha\alpha'}(\overline{r}, \overline{r}_1) \rangle \\ \cdot \langle G_{\beta\beta'}^*(\overline{r}', \overline{r}_1') \rangle Q(\overline{r}_1' - \overline{r}_1) \langle E_{\alpha'}(\overline{r}_1) E_{\beta'}^*(\overline{r}_1') \rangle \end{aligned} \tag{7.3.6}$$

where

$$Q(\overline{r}) = \sum_{s_j=1}^{L} \sum_{s_l=1}^{L} \left[n_{s_j} \delta_{s_j s_l} \delta(\overline{r}) + n_{s_j} n_{s_l} h_{s_j s_l}(\overline{r}) \right] C_{s_l}^{(0)} C_{s_j}^{(0)*} \tag{7.3.7}$$

From (7.3.6), it follows that the covariance of the incoherent fields at $\overline{r}$ and $\overline{r}'$ comes from scattering at $\overline{r}_1$ and $\overline{r}_1'$ with $\overline{r}_1$ and $\overline{r}_1'$ integrated over the entire lower half-space. The contributions of the integral come from summing up all the scattering from many scatterers over the lower half-space. Thus we need to identify constructive interference terms in the integrand of (7.3.6) that will accumulate over the integration region. For the terms that do not have constructive interference, they exhibit phase variations and will "average out" on integration in (7.3.6). Since the major contributions come from integration over many (wavelengths)^3 of space and do not come from $\overline{r}$ close to $\overline{r}_1$ nor $\overline{r}'$ close to $\overline{r}_1'$, the singularity of the Green's dyadic can be neglected in (7.3.6). The function $h_{s_j s_l}(\overline{r})$ is only nonzero over a distance range of several particle diameters. Thus the function $Q(\overline{r}_1' - \overline{r}_1)$ is short-ranged.

The rest of the integrand in (7.3.6) is varying on the wavelength scale. Since $ka_{s_j} \ll 1$, the function $Q(\overline{r}_1' - \overline{r}_1)$ is sharply peaked in the integrand of (7.3.6). Hence in (7.3.6), the rest of the integrand can be approximated by $\overline{r}_1' \simeq \overline{r}_1$. Thus

$$\langle \mathcal{E}_\alpha(\overline{r})\mathcal{E}_\beta^*(\overline{r}')\rangle \simeq \sum_{\alpha'\beta'} \int_{z_1<0} d\overline{r}_1 \langle G_{\alpha\alpha'}(\overline{r},\overline{r}_1)\rangle\langle G_{\beta\beta'}^*(\overline{r}',\overline{r}_1')\rangle$$
$$\int d\overline{r}_1' Q(\overline{r}_1' - \overline{r}_1)\langle E_{\alpha'}(\overline{r}_1)E_{\beta'}^*(\overline{r}_1)\rangle \tag{7.3.8a}$$

This gives

$$\langle \overline{\mathcal{E}}(\overline{r})\overline{\mathcal{E}}^*(\overline{r}')\rangle = Q_o \int_{z_1<0} d\overline{r}_1 \langle \overline{\overline{G}}(\overline{r},\overline{r}_1)\rangle \cdot \langle \overline{E}(\overline{r}_1)\overline{E}^*(\overline{r}_1)\rangle \cdot \langle \overline{\overline{G}}^{*t}(\overline{r}',\overline{r}_1)\rangle \tag{7.3.8b}$$

where the superscript t denotes transpose, and

$$Q_o = \int_{-\infty}^{\infty} d\overline{r}\; Q(\overline{r})$$
$$= \frac{12\pi|K|^4}{|D|^2} \sum_{s_l=1}^{L} f_{s_l} y_{s_l} \left\{ a_{s_l}^3 y_{s_l}^* + \sum_{s_j=1}^{L} n_{s_j} a_{s_j}^3 y_{s_j}^* 8\pi^3 H_{s_j s_l}(\overline{p}=0) \right\} \tag{7.3.9}$$

The quantity Q_o is real because $H_{s_j s_l} = H_{s_l s_j}$.

To solve (7.3.8), we make a spectral decomposition of the incoherent field

$$\overline{\mathcal{E}}(\overline{r}) = \int_{-\infty}^{\infty} d\overline{k}_\perp e^{i\overline{k}_\perp\cdot\overline{r}_\perp} \left\{ \overline{\mathcal{E}}_u(z,\overline{k}_\perp) e^{iK_z' z} + \overline{\mathcal{E}}_d(z,\overline{k}_\perp) e^{-iK_z' z} \right\} \tag{7.3.10}$$

where $\overline{k}_\perp$ and $\overline{r}_\perp$ denote transverse wave vector and transverse coordinate vector, respectively, $K_z' = \mathrm{Re}(K_z)$, $\overline{\mathcal{E}}_u(z,\overline{k}_\perp)$ is the upward propagating incoherent envelope amplitude with transverse wave vector $\overline{k}_\perp$, and $\overline{\mathcal{E}}_d(z,\overline{k}_\perp)$ is the corresponding downward propagating incoherent envelope amplitude. Further decomposition can be made in terms of TM and TE waves.

$$\overline{\mathcal{E}}_u(z,\overline{k}_\perp) = \mathcal{E}_{hu}(z,\overline{k}_\perp)\hat{h}(K_z') + \mathcal{E}_{eu}(z,\overline{k}_\perp)\hat{e}(K_z') \tag{7.3.11}$$

$$\overline{\mathcal{E}}_d(z,\overline{k}_\perp) = \mathcal{E}_{hd}(z,\overline{k}_\perp)\hat{h}(-K_z') + \mathcal{E}_{ed}(z,\overline{k}_\perp)\hat{e}(-K_z') \tag{7.3.12}$$

We note that $K_z'' = \mathrm{Im}(K_z)$, and since $K_z' \gg K_z''$, we have included the K_z'' dependence of propagation in the envelope functions of $\overline{\mathcal{E}}_u$ and $\overline{\mathcal{E}}_d$. Note that incoherent waves build up slowly on distance scale much bigger than a wavelength. It builds up on the distance scale of the mean free path. Thus it is reasonable to assume that evanescent waves in the plane wave spectrum of (7.3.10) have negligible contribution to the incoherent waves. Since the main contributions come from radiating waves, we have $K_z' \gg K_z''$, so that we can replace K_z in the argument of $\hat{h}$ and $\hat{e}$ by K'_z. In deriving the dense media radiative transfer equation, we distinguish two distance scales. The long-distance scale is characterized by the mean free path l_{mfp}, with $l_{mfp} = 1/(2K'')$, $K'' = \text{Im}(K)$. The short distance scale is characterized by the wavelength scale $\lambda_e = 2\pi/K'$ with $K' = \text{Re}(K)$. Transport-type equations are derived based on the assumption that $l_{mfp} \gg \lambda_e$, $l_{mfp} \gg a_{s_j}$. That is, the mean free path is much larger than the wavelength and the particle size. The condition is generally met for many problems involving multiple scattering except in the case of strong localization. Thus the envelope amplitude $\overline{\mathcal{E}}_u$ and $\overline{\mathcal{E}}_d$ in (7.3.11) and (7.3.12) are varying on the l_{mfp} scale.

For z and z' separated on the λ_e scale, we assume the trial solution

$$\langle \overline{\mathcal{E}}_m(z, \overline{k}_\perp) \overline{\mathcal{E}}_n^*(z', \overline{k}_{\perp 1}) \rangle = \delta\left(\overline{k}_\perp - \overline{k}_{\perp 1}\right) \overline{\overline{\Lambda}}^{mn}\left(z, z', \overline{k}_\perp\right) \tag{7.3.13a}$$

where $m, n = u, d$. The assumption of (7.3.13a) is that incoherent fields with different transverse wave vectors $\overline{k}_\perp$ of propagation are uncorrelated. For the same transverse wave vector, there is nonzero correlation. The upward and downward propagation waves with the same transverse $\overline{k}_\perp$ wave vector are correlated because of the effective reflecting interface at $z = 0$. This correlation is represented by $\overline{\overline{\Lambda}}^{ud}$. Using (7.3.13$a$) and (7.3.10), we have

$$\begin{aligned} &\langle \overline{\mathcal{E}}(\overline{r}) \overline{\mathcal{E}}^*(\overline{r}') \rangle \\ &= \int_{-\infty}^{\infty} d\overline{k}_\perp e^{i\overline{k}_\perp \cdot (\overline{r}_\perp - \overline{r}'_\perp)} \left\{ \overline{\overline{\Lambda}}^{uu}(z, z', \overline{k}_\perp) e^{iK'_z(z-z')} + \overline{\overline{\Lambda}}^{dd}(z, z', \overline{k}_\perp) e^{-iK'_z(z-z')} \right. \\ &\left. + \overline{\overline{\Lambda}}^{ud}(z, z', \overline{k}_\perp) e^{iK'_z(z+z')} + \overline{\overline{\Lambda}}^{du}(z, z', \overline{k}_\perp) e^{-iK'_z(z+z')} \right\} \end{aligned} \tag{7.3.13b}$$

For $\overline{r} \neq \overline{r}'$, the mean Green's function of (7.2.2), (7.2.3), and (7.2.18) can be written in the form

$$\langle \overline{\overline{G}}(\overline{r}, \overline{r}') \rangle = \int_{-\infty}^{\infty} d\overline{k}_\perp e^{i\overline{k}_\perp \cdot (\overline{r}_\perp - \overline{r}'_\perp)} \left\{ \overline{\overline{g}}_P\left(\overline{k}_\perp, z, z'\right) + \overline{\overline{g}}_R\left(\overline{k}_\perp, z, z'\right) \right\} \tag{7.3.14}$$

where

$$\overline{\overline{g}}_P\left(\overline{k}_\perp, z, z'\right) = \begin{cases} \overline{\overline{\sigma}}_>(K_z)\, e^{iK_z(z-z')} & \text{for } z > z' \\ \overline{\overline{\sigma}}_<(K_z)\, e^{-iK_z(z-z')} & \text{for } z < z' \end{cases} \tag{7.3.15}$$

$$\overline{\overline{g}}_R\left(\overline{k}_\perp, z, z'\right) = \overline{\overline{\sigma}}_R(K_z)\, e^{-iK_z(z+z')} \tag{7.3.16}$$

with

$$\overline{\overline{\sigma}}_>(K_z) = \frac{i}{8\pi^2 K_z} \left[\hat{e}(K_z)\, \hat{e}(K_z) + \hat{h}(K_z)\, \hat{h}(K_z) \right] \tag{7.3.17}$$

$$\overline{\overline{\sigma}}_<(K_z) = \frac{i}{8\pi^2 K_z} \left[\hat{e}(-K_z)\, \hat{e}(-K_z) + \hat{h}(-K_z)\, \hat{h}(-K_z) \right] \tag{7.3.18}$$

$$\overline{\overline{\sigma}}_R(K_z) = \frac{i}{8\pi^2 K_z}\left[R_e(\overline{k}_\perp)\hat{e}(-K_z)\hat{e}(K_z) + R_h(\overline{k}_\perp)\hat{h}(-K_z)\hat{h}(K_z)\right] \tag{7.3.19}$$

In (7.3.8*b*) the integrand includes the correlation of the total field which is a sum of the correlation of the incoherent field. From (7.2.26) on the coherent field, we have the correlation of the coherent field

$$\langle \overline{E}(\overline{r}_1)\rangle\langle \overline{E}^*(\overline{r}_1)\rangle = \overline{\overline{\lambda}}(K_{iz})e^{2K''_{iz}z_1} \tag{7.3.20}$$

with

$$\overline{\overline{\lambda}}(K_{iz}) = \left[E_{ih}T_{ih}\hat{h}(-K_{iz}) + E_{ie}T_{ie}\hat{e}(-K_{iz})\right] \cdot \left[E^*_{ih}T^*_{ih}\hat{h}(-K_{iz}) + E^*_{ie}T^*_{ie}\hat{e}(-K_{iz})\right] \tag{7.3.21}$$

In (7.3.20), $K''_{iz} = \mathrm{Im}(K_{iz})$. Thus

$$\langle \overline{E}(\overline{r}_1)\overline{E}^*(\overline{r}_1)\rangle = \overline{\overline{\lambda}}(K_{iz})\exp(2K''_{iz}z_1) + \langle \overline{\mathcal{E}}(\overline{r}_1)\overline{\mathcal{E}}^*(\overline{r}_1)\rangle \tag{7.3.22}$$

Note that we have assumed the wave dependence of $\langle \overline{\mathcal{E}}(\overline{r}_1)\overline{\mathcal{E}}^*(\overline{r}_1)\rangle$ in (7.3.13*b*). Substituting (7.3.20), (7.3.22), (7.3.14), and (7.3.13) into (7.3.8*b*), the objective is to balance the wave dependences based on the phase dependence of the various terms so that equations can be obtained for the covariance $\overline{\overline{\Lambda}}^{mn}$ of the incoherent field. Without loss generality, let $z > z'$ in (7.3.8). Then

$$\begin{aligned}
&\overline{\overline{\Lambda}}^{uu}(z,z',\overline{k}_\perp)e^{iK'_z(z-z')} + \overline{\overline{\Lambda}}^{dd}(z,z',\overline{k}_\perp)e^{-iK'_z(z-z')} \\
&+ \overline{\overline{\Lambda}}^{ud}(z,z',\overline{k}_\perp)e^{iK'_z(z+z')} + \overline{\overline{\Lambda}}^{du}(z,z',\overline{k}_\perp)e^{-iK'_z(z+z')} \\
&= Q_o 4\pi^2\Bigg\{\int_{-\infty}^{z'} dz_1\left[\overline{\overline{\sigma}}_>(K_z)e^{iK_z(z-z_1)} + \overline{\overline{\sigma}}_R(K_z)e^{-iK_z(z+z_1)}\right] \\
&\quad\cdot\left[\overline{\overline{\lambda}}(K_{iz})e^{2K''_{iz}z_1} + \langle \overline{\mathcal{E}}(\overline{r}_1)\overline{\mathcal{E}}^*(\overline{r}_1)\rangle\right] \\
&\quad\cdot\left[\overline{\overline{\sigma}}^{t*}_>(K_z)e^{-iK^*_z(z'-z_1)} + \overline{\overline{\sigma}}^{t*}_R(K_z)e^{iK^*_z(z'+z_1)}\right] \\
&\quad+\int_{z'}^{z} dz_1\left[\overline{\overline{\sigma}}_>(K_z)e^{iK_z(z-z_1)} + \overline{\overline{\sigma}}_R(K_z)e^{-iK_z(z+z_1)}\right] \\
&\quad\cdot\left[\overline{\overline{\lambda}}(K_{iz})e^{2K''_{iz}z_1} + \langle \overline{\mathcal{E}}(\overline{r}_1)\overline{\mathcal{E}}^*(\overline{r}_1)\rangle\right] \\
&\quad\cdot\left[\overline{\overline{\sigma}}^{*t}_<(K_z)e^{iK^*_z(z'-z_1)} + \overline{\overline{\sigma}}^{*t}_R(K_z)e^{iK^*_z(z'+z_1)}\right]
\end{aligned}$$

$$+\int_{z}^{0} dz_1 \left[\overline{\overline{\sigma}}_{<}(K_z)\, e^{-iK_z(z-z_1)} + \overline{\overline{\sigma}}_R(K_z)\, e^{-iK_z(z+z_1)}\right]$$
$$\cdot \left[\overline{\overline{\lambda}}(K_{iz})\, e^{2K''_{iz}z_1} + \langle \overline{\mathcal{E}}(\overline{r}_1)\, \overline{\mathcal{E}}^*(\overline{r}_1)\rangle\right]$$
$$\cdot \left[\overline{\overline{\sigma}}^{*t}_{<}(K_z)\, e^{iK^*_z(z'-z_1)} + \overline{\overline{\sigma}}^{*t}_R(K_z)\, e^{iK^*_z(z'+z_1)}\right]\Bigg\} \tag{7.3.23}$$

We shall let $z > z'$, but with $|z - z'|$ on the wavelength scale. We also note that incoherent field comes from integration over distance of mean free path scale. Thus we can neglect the $\int_{z'}^{z} dz_1$ term in (7.3.23) which integrates only over distance of wavelength scale. Also using (7.3.13b) in (7.3.23), we have

$$\overline{\overline{\Lambda}}^{uu}\left(z, z', \overline{k}_\perp\right) e^{iK'_z(z-z')} + \overline{\overline{\Lambda}}^{dd}\left(z, z', \overline{k}_\perp\right) e^{-iK'_z(z-z')}$$
$$+\overline{\overline{\Lambda}}^{ud}\left(z, z', \overline{k}_\perp\right) e^{iK'_z(z+z')} + \overline{\overline{\Lambda}}^{du}\left(z, z', \overline{k}_\perp\right) e^{-iK'_z(z+z')}$$
$$= Q_o 4\pi^2 \Bigg\{ \int_{-\infty}^{z'} dz_1 \left[\overline{\overline{\sigma}}_{>}(K_z)\, e^{iK_z(z-z_1)} + \overline{\overline{\sigma}}_R(K_z)\, e^{-iK_z(z+z_1)}\right]$$
$$\cdot \Bigg\{\overline{\overline{\lambda}}(K_{iz}) e^{2K''_{iz}z_1} + \int_{-\infty}^{\infty} d\overline{k}_\perp \Bigg[\overline{\overline{\Lambda}}^{uu}(z_1, z_1, \overline{k}_\perp) + \overline{\overline{\Lambda}}^{dd}(z_1, z_1, \overline{k}_\perp)$$
$$+ \overline{\overline{\Lambda}}^{ud}(z_1, z_1, \overline{k}_\perp) e^{2iK'_z z_1} + \overline{\overline{\Lambda}}^{du}(z_1, z_1, \overline{k}_\perp) e^{-2iK'_z z_1}\Bigg]\Bigg\}$$
$$\cdot \left[\overline{\overline{\sigma}}^{t*}_{>}(K_z)\, e^{-iK^*_z(z'-z_1)} + \overline{\overline{\sigma}}^{t*}_R(K_z)\, e^{iK^*_z(z'+z_1)}\right]$$
$$+ \int_{z}^{0} dz_1 \left[\overline{\overline{\sigma}}_{<}(K_z)\, e^{-iK_z(z-z_1)} + \overline{\overline{\sigma}}_R(K_z)\, e^{-iK_z(z+z_1)}\right]$$
$$\cdot \Bigg\{\overline{\overline{\lambda}}(K_{iz}) e^{2K''_{iz}z_1} + \int_{-\infty}^{\infty} d\overline{k}_\perp \Bigg[\overline{\overline{\Lambda}}^{uu}(z_1, z_1, \overline{k}_\perp) + \overline{\overline{\Lambda}}^{dd}(z_1, z_1, \overline{k}_\perp)$$
$$+ \overline{\overline{\Lambda}}^{ud}(z_1, z_1, \overline{k}_\perp) e^{2iK'_z z_1} + \overline{\overline{\Lambda}}^{du}(z_1, z_1, \overline{k}_\perp) e^{-2iK'_z z_1}\Bigg]\Bigg\}$$
$$\cdot \left[\overline{\overline{\sigma}}^{*t}_{<}(K_z)\, e^{iK^*_z(z'-z_1)} + \overline{\overline{\sigma}}^{*t}_R(K_z)\, e^{iK^*_z(z'+z_1)}\right]\Bigg\} \tag{7.3.24}$$

We first note that the term $\overline{\overline{\lambda}}(K_{iz}) \exp(2K''_{iz}z_1) + \langle \overline{\mathcal{E}}(\overline{r}_1)\overline{\mathcal{E}}^*(\overline{r}_1)\rangle$ varies on the mean free path scale and does not vary on wavelength scale. The magnitudes of the various terms on the right-hand side of (7.3.24) can be estimated. We distinguish between constructive interference terms and destructive interference terms. For example, a typical constructive interference term is

$$\int dz_1 e^{-iK_z(z-z_1)} e^{iK^*_z(z'-z_1)} = \mathcal{O}\left(\frac{1}{K''_z}\right) \tag{7.3.25}$$

where $\mathcal{O}$ represents the order and $K_z'' = \text{Im}(K_z)$. Note that the terms in the integrand of (7.3.25) balance the phases so that the integrand in (7.3.25) has no phase variation with integration variable z_1. On the other hand, a typical destructive interference term is

$$\int dz_1 e^{-iK_z(z-z_1)+iK_z^*(z'+z_1)} = \mathcal{O}\left(\frac{1}{K_z'}\right) \tag{7.3.26}$$

where $K_z' = \text{Re}(K_z)$. Note that the terms in the integrand of (7.3.26) do not balance the phases and the integrand has phase variation with integration variable z_1. Generally, $K_z'' \ll K_z'$. Terms of the type of (7.3.25) are of the constructive interference type because the phases of the integrand balance each other so that the integrand has little phase variation over the range of integration. Terms of the type of (7.3.26) are of the destructive interference type because the phases do not balance each other. Since $K_z' \gg K_z''$, terms of the destructive interference type are much smaller than that of the constructive interference type. Physically this corresponds to the fact that scattering due to products of fields travelling with different phases will average out to a small number after several wavelengths. We shall only retain terms of the constructive interference type. Let

$$\overline{\overline{J}}(z_1) = \int_{-\infty}^{\infty} d\overline{\gamma}_\perp \left\{ \overline{\overline{\Lambda}}^{uu}(z_1, z_1, \overline{\gamma}_\perp) + \overline{\overline{\Lambda}}^{dd}(z_1, z_1, \overline{\gamma}_\perp) \right\} \tag{7.3.27}$$

The next step is to balance terms on both sides of (7.3.23) that have the same phase dependence of z and z'. There are four types of phase dependence in (7.3.24): $\exp(iK_z'(z-z'))$, $\exp(-iK_z'(z-z'))$, $\exp(iK_z'(z+z'))$, and $\exp(-iK_z'(z+z'))$. Balancing terms with phase dependence $\exp(iK_z'(z-z'))$ gives

$$\begin{aligned} \overline{\overline{\Lambda}}^{uu}(z, z', \overline{k}_\perp)\, e^{iK_z'(z-z')} = {} & 4\pi^2 Q_o e^{iK_z z - iK_z^* z'} \int_{-\infty}^{z'} dz_1 \overline{\overline{\sigma}}_>(K_z)\, e^{2K_z'' z_1} \\ & \cdot \left[\overline{\overline{\lambda}}(K_{iz})\, e^{2K_{iz}'' z_1} + \overline{\overline{J}}(z_1) \right] \cdot \overline{\overline{\sigma}}_>^{t*}(K_z) \end{aligned} \tag{7.3.28}$$

Balancing terms with phase dependence $\exp(-iK_z'(z-z'))$ gives

$$\begin{aligned} \overline{\overline{\Lambda}}^{dd}(z, z', \overline{k}_\perp) e^{-iK_z'(z-z')} = {} & 4\pi^2 Q_o e^{-iK_z z + iK_z^* z'} \int_{-\infty}^{z'} dz_1 \overline{\overline{\sigma}}_R(K_z) e^{2K_z'' z_1} \\ & \cdot \left[\overline{\overline{\lambda}}(K_{iz}) e^{2K_{iz}'' z_1} + \overline{\overline{J}}(z_1) \right] \cdot \overline{\overline{\sigma}}_R^{t*}(K_z) \\ & + 4\pi^2 Q_o e^{-iK_z z + iK_z^* z'} \int_z^0 dz_1 \overline{\overline{\sigma}}_<(K_z) e^{-2K_z'' z_1} \\ & \cdot \left[\overline{\overline{\lambda}}(K_{iz}) e^{2K_{iz}'' z_1} + \overline{\overline{J}}(z_1) \right] \cdot \overline{\overline{\sigma}}_<^{*t}(K_z) \end{aligned}$$

$$+ 4\pi^2 Q_o e^{-iK_z z + iK_z^* z'} \int_z^0 dz_1 \overline{\overline{\sigma}}_R(K_z) e^{2K_z'' z_1}$$

$$\cdot \left[\overline{\overline{\lambda}}(K_{iz}) e^{2K_{iz}'' z_1} + \overline{\overline{J}}(z_1) \right] \cdot \overline{\overline{\sigma}}_R^{*t}(K_z) \qquad (7.3.29)$$

Balancing terms with phase dependence $e^{iK_z'(z+z')}$ gives

$$\overline{\overline{\Lambda}}^{ud}(z, z', \overline{k}_\perp) e^{iK_z'(z+z')} = 4\pi^2 Q_o e^{iK_z z} e^{iK_z^* z'} \int_{-\infty}^{z'} dz_1 \overline{\overline{\sigma}}_>(K_z) e^{2K_z'' z_1}$$

$$\cdot \left[\overline{\overline{\lambda}}(K_{iz}) e^{2K_z'' z_1} + \overline{\overline{J}}(z_1) \right] \cdot \overline{\overline{\sigma}}_R^{t*}(K_z) \qquad (7.3.30)$$

Balancing terms with phase dependence $e^{-iK_z'(z+z')}$, we have an equation that is the complex conjugate of (7.3.30). Next we let z' approach z in (7.3.28)–(7.3.30). From (7.3.28) we get

$$\overline{\overline{\Lambda}}^{uu}(z, z, \overline{k}_\perp) = 4\pi^2 Q_o \int_{-\infty}^{z} dz_1 \overline{\overline{\sigma}}_>(K_z) e^{-2K_z''(z - z_1)}$$

$$\cdot \left[\overline{\overline{\lambda}}(K_{iz}) e^{2K_{iz}'' z_1} + \overline{\overline{J}}(z_1) \right] \cdot \overline{\overline{\sigma}}_>^{t*}(K_z) \qquad (7.3.31)$$

From (7.3.29), we obtain the following by letting $z' \to z$:

$$\overline{\overline{\Lambda}}^{dd}(z, z, \overline{k}_\perp)$$

$$= 4\pi^2 Q_o \int_{-\infty}^{z} dz_1 \overline{\overline{\sigma}}_R(K_z) e^{2K_z''(z+z_1)} \cdot \left[\overline{\overline{\lambda}}(K_{iz}) e^{2K_{iz}'' z_1} + \overline{\overline{J}}(z_1) \right] \cdot \overline{\overline{\sigma}}_R^{t*}(K_z)$$

$$+ 4\pi^2 Q_o \int_z^0 dz_1 \Big\{ \overline{\overline{\sigma}}_<(K_z) e^{-2K_z''(z_1 - z)} \cdot \left[\overline{\overline{\lambda}}(K_{iz}) e^{2K_{iz}'' z_1} + \overline{\overline{J}}(z_1) \right] \cdot \overline{\overline{\sigma}}_<^{*t}(K_z)$$

$$+ \overline{\overline{\sigma}}_R(K_z) e^{2K_z''(z+z_1)} \cdot \left[\overline{\overline{\lambda}}(K_{iz}) e^{2K_{iz}'' z_1} + \overline{\overline{J}}(z_1) \right] \cdot \overline{\overline{\sigma}}_R^{*t}(K_z) \Big\}$$

$$= 4\pi^2 Q_o \int_{-\infty}^{0} dz_1 \overline{\overline{\sigma}}_R(K_z) e^{2K_z''(z+z_1)} \cdot \left[\overline{\overline{\lambda}}(K_{iz}) e^{2K_{iz}'' z_1} + \overline{\overline{J}}(z_1) \right] \cdot \overline{\overline{\sigma}}_R^{t*}(K_z)$$

$$+ 4\pi^2 Q_o \int_z^0 dz_1 \overline{\overline{\sigma}}_<(K_z) e^{-2K_z''(z_1 - z)} \cdot \left[\overline{\overline{\lambda}}(K_{iz}) e^{2K_{iz}'' z_1} + \overline{\overline{J}}(z_1) \right] \cdot \overline{\overline{\sigma}}_<^{t*}(K_z)$$

$$(7.3.32)$$

Let $z' \to z$ in (7.3.30), we have

$$\overline{\overline{\Lambda}}^{ud}(z, z, \overline{k}_\perp) = 4\pi^2 Q_o \int_{-\infty}^{z} dz_1 \, \overline{\overline{\sigma}}_>(K_z) e^{2K_z'' z_1}$$

$$\cdot \left[\overline{\overline{\lambda}}(K_{iz}) e^{2K_{iz}'' z_1} + \overline{\overline{J}}(z_1) \right] \cdot \overline{\overline{\sigma}}_R^{t*}(K_z) \qquad (7.3.33)$$

Equations (7.3.31) to (7.3.33) are the integral equations for the covariance dyads. Since $\overline{\overline{J}}$ is a function of $\overline{\overline{\Lambda}}^{uu}$ and $\overline{\overline{\Lambda}}^{dd}$, we note from (7.3.31) and (7.3.32) that $\overline{\overline{\Lambda}}^{uu}$ and $\overline{\overline{\Lambda}}^{dd}$ are dependent on each other, while $\overline{\overline{\Lambda}}^{ud}$, the correlation between the upward and downward waves with the same $\overline{k}_\perp$, is dependent on $\overline{\overline{\Lambda}}^{uu}$ and $\overline{\overline{\Lambda}}^{dd}$ through (7.3.33). We can convert (7.3.31)–(7.3.33) into integro-differential equations with boundary conditions.

Differentiating (7.3.31), (7.3.32), and (7.3.33) with respect to z gives, respectively,

$$\begin{aligned}\frac{d}{dz}\overline{\overline{\Lambda}}^{uu}(z,z,\overline{k}_\perp) = &-2K_z''\overline{\overline{\Lambda}}^{uu}(z,z,\overline{k}_\perp) + 4\pi^2 Q_o \overline{\overline{\sigma}}_>(K_z)\\ &\cdot\left[\overline{\overline{\lambda}}(K_{iz})\, e^{2K_{iz}''z} + \overline{\overline{J}}(z)\right]\cdot \overline{\overline{\sigma}}_>^{t*}(K_z)\end{aligned} \tag{7.3.34}$$

$$\begin{aligned}\frac{d}{dz}\overline{\overline{\Lambda}}^{dd}(z,z,\overline{k}_\perp) = &\,2K_z''\overline{\overline{\Lambda}}^{dd}(z,z,\overline{k}_\perp) - 4\pi^2 Q_o \overline{\overline{\sigma}}_<(K_z)\\ &\cdot\left[\overline{\overline{\lambda}}(K_{iz})\, e^{2K_{iz}''z} + \overline{\overline{J}}(z)\right]\cdot \overline{\overline{\sigma}}_<^{*t}(K_z)\end{aligned} \tag{7.3.35}$$

$$\begin{aligned}\frac{d}{dz}\overline{\overline{\Lambda}}^{ud}(z,z,\overline{k}_\perp) = &\,2K_z''\overline{\overline{\Lambda}}^{ud}(z,z,\overline{k}_\perp) + 4\pi^2 Q_o \overline{\overline{\sigma}}_>(K_z)\, e^{2K_z''z}\\ &\cdot\left[\overline{\overline{\lambda}}(K_{iz})\, e^{2K_{iz}''z} + \overline{\overline{J}}(z)\right]\cdot \overline{\overline{\sigma}}_R^{t*}(K_z)\end{aligned} \tag{7.3.36}$$

The boundary conditions for $\overline{\overline{\Lambda}}^{uu}$ and $\overline{\overline{\Lambda}}^{dd}$ can be obtained by setting $z=0$ in (7.3.31) and (7.3.32). From (7.3.31) we have

$$\begin{aligned}\overline{\overline{\Lambda}}^{uu}(0,0,\overline{k}_\perp) = &\,4\pi^2 Q_o \int_{-\infty}^{0} dz_1 \frac{i}{8\pi^2 K_z}\left[\hat{e}(K_z)\hat{e}(K_z) + \hat{h}(K_z)\hat{h}(K_z)\right]\\ &\cdot e^{2K_z''z_1}\cdot\left[\overline{\overline{\lambda}}(K_{iz})e^{2K_{iz}''z_1} + \overline{\overline{J}}(z_1)\right]\cdot\left(\frac{-i}{8\pi^2 K_z^*}\right)\\ &\cdot\left[\hat{e}(K_z)\hat{e}(K_z) + \hat{h}(K_z)\hat{h}(K_z)\right]\end{aligned} \tag{7.3.37a}$$

On the other hand, from (7.3.32) we obtain

$$\begin{aligned}&\overline{\overline{\Lambda}}^{uu}(0,0,\overline{k}_\perp)\\ &= 4\pi^2 Q_o \int_{-\infty}^{0} dz_1 \frac{i}{8\pi^2 K_z}\left[R_e(\overline{k}_\perp)\hat{e}(-K_z)\hat{e}(K_z) + R_h(\overline{k}_\perp)\hat{h}(-K_z)\hat{h}(K_z)\right]\\ &\quad\cdot e^{2K_z''z_1}\cdot\left[\overline{\overline{\lambda}}(K_{iz})e^{2K_{iz}''z_1} + \overline{\overline{J}}(z_1)\right]\cdot\left(\frac{-i}{8\pi^2 K_z^*}\right)\\ &\quad\cdot\left[R_e^*(\overline{k}_\perp)\hat{e}(K_z)\hat{e}(-K_z) + R_h^*(\overline{k}_\perp)\hat{h}(K_z)\hat{h}(-K_z)\right]\end{aligned} \tag{7.3.37b}$$

Let

$$\overline{\overline{R}}_c(K_z) = R_e\left(\overline{k}_\perp\right)\hat{e}(-K_z)\,\hat{e}(K_z) + R_h\left(\overline{k}_\perp\right)\hat{h}(-K_z)\,\hat{h}(K_z) \tag{7.3.38}$$

Then from (7.3.37*a*) and (7.3.37*b*) we have the following boundary condition at $z = 0$ for the covariance functions:

$$\overline{\overline{\Lambda}}^{dd}\left(z, z, \overline{k}_\perp\right)\Big|_{z=0} = \overline{\overline{R}}_c(K_z)\cdot\overline{\overline{\Lambda}}^{uu}\left(z, z, \overline{k}_\perp\right)\Big|_{z=0}\cdot\overline{\overline{R}}_c^{t*}(K_z) \tag{7.3.39}$$

Equations (7.3.34) and (7.3.35) with boundary condition of (7.3.39) are the transport equations for the covariance dyads. They are in the form that resembles the radiative transfer equation. They can be cast into the more standard form of radiative transfer equations in terms of angular variable (θ, ϕ). Set

$$\overline{k}_\perp = k_x\hat{x} + k_y\hat{y}$$
$$k_x = K'\sin\theta\cos\phi$$
$$k_y = K'\sin\theta\sin\phi$$
$$\int_{-\infty}^{\infty} d\overline{k}_\perp \simeq \int_0^{\pi/2} d\theta\sin\theta\cos\theta\int_0^{2\pi} d\phi K'^2$$

Only the radiating waves can contribute to constructive interference terms. Constructive interference terms arise from scattering from waves integrated over a sizable region of many wavelengths and evanescent waves do not fall into that category. Next we put (7.3.35), (7.3.36), and (7.3.39) in 2×2 matrix form. Let

$$\Lambda_{11}^{uu}(z, z, \overline{k}_\perp) = \hat{h}(K_z')\cdot\overline{\overline{\Lambda}}^{uu}(z, z, \overline{k}_\perp)\cdot\hat{h}(K_z')$$
$$\Lambda_{12}^{uu}(z, z, \overline{k}_\perp) = \hat{h}(K_z')\cdot\overline{\overline{\Lambda}}^{uu}(z, z, \overline{k}_\perp)\cdot\hat{e}(K_z')$$
$$\Lambda_{12}^{dd}(z, z, \overline{k}_\perp) = \hat{h}(-K_z')\cdot\overline{\overline{\Lambda}}^{dd}(z, z, \overline{k}_\perp)\cdot\hat{e}(-K_z')$$

etc., to define $\Lambda_{ij}^{uu}(z, z, \overline{k}_\perp)$ and $\Lambda_{ij}^{dd}(z, z, \overline{k}_\perp)$, $i, j = 1, 2$. Then we define the 2×2 matrices for $0 \le \theta < \pi/2$, $0 \le \phi < 2\pi$ by

$$\underline{\underline{\Lambda}}^{uu}\left(z, z, \overline{k}_\perp\right) = \begin{bmatrix}\Lambda_{11}^{uu}(z, z, \overline{k}_\perp) & \Lambda_{12}^{uu}(z, z, \overline{k}_\perp)\\ \Lambda_{21}^{uu}(z, z, \overline{k}_\perp) & \Lambda_{22}^{uu}(z, z, \overline{k}_\perp)\end{bmatrix} = \underline{\underline{\Lambda}}(z, \theta, \phi) \tag{7.3.40}$$

$$\underline{\underline{\Lambda}}^{dd}\left(z, z, \overline{k}_\perp\right) = \begin{bmatrix}\Lambda_{11}^{dd}(z, z, \overline{k}_\perp) & \Lambda_{12}^{dd}(z, z, \overline{k}_\perp)\\ \Lambda_{21}^{dd}(z, z, \overline{k}_\perp) & \Lambda_{22}^{dd}(z, z, \overline{k}_\perp)\end{bmatrix} = \underline{\underline{\Lambda}}(z, \pi-\theta, \phi) \tag{7.3.41}$$

For $0 < \theta < \pi/2$, $0 < \phi < 2\pi$, let

$$\underline{\underline{F}}\left(K_z', K_{z\gamma}'\right) = \begin{bmatrix} f_{11}(K_z', K_{z\gamma}') & f_{12}(K_z', K_{z\gamma}')\\ f_{21}(K_z', K_{z\gamma}') & f_{22}(K_z', K_{z\gamma}')\end{bmatrix} = \underline{\underline{F}}(\theta, \phi; \theta_\gamma, \phi_\gamma) \tag{7.3.42}$$

where $K_{z\gamma} = (K^2 - \gamma_\perp^2)^{1/2}$, $K'_{z\gamma} = \mathrm{Re}(K_{z\gamma})$,

$$f_{11}\left(K'_z, K'_{z\gamma}\right) = \hat{h}\left(K'_z\right) \cdot \hat{h}\left(K'_{z\gamma}\right) \tag{7.3.43}$$

$$f_{12}\left(K'_z, K'_{z\gamma}\right) = \hat{h}\left(K'_z\right) \cdot \hat{e}\left(K'_{z\gamma}\right) \tag{7.3.44}$$

$$f_{21}\left(K'_z, K'_{z\gamma}\right) = \hat{e}\left(K'_z\right) \cdot \hat{h}\left(K'_{z\gamma}\right) \tag{7.3.45}$$

$$f_{22}\left(K'_z, K'_{z\gamma}\right) = \hat{e}\left(K'_z\right) \cdot \hat{e}\left(K'_{z\gamma}\right) \tag{7.3.46}$$

Also

$$\underline{\underline{F}}\left(-K'_z, K'_{z\gamma}\right) = \underline{\underline{F}}\left(\pi - \theta, \phi; \theta_\gamma, \phi_\gamma\right) \tag{7.3.47}$$

etc. Then, (7.3.34) and (7.3.35) assume the following matrix form, respectively:

$$\begin{aligned}\frac{d}{dz}\underline{\underline{\Lambda}}(z,\theta,\phi) = & -\kappa_e \sec\theta \underline{\underline{\Lambda}}(z,\theta,\phi) + \frac{Q_o \sec^2\theta}{16\pi^2 K'^2}\underline{\underline{F}}(\theta,\phi;\pi-\theta_i,\phi_i) \\ & \cdot \underline{\underline{\lambda}}(K_{iz}) \cdot \underline{\underline{F}}(\pi-\theta_i,\phi_i;\theta,\phi)\, e^{\kappa_e z \sec\theta_i} \\ & + \frac{Q_o \sec^2\theta}{16\pi^2}\int_0^\pi d\theta_\gamma \sin\theta_\gamma |\cos\theta_\gamma| \int_0^{2\pi} d\phi_\gamma \\ & \cdot \left\{\underline{\underline{F}}(\theta,\phi;\theta_\gamma,\phi_\gamma) \cdot \underline{\underline{\Lambda}}(z,\theta_\gamma,\phi_\gamma) \cdot \underline{\underline{F}}(\theta_\gamma,\phi_\gamma;\theta,\phi)\right\}\end{aligned} \tag{7.3.48}$$

$$\begin{aligned}\frac{d}{dz}\underline{\underline{\Lambda}}(z,\pi-\theta,\phi) = & \,\kappa_e \sec\theta \underline{\underline{\Lambda}}(z,\pi-\theta,\phi) - \frac{Q_o \sec^2\theta}{16\pi^2 K'^2}\underline{\underline{F}}(\pi-\theta,\phi;\pi-\theta_i,\phi_i) \\ & \cdot \underline{\underline{\lambda}}(K_{iz}) \cdot \underline{\underline{F}}(\pi-\theta_i,\phi_i;\pi-\theta,\phi)\, e^{\kappa_e z \sec\theta_i} \\ & - \frac{Q_o \sec^2\theta}{16\pi^2}\int_0^\pi d\theta_\gamma \sin\theta_\gamma |\cos\theta_\gamma| \int_0^{2\pi} d\phi_\gamma \\ & \cdot \left\{\underline{\underline{F}}(\pi-\theta,\phi;\theta_\gamma,\phi_\gamma) \cdot \underline{\underline{\Lambda}}(z,\theta_\gamma,\phi_\gamma) \cdot \underline{\underline{F}}(\theta_\gamma,\phi_\gamma;\pi-\theta,\phi)\right\}\end{aligned} \tag{7.3.49}$$

where

$$\kappa_e = 2K'' \tag{7.3.50}$$

is the extinction rate and

$$\underline{\underline{\lambda}}(K_{iz}) = \begin{bmatrix} |E_{ih}|^2|T_{ih}|^2 & E_{ih}E_{ie}^*T_{ih}T_{ie}^* \\ E_{ie}E_{ih}^*T_{ie}T_{ih}^* & |E_{ie}|^2|T_{ie}|^2 \end{bmatrix} \tag{7.3.51}$$

Note that $K'_{iz} = K' \sin\theta_i$, and the angles θ_i and θ_{oi} are related by Snell's law $\theta_i = \sin^{-1}(k \sin\theta_{oi}/K')$. In deriving (7.3.48) and (7.3.49), we have also made use of the property that $K''_z = K'' \sec\theta$. Equations (7.3.48)–(7.3.49) are the radiative transfer equations for the field covariance matrix $\underline{\underline{\Lambda}}(z,\theta,\phi)$. The boundary condition is obtained from (7.3.38)–(7.3.39) and is, for $0 \leq$

$\theta < \pi/2$, $0 \leq \phi < 2\pi$,

$$\underline{\underline{\Lambda}}(z = 0, \pi - \theta, \phi) = \begin{bmatrix} |R_h(\overline{k}_\perp)|^2 \Lambda_{11}(z = 0, \theta, \phi) & R_h(\overline{k}_\perp) R_e^*(\overline{k}_\perp) \Lambda_{12}(z = 0, \theta, \phi) \\ R_h^*(\overline{k}_\perp) R_e(\overline{k}_\perp) \Lambda_{21}(z = 0, \theta, \phi) & |R_e(\overline{k}_\perp)|^2 \Lambda_{22}(z = 0, \theta, \phi) \end{bmatrix} \tag{7.3.52}$$

Equations (7.3.48)–(7.3.52) can be put in the standard form using the four Stokes parameters by defining

$$\mathcal{I}_1(z, \theta, \phi) = \frac{K'^3}{k} |\cos\theta| \Lambda_{11}(z, \theta, \phi) \tag{7.3.53}$$

$$\mathcal{I}_2(z, \theta, \phi) = \frac{K'^3}{k} |\cos\theta| \Lambda_{22}(z, \theta, \phi) \tag{7.3.54}$$

$$\mathcal{I}_3(z, \theta, \phi) = -2\frac{K'^3}{k} |\cos\theta| \mathrm{Re}\,(\Lambda_{21}(z, \theta, \phi)) \tag{7.3.55}$$

$$\mathcal{I}_4(z, \theta, \phi) = -2\frac{K'^3}{k} |\cos\theta| \mathrm{Im}\,(\Lambda_{21}(z, \theta, \phi)) \tag{7.3.56}$$

and the 4×1 Stokes vector is

$$\underline{\mathcal{I}}(z, \theta, \phi) = \begin{bmatrix} \mathcal{I}_1(z, \theta, \phi) \\ \mathcal{I}_2(z, \theta, \phi) \\ \mathcal{I}_3(z, \theta, \phi) \\ \mathcal{I}_4(z, \theta, \phi) \end{bmatrix} \tag{7.3.57}$$

The albedo will then be

$$\tilde{\omega} = \frac{Q_o}{6\pi\kappa_e} \tag{7.3.58}$$

The scattering coupling coefficients for Stokes parameters are the phase matrix elements, which, in view of (7.3.48)–(7.3.49), can be expressed in terms of products of $\underline{\underline{F}}(\theta, \phi; \theta_\gamma, \phi_\gamma)$ elements. The resulting radiative transfer equation for the Stokes parameters and the phase matrix elements are given in the Sections 4 and 6, respectively, for active and passive remote sensing.

4 Dense Media Radiative Transfer Equations for Active Remote Sensing

In this section we summarize the results concerning the dense media radiative transfer equation as derived for active remote sensing. The incident electromagnetic wave is in the direction $(\pi - \theta_{oi}, \phi_i)$. The incident electric field is given by (7.2.21), with E_{ih} and E_{ie} being the amplitudes of TM and TE wave, respectively.

Step 1: Effective Propagation Constants, Extinction Rates, and Coherent Transmitted Wave

As shown in Section 3, the lower half-space of region 1 can be represented by an effective propagation constant K, where K obeys the equation

$$\begin{aligned} K^2 &\approx K'^2 + i2K'K'' \\ &= k^2 + \frac{3K_o^2}{D(K_o)} \sum_{s_l=1}^{L} f_{s_l} y_{s_l}(K_o) \left\{ 1 + i \frac{2K_o^3}{3D(K_o)} \right. \\ &\left. \cdot \left[a_{s_l}^3 y_{s_l}(K_o) + \sum_{s_j=1}^{L} y_{s_j}(K_o) a_{s_j}^3 n_{s_j} 8\pi^3 H_{s_j s_l}(p=0) \right] \right\} \end{aligned} \quad (7.4.1)$$

where K_o^2 represents the zeroth-order solution and satisfies the following equation

$$K_o^2 = k^2 + \frac{3K_o^2}{D(K_o)} \sum_{s_l=1}^{L} f_{s_l} y_{s_l}(K_o) \quad (7.4.2)$$

and

$$D(K_o) = 1 - \sum_{s_j=1}^{L} f_{s_j} y_{s_j}(K_o) \quad (7.4.3)$$

$$y_{s_j}(K_o) = \frac{k_{s_j}^2 - k^2}{3K_o^2 + (k_{s_j}^2 - k^2)} \quad (7.4.4)$$

To calculate numerical results of the nonlinear equation of K in (7.4.1), first solve (7.4.2) for K_o. Equation (7.4.2) with (7.4.3)–(7.4.4) is a polynomial equation for K_o, and only one solution has the correct physical meaning of effective propagation constant. That solution is then substituted back in (7.4.1) to calculate K. The extinction rate is $\kappa_e = 2K''$, where $K'' = \text{Im}(K)$. Snell's law can then be applied so that the transmitted angle of the coherent wave is $\theta_i = \sin^{-1}(k \sin\theta_{oi}/K')$, where $K' = \text{Re}(K)$. The coherent transmission wave amplitudes for TM and TE waves are given, respectively, by (7.2.24) and (7.2.25) with $k_{iz} = k\cos\theta_{oi}$ and $K_{iz} = (K^2 - k^2 \sin^2\theta_{oi})^{1/2}$.

Step 2: Dense Media Radiative Transfer Equation for the Incoherent Wave

By using the definitions of Stokes parameters of (7.3.53)–(7.3.56) in (7.3.48)–(7.3.49), it follows after some algebraic manipulations that the dense media radiative transfer equation for the incoherent 4×1 Stokes vector

$\underline{\mathcal{I}}(z,\theta,\phi)$ in region 1 is, for $0 \le \theta \le \pi$, and $0 \le \phi \le 2\pi$,

$$\cos\theta \frac{d}{dz}\underline{\mathcal{I}}(z,\theta,\phi) = -\kappa_e \underline{\mathcal{I}}(z,\theta,\phi) + \frac{\kappa_e \tilde{\omega}}{4\pi}\frac{K'}{k}\underline{\underline{P}}(\theta,\phi;\pi-\theta_i,\phi_i)\,\underline{\lambda}(K_{iz})\,e^{\kappa_e z \sec\theta_i}$$
$$+\frac{\kappa_e\tilde{\omega}}{4\pi}\int_0^{\pi} d\theta_\gamma \sin\theta_\gamma \int_0^{2\pi} d\phi_\gamma \underline{\underline{P}}(\theta,\phi;\theta_\gamma,\phi_\gamma)\,\underline{\mathcal{I}}(z,\theta_\gamma,\phi_\gamma) \tag{7.4.5}$$

where $\tilde{\omega}$ is

$$\tilde{\omega} = \frac{2|K_o|^4}{\kappa_e |D(K_o)|^2}\sum_{s_l=1}^{L} f_{s_l} y_{s_l}(K_o)$$
$$\cdot \left\{ a_{s_l}^3 y_{s_l}^*(K_o) + \sum_{s_j=1}^{L} n_{s_j} a_{s_j}^3 y_{s_j}^*(K_o) 8\pi^3 H_{s_j s_l}(\overline{p}=0) \right\} \tag{7.4.6}$$

and the coherent transmitted Stokes vector, $\underline{\lambda}(K_{iz})$, is

$$\underline{\lambda}(K_{iz}) = \begin{bmatrix} T_{ih}^2 |E_{ih}|^2 \\ T_{ie}^2 |E_{ie}|^2 \\ -2T_{ie}T_{ih}\mathrm{Re}(E_{ih}^* E_{ie}) \\ -2T_{ie}T_{ih}\mathrm{Im}(E_{ih}^* E_{ie}) \end{bmatrix} \tag{7.4.7}$$

The quantities T_{ih} and T_{ie} are practically real since $K_{iz}'' \ll K_{iz}'$; and for convenience, we shall just take their respective real parts when calculating (7.4.7). The second term in (7.4.5) is the source term for the incoherent wave and corresponds to a single scattering of the coherent wave into incoherent wave. The elements of the 4×4 phase matrix $\underline{\underline{P}}(\theta,\phi;\theta_\gamma,\phi_\gamma)$ are

$$P_{11}(\theta,\phi;\theta_\gamma,\phi_\gamma) = \frac{3}{2}\left[\sin\theta\sin\theta_\gamma + \cos\theta\cos\theta_\gamma\cos(\phi-\phi_\gamma)\right]^2 \tag{7.4.8}$$

$$P_{12}(\theta,\phi;\theta_\gamma,\phi_\gamma) = \frac{3}{2}\cos^2\theta\sin^2(\phi-\phi_\gamma) \tag{7.4.9}$$

$$P_{13}(\theta,\phi;\theta_\gamma,\phi_\gamma) = -\frac{3}{2}\left[\sin\theta\sin\theta_\gamma + \cos\theta\cos\theta_\gamma\cos(\phi-\phi_\gamma)\right]$$
$$\cdot\cos\theta\sin(\phi-\phi_\gamma) \tag{7.4.10}$$

$$P_{21}(\theta,\phi;\theta_\gamma,\phi_\gamma) = \frac{3}{2}\cos^2\theta_\gamma\sin^2(\phi-\phi_\gamma) \tag{7.4.11}$$

$$P_{22}(\theta,\phi;\theta_\gamma,\phi_\gamma) = \frac{3}{2}\cos^2(\phi-\phi_\gamma) \tag{7.4.12}$$

$$P_{23}(\theta,\phi;\theta_\gamma,\phi_\gamma) = \frac{3}{2}\cos\theta_\gamma\sin(\phi-\phi_\gamma)\cos(\phi-\phi_\gamma) \tag{7.4.13}$$

$$P_{31}(\theta,\phi;\theta_\gamma,\phi_\gamma) = 3\cos\theta_\gamma \sin(\phi-\phi_\gamma) \cdot [\sin\theta\sin\theta_\gamma + \cos\theta\cos\theta_\gamma\cos(\phi-\phi_\gamma)] \qquad (7.4.14)$$

$$P_{32}(\theta,\phi;\theta_\gamma,\phi_\gamma) = -3\cos\theta\cos(\phi-\phi_\gamma)\sin(\phi-\phi_\gamma) \qquad (7.4.15)$$

$$P_{33}(\theta,\phi;\theta_\gamma,\phi_\gamma) = \frac{3}{2}[\sin\theta\sin\theta_\gamma\cos(\phi-\phi_\gamma) + \cos\theta\cos\theta_\gamma\cos 2(\phi-\phi_\gamma)] \qquad (7.4.16)$$

$$P_{44}(\theta,\phi;\theta_\gamma,\phi_\gamma) = \frac{3}{2}[\sin\theta\sin\theta_\gamma\cos(\phi-\phi_\gamma) + \cos\theta\cos\theta_\gamma] \qquad (7.4.17)$$

$$P_{14} = P_{24} = P_{34} = P_{41} = P_{42} = P_{43} = 0 \qquad (7.4.18)$$

The phase matrix is identical to the Rayleigh phase matrix of the conventional radiative transfer theory. This is because of the small-particle assumption and because the pair distribution functions are only correlated for the range of a few diameters.

At $z = 0$ the boundary condition for the Stokes vector is, from (7.3.52)–(7.3.57)

$$\underline{I}(z,\pi-\theta,\phi) = \underline{\underline{R}}(\theta)\underline{I}(z,\theta,\phi) \qquad (7.4.19)$$

where

$$\underline{\underline{R}}(\theta) = \begin{bmatrix} |R_h(\theta)|^2 & 0 & 0 & 0 \\ 0 & |R_e(\theta)|^2 & 0 & 0 \\ 0 & 0 & \mathrm{Re}\Big(R_h(\theta)R_e^*(\theta)\Big) & -\mathrm{Im}\Big(R_h(\theta)R_e^*(\theta)\Big) \\ 0 & 0 & \mathrm{Im}\Big(R_h(\theta)R_e^*(\theta)\Big) & \mathrm{Re}\Big(R_h(\theta)R_e^*(\theta)\Big) \end{bmatrix} \qquad (7.4.20)$$

where $R_e(\theta)$ and $R_h(\theta)$ are given by (7.2.19)–(7.2.20), respectively, with $K_z = K\cos\theta$ and $k_z = (k^2 - K^2\sin^2\theta)^{1/2}$.

The dense media radiative transfer equation (7.4.5) is to be solved subject to the boundary condition (7.4.19). After the solution is calculated, the incoherent Stokes vector that is transmitted back into region 0, $\underline{I}_o(\theta_o,\phi)$, with $\theta_o = \sin^{-1}(K'\sin\theta/k)$ according to Snell's law, is

$$\underline{I}_o(\theta_o,\phi_o) = T(\theta)\underline{I}(z=0,\theta,\phi) \qquad (7.4.21)$$

For θ less than critical angle $\theta_c \left(= \sin^{-1}(k/K')\right)$ we have

$$\underline{\underline{T}}(\theta) = \frac{k^2}{K'^2}\begin{bmatrix} 1-|R_h(\theta)|^2 & 0 & 0 & 0 \\ 0 & 1-|R_e(\theta)|^2 & 0 & 0 \\ 0 & 0 & \frac{\cos\theta_o}{\cos\theta}\mathrm{Re}\Big(T_h(\theta)T_e^*(\theta)\Big) & -\frac{\cos\theta_o}{\cos\theta}\mathrm{Im}\Big(T_h(\theta)T_e^*(\theta)\Big) \\ 0 & 0 & \frac{\cos\theta_o}{\cos\theta}\mathrm{Im}\Big(T_h(\theta)T_e^*(\theta)\Big) & \frac{\cos\theta_o}{\cos\theta}\mathrm{Re}\Big(T_h(\theta)T_e^*(\theta)\Big) \end{bmatrix} \qquad (7.4.22)$$

$T_h(\theta) = 1 + R_h(\theta)$ and $T_e(\theta) = 1 + R_e(\theta)$. For θ greater than the critical angle θ_c, $\underline{\underline{T}}(\theta) = 0$.

The dense media radiative transfer equation (7.4.5) with boundary conditions (7.4.19)–(7.4.20) resembles that of conventional radiative transfer equation of independent scattering except that the expressions for κ_e and $\tilde{\omega}$ in terms of physical parameters are different from those of the conventional theory because of correlated scattering and the fact that the effective propagation constant K has been taken into account. The procedure of numerical solutions of the dense media transfer equation follow the same procedure as that of the conventional ones as described in Volume I.

5 General Relation between Active and Passive Remote Sensing with Temperature Distribution

Relations exist between the brightness temperatures of passive remote sensing and the bistatic coefficients of active remote sensing when the medium is of uniform temperature. The expressions were derived in Volume I, Chapter 3, Section 5. In this section, we derive a similar expression for nonuniform temperature distribution. An expression will be derived relating the thermal radiated field intensity of a medium with nonuniform temperature distribution to the electromagnetic fields in the problem of active remote sensing. Specifically, it is shown that the thermal radiated field intensity is equal to a constant times the integration of the product of the temperature distribution and the divergence of the Poynting's vector for the problem of active remote sensing. The derivation is based on the field equations of a medium of discrete scatterers and the fluctuation-dissipation theorem. The key quantity to utilize in the derivation is the dyadic Green's function of the random medium, which is common to both problems and is also reciprocal. In the following, active and passive remote sensing problems will be labeled as Problems A and B, respectively.

Problem A:

Consider a plane electromagnetic wave of unit amplitude with polarization $\hat{\alpha}$ and incident angle $(\pi - \theta_{oi}, \phi_i)$ impinging upon a half-space medium of N dielectric scatterers of permittivity ϵ_s and wavenumber k_s (Fig. 7.5.1). The particles occupy volumes $V_1, V_2, \ldots V_N$, respectively. The background permittivity in the lower half-space as well as the upper half-space is ϵ. Let $\overline{E}^{(A)}$ and $\overline{S}^{(A)}$ denote respectively the electric field and time-averaged Poynting's vector, respectively, and the superscript (A) is used to denote

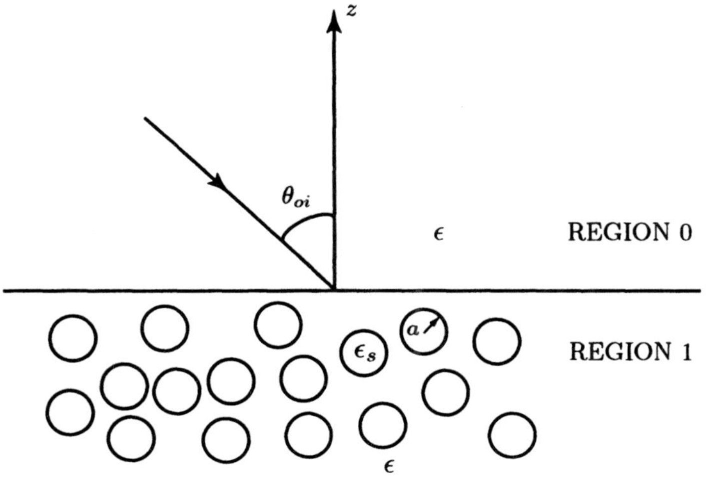

Figure 7.5.1 Problem A: Active remote sensing with a plane electromagnetic wave of unit amplitude, polarization $\hat{\alpha}$, and incident angle $(\pi - \theta_{oi}, \phi_i)$ impinging upon a half-space of particles. Electrical field and Poynting's vector are denoted respectively by $\overline{E}^{(A)}$ and $\overline{S}^{(A)}$.

Problem A. The electric field $\overline{E}^{(A)}$ obeys the vector wave equation,

$$\nabla \times \nabla \times \overline{E}^{(A)}(\overline{r}) - k^2 \overline{E}^{(A)}(\overline{r}) - \sum_{l=1}^{N} \left(k_s^2 - k^2\right) V_l(\overline{r}) \overline{E}^{(A)}(\overline{r}) = 0 \qquad (7.5.1)$$

where $V_l(\overline{r}) = 1$ for $\overline{r}$ in V_l and is zero otherwise. The time-averaged Poynting's vector $\overline{S}^{(A)}$ is

$$\overline{S}^{(A)} = \frac{1}{4i\omega\mu} \left\{ \overline{E}^{(A)*} \times (\nabla \times \overline{E}_A) - \overline{E}^{(A)} \times \left(\nabla \times E^{(A)*}\right) \right\} \qquad (7.5.2)$$

It follows from (7.5.1) and (7.5.2) that

$$\nabla \cdot \overline{S}^{(A)}(\overline{r}) = -\frac{\omega}{2} \sum_{l=1}^{N} \epsilon_s'' V_l(\overline{r}) \left| \overline{E}^{(A)}(\overline{r}) \right|^2 \qquad (7.5.3)$$

where $k_s^2 = \omega^2 \mu \epsilon_s$ and $\epsilon_s'' = \mathrm{Im}(\epsilon_s)$.

Problem B:

In Problem B of thermal emission, let $T^{(B)}(\overline{r})$ be the temperature distribution in the lower half-space which is the same medium as in Problem A (Fig. 7.5.2). Then, from the fluctuation-dissipation theorem, the current $\overline{J}^{(B)}$

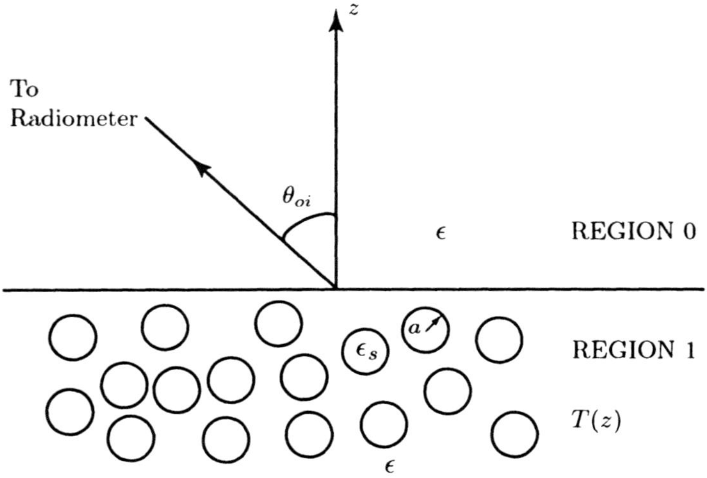

Figure 7.5.2 Problem B: Passive remote sensing with observation of thermal emission of same medium in direction $(\theta_{oi}, \pi + \phi_{0i})$. Temperature distribution of lower half-space is $T^{(B)}(\overline{r})$. Electric field and Poynting's vector are denoted respectively by $\overline{E}^{(B)}$ and $\overline{S}^{(B)}$.

obeys the thermodynamic average (denoted by subscript th)

$$\left\langle \overline{J}^{(B)}(\overline{r},\omega)\overline{J}^{(B)*}(\overline{r}',\omega')\right\rangle_{th} = \frac{4}{\pi}\omega\epsilon''(\overline{r})K_b T^{(B)}(\overline{r})\overline{\overline{I}}\delta(\omega-\omega')\delta(\overline{r}-\overline{r}') \quad (7.5.4)$$

where K_b is Boltzmann's constant and $\epsilon''(\overline{r})$ is the imaginary part of the permittivity in the lower half-space. Thus $\epsilon''(\overline{r}) = \epsilon_s''$ for $\overline{r}$ in $V_1, V_2 \ldots V_N$ and $\epsilon''(\overline{r}) = 0$ otherwise.

The right-hand side of equation (7.5.4) is proportional to $\delta(\omega - \omega')$ implying zero correlation between two different frequencies. Measurement is usually made for a bandwidth of $\Delta\omega$ about ω. Thus the quantity to calculate in Problem B is

$$\int_{\Delta\omega} d\omega \left\langle \overline{E}^{(B)}(\overline{r},\omega)\cdot\overline{E}^{(B)*}(\overline{r},\omega')\right\rangle_{th}$$

in terms of which the brightness temperature can be expressed. The vector wave equation governing $\overline{E}^{(B)}(\overline{r},\omega)$ is

$$\nabla\times\nabla\times\overline{E}^{(B)}(\overline{r}) - k^2\overline{E}^{(B)}(\overline{r}) - \sum_{l=1}^{N}\left(k_s^2 - k^2\right)V_l(\overline{r})\overline{E}^{(B)}(\overline{r})$$
$$= i\omega\mu\sum_{l=1}^{N}V_l(\overline{r})\overline{J}^{(B)}(\overline{r}) \quad (7.5.5)$$

The dyadic Green's function for the medium, $\overline{\overline{G}}(\overline{r},\overline{r}')$, is common to both Problems A and B.

$$\nabla\times\nabla\times\overline{\overline{G}}(\overline{r},\overline{r}')-k^2\overline{\overline{G}}(\overline{r},\overline{r}')=\sum_{l=1}^{N}\left(k_s^2-k^2\right)V_l(\overline{r})\overline{\overline{G}}(\overline{r},\overline{r}')+\overline{\overline{I}}\delta(\overline{r}-\overline{r}') \quad (7.5.6)$$

In terms of dyadic Green's function, the electric field $\overline{E}^{(B)}(\overline{r})$ is

$$\overline{E}^{(B)}(\overline{r})=i\omega\mu\int d\overline{r}'\overline{\overline{G}}(\overline{r},\overline{r}')\cdot\sum_{l=1}^{N}V_l(\overline{r}')\overline{J}^{(B)}(\overline{r}') \quad (7.5.7)$$

By making use of (7.5.4), (7.5.7) and the symmetry relation of Green's function

$$\overline{\overline{G}}(\overline{r},\overline{r}')=\overline{\overline{G}}^t(\overline{r},\overline{r}') \quad (7.5.8)$$

with superscript t indicating transpose, it follows that

$$\int_{\Delta\omega}d\omega\left\langle\left|\hat{\alpha}\cdot\overline{E}^{(B)}(\overline{r},\omega)\right|^2\right\rangle_{th}$$
$$=\int d\overline{r}'\left|\overline{\overline{G}}(\overline{r},\overline{r}')\cdot\hat{\alpha}\right|^2\omega^2\mu^2\sum_{l=1}^{N}V_l(\overline{r}')\frac{4}{\pi}\omega\epsilon_s''K_bT^{(B)}(\overline{r}') \quad (7.5.9)$$

Note that by virtue of the symmetry relations of (7.5.8), source and observation locations have been interchanged in (7.5.9). The observation point $\overline{r}$ in Problem B is usually in the far field region away from the medium of particles. Let it be in the observation direction $(\theta_{oi},\pi+\phi_i)$. Then $\overline{\overline{G}}(\overline{r}',\overline{r})\cdot\hat{\alpha}$ in (7.5.9) corresponds to the electric field in the particles due to an incident field of polarization $\hat{\alpha}$ impinging at an angle $(\pi-\theta_{oi},\phi_i)$ upon the lower half-space. This electric field is proportional to the electric field of Problem A. The exact relation is

$$\lim_{r\to\infty}\overline{\overline{G}}(\overline{r}',\overline{r})\cdot\hat{\alpha}=\frac{e^{ikr}}{4\pi r}\overline{E}^{(A)}(\overline{r}') \quad (7.5.10)$$

Substituting (7.5.10) into (7.5.9) indicates that the integrand of (7.5.9) is proportional to the right-hand side of (7.5.5),

$$\lim_{r\to\infty}\int_{\Delta\omega}d\omega\left\langle\left|\hat{\alpha}\cdot\overline{E}^{(B)}(\overline{r},\omega)\right|^2\right\rangle_{th}$$
$$=-\frac{\omega^2\mu^2}{2\pi^3r^2}\int_{z'<0}d\overline{r}'K_bT^{(B)}(\overline{r}')\nabla'\cdot\overline{S}^{(A)}(\overline{r}') \quad (7.5.11)$$

The thermal radiation intensity in Problem B is therefore proportional to the integration of the product of the temperature distribution and the divergence of Poynting's vector for Problem A. Hence, if the field of Problem A is solved, the radiation intensity of Problem B can be calculated by using (7.5.11).

In Sections 3 and 4, Problem A was treated by using analytic wave theory based on QCA-CP and modified ladder approximation. Using the results, it follows that the field can be decomposed into coherent and incoherent waves, $\overline{E}^{(A)} = \langle \overline{E} \rangle + \overline{\mathcal{E}}$. Thus the divergence of the Poynting's vector is

$$\left\langle \nabla \cdot \overline{S}^{(A)} \right\rangle = \nabla \cdot S_c^{(A)} + \left\langle \nabla \cdot \overline{\mathcal{S}}^{(A)} \right\rangle \tag{7.5.12}$$

where $\langle\ \rangle$ stands for configuration average, $\overline{S}_c^{(A)}$ stands for the Poynting's vector for the coherent wave $\langle \overline{E} \rangle$, and $\overline{\mathcal{S}}^{(A)}$ the Poynting's vector for incoherent wave $\overline{\mathcal{E}}$. We also have, for the coherent wave,

$$\nabla \cdot \overline{S}_c^{(A)} = -\frac{K'K''}{\omega\mu} e^{\kappa_e z \sec\theta_i} \left[|E_{iv} T_v|^2 + |E_{ih} T_h|^2 \right] \tag{7.5.13}$$

where $E_{iv} = 1$ for incident vertical polarized waves $\hat{\alpha} = \hat{v}$ and $E_{ih} = 1$ for incident horizontal polarized waves $\hat{\alpha} = \hat{h}$, and T_v and T_h are respectively transmission coefficients at the $z = 0$ boundary for vertical and horizontal polarized waves. For the incoherent waves, only the scattering terms of the constructive interference type will be included since integration over the lower half-space is generally implied in evaluating the scattering contributions. It then follows that

$$\left\langle \nabla \cdot \overline{\mathcal{S}}^{(A)} \right\rangle = \frac{k}{2\omega\mu} \int_0^\pi d\theta \sin\theta \int_0^{2\pi} d\phi \cos\theta \frac{d}{dz} \left[I_1^{(A)}(z,\theta,\phi) + I_2^{(A)}(z,\theta,\phi) \right] \tag{7.5.14}$$

where $I_1^{(A)}$ and $I_2^{(A)}$ are the first two components of the incoherent Stokes vector that obey the radiative wave equation for Problem A. The brightness temperature of Problem B, $T_{B\alpha}^{(B)}(\theta_{oi})$, can be expressed in terms of the configuration average of the left-hand side of (7.5.11). It then follows from (7.5.11) and (7.5.12)–(7.5.14) that

$$\begin{aligned} T_{B\alpha}^{(B)}(\theta_{oi}) = \frac{1}{\cos\theta_{oi}} \int_{-\infty}^0 dz T^{(B)}(z) \\ \Bigg\{ \kappa_e e^{\kappa_e z \sec\theta_i} \left[\frac{K'}{k} |T_v E_{iv}|^2 + \frac{K'}{k} |T_h E_{ih}|^2 \right] \\ - \int_0^\pi d\theta \sin\theta \int_0^{2\pi} d\phi \cos\theta \frac{d}{dz} \left(I_1^{(A)}(z,\theta,\phi) + I_2^{(A)}(z,\theta,\phi) \right) \Bigg\} \end{aligned} \tag{7.5.15}$$

with $\alpha = v, h$.

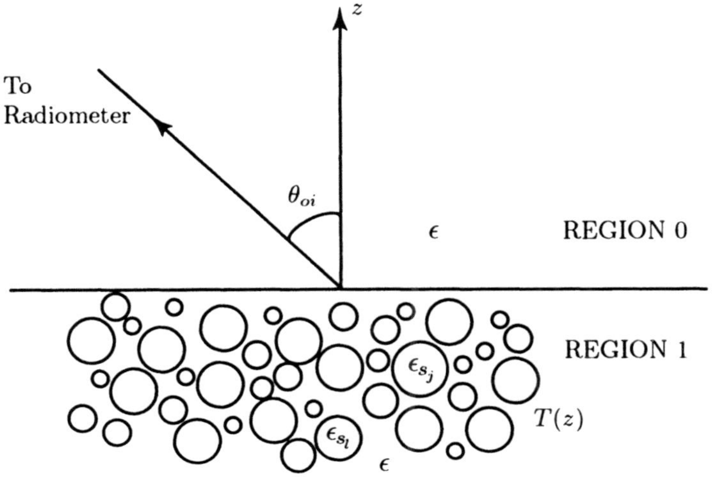

Figure 7.6.1 Passive remote sensing with a radiometer measuring the thermal emission of a half-space discrete random medium of multiple species in the direction θ_{oi}.

The Stokes vector can be defined to be the sum of coherent and incoherent Stokes vector

$$\overline{J}^{(A)}(z,\theta,\phi) = \overline{I}^{(A)}(z,\theta,\phi) + \frac{K'}{k}\overline{T}_0 e^{\kappa_e z \sec\theta_i}\delta(\cos\theta - \cos(\pi-\theta_i))\delta(\phi-\phi_i) \tag{7.5.16}$$

where $\overline{I}^{(A)}$ is the 4×1 Stokes column vector for the incoherent wave and $\overline{T}_0$ is a quantity that is proportional to the 4×1 Stokes column vector of the coherent wave. Using (7.5.15)–(7.5.16) gives,

$$T_{B\alpha}^{(B)}(\theta_{oi}) = -\frac{1}{\cos\theta_{oi}}\int_{-\infty}^{0} dz T^{(B)}(z)\sum_{j=1}^{2}\int_{0}^{4\pi} d\Omega(\hat{s}\cdot\hat{z})\frac{dJ_j^{(A)}(z,\hat{s})}{dz} \tag{7.5.17}$$

where $\hat{s}$ represents the angular direction (θ,ϕ) and $J_1^{(A)}$ and $J_2^{(A)}$ are the first two components of $\overline{J}^{(A)}$. The right-hand side of (7.5.17) is of a form such that radiative wave equations can be set up for Problem B to give results equivalent to (7.5.17).

6 Dense Media Radiative Transfer Equations for Passive Remote Sensing

By using the relation between active and passive microwave remote sensing established in Section 5 for random medium with nonuniform temperature distribution, we can derive DMRT for passive remote sensing. We consider the problem that the temperature profile $T(z)$ is only a function of z (Fig. 7.6.1). We can also make use of the azimuthal symmetry of the problem so that $\mathcal{I}_1$ and $\mathcal{I}_2$ are independent of ϕ and $\mathcal{I}_3 = \mathcal{I}_4 = 0$. Integration over ϕ can also be performed. Thus the dense media radiative transfer equations for passive remote sensing assume the following matrix form of dimension 2. For $0 \leq \theta \leq \pi$ and in region 1, we have

$$\cos\theta \frac{\partial}{\partial z}\underline{\mathcal{I}}(z,\theta) = -\kappa_e \underline{\mathcal{I}}(z,\theta) + \kappa_e(1-\tilde{\omega})CT(z)\begin{bmatrix}1\\1\end{bmatrix} + \frac{3}{8}\kappa_e\tilde{\omega}\int_0^\pi d\theta' \sin\theta' \underline{\underline{p}}(\theta,\theta') \cdot \underline{\mathcal{I}}(z,\theta') \tag{7.6.1}$$

where

$$\underline{\mathcal{I}}(z,\theta) = \begin{bmatrix}\mathcal{I}_1(z,\theta)\\ \mathcal{I}_2(z,\theta)\end{bmatrix} \tag{7.6.2}$$

and $\mathcal{I}_1$ is the TM specific intensity and $\mathcal{I}_2$ is the TE specific intensity. Also in (7.6.1) we have $C = K_b K'^2/(\lambda^2 k^2)$, and

$$\underline{\underline{p}}(\theta,\theta') = \begin{bmatrix}p_{11}(\theta,\theta') & p_{12}(\theta,\theta')\\ p_{21}(\theta,\theta') & p_{22}(\theta,\theta')\end{bmatrix} \tag{7.6.3}$$

where

$$p_{11}(\theta,\theta') = 2\sin^2\theta\sin^2\theta' + \cos^2\theta\cos^2\theta' \tag{7.6.4}$$

$$p_{12}(\theta,\theta') = \cos^2\theta \tag{7.6.5}$$

$$p_{21}(\theta,\theta') = \cos^2\theta' \tag{7.6.6}$$

$$p_{22}(\theta,\theta') = 1 \tag{7.6.7}$$

The boundary conditions for (7.6.1) are, for $0 \leq \theta \leq \pi/2$,

$$\mathcal{I}_1(z=0,\pi-\theta) = |R_h(\theta)|^2 \mathcal{I}_1(z=0,\theta) \tag{7.6.8}$$

$$\mathcal{I}_2(z=0,\pi-\theta) = |R_e(\theta)|^2 \mathcal{I}_2(z=0,\theta) \tag{7.6.9}$$

where $R_h(\theta)$ and $R_e(\theta)$ are TM and TE reflection coefficients, respectively, and are given by (7.2.19) and (7.2.20) with $K_z = K\cos\theta$ and $k_z = (k^2 - K^2\sin^2\theta)^{1/2}$.

After (7.6.1) is solved subject to the boundary conditions of (7.6.8)–(7.6.9), the brightness temperatures in the direction $\theta_o = \sin^{-1}(K' \sin\theta/k)$ for TM and TE polarizations are given by

$$\begin{bmatrix}T_{B1}(\theta_o)\\ T_{B2}(\theta_o)\end{bmatrix} = \frac{1}{C}\begin{bmatrix}\left(1-|R_h(\theta)|^2\right)\mathcal{I}_1(z=0,\theta)\\ \left(1-|R_e(\theta)|^2\right)\mathcal{I}_2(z=0,\theta)\end{bmatrix} \tag{7.6.10}$$

7 Numerical Illustrations of Active and Passive Remote Sensing

In this section we illustrate the numerical results of the dense media radiative transfer equations for active and passive remote sensing. Given the physical parameters of sizes, concentrations, and permittivities, we first calculate the pair distribution functions of multiple species of particles and their Fourier transforms, the effective propagation constant K, the extinction rate κ_e, and the effective permittivity $\epsilon_{\text{eff}} = K^2/\omega^2\mu$. Besides the effective permittivity, we can also define a mixing formula effective permittivity $\epsilon_{\text{eff}}^{(m)}$ which ignores the scattering contributions of (7.2.17) that are size dependent as represented by the terms in the square bracket. Thus

$$\epsilon_{\text{eff}}^{(m)} = \frac{K_o^2}{\omega^2\mu} \tag{7.7.1}$$

where, from (7.2.17), K_o obeys the equation

$$K_o^2 = k^2 + \frac{3K_o^2}{D(K_o)} \sum_{s_l=1}^{L} f_{s_l} y_{s_l}(K_o) \tag{7.7.2}$$

The albedo $\tilde{\omega}$ is then calculated by using (7.4.6). We then solve DMRT equations and calculate bistatic scattering cross sections for active remote sensing and brightness temperatures for passive remote sensing.

We first compare the results of the dense media radiative transfer equations with those of the conventional radiative transfer equations. The conventional radiative transfer equations are valid for sparse concentration of particles. The relations governing the parameters of the conventional transfer equation and the physical parameters of the medium are based on independent scattering. They are as follows:

$$\kappa_e = \kappa_a + \kappa_s \tag{7.7.3}$$

$$\kappa_a = \sum_{s_j=1}^{L} \frac{\epsilon''_{s_j}}{\epsilon} \left| \frac{3\epsilon}{\epsilon_{s_j} + 2\epsilon} \right|^2 f_{s_j} k \tag{7.7.4}$$

$$\kappa_s = \sum_{s_j=1}^{L} 2 f_{s_j} k^4 a_{s_j}^3 \left| \frac{\epsilon_{s_j} - \epsilon}{\epsilon_{s_j} + 2\epsilon} \right|^2 \tag{7.7.5}$$

$$\tilde{\omega} = \frac{\kappa_s}{\kappa_e} \tag{7.7.6}$$

$$K' = k \tag{7.7.7}$$

and $f_{s_j} = 4\pi n_{s_j} a_{s_j}^3/3$ is the fractional volume occupied by species s_j. In

Fig. 7.7.1 the results of the brightness temperatures at 18 GHz between the two approaches are compared for a given set of physical parameters that are characteristic of snow . For the sake of illustration, we have assumed only one species of particles in Fig. 7.7.1. The presence of scattering prevents the thermal emission from reaching the radiometer giving rise to lower brightness temperatures. In this chapter, we only treat non-sticky particles. Since conventional transfer equations predict a much larger albedo, the calculated brightness temperatures are much lower than those of the dense media radiative transfer equations.

In Fig. 7.7.2 the theoretical results of the dense media transfer equations for one particle species are compared with experimental data of brightness temperature measurements over snow. Only the input physical parameters are specified to match the experimental data. A physical fractional volume of 30% is used in calculating the theoretical result. The permittivity of particles $\epsilon_s = (3.2 + i0.016)\epsilon_o$ represents typical permittivity of ice in snow and the calculated mixing formula effective permittivity $\epsilon_{\text{eff}}^{(m)}$ according to (7.7.1)–(7.7.2) is $(1.49 + i0.0029)\epsilon_o$ and represents typical permittivity of the overall snow medium.

In the studies of remote sensing of geophysical terrain, very often the ground truth data are given in terms of a histogram of particle size distribution. We next use the modified gamma distribution

$$n(a) = K_1 a^P \exp\left(-K_2 a^Q\right) \tag{7.7.8}$$

where a is the size of particle and $n(a)da$ gives the number of particles per unit volume having sizes between a and $a + da$.

The total fractional volume f_{tot} is obtained by multiplying $n(a)$ by $4\pi a^3/3$ and integrating over the entire range of size from 0 to infinity

$$f_{tot} = \frac{4\pi}{3} \frac{K_1}{Q} K_2^{-\frac{P+4}{Q}} \Gamma\left(\frac{P+4}{Q}\right) \tag{7.7.9}$$

where Γ is gamma function. The parameter K_2 is related to the mode size a_c. The mode radius a_c is such that $n(a_c)$ is a maximum in the distribution. The mode radius a_c can be found by setting the derivative of (7.7.8) with respect to a equal to zero. Then

$$a_c = \left(\frac{P}{K_2 Q}\right)^{\frac{1}{Q}} \tag{7.7.10}$$

Instead of using K_1, K_2, P, and Q to characterize the distribution, we will use f_{tot}, a_c, P, and Q.

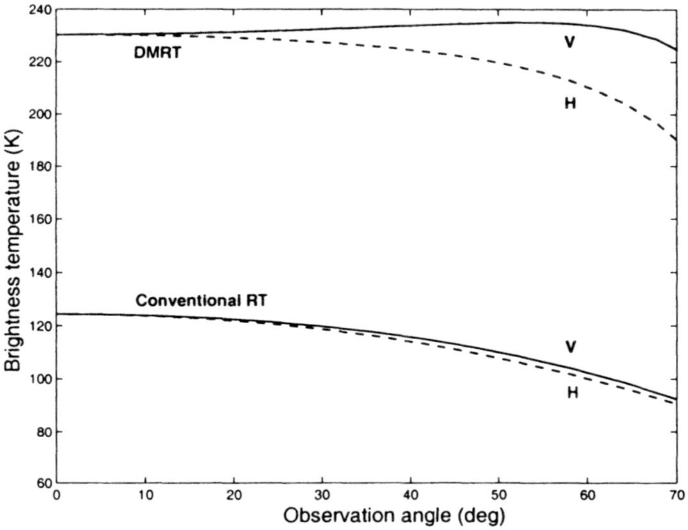

Figure 7.7.1 Comparison of brightness temperature of vertical V polarization (TM) and horizontal H polarization (TE) as a function of angle θ_o as computed by conventional radiative transfer theory and dense media radiative transfer theory. The input physical parameters are for a half-space medium with one particle species with frequency = 18 GHz, $f_1 = 0.3$, $\epsilon_1 = (3.2+i0.016)\epsilon_o$, $a_1 = 0.175$ cm, and $T = 272$ K. The background medium permittivity is ϵ_o. The calculated mixing formula effective permittivity $\epsilon_{\text{eff}}^{(m)}$ is $(1.49+i0.0029)\epsilon_o$. Parameters represent dry snow.

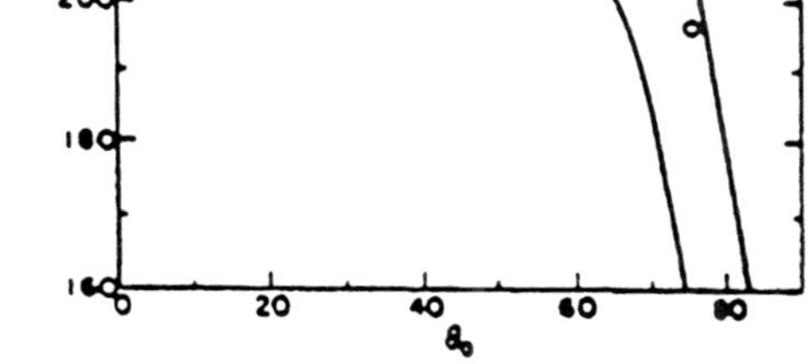

Figure 7.7.2 Comparison of snow field brightness temperature experimental measurements [Shiue et al. 1978; Kong et al. 1979] with theoretical results of dense media radiative transfer equations using the input physical parameters of Fig. 7.7.1. V stands for vertical polarization and H stands for horizontal polarization.

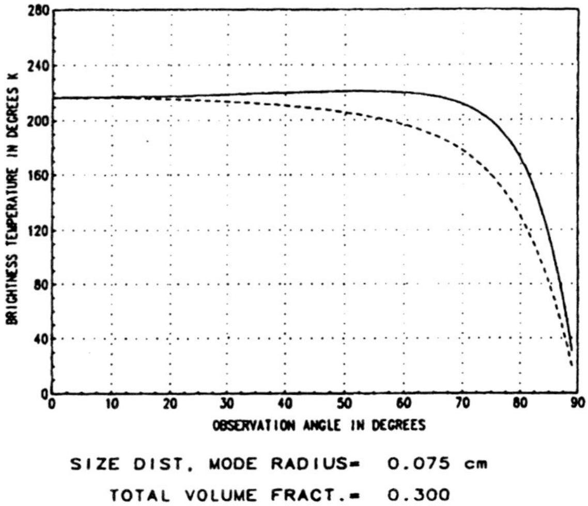

Figure 7.7.3 Brightness temperature of vertical polarization (solid line) and horizontal polarization (dashed line) as a function of observation angle θ_o at frequency = 18 GHz, for a half-space medium with a gamma size distribution of particles with $P = 6$, $Q = 2$, $a_c = 0.075$ cm, $f_{tot} = 0.3$, $\epsilon_s = (3.2 + i0.002)\epsilon_o$ and $T = 272$ K. The computed values are: effective propagation constant $K = (4.60 + i0.2865 \times 10^{-2})$ cm^{-1}, $\kappa_e = 0.00573$ cm^{-1}, and $\tilde{\omega} = 0.7573$.

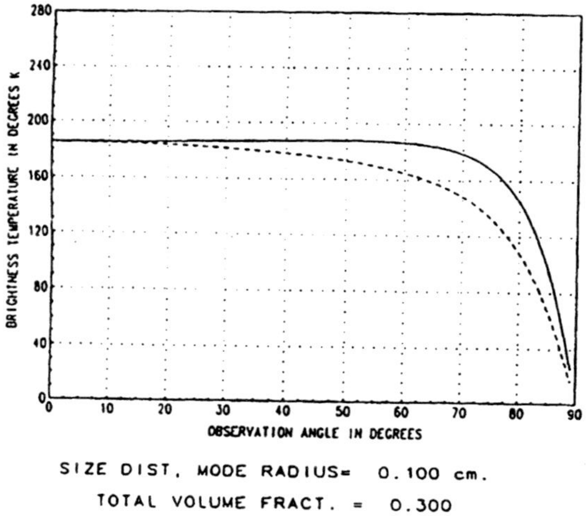

Figure 7.7.4 Brightness temperature of vertical polarization (solid line) and horizontal polarization (dashed line) as a function of observation angle θ_o at frequency = 18 GHz, for a half-space medium with a gamma size distribution of particles with $P = 6$, $Q = 2$, $a_c = 0.1$ cm, $f_{tot} = 0.3$, $\epsilon_s = (3.2 + i0.002)\epsilon_o$ and $T = 272$ K. The computed values are: effective propagation constant $K = (4.60 + i0.584 \times 10^{-2})$ cm^{-1}, $\kappa_e = 0.0117$ cm^{-1}, and $\tilde{\omega} = 0.881$.

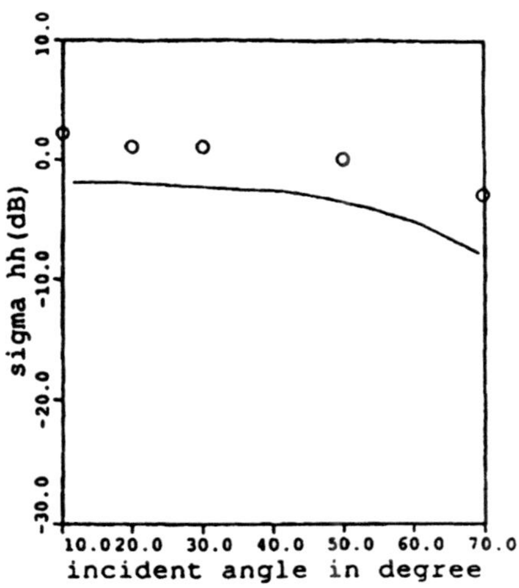

Figure 7.7.5 Comparison of dense medium radiative transfer theory with backscattering data from snow. The input parameters are frequency = 17 GHz, $\epsilon_s = (3.2 + i0.002)\epsilon_o$, d = layer thickness = 27 cm, ϵ_2 = permittivity of homogeneous half-space below scattering layer = $(6 + i0.6)\epsilon_o$. The gamma size distribution parameters are $P = 6$, $Q = 2$, $a_c = 0.125$ cm, and $f_{tot} = 0.2$. Based on these input parameters, the computed values are $K = (4.064 + i0.00941)$ cm^{-1}, $\kappa_e = 0.0188$ cm^{-1}, and $\tilde{\omega} = 0.961$.

In Figs. 7.7.3 and 7.7.4 we plot the brightness temperatures for a half-space medium with particles obeying a gamma size distribution for two different mode radii of 0.075 cm and 0.1 cm, respectively. The permittivity of the particles is $\epsilon_s = (3.2 + i0.002)\epsilon_o$. From the figures, it can be seen that the larger particles which correspond to the tail of the gamma size distribution can contribute significantly to scattering and scattering-induced decreasing of brightness temperatures. Figure 7.7.4 has a lower brightness temperature than Fig. 7.7.3 because of a larger mode radius.

In Fig. 7.7.5 we compare the result of backscattering coefficient in active remote sensing with experimental data at 17 GHz [Stiles and Ulaby, 1980]. The parameters are a two-layer medium with the uppermost layer being air. The middle layer is a slab of scatterers with $\epsilon_s = (3.2 + i0.002)\epsilon_o$ of thickness $d = 27$ cm overlying the lowermost layer of a homogeneous half-space of $\epsilon_2 = (6 + i0.6)\epsilon_o$. The particles in the scattering layer obey gamma size distribution with $P = 6$, $Q = 2$, and $a_c = 0.125$ cm. The backscattering coefficient is calculated by solving (7.4.5) subject to boundary conditions at the top and the lower boundary. The backscattering coefficient $\sigma = 4\pi \cos\theta_{oi} \cdot \mathcal{I}_o(\theta_{oi}, \pi + \phi_{oi})$ for HH polarization is illustrated in Fig. 7.7.5 as a function of incidence angle θ_{oi}. Figure 7.7.5 shows that by using a broad size distribution with $P = 6$ and $Q = 2$, one can achieve a reasonable match of the backscattering measurements with a mode radius of $a_c = 0.125$ cm.

REFERENCES AND ADDITIONAL READINGS

Barabanenkov, Yu. N. and M. Yu. Barabanenkov (1996), Weak time dispersion in scalar theory of radiative transfer for pulse propagation in dense strongly scattering media, *Phys. Lett. A*, **221**, 421–426.

Barabanenkov, Yu. N. and M. Yu. Barabanenkov (1997), Radiative transfer theory with time delay for the effect of pulse entrapping in a resonant random medium: general transfer equation and point-like scatterer model, *Waves in Random Media*, **7**, 607–633.

Barabanenkov, Yu. N., Yu. A. Kravtsov, V. D. Ozrin, and A. I. Saichev (1991a), Enhanced backscattering: the universal wave phenomenon, *Proc. IEEE*, **79(10)**, 1367–1370.

Barabanenkov, Yu. N., Yu. A. Kravtsov, V. D. Ozrin, and A. I. Saichev (1991b), Enhanced backscattering in optics, *Progress in Optics*, **29**, 65–197.

Barabanenkov, Yu. N. and D. I. Kryukov (1992), Functional Fokker-Planck formalism for wave propagation in random media under delocalized and weak localized regions, *Waves in Random Media*, **2(1)**, 1–6.

Barabanenkov, Yu. N., L. M. Zurk, M. Yu. Barabanenkov (1995), Poynting's theorem and electromagnetic wave multiple scattering in dense media near resonance: modified radiative transfer equation, *J. Electromag. Waves and Appl.*, **9**, 1393–1420.

Chandrasekhar, S. (1960), *Radiative Transfer*, Dover, New York.

Chang, A., P. Gloersen, T. Schmugge, T. Wilheit, and H. Zwally (1976), Microwave emission from snow and glacier ice, *J. of Glaciology*, **16**, 23–39.

Chuah, H.-T., S. Tjuatja, A. K. Fung, and J. W. Bredow (1996), A phase matrix for a dense discrete random medium: evaluation of volume scattering coefficient, *IEEE Trans. Geosci. Remote Sens.*, **34(5)**, 1137–1143.

Davis, R. E., J. Dozier, and A. T. C. Chang (1987), Snow property measurement correlative to microwave emission at 35 GHz, *IEEE Trans. Geosci. Remote Sens.*, **25(6)**, 751–757.

Ding, K. H. and L. Tsang (1988), Effective propagation constants of dense nontenuous media with multi-species of particles, *J. Electromag. Waves and Appl.*, **2(8)**, 757–777.

Ding, K. H. and L. Tsang (1989), Effective propagation constants in media with densely distributed dielectric particles of multiple sizes and permittivities, *Progress in Electromag. Res.*, **PIER 1**, edited by J. A. Kong, Chapter 3, 241–295, Elsevier Science Publishing Co., New York.

Kong, J. A., R. Shin, J. C. Shiue, and L. Tsang (1979), Theory and experiment for passive microwave remote sensing of snowpacks, *J. Geophys. Res.*, **84(B10)**, 5669–5673.

Sheng, P., Ed. (1990), *Scattering and Localization of Classical Waves in Random Media*, World Scientific, Singapore.

Sheng, P. (1995), *Introduction to Wave Scattering, Localization, and Mesoscopic Phenomena*, Academic, San Diego.

Shiue, J. C., A. T. C. Chang, H. Boyne, and D. Ellerbruch (1978), Remote sensing of snowpack with microwave radiometers for hydrologic applications, *Proc. 12th Int. Symp. on Remote Sensing of the Environment*, **2**, 877–886, Univ. of Michigan, Ann Arbor, MI.

Stiles, W. H., F. Ulaby (1980), The active and passive microwave response to snow parameters I: Wetness, *J. Geophys. Res.*, **85(C2)**, 1037–1044.

Tsang, L. (1987), Passive remote sensing of dense nontenuous media, *J. Electromag. Waves and Appl.*, **1(2)**, 159–173.

Tsang, L. (1991), Polarimetric passive microwave remote sensing of random discrete scatterers and rough surfaces, *J. Electromag. Waves and Appl.*, **5(1)**, 41–57.

Tsang, L. (1992), Dense media radiative transfer theory for dense discrete random media with spherical particles of multiple sizes and permittivities, *Progress in Electromag. Res.*, **PIER 6**, Dielectric Properties of Heterogeneous Materials, edited by A. Priou, ed., Chapter 5, 181–230, Elsevier, New York.

Tsang, L., K. H. Ding, and B. Wen (1990), Polarimetric signatures of random discrete scatterers based on vector radiative transfer theory. Part I. Nonspherical particles and Part II. Dense media, *Progress in Electromag. Res.*, **PIER 3**, Polarimetric Remote Sensing, edited by J. A. Kong, Chapter 2, 75–142.

Tsang, L. and A. Ishimaru (1984), Backscattering enhancement of random discrete scatterers, *J. Opt. Soc. Am. A*, **1**, 836–839.

Tsang, L. and A. Ishimaru (1985), Radiative wave and cyclical transfer equations for dense nontenuous media, *J. Opt. Soc. Am. A*, **2**, 2187–2194.

Tsang, L. and A. Ishimaru (1987), Radiative wave equations for vector electromagnetic propagation in dense nontenuous media, *J. Electromag. Waves and Appl.*, **1**, 59–72.

Tsang, L. and J. A. Kong (1980), Multiple scattering of electromagnetic waves by random distribution of discrete scatterers with coherent potential and quantum mechanical formulism, *J. Appl. Phys.*, **51**, 3465–3485.

Tsang, L. and J. A. Kong (1982), Effective propagation constants for coherent electromagnetic wave propagation in media embedded with dielectric scatterers, *J. Appl. Phys.*, **53**, 7162–7173.

Tsang, L., J. A. Kong, and R. T. Shin (1985), *Theory of Microwave Remote Sensing*, Wiley-Interscience, New York.

Tsolakis, A. I., I. M. Besieris, and W. E. Kohler (1985), Two-frequency radiative transfer equation for scalar wave in a random distribution of discrete scatterers with pair correlation, *Radio Sci.*, **20**, 1037–1052.

Wen, B., L. Tsang, D. P. Winebrenner, and A. Ishimaru (1990), Dense medium radiative transfer theory: Comparison with experiment and applications to microwave remote sensing and polarimetry, *IEEE Trans. Geosci. Remote Sens.*, **28**, 46–59.

Wiesman, A. and C. Mätzler (1999), Microwave emission model of layered snowpacks, *Remote Sens. Environ.*, **70**, 307–316.

Chapter 8

BACKSCATTERING ENHANCEMENT

1 Introduction

Backscattering enhancement denotes the phenomenon that the bistatic scattering has a maximum in the backscattering direction. Thus, when the bistatic scattering coefficient is plotted as a function of scattered angle for a fixed incident angle, the coefficient exhibits a sharp peak in the backscattering direction of $(\theta_s = \theta_i, \phi_s = \pi + \phi_i)$. Such a phenomenon has been observed in laboratory-controlled experiments [Kuga and Ishimaru, 1984]. The interesting behavior is that the angular width of the peak is small and can be as small as a fraction of a degree. In this chapter we will analyze this phenomenon based on multiple scattering theory. We distinguish the backscattering enhancement from the usual glory effect which also has maximum intensity in the backscattering direction. The usual glory effect is attributed to the Mie scattering as a result of single scattering by particles and not as a result of multiple scattering.

The backscattering enhancement is a surprising phenomenon because it is a result of coherent multiple scattering. In volume scattering, the particles have random positions that will result in random phases of the scattered waves from many particles. It was believed that because of random phase, all phase information would be lost after multiple scattering. Thus the radiative transfer theory is built on the fact that since phase information is lost by the wave, the intensities will add on the average. However backscattering enhancement is observed in strong multiple scattering situations. In the laboratory experiment performed at the University of Washington [Kuga and Ishimaru, 1984; Tsang and Ishimaru, 1984], the optical thickness of the medium was of the order of thousands. That means the wave goes through thousands of orders of multiple scattering by particles. Nevertheless, the wave remembers its phase in the backscattering direction.

In physics, the backscattering enhancement is associated with weak Anderson localization [Anderson, 1958]. Anderson predicts the coherent multiple scattering effects due to randomly placed impurities that will lead to an enhancement of scattering cross section. The phenomenon was difficult to demonstrate for electron waves because electron waves in condensed matter have a distribution in momentum-vectors. It is easier to observe the phenomenon in light waves experiment because light waves have well defined incident and scattered wave vectors. The experiment was subsequently repeated and confirmed by physicists [Wolf and Maret, 1985; Akkermans et al. 1986; van Albada et al. 1987]. The related phenomenon of photonic band gap was also shown [Yablonovitch, 1987, 1993; Joannopoulos et al. 1995] and has led to important technological applications. The backscattering enhance-

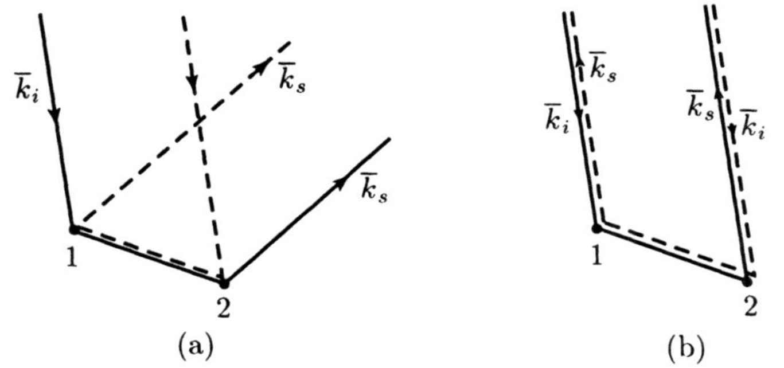

Figure 8.1.1 Physical processes for double scattering involving two particles: (a) Scattered direction $\overline{k}_s$ is not in the backscattering direction. E_{21} is solid line and E_{12} is dotted line. (b) Scattered direction $\overline{k}_s$ is in the backscattering direction, that is, $\overline{k}_s = -\overline{k}_i$. E_{21} is solid line and E_{12} is dotted line. $E_{21} = E_{12}$ by reciprocity.

ment was later observed in random rough surfaces [Mendez and O'Donnell, 1987; Gu et al. 1989; Kim et al. 1990]. Recently the coherent multiple scattering effects were also observed in gain medium with randomly placed inhomegenities or impurities [Lawandy et al. 1994; Wiersma et al. 1995; Wiersma and Lagendijk, 1996, 1997]. Random lasers may become possible with a substantial lowering of the lasing threshold gain.

1.1 Volume Scattering

To illustrate backscattering enhancement, consider double scattering involving particles 1 and 2. The scattered electric field E is a sum of two fields E_{12} and E_{21} (Fig. 8.1.1)

$$E = E_{21} + E_{12} \tag{8.1.1}$$

The field E_{21} is the scattered field that originates from the incident wave scattered by particle 1, traveling from particle 1 to particle 2 and finally being scattered by particle 2. The field E_{12} corresponds to the wave which is scattered by particle 2 before scattered by particle 1 (Fig. 8.1.1). The averaged scattered intensity is

$$\langle |E|^2 \rangle = \langle |E_{21}|^2 \rangle + \langle |E_{12}|^2 \rangle + 2\mathrm{Re}\langle E_{12} E_{21}^* \rangle \tag{8.1.2}$$

where the ensemble average is to be taken over particle positions. It can be shown from Fig. 8.1.1a that the phase difference between E_{12} and E_{21} is $(\overline{k}_i + \overline{k}_s) \cdot \overline{r}_{12}$, where $\overline{r}_{12}$ is the vector joining particles 1 and 2. We consider the following two cases:

1. $\overline{k}_s$ is not in the backscattering direction, so that $\overline{k}_i + \overline{k}_s$ is not equal to zero (Fig. 8.1.1a). In this case, the phase difference between E_{12} and E_{21} is not equal to zero and is given by $(\overline{k}_i + \overline{k}_s) \cdot \overline{r}_{12}$ that depends on the relative particle positions. Since the particle positions are random, the phase difference will also be random. On averaging over particle positions, $\langle E_{12}E_{21}^*\rangle$ is equal to zero, and from (8.1.2) we have

$$\langle |E|^2\rangle = \langle |E_{21}|^2\rangle + \langle |E_{12}|^2\rangle \tag{8.1.3}$$

Thus, the intensities add. This is in agreement with the assumption of radiative transfer theory that scattered fields are uncorrelated so that the intensities add.

2. $\overline{k}_i + \overline{k}_s = 0$. This case corresponds to backscattering and is illustrated in Fig. 8.1.1b. The phase difference vanishes. The two fields E_{12} and E_{21} are always in phase in spite of the randomness of particle positions. Thus, $\langle E_{12}E_{21}^*\rangle$ is not equal to zero. From reciprocity, we actually have

$$E_{12} = E_{21} \tag{8.1.4}$$

Then from (8.1.2) we have

$$\langle |E|^2\rangle = \langle |E_{21}|^2\rangle + \langle |E_{12}|^2\rangle + 2\langle |E_{12}|^2\rangle = 4\langle |E_{21}|^2\rangle \tag{8.1.5}$$

The result in (8.1.5) is twice that of (8.1.3). This phenomenon is not accounted for in the radiative transfer theory. We note that the result in (8.1.5) only occurs in the backscattering direction. For other directions, (8.1.3) is a good approximation. Thus, when the bistatic scattering coefficient is plotted as a function of scattered angle at a fixed incident angle, there is a sharp peak in the backscattering direction. Note that this effect does not occur in single scattering.

The phenomenon can be extended to higher-order multiple scattering. Consider scattering by six scatterers (Fig. 8.1.2). The wave that is scattered by particle $1, 2, 3, \ldots, 6$ (E_{654321}) and the wave that is scattered by particles $6, 5, 4, 3, 2, 1 (E_{123456})$ are exactly in phase in the backscattering direction. They are in phase no matter how random the particle positions are. In other scattered directions, the relative phase of the two waves depends on the positions of the six particles which are assumed to be random. Random phase sets in on averaging over particle positions. The two waves will be uncorrelated in such a case.

1.2 Volume Scattering in the Presence of Reflective Boundary

In this section, it will be shown that backscattering enhancement occurs in the single volume-scattering in the presence of a reflective boundary.

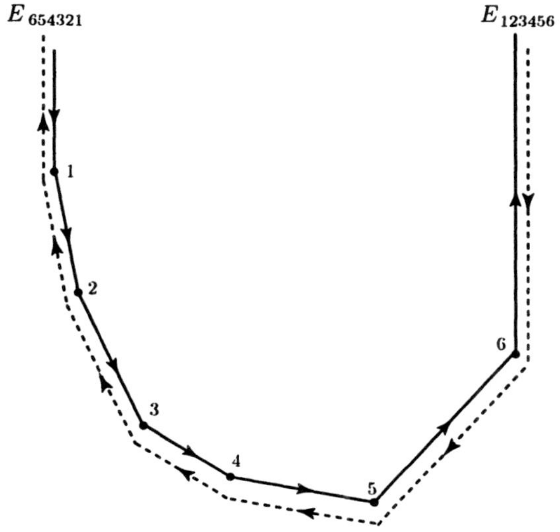

Figure 8.1.2 Scattering by six particles: two waves in phase in the backscattering direction.

The reflective surface can be smooth or rough. Consider a layer of inhomogeneities above a reflective boundary. Single scattering by a particle from incident direction $\overline{k}_i$ into scattering direction $\overline{k}_s$ consists of four contributions (Fig. 8.1.3). The electric field E is

$$E = E_p + E_{pb} + E_{bp} + E_{bpb} \tag{8.1.6}$$

where E_p is the field caused by scattering of the incident wave along $\overline{k}_i$ by the particle into the direction $\overline{k}_s$ (Fig. 8.1.3a). The field E_{bp} is the wave that is scattered by the particle and then reflected by the boundary into the direction $\overline{k}_s$ (Fig. 8.1.3c). The field E_{pb} is the wave which is first scattered by the boundary before being scattered by the particle (Fig. 8.1.3c). Finally the field E_{bpb} is the field that is scattered by the boundary, then the particle, and then the boundary (Fig. 8.1.3b). All four terms E_p, E_{pb}, E_{bp}, and E_{bpb} are a result of single volume scattering. The averaged intensity is

$$\begin{aligned}\langle |E|^2\rangle = {} & \langle |E_p|^2\rangle + \langle |E_{bpb}|^2\rangle + \langle |E_{bp}|^2\rangle + \langle |E_{pb}|^2\rangle + 2\mathrm{Re}\langle E_p E_{bpb}^*\rangle + 2\mathrm{Re}\langle E_p E_{bp}^*\rangle \\ & + 2\mathrm{Re}\langle E_p E_{pb}^*\rangle + 2\mathrm{Re}\langle E_{bpb} E_{bp}^*\rangle + 2\mathrm{Re}\langle E_{bpb} E_{pb}^*\rangle + 2\mathrm{Re}\langle E_{bp} E_{pb}^*\rangle\end{aligned} \tag{8.1.7}$$

Consider the two cases:

1. Scattered direction not in backscattering direction. There are nonzero phase differences among the four field quantities as can be seen from Fig. 8.1.3. The phase differences depend on the particle position. Hence,

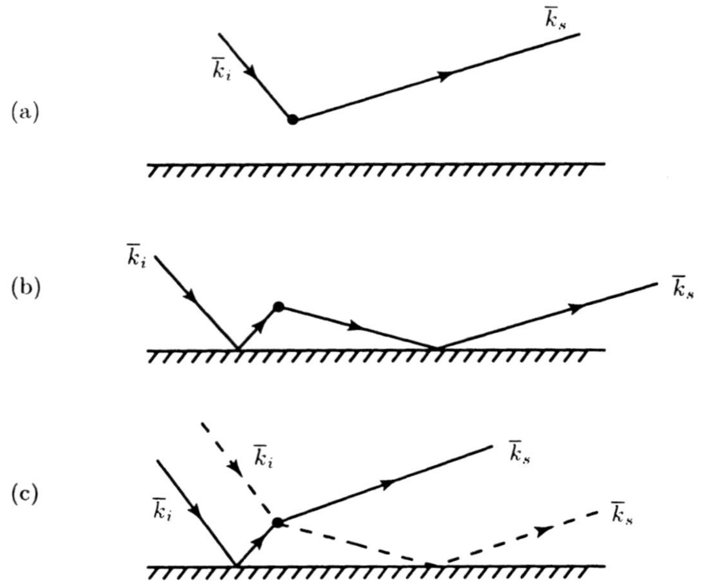

Figure 8.1.3 Physical processes for single-volume scattering in the presence of reflection boundary. Scattered direction $\overline{k}_s$ is not in backscattering direction. (a) E_p; (b) E_{bpb}; (c) solid line, E_{pb}; dotted line, E_{bp}.

on averaging over particle positions that are assumed to be random, the cross terms in (8.1.7) vanish, and

$$\langle |E|^2 \rangle = \langle |E_p|^2 \rangle + \langle |E_{bpb}|^2 \rangle + \langle |E_{bp}|^2 \rangle + \langle |E_{pb}|^2 \rangle \tag{8.1.8}$$

The total intensity is the sum of the intensities of each field quantity. This is in agreement with the assumption of radiative transfer theory.

2. Scattered direction along backscattering direction. The physical processes associated with the four field quantities are depicted in Fig. 8.1.4. The phase difference between E_{bp} and E_{pb} vanishes. By reciprocity we have

$$E_{bp} = E_{pb} \tag{8.1.9}$$

The nonzero phase differences still exist between E_p and the rest of the field quantities and between E_{bpb} and the rest of the field quantities. Hence, from (8.1.7) we obtain

$$\langle |E|^2 \rangle = \langle |E_p|^2 \rangle + \langle |E_{bpb}|^2 \rangle + 2\langle |E_{bp}|^2 \rangle + 2\langle |E_{pb}|^2 \rangle \tag{8.1.10}$$

A comparison between (8.1.8) and (8.1.10) shows that there is backscattering enhancement which is thus a first-order volume scattering

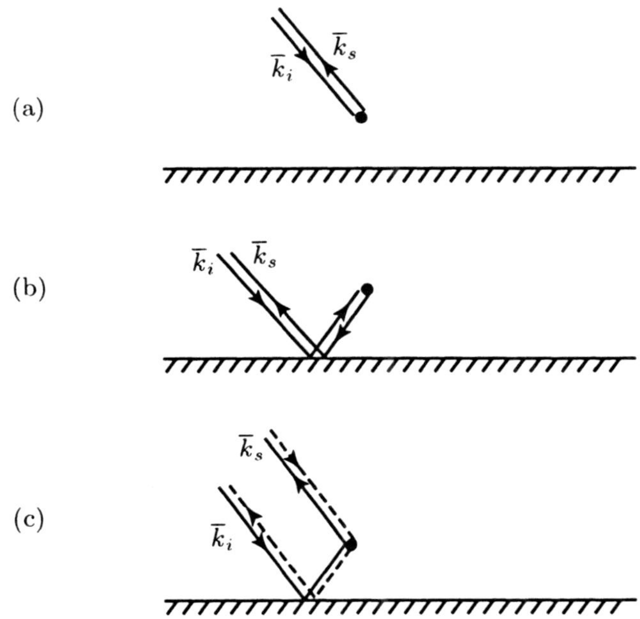

Figure 8.1.4 Physical processes for single volume scattering in the presence of reflective boundary. Scattered direction $\overline{k}_s$ is in backscattering direction, $\overline{k}_s = -\overline{k}_i$. (a) E_p; (b) E_{bpb}; (c) Solid line, E_{pb}; dotted line, E_{bp}; $E_{bp} = E_{pb}$ by reciprocity.

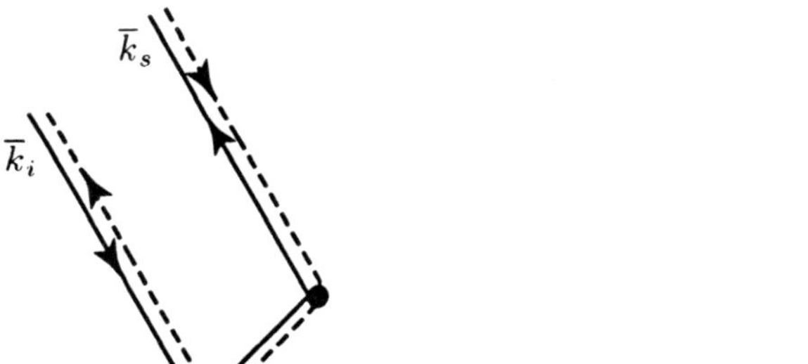

Figure 8.1.5 Single-volume scattering in the presence of a rough surface. E_{bp} is still equal to E_{pb} by reciprocity. Solid line, E_{pb}; dotted line, E_{bp}.

in the presence of a reflective boundary. Backscattering enhancement still exists even if the reflective boundary is a rough surface. The physical processes for E_{pb} and E_{bp} for the case of volume scattering in the presence of a rough boundary are illustrated in Fig. 8.1.5. By reciprocity, we still have $E_{bp} = E_{pb}$. Thus, although rough surfaces can convert a coherent wave into incoherent scattered energy, yet the backscattering enhancement phenomenon is preserved.

2 Second-Order Volume Scattering Theory of Isotropic Point Scatterers

In this section, we illustrate backscattering enhancement by considering the second order theory of volume scattering. To simplify the treatment, we use the model of isotropic point scatterers. Consider a plane wave incident upon a half-space of point scatterers in the direction θ_i (Fig. 8.2.1). Let k be the wavenumber of the upper half-space (region 0) and the wavenumber of the background medium of the lower half-space (region 1). The effective wavenumber K for region 1 is given by, using Foldy's approximation,

$$K = k + \frac{2\pi n_0 f}{k} = K' + iK'' \tag{8.2.1a}$$

where n_0 is the number of particles per unit volume and f is the complex scattering amplitude. The imaginary part of f contributes to the attenuation constant K'' of the effective wave number. It is assumed that the concentration of particles is low so that K' is close to k. Thus

$$K = k + iK'' \tag{8.2.1b}$$

where

$$K'' = \frac{2\pi n_0}{k} \mathrm{Im} f \tag{8.2.1c}$$

The mean field in region 1 is

$$\psi_m(\overline{r}) = \exp(i\overline{K}_i \cdot \overline{r}) \tag{8.2.2}$$

where

$$\overline{K}_i = K_{ix}\hat{x} + K_{iy}\hat{y} - K_{iz}\hat{z} \tag{8.2.3}$$

$$K_{ix} = k \sin\theta_i \cos\phi_i; \qquad K_{iy} = k \sin\theta_i \sin\phi_i \tag{8.2.4a}$$

$$K_{iz} = (K^2 - k^2 \sin^2\theta_i)^{1/2} = K'_{iz} + iK''_{iz} \tag{8.2.4b}$$

In what follows we shall use the second-order multiple-scattering theory to calculate the scattered field in region 0. Diagrammatically, the second-order multiple-scattering theory is given by

$E(\psi\psi^*) =$ 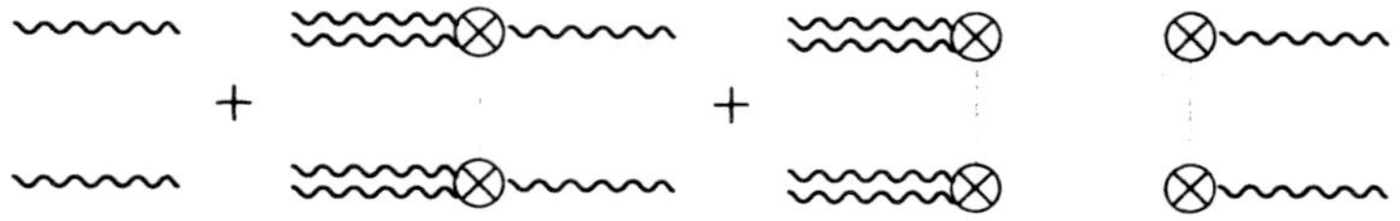

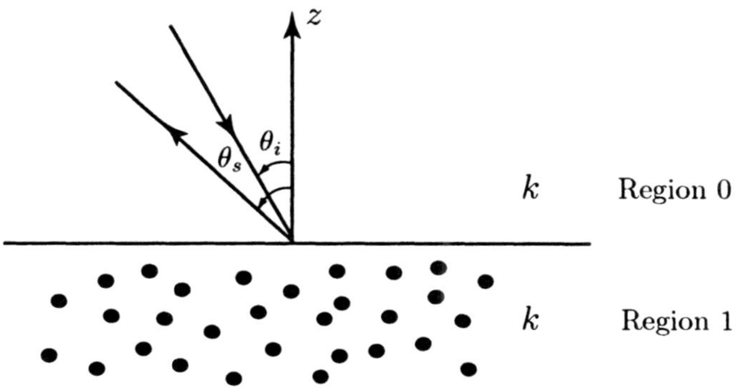

Figure 8.2.1 Scattering of waves by a half-space of point scatterers.

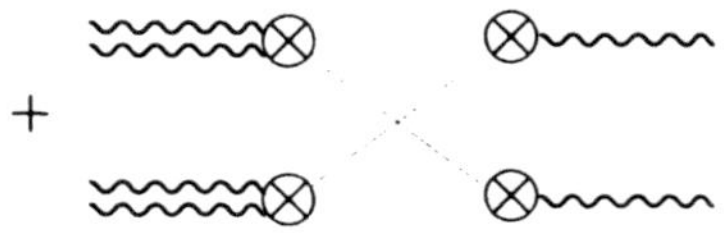

(8.2.5)

where

 $= \psi_m$ (8.2.5*a*)

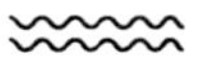 $= G_{01m}$ (8.2.5*b*)

$= G_{11m}$ (8.2.5*c*)

The first term in (8.2.5) is the intensity of the mean field; the second term (denoted by L_1) corresponds to single scattering of the mean field and is the result of first-order multiple-scattering theory. The third term in (8.2.5) is called the second-order ladder term. We denote it by L_2. The last term in (8.2.5) is called the second-order cyclical term and will be denoted by C_2. The second-order ladder term is included in the ladder approximation, whereas the cyclical term contribution is comparable with the ladder term for observation in the vicinity of the backscattering direction.

The mean Green's functions in (8.2.5) are half-space Green's functions. Since observation of the scattered field is in region 0, the leftmost Green's functions in each diagram are G_{01m}, and the rest of the mean Green's functions in the diagram are G_{11m}. The mean Green's function G_{11m} has both source and observation points in region 1. It obeys the wave equation

$$(\nabla^2 + K^2)G_{11m}(\overline{r}, \overline{r}') = -\delta(\overline{r} - \overline{r}') \quad (8.2.6)$$

It is given by the expression

$$G_{11m}(\overline{r},\overline{r}') = \frac{\exp(iK|\overline{r}-\overline{r}'|)}{4\pi|\overline{r}-\overline{r}'|} \tag{8.2.7}$$

The mean Green's function G_{01m} has a source in region 1 and an observation point in region 0. It obeys the wave equation

$$(\nabla^2 + k^2)G_{01m}(\overline{r},\overline{r}') = 0 \tag{8.2.8}$$

Since the observation point in region 0 is far away from the scattering layer, the far-field approximation can be made on G_{01m} :

$$G_{01m}(\overline{r},\overline{r}') = \frac{\exp(ikr)}{4\pi r}\exp(-i\overline{K}_s\cdot\overline{r}') \tag{8.2.9}$$

where

$$\overline{K}_s = K_{sx}\hat{x} + K_{sy}\hat{y} + K_{sz}\hat{z} \tag{8.2.10}$$
$$K_{sx} = k\sin\theta_s\cos\phi_s \tag{8.2.11}$$
$$K_{sy} = k\sin\theta_s\sin\phi_s \tag{8.2.12}$$
$$K_{sz} = (K^2 - k^2\sin^2\theta_s)^{1/2} = K'_{sz} + iK''_{sz} \tag{8.2.13}$$

and (θ_s, ϕ_s) is the scattered direction. In the backscattering direction, $\overline{K}_s = -\overline{K}_i$ so that $\theta_s = \theta_i$ and $\phi_s = \pi + \phi_i$.

The first-order scattering term L_1 is, from (8.2.5),

$$L_1 = n_0 \int d\overline{r}_j d\overline{r}_1 d\overline{r}_2 d\overline{r}_3 d\overline{r}_4 G_{01m}(\overline{r},\overline{r}_1)T_j(\overline{r}_1,\overline{r}_2) \cdot \psi_m(\overline{r}_2)G^*_{01m}(\overline{r},\overline{r}_3)T^*_j(\overline{r}_3,\overline{r}_4)\psi^*_m(\overline{r}_4) \tag{8.2.14}$$

where T_j is the transition operator of the jth particle and $\overline{r}_j$ is its position. The integrations of the space variables are carried out over region 1.

For point scatterers,

$$T_j(\overline{r}_1,\overline{r}_2) = 4\pi f\delta(\overline{r}_1 - \overline{r}_j)\delta(\overline{r}_2 - \overline{r}_j). \tag{8.2.15}$$

Substituting (8.2.2), (8.2.9), and (8.2.15) into (8.2.14), the four Dirac delta functions in $T_jT_j^*$ cancel the 4-fold $d\overline{r}_1 d\overline{r}_2 d\overline{r}_3 d\overline{r}_4$ integrations. We obtain

$$L_1 = n_o \int d\overline{r}_j (4\pi)^2 |f|^2 |G_{01m}(\overline{r},\overline{r}_j)\psi_m(\overline{r}_j)|^2 = \frac{n_o|f|^2}{r^2}\int d\overline{r}_{\perp j}\int_{-\infty}^{0} dz_j e^{(2K''_{sz}+2K''_{iz})z_j} \tag{8.2.16}$$

Integrating (8.2.16) gives

$$L_1 = \frac{|f|^2 n_0 A}{r^2}\frac{1}{2(K''_{sz} + K''_{iz})}, \tag{8.2.17}$$

where A is the illuminated area of the layer of scatterers. Note from (8.2.4b)

$$K'^2 - k^2 \sin^2 \theta_i = K'^2_{iz} - K''^2_{iz} \tag{8.2.18a}$$

$$2K'K'' = 2K'_{iz}K''_{iz} \tag{8.2.18b}$$

For this case $K' \simeq k$ and $K''_{iz} \ll K'_{iz}$. Then

$$K'_{iz} \simeq K' \cos\theta_i \tag{8.2.18c}$$

Since the attenuation rate is much smaller than the wavenumber, we have the following approximate relation from (8.2.18b) and (8.2.18c):

$$K''_{iz} = K''/\mu_i \tag{8.2.19}$$

where $\mu_i = \cos\theta_i$. Similarly,

$$K''_{sz} = K''/\mu_s \tag{8.2.20}$$

where $\mu_s = \cos\theta_s$. Hence

$$L_1 = \frac{\tilde{\omega}}{4\pi}\frac{A}{r^2}\frac{1}{\left(\frac{1}{\mu_s} + \frac{1}{\mu_i}\right)} \tag{8.2.21}$$

where

$$\tilde{\omega} = \frac{4\pi n_o |f|^2}{2K''} \tag{8.2.22}$$

is the albedo of isotropic point scatterers.

The second-order ladder term is, from (8.2.5)

$$L_2 = n_o^2 \int d\overline{r}_j d\overline{r}_l d\overline{r}_1 d\overline{r}_2 d\overline{r}_3 d\overline{r}_4 d\overline{r}_5 d\overline{r}_6 d\overline{r}_7 d\overline{r}_8 G_{01m}(\overline{r},\overline{r}_1) T_j(\overline{r}_1,\overline{r}_2) G_{11m}(\overline{r}_2,\overline{r}_3)$$
$$\cdot\, T_l(\overline{r}_3,\overline{r}_4)\psi_m(\overline{r}_4) \cdot G^*_{01m}(\overline{r},\overline{r}_5) T^*_j(\overline{r}_5,\overline{r}_6) G^*_{11m}(\overline{r}_6,\overline{r}_7) T^*_l(\overline{r}_7,\overline{r}_8)\psi^*_m(\overline{r}_8) \tag{8.2.23}$$

Substituting in the expressions for T_j and T_l, we get

$$L_2 = n_0^2 \int d\overline{r}_j \int d\overline{r}_l (4\pi)^4 |f|^4 |G_{01m}(\overline{r},\overline{r}_j)|^2 |G_{11m}(\overline{r}_j,\overline{r}_l)|^2 |\psi_m(\overline{r}_l)|^2 \tag{8.2.24}$$

From (8.2.24) and using the expressions for G_{01m} and G_{11m}, we have

$$L_2 = \frac{n_o^2 |f|^4}{r^2} \int_{z_j \le 0} d\overline{r}_j \int_{z_l \le 0} d\overline{r}_l \frac{e^{-2K''|\overline{r}_j - \overline{r}_l|}}{|\overline{r}_j - \overline{r}_l|^2} e^{2K''_{sz} z_j} e^{2K''_{iz} z_l} \tag{8.2.25}$$

To perform the integral in (8.2.25), let

$$\overline{r}_d = \overline{r}_j - \overline{r}_l \tag{8.2.26a}$$

$$\overline{r}_A = \frac{\overline{r}_j + \overline{r}_l}{2} \tag{8.2.26b}$$

Then

$$z_A = \frac{z_j + z_l}{2} \tag{8.2.26c}$$

$$z_d = z_j - z_l \tag{8.2.26d}$$

Since z_j and z_l only extend over the lower half-space, we have

$$\int_{-\infty}^{0} dz_j \int_{-\infty}^{0} dz_l = \int_{0}^{\infty} dz_d \int_{-\infty}^{-\frac{z_d}{2}} dz_A + \int_{-\infty}^{0} dz_d \int_{-\infty}^{\frac{z_d}{2}} dz_A \tag{8.2.27}$$

Using spherical coordinates for $\overline{r}_d$, $x_d = r_d \sin \theta d \cos \phi_d$, $y_d = r_d \sin \theta_d \sin \phi_d$, $z_d = r_d \cos \theta_d$, we have

$$L_2 = \frac{n_o^2 |f|^4}{r^2} \int_{-\infty}^{\infty} d\overline{r}_{A\perp} \int_0^{2\pi} d\phi_d \int_0^{\infty} dr_d r_d{}^2 \Bigg(\int_0^{\pi/2} d\theta_d \sin\theta_d \int_{-\infty}^{-r_d \cos\ \theta_d/2} dz_A$$

$$+ \int_{\pi/2}^{\pi} d\theta_d \sin\theta_d \int_{-\infty}^{r_d \cos\ \theta_d/2} dz_A \Bigg) \frac{e^{-2K'' r_d}}{r_d^2} e^{(2K''_{sz}+2K''_{iz})z_A} e^{(K''_{sz}-K''_{iz}) r_d \cos\theta_d}$$

$$= \frac{2\pi n_o^2 |f|^4 A}{r^2} \cdot \frac{1}{2(K''_{sz}+K''_{iz})} \int_0^{\infty} dr_d \Bigg(\int_0^{\pi/2} d\theta_d \sin\theta_d e^{-2K'' r_d - 2K''_{iz} r_d \cos\theta_d}$$

$$+ \int_{\pi/2}^{\pi} d\theta_d \sin\theta_d e^{-2K'' r_d + 2K''_{sz} r_d \cos\theta_d} \Bigg) \tag{8.2.28}$$

Carrying out the integrals in (8.2.28), gives

$$L_2 = \frac{n_0^2 |f|^4 A}{r^2} \frac{2\pi}{2K''_{sz}+2K''_{iz}} \int_0^1 \frac{d\mu_d}{\mu_d} \frac{\frac{4K''}{\mu_d} + 2K''_{iz} + 2K''_{sz}}{\left(\frac{2K''}{\mu_d} + 2K''_{iz}\right)\left(\frac{2K''}{\mu_d} + 2K''_{sz}\right)} \tag{8.2.29}$$

By using the relations of (8.2.19) and (8.2.20) and integrating the expression in (8.2.29), we have

$$L_2 = \left(\frac{\tilde{\omega}}{4\pi}\right)^2 \frac{2\pi A}{r^2} \frac{1}{\left(\frac{1}{\mu_s} + \frac{1}{\mu_i}\right)} \left[\mu_i \ln\left(1+\frac{1}{\mu_i}\right) + \mu_s \ln\left(1+\frac{1}{\mu_s}\right)\right] \tag{8.2.30}$$

The sum of the first two ladder terms L_1 and L_2 is identical to the result of a second iteration of the radiative-transfer equation for a half-space of point particles.

The second-order cyclical term C_2 is, on making use of the expressions of T_j and T_l from (8.2.15),

$$C_2 = n_0^2 \int d\overline{r}_j \int d\overline{r}_l (4\pi)^4 |f|^4 G_{01m}(\overline{r}, \overline{r}_j) G_{11m}(\overline{r}_j, \overline{r}_l) \psi_m(\overline{r}_l)$$

$$\cdot\, G^*_{01m}(\overline{r}, \overline{r}_l) G^*_{11m}(\overline{r}_l, \overline{r}_j) \psi^*_m(\overline{r}_j) \tag{8.2.31}$$

Comparing (8.2.24) and (8.2.31), we note the difference between the second-order ladder term and the second-order cyclical term. In the second-order cyclical term C_2, the argument of ψ_m is $\overline{r}_l$ and the argument of ψ_m^* is $\overline{r}_j$. Also G_{01m} and G_{01m}^* have different scatterer arguments $\overline{r}_j$ and $\overline{r}_l$, respectively. These different arguments are due to the cyclical arrangement of the scatterers.

Although the integrand for L_2 contains no phase terms, the integrand for C_2 contains a phase dependence that results from this cyclical arrangement of the scatterers. If we use the expressions of G_{01m} and G_{11m} and make the substitutions, we have

$$C_2 = \frac{n_o^2 |f|^4 A}{r^2} \int dz_A \int d\overline{r}_d \frac{e^{-2K'' r_d}}{r_d^2} e^{-i\overline{K}_s \cdot \overline{r}_j + i\overline{K}_s^* \cdot \overline{r}_l} e^{i\overline{K}_i \cdot \overline{r}_l - i\overline{K}_i^* \cdot \overline{r}_j} \tag{8.2.32}$$

Let $\overline{K}_s'$ and $\overline{K}_i'$ denote the real parts of $\overline{K}_s$ and $\overline{K}_i$, respectively. Then

$$C_2 = \frac{n_0^2 |f|^4 A}{r^2} \int dz_A \int d\overline{r}_d \frac{\exp(-2K'' r_d)}{r_d^2} \times \exp(-i\overline{K}_d \cdot \overline{r}_d) \exp[2(K_{iz}'' + K_{sz}'') z_A] \tag{8.2.33}$$

where

$$\overline{K}_d = \overline{K}_s' + \overline{K}_i' = (K_{ix} + K_{sx})\,\hat{x} + (K_{iy} + K_{sy})\,\hat{y} + \left(-K_{iz}' + K_{sz}'\right)\hat{z} \tag{8.2.34}$$

In the backscattering direction, $\overline{K}_d = 0$. We note that if attenuation is ignored, i.e., set $K'' = 0$ in the integrand of (8.2.33), then C_2 will be a delta function at $\overline{K}_d = 0$, giving a sharp peak with no angular width. This will be the result of using the second order Born approximation, in which attenuation is neglected. If attenuation is included, the delta function in angles will become a peak with a finite angular width. We note that $\overline{K}_d \cdot \overline{r}_d$ in the integrand of (8.2.33) corresponds physically to the phase difference between the field and its conjugate in double scattering (Fig. 8.2.2). In the backscattering direction, the phase difference is zero and gives a maximum for the cyclical term C_2. From (8.2.33) and making transformation to spherical coordinates similar to what we done in (8.2.28) we obtain

$$C_2 = \frac{n_o^2 |f|^4 A}{r^2} \int_0^{2\pi} d\phi_d \int_0^\infty dr_d \left(\int_0^{\pi/2} d\theta_d \sin\theta_d \int_{-\infty}^{-r_d \cos\theta_d/2} dz_A \right.$$
$$\left. + \int_{\frac{\pi}{2}}^{\pi} d\theta_d \sin\theta_d \int_{-\infty}^{r_d \cos\theta_d/2} dz_A \right) e^{-2K_d'' r_d} e^{-i\overline{K}_d \cdot \overline{r}_d} e^{2(K_{iz}'' + K_{sz}'') z_A} \tag{8.2.35}$$

Let

$$\overline{K}_{d\perp} = (K_{ix} + K_{sx})\,\hat{x} + (K_{iy} + K_{sy})\,\hat{y} \tag{8.2.36}$$

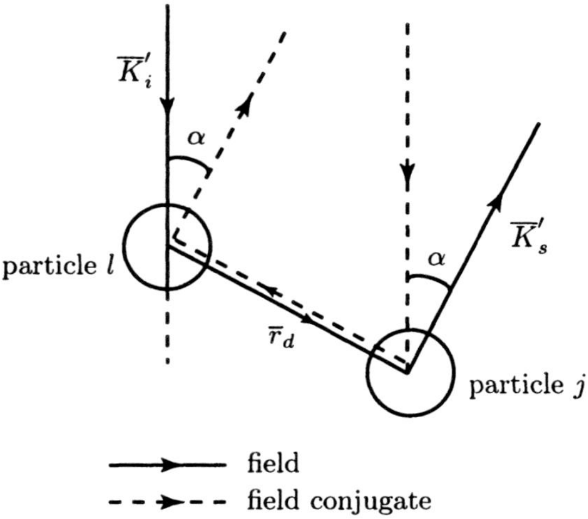

Figure 8.2.2 Path difference between field and its conjugate in double scattering.

so that

$$\overline{K}_d = \overline{K}_{d\perp} + \left(K'_{sz} - K'_{iz}\right)\hat{z} \tag{8.2.37}$$

Also let $\overline{K}_{d\perp} = K_{d\perp}\cos\phi_{K_d}\ \hat{x} + K_{d\perp}\sin\phi_{K_d}\ \hat{y}$. Then

$$\overline{K}_d \cdot \overline{r}_d = K_{d\perp} r_d \sin\theta_d \cos(\phi_d - \phi_{K_d}) + \left(K'_{sz} - K'_{iz}\right) r_d \cos\theta_d \tag{8.2.38}$$

Put (8.2.36) and (8.2.38) into (8.2.35), and integrating over $d\phi_d$ and dz_A gives

$$C_2 = \frac{\pi n_o^2 |f|^4 A}{\left(K''_{sz} + K''_{iz}\right) r^2} \int_0^\infty dr_d$$
$$\left\{ \int_0^{\pi/2} d\theta_d \sin\theta_d e^{-2K'' r_d} e^{-i(K'_{sz}-K'_{iz}) r_d \cos\theta_d} e^{-(K''_{sz}+K''_{iz}) r_d \cos\theta_d} J_o\left(K_{d\perp} r_d \sin\theta_d\right) \right.$$
$$\left. + \int_{\frac{\pi}{2}}^{\pi} d\theta_d \sin\theta_d e^{-2K'' r_d} e^{-i(K'_{sz}-K'_{iz}) r_d \cos\theta_d} e^{(K''_{sz}+K''_{iz}) r_d \cos\theta_d} J_o\left(K_{d\perp} r_d \sin\theta_d\right) \right\} \tag{8.2.39}$$

It can be shown that the two terms in the curly brackets of (8.2.39) are complex conjugates of each other. Next, we define

$$\gamma = 1 + \frac{i(K'_{sz} - K'_{iz})\cos\theta_d}{2K''} + \frac{\cos\theta_d}{2}\left(\frac{1}{\mu_i} + \frac{1}{\mu_s}\right) \tag{8.2.40}$$

$$\beta = \frac{\sin\theta_d}{2K''}\left[K_d^2 - (K'_{sz} - K'_{iz})^2\right]^{1/2} = \frac{K_{d\perp}\sin\theta_d}{2K''} \tag{8.2.41}$$

Then

$$C_2 = \frac{2\pi n_o^2 |f|^4 A}{(K''_{sz} + K''_{iz})r^2} \text{Re} \int_0^\infty dr_d \int_0^{\pi/2} d\theta_d \sin\theta_d e^{-2K'' r_d \gamma} J_o(2K''\beta r_d) \tag{8.2.42}$$

Using the integral identity of Bessel's function,

$$\int_0^\infty dx e^{-ax} J_o(bx) = \frac{1}{\sqrt{a^2 + b^2}} \tag{8.2.43}$$

we obtain

$$C_2 = \left(\frac{\tilde{\omega}}{4\pi}\right)^2 \frac{4\pi A}{r^2} \frac{1}{\left(\frac{1}{\mu_i} + \frac{1}{\mu_s}\right)} \int_0^{\pi/2} d\theta_d \sin\theta_d \text{Re}\left(\frac{1}{(\beta^2 + \gamma^2)^{1/2}}\right) \tag{8.2.44}$$

Let α be the angle of separation between $\overline{K}'_i$ and $\overline{K}'_s$ (Fig. 8.2.2), then $K_d = 2K' \sin(\alpha/2)$ denotes the magnitude of the deviation of the scattering observation direction from the backscattering direction. In the backscattering direction, $K_d = 0$, and it can readily be shown that the cyclical term C_2 is equal to the ladder term L_2. For $K_d \gg 2K''$, it can be shown that C_2 is much less than L_2. Thus the cyclical term C_2 exhibits a peak in the backscattering direction with angular width of the order of K''/K'.

In Fig. 8.2.3 we plot the bistatic scattering coefficient $\sigma(\theta_s, \phi_s; \theta_i, \phi_i = 0) = r^2 4\pi(L_1 + L_2 + C_2)/(A\cos\theta_i)$ as a function of scattered angle θ_s. The angle ϕ_s is chosen to be 180°, and θ_s is close to θ_i so that the scattered direction is in the vicinity of the backscattering direction. The incidence angle is 5°. It can be seen from Fig. 8.2.3 that the bistatic scattering coefficient exhibits a sharp peak in the backscattering direction. The angular width of the sharp peak is of the order of attenuation rate divided by wavenumber. Since the peak arises from the second-order cyclical term, backscattering enhancement is not observable if single scattering is the dominant response.

In the laboratory-controlled experiment [Kuga and Ishimaru, 1984], the observed angular width of the backscattering peak is between 0.1° and 0.5°, and the measured attenuation rate $2K''$ is about 50 to 100 times less than the wavenumber K'. Hence the angular width is of the order K''/K' and is in agreement with the theoretical prediction despite the simple model of second order scattering of isotropic point scatterers that is used in the analysis in this section.

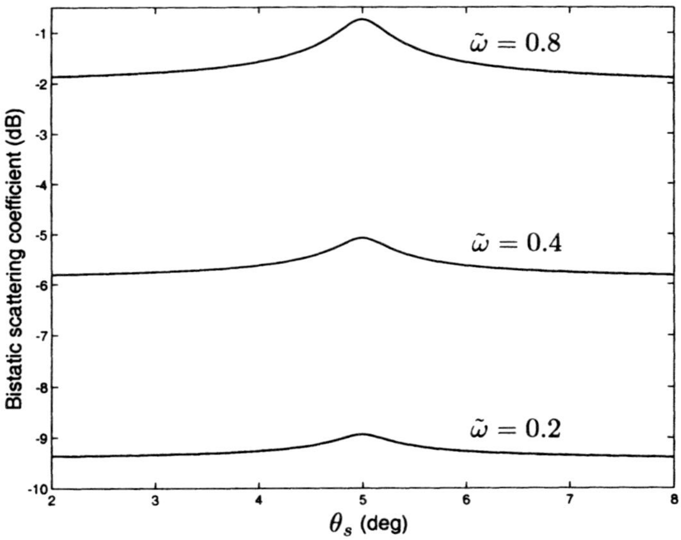

Figure 8.2.3 Bistatic scattering coefficients as a function of θ_s for different albedos with $\theta_i = 5°, \phi_s = 180°$, and $2K''/K' = 3.18 \times 10^{-3}$. The second-order theory is used.

3 Summation of Ladder Terms and Cyclical Terms for Isotropic Point Scatterers

The second-order theory includes the first-order scattering, the second-order ladder term, and the second-order cyclical term. However, the second-order solution is not adequate when the albedo and the optical thickness of the scattering medium become appreciable. In this section, by using the model of isotropic point scatterers, we investigate the multiple-scattering phenomenon by summing all the ladder and cyclical terms. As shown in Chapter 5, Section 5 the summation of all the ladder terms leads to the classical Schwarzschild-Milne integral equation with the exponential integral as the kernel. On the other hand, we shall show in this section that the summation of the cyclical terms leads to a two-variable cyclical transfer integral equation with a more general kernel. The integral equation is then solved numerically. The results thus include all the multiple scattering associated with the cyclical terms. Results of multiple scattering theory differ significantly from the second-order theory for appreciable optical thickness and albedo. The multiple-scattering solution also gives a sharper backscattering enhancement peak than the second-order solution.

3.1 Formulation

Consider a plane wave incident upon a slab of point scatterers (Fig. 8.3.1). The mean field $\psi_m(\overline{r})$, the mean Green's functions of G_{01m}, G_{11m}, the effective propagation constant K, and the scattered wave vector $\overline{K}_s$ are as given in Section 2. We also let

$$\overline{K}_i = k \sin\theta_i \cos\phi_i \hat{x} + k \sin\theta_i \sin\phi_i \hat{y} - K_{iz}\hat{z} \tag{8.3.1}$$

$$K_{iz} = (K^2 - k^2 \sin^2\theta_i)^{1/2} \simeq k\cos\theta_i + \frac{iK''}{\mu_i} \tag{8.3.2}$$

Let $I_C(\overline{r})$ denote the intensity in region 0 that is a summation of all the cyclical terms. The total intensity is

$$I(\overline{r}) = I_L(\overline{r}) + I_C(\overline{r}) \tag{8.3.3}$$

In (8.3.3), $I_L(\overline{r})$ is the ladder intensity as determined in Section 5 of Chapter 5. It consists of summation of all ladder terms. In Section 5 of Chapter 5, we show that the summation of all ladder terms gives a result that is equivalent to radiative transfer theory. Using the same diagrametric notation as Section 5 of Chapter 5 and Section 2 of this chapter, we have

$I_C(\overline{r}) =$  (8.3.4)

Next define the cyclical operator  such that

$I_C(\overline{r}) =$ 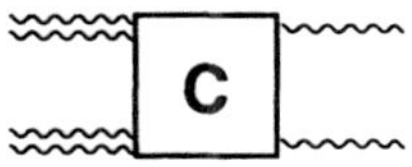 (8.3.5)

Then from (8.3.4) and (8.3.5), the cyclical operator obeys the diagrammatic equation

 (8.3.6)

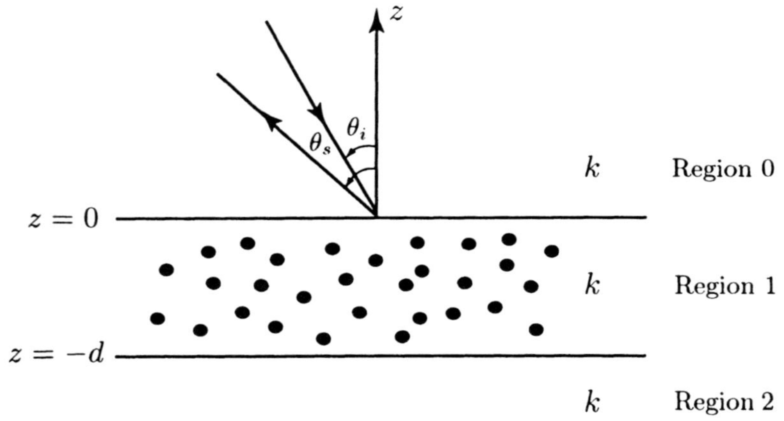

Figure 8.3.1 Scattering of waves by a slab of isotropic point scatterers.

Equations (8.3.5) and (8.3.6) include all the multiple-scattering effects in the cyclical terms. The objective is to put (8.3.6) in a numerically tractable form. By solving the equations numerically, the results include all the cyclical terms. The bistatic scattering coefficient for the cyclical intensity is denoted by $\gamma_C(\mu_s, \phi_s; \mu_i, \phi_i)$ and is defined as

$$\gamma_C(\mu_s, \phi_s; \mu_i, \phi_i) = \frac{4\pi r^2}{A \cos\theta_i} I_C(\overline{r}) \tag{8.3.7}$$

where A is the area of the layer of random medium under observation. The total bistatic scattering coefficient is

$$\gamma(\mu_s, \phi_s; \mu_i, \phi_i) = \gamma_L(\mu_s, \phi_s; \mu_i, \phi_i) + \gamma_C(\mu_s, \phi_s; \mu_i, \phi_i) \tag{8.3.8}$$

The bistatic scattering coefficient consisting of all the ladder terms is calculated in (5.5.53) by solving the radiative transfer equations. To study the cyclical equation of (8.3.6), let

$$C(\overline{r}_1, \overline{r}_2; \overline{r}_3, \overline{r}_4) = \boxed{\mathbf{C}}\ \ \text{(corners: 1, 2 top; 3, 4 bottom)} \tag{8.3.9}$$

Substituting (8.3.9) into (8.3.6), we have

$$\begin{aligned} C(\overline{r}_1, \overline{r}_2, \overline{r}_3, \overline{r}_4) &= n_o^2 \int d\overline{r}_j d\overline{r}_l d\overline{r}_5 d\overline{r}_6 d\overline{r}_7 d\overline{r}_8 \\ &\cdot T_j(\overline{r}_1, \overline{r}_5) G_{11m}(\overline{r}_5, \overline{r}_6) T_l(\overline{r}_6, \overline{r}_2) T_l^*(\overline{r}_3, \overline{r}_7) G_{11m}(\overline{r}_7, \overline{r}_8) T_j^*(\overline{r}_8, \overline{r}_4) \\ &+ n_o^3 \int d\overline{r}_j d\overline{r}_l d\overline{r}_m d\overline{r}_5 d\overline{r}_6 d\overline{r}_7 d\overline{r}_8 d\overline{r}_9 d\overline{r}_{10} d\overline{r}_{11} d\overline{r}_{12} \\ &\cdot T_j(\overline{r}_1, \overline{r}_5) G_{11m}(\overline{r}_5, \overline{r}_6) T_l(\overline{r}_6, \overline{r}_7) G_{11m}(\overline{r}_7, \overline{r}_8) T_m(\overline{r}_8, \overline{r}_2) \end{aligned}$$

$$\cdot T_m^*(\overline{r}_3,\overline{r}_9)G_{11m}^*(\overline{r}_9,\overline{r}_{10})T_l^*(\overline{r}_{10},\overline{r}_{11})G_{11m}^*(\overline{r}_{11},\overline{r}_{12})T_j^*(\overline{r}_{12},\overline{r}_4)$$
$$+ n_o^2 \int d\overline{r}_j d\overline{r}_l d\overline{r}_5 d\overline{r}_6 d\overline{r}_7 d\overline{r}_8 d\overline{r}_9 d\overline{r}_{10} d\overline{r}_{11} d\overline{r}_{12}$$
$$\cdot T_j(\overline{r}_1,\overline{r}_5)G_{11m}(\overline{r}_5,\overline{r}_6)G_{11m}(\overline{r}_7,\overline{r}_8)T_l(\overline{r}_8,\overline{r}_2)$$
$$\cdot T_l(\overline{r}_3,\overline{r}_9)G_{11m}^*(\overline{r}_9,\overline{r}_{10})G_{11m}^*(\overline{r}_{11},\overline{r}_{12})T_j(\overline{r}_{12},\overline{r}_4)C(\overline{r}_6,\overline{r}_7;\overline{r}_{10},\overline{r}_{11}) \quad (8.3.10)$$

Since

$$T_j(\overline{r}_1,\overline{r}_2) = 4\pi f \delta(\overline{r}_1 - \overline{r}_j)\delta(\overline{r}_2 - \overline{r}_j) \quad (8.3.11)$$

We have

$$\begin{aligned} C(\overline{r}_1,\overline{r}_2;\overline{r}_3,\overline{r}_4) = {} & n_o^2|4\pi f|^4 G_{11m}(\overline{r}_1,\overline{r}_2)G_{11m}^*(\overline{r}_2,\overline{r}_1)\delta(\overline{r}_1-\overline{r}_4)\delta(\overline{r}_2-\overline{r}_3) \\ & + n_o^3|4\pi f|^6 \int d\overline{r}_l G_{11m}(\overline{r}_1,\overline{r}_l)G_{11m}(\overline{r}_l,\overline{r}_2) \\ & \cdot G_{11m}^*(\overline{r}_2,\overline{r}_l)G_{11m}^*(\overline{r}_l,\overline{r}_1)\delta(\overline{r}_1-\overline{r}_4)\delta(\overline{r}_2-\overline{r}_3) \\ & + n_o^2|4\pi f|^4 \delta(\overline{r}_1-\overline{r}_4)\delta(\overline{r}_2-\overline{r}_3)\int d\overline{r}_6 d\overline{r}_7 d\overline{r}_{10} d\overline{r}_{11} \\ & \cdot G_{11m}(\overline{r}_1,\overline{r}_6)G_{11m}(\overline{r}_7,\overline{r}_2)G_{11m}^*(\overline{r}_2,\overline{r}_{10}) \\ & \cdot G_{11m}^*(\overline{r}_{11},\overline{r}_1)C(\overline{r}_6,\overline{r}_7;\overline{r}_{10},\overline{r}_{11}) \end{aligned} \quad (8.3.12)$$

To solve (8.3.12), let

$$C(\overline{r}_1,\overline{r}_2;\overline{r}_3,\overline{r}_4) = A(\overline{r}_1,\overline{r}_2)\delta(\overline{r}_1-\overline{r}_4)\delta(\overline{r}_2-\overline{r}_3) \quad (8.3.13)$$

Substituting (8.3.13) into (8.3.12), we have

$$\begin{aligned} A(\overline{r}_1,\overline{r}_2) = {} & n_o^2(4\pi)^2|f|^4 \frac{e^{-\kappa_e|\overline{r}_1-\overline{r}_2|}}{|\overline{r}_1-\overline{r}_2|^2} + n_o^3(4\pi)^2|f|^6 \int_{V_1} d\overline{r}' \frac{e^{-\kappa_e|\overline{r}_1-\overline{r}'|}}{|\overline{r}_1-\overline{r}'|^2}\frac{e^{-\kappa_e|\overline{r}_2-\overline{r}'|}}{|\overline{r}_2-\overline{r}'|^2} \\ & + n_o^2|f|^4 \int_{V_1} d\overline{r}' \int_{V_1} d\overline{r}'' \frac{e^{-\kappa_e|\overline{r}_1-\overline{r}'|}}{|\overline{r}_1-\overline{r}'|^2}\frac{e^{-\kappa_e|\overline{r}_2-\overline{r}''|}}{|\overline{r}_2-\overline{r}''|^2} A(\overline{r}',\overline{r}'') \end{aligned} \quad (8.3.14)$$

where V_1 is the volume in the region between $z = 0$ and $z = -d$. Because of translational invariance in the horizontal plane, let

$$A(\overline{r}_1,\overline{r}_2) = \int_{-\infty}^{\infty} d\overline{k}_{D\perp}\, e^{i\overline{k}_{D\perp}\cdot(\overline{r}_{1\perp}-\overline{r}_{2\perp})}\, \widetilde{A}(\overline{k}_{D\perp};z_1,z_2) \quad (8.3.15)$$

From (5.5.37), (5.5.38) and (5.5.39), we have the identity

$$\frac{e^{-\kappa_e|\overline{r}_1-\overline{r}_2|}}{|\overline{r}_1-\overline{r}_2|^2} = \frac{1}{4\pi^2}\int_{-\infty}^{\infty} d\overline{k}_{A\perp}\int_{-\infty}^{\infty} d\overline{k}_{D\perp}\frac{e^{i\overline{k}_{D\perp}\cdot(\overline{r}_{1\perp}-\overline{r}_{2\perp})}}{K_{z(A+D)}K_{z(A-D)}^*} \cdot e^{i(K_{z(A+D)}-K_{z(A-D)}^*)|z_1-z_2|} \quad (8.3.16)$$

where

$$K_{z(A+D)} = \left[K^2 - \left|\overline{k}_{A\perp} + \frac{\overline{k}_{D\perp}}{2}\right|^2\right]^{1/2} \tag{8.3.17}$$

$$K_{z(A-D)} = \left[K^2 - \left|\overline{k}_{A\perp} - \frac{\overline{k}_{D\perp}}{2}\right|^2\right]^{1/2} \tag{8.3.18}$$

We also let

$$\tau = \kappa_e z \tag{8.3.19}$$

$$\alpha = \frac{|\overline{k}_{d\perp}|}{\kappa_e} \tag{8.3.20}$$

From (5.5.42) and (5.5.46)

$$\begin{aligned} W(\tau;\alpha) &= \frac{1}{2\pi}\int_{-\infty}^{\infty} d\overline{k}_{A\perp}\,\frac{e^{i(K_{z(A+D)}-K^*_{z(A-D)})\tau/\kappa_e}}{K_{z(A+D)}K^*_{z(A-D)}} \\ &= \int_1^{\infty} dt\,\frac{e^{-\tau(\alpha^2+t^2)^{1/2}}}{(\alpha^2+t^2)^{1/2}} \end{aligned} \tag{8.3.21}$$

Using (8.3.21) in (8.3.16)

$$\frac{e^{-\kappa_e|\overline{r}_1-\overline{r}_2|}}{|\overline{r}_1-\overline{r}_2|^2} = \frac{1}{2\pi}\int_{-\infty}^{\infty} d\overline{k}_{D\perp} e^{i\overline{k}_{D\perp}\cdot(\overline{r}_{1\perp}-\overline{r}_{2\perp})} W\left(|z_1-z_2|\,\kappa_e; \frac{|\overline{k}_{D\perp}|}{\kappa_e}\right) \tag{8.3.22}$$

We also let the albedo $\tilde{\omega} = 4\pi n_o|f|^2/\kappa_e$. Substituting (8.3.15) and (8.3.22) into (8.3.14), we obtain

$$\begin{aligned} \widetilde{A}(\overline{k}_{D\perp}; z_1, z_2) &= \frac{\kappa_e^2\tilde{\omega}^2}{2\pi} W(|\tau_1-\tau_2|;\alpha) \\ &+ \frac{\tilde{\omega}^3\kappa_e^2}{4\pi}\int_{-\tau_d}^{0} d\tau W(|\tau_1-\tau|;\alpha)W(|\tau_2-\tau|;\alpha) \\ &+ \frac{\tilde{\omega}^2}{4}\int_{-\tau_d}^{0} d\tau'\int_{-\tau_d}^{0} d\tau'' W(|\tau_1-\tau'|;\alpha) \\ &\cdot W(|\tau_2-\tau''|;\alpha)\widetilde{A}(\overline{k}_{D\perp}; z', z'') \end{aligned} \tag{8.3.23}$$

To solve (8.3.23), we further let

$$\widetilde{A}(\overline{k}_{D\perp}; z_1, z_2) = \frac{\kappa_e^2}{\pi}\left[\Gamma(\tau_1,\tau_2;\alpha) + \frac{\tilde{\omega}^2}{2}W(|\tau_1-\tau_2|;\alpha)\right] \tag{8.3.24}$$

Substituting (8.3.24) into (8.3.23) gives the following cyclical transfer integral equation for Γ.

$$\Gamma(\tau_1, \tau_2; \alpha) = \frac{\tilde{\omega}^3}{4} \int_{-\tau_d}^{0} d\tau W(|\tau_1 - \tau|; \alpha)\, W(|\tau - \tau_2|; \alpha)$$
$$+ \frac{\tilde{\omega}^4}{8} \int_{-\tau_d}^{0} d\tau \int_{-\tau_d}^{0} d\tau' W(|\tau_1 - \tau|; \alpha)\, W(|\tau' - \tau_2|; \alpha) W(|\tau - \tau'|; \alpha)$$
$$+ \frac{\tilde{\omega}^2}{4} \int_{-\tau_d}^{0} d\tau \int_{-\tau_d}^{0} d\tau' W(|\tau_1 - \tau|; \alpha)\, W(|\tau' - \tau_2|; \alpha) \Gamma(\tau, \tau'; \alpha) \quad (8.3.25)$$

By using (8.3.7), (8.3.9), and (8.3.13) and expressions of G_{01m} and ψ_m of Section 2, we have

$$\gamma_C(\mu_s, \phi_s; \mu_i, \phi_i) = \frac{4\pi r^2}{A \cos\theta_i} \int d\overline{r}_1 d\overline{r}_2 d\overline{r}_3 d\overline{r}_4 C(\overline{r}_1, \overline{r}_2; \overline{r}_3, \overline{r}_4)$$
$$\cdot\, G_{01m}(\overline{r}, \overline{r}_1) \psi_m(\overline{r}_2) G^*_{01m}(\overline{r}, \overline{r}_3) \psi^*_m(\overline{r}_4) \quad (8.3.26)$$

Following Section 2 of the second-order theory, we define

$$\overline{K}_d = \overline{K}'_i + \overline{K}'_s \quad (8.3.27)$$

to be the deviation from the backscattering direction. Then

$$\gamma_C(\mu_s, \phi_s; \mu_i, \phi_i) = \frac{1}{4\pi A \cos\theta_i} \int d\overline{r}_1 d\overline{r}_2 A(\overline{r}_1, \overline{r}_2) e^{(K''_{iz} + K''_{sz})(z_1 + z_2)} e^{-i\overline{K}_d \cdot (\overline{r}_1 - \overline{r}_2)} \quad (8.3.28)$$

Using (8.3.25) in (8.3.28), we have

$$\gamma_C(\mu_s, \phi_s; \mu_i, \phi_i) = \frac{\pi}{\cos\theta_i} \int_{-d}^{0} dz_1 \int_{-d}^{0} dz_2 \tilde{A}(\overline{K}_{d\perp}; z_1, z_2)$$
$$\cdot\, e^{(K''_{iz} + K''_{sz})(z_1 + z_2)} e^{-i(K'_{iz} - K'_{sz})(z_1 - z_2)} \quad (8.3.29)$$

Substituting (8.3.24) into (8.3.29) and noting that $\Gamma(\tau_1, \tau_2; \alpha)$ is symmetric in τ_1 and τ_2, we have

$$\gamma_C(\mu_s, \phi_s; \mu_i, \phi_i) = \frac{1}{\mu_i} \int_{-\tau_d}^{0} d\tau_1 \int_{-\tau_d}^{0} d\tau_2 \left[\frac{\tilde{\omega}^2}{2} W(|\tau_1 - \tau_2|; \alpha) + \Gamma(\tau_1, \tau_2; \alpha) \right]$$
$$\times \cos\left[\frac{k}{\kappa_e} (\mu_s - \mu_i)(\tau_1 - \tau_2) \right] e^{(1/\mu_s + 1/\mu_i)(\tau_1 + \tau_2)/2} \quad (8.3.30)$$

where

$$\alpha = \frac{|\overline{K}_{D\perp}|}{\kappa_e} = \frac{k}{\kappa_e} \Big[(\sin\theta_i \cos\phi_i + \sin\theta_s \cos\phi_s)^2$$
$$+ (\sin\theta_i \sin\phi_i + \sin\theta_s \sin\phi_s)^2 \Big]^{1/2} \quad (8.3.31)$$

denotes the deviation from the backscattering direction.

The kernel of the ladder equation is given in (5.5.49) with the integral equation of (5.5.50). The kernel depends on $W(\tau; \alpha)$ with $\alpha = 0$. On the

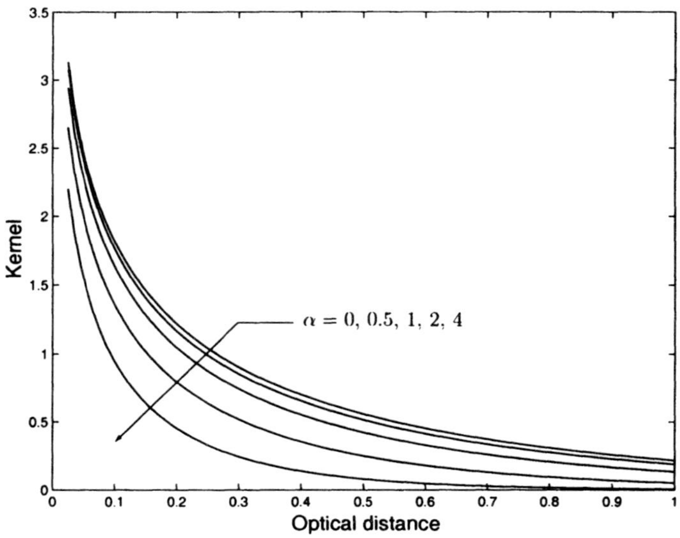

Figure 8.3.2 The kernel $W(\tau;\alpha)$ as a function of τ for $\alpha = 0$, 0.5, 1, 2 and 4.

other hand, the cyclical equation (8.3.25) has the same kernel $W(\tau;\alpha)$ but with α generally not equal to zero.

To calculate the cyclical bistatic coefficient requires a solution of the integral equation in (8.3.25). Once Γ is solved in (8.3.25), the cyclical bistatic coefficient is calculated by using (8.3.30). Equation (8.3.25) is a two-variable integral equation with independent variables τ_1 and τ_2. Note that Γ in (8.3.25) is only coupled to the same α.

3.2 Numerical Illustrations

We next calculate numerically the total bistatic scattering coefficient, which is a sum of the ladder and the cyclical bistatic scattering coefficients according to (8.3.8). The task is to solve the integral equation (8.3.25) numerically and calculate the cyclical bistatic coefficient. The ladder bistatic coefficient can be obtained by solving the radiative transfer equation and using (5.5.53).

In Fig. 8.3.2 we plot the kernel $W(\tau;\alpha)$ as a function of τ for different α's. The $W(\tau;\alpha)$ is calculated by numerically evaluating the integral in (8.3.21). The following small argument approximation can also be used for $\tau \ll 1$ and $\alpha\tau \ll 1$

$$W(\tau;\alpha) \simeq -\ln\tau - U - \ln\left(\frac{1+\sqrt{1+\alpha^2}}{2}\right) \tag{8.3.32}$$

where $U = 0.57721$ is Euler's constant. We note from Fig. 8.3.2 that $W(\tau;\alpha)$ decreases exponentially with τ and also decreases with α. This implies that there is less cyclical scattering with increasing α consistent with the backscattering enhancement phenomenon. The kernel also has a logarithmic singularity at $\tau = 0$.

The integral equation of (8.3.25) is solved by the method of moments by discretizing τ_1 and τ_2 into intervals of width $\Delta\tau$. Pulse functions of width $\Delta\tau$ are used to expand the unknown variable $\Gamma(\tau_1, \tau_2; \alpha)$, and point matching is used to obtain the final set of matrix equations. To calculate the first two terms on the right-hand side of (8.3.25) and to evaluate the matrix elements for the third term, we perform a numerical integration over the patch. In view of the fast variation of W for all values of τ, a much smaller patch $\Delta\tau_s \ll \Delta\tau$ is chosen for this numerical integration procedure. In view of the logarithmic singularity of the kernel, the self-patch has a large contribution. Computation of the self-patch contribution is accomplished by using the small argument approximation of (8.3.32) and choosing a small enough $\Delta\tau_s$ such that $\alpha\Delta\tau_s \ll 1$. For example,

$$\int_{\tau_m-\Delta\tau_s}^{\tau_m+\Delta\tau_s} d\tau\, W(|\tau_m - \tau|) \simeq 2\Delta\tau_s P \tag{8.3.33}$$

where

$$P = -\ln\Delta\tau_s + 1 - U - \ln\left(\frac{1+\sqrt{1+\alpha^2}}{2}\right) \tag{8.3.34}$$

The symmetry property

$$\Gamma(\tau_1, \tau_2; \alpha) = \Gamma(\tau_2, \tau_1; \alpha) \tag{8.3.35}$$

is also used to reduce the number of unknowns in the integral equation and the associated matrix equation. Thus the number of basis functions for Γ is $N = (\tau_d/\Delta\tau)(\tau_d/\Delta\tau + 1)/2$. The two-variable integral equation is thus cast into a matrix equation of dimension N which is then solved numerically by standard techniques.

In the following we illustrate numerical results for $\kappa_e/k = 3.18 \times 10^{-3}$, $\theta_i = 0$, $\phi_i = 0$, $\phi_s = \pi$, θ_s from 0° to 1° and $\tau_d = 2$ and 4, as well as for albedos $\tilde{\omega}$ ranging from 0.4 to 0.95. For this range of angles, the values of α range from 0 to 5.49. For these values of α, it is sufficient to use $\Delta\tau = 0.25$ and $\Delta\tau_s = 0.025$. This corresponds to $N = 36$ for $\tau_d = 2$ and $N = 136$ for $\tau_d = 4$. It is found that for $\tilde{\omega} \le 0.85$, a value of $\tau_d = 4$ approximates well the half-space solution.

For comparison, we also plot the single-scattering solution $\gamma^{(1)}$ and the result for the second-order multiple scattering theory $\gamma^{(2)}$ calculated in Section 2. We have, for an optical thickness of τ_d,

$$\gamma^{(1)}(\mu_s, \mu_i) = \frac{\tilde{\omega}}{\mu_i} \frac{1 - e^{-\tau_d(1/\mu_s + 1/\mu_i)}}{1/\mu_s + 1/\mu_i} \tag{8.3.36}$$

$$\gamma^{(2)}(\mu_s, \phi_s; \mu_i, \phi_i) = \gamma^{(1)}(\mu_s, \mu_i) + \gamma_L^{(2)}(\mu_s, \mu_i) + \gamma_C^{(2)}(\mu_s, \phi_s; \mu_i, \phi_i) \tag{8.3.37}$$

The second-order theory is the sum of the first-order ladder term, the second-order ladder term, and the second-order cyclical term. We have that

$$\begin{aligned} \gamma_L^{(2)}(\mu_s, \mu_i) = \frac{\tilde{\omega}^2}{2\mu_i} \int_0^1 d\mu \Bigg\{ \frac{1}{1 + \mu/\mu_i} \Bigg[& \frac{1 - e^{-\tau_d(1/\mu_s + 1/\mu_i)}}{1/\mu_s + 1/\mu_i} \\ & - \frac{e^{-\tau_d(1/\mu + 1/\mu_i)} - e^{-\tau_d(1/\mu_s + 1/\mu_i)}}{1/\mu_s - 1/\mu} \Bigg] + \frac{1}{\mu/\mu_i - 1} \\ & \cdot \Bigg[\frac{1 - e^{-\tau_d(1/\mu_s + 1/\mu)}}{1/\mu_s + 1/\mu} - \frac{1 - e^{-\tau_d(1/\mu_s + 1/\mu_i)}}{1/\mu_s + 1/\mu_i} \Bigg] \Bigg\} \end{aligned} \tag{8.3.38}$$

is the second-order ladder term, and

$$\gamma_C^{(2)}(\mu_s, \phi_s; \mu_i, \phi_i) = \frac{\tilde{\omega}^2}{2\mu_i} \int_{-\tau_d}^{0} d\tau_1 \int_{-\tau_d}^{0} d\tau_2 \, W(|\tau_1 - \tau_2|; \alpha) \tag{8.3.39}$$

is the second-order cyclical term.

It is to be noted that in the backscattering direction of $\alpha = 0$ we have

$$\gamma_C = \gamma_L - \gamma^{(1)} \tag{8.3.40}$$

The equality of (8.3.40) is proved by showing that the nth-order ladder term ($n \geq 2$) is equal to the nth-order cyclical term at $\alpha = 0$. Diagrammatically, the nth-order ladder term is

$I_L^{(n)}(\bar{r}) =$ 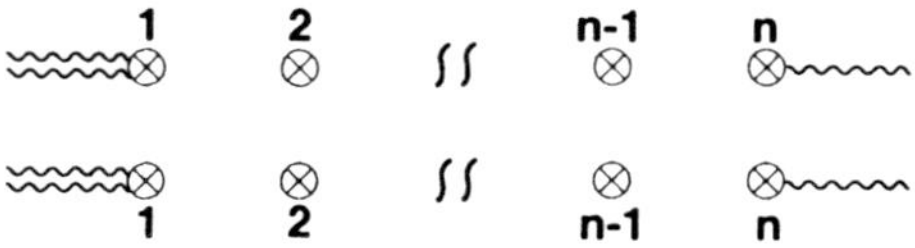 (8.3.41)

The nth-order cyclical term is

$I_C^{(n)}(\bar{r}) =$  (8.3.42)

Hence using the expressions for G_{01m}, G_{11m}, $\psi_m(\overline{r})$, and the reciprocity property of G_{11m}, it follows that at $\alpha = 0$,

$$I_L^{(n)}(\overline{r}) = \frac{n_o^n(4\pi)^{2(n-1)}|f|^{2n}}{r^2}\int_{V_1} d\overline{r}_1 d\overline{r}_2 \cdots d\overline{r}_n \times e^{2K''_{sz}z_1}|G_{11m}(\overline{r}_1,\overline{r}_2)|^2 \cdots |G_{11m}(\overline{r}_{n-1},\overline{r}_n)|^2 e^{2K''_{iz}z_n} \quad (8.3.43)$$

$$I_C^{(n)}(\overline{r}) = \frac{n_o^n(4\pi)^{2(n-1)}|f|^{2n}}{r^2}\int_{V_1} d\overline{r}_1 d\overline{r}_2 \cdots d\overline{r}_n\, e^{-i\overline{K}_s\cdot\overline{r}_1 + i\overline{K}_s^*\cdot\overline{r}_n} \times |G_{11m}(\overline{r}_1,\overline{r}_2)|^2 \cdots |G_{11m}(\overline{r}_{n-1},\overline{r}_n)|^2 e^{i\overline{K}_i\cdot\overline{r}_n - i\overline{K}_i^*\cdot\overline{r}_1} \quad (8.3.44)$$

In the backscattering direction, $\alpha = 0$, we have $\overline{K}_{i\perp} + \overline{K}_{s\perp} = 0$ and $K_{iz} = K_{sz}$. Making these substitutions in (8.3.44) and comparing with (8.3.43) give

$$I_C^{(n)}(\overline{r}) = I_L^{(n)}(\overline{r}) \quad (8.3.45)$$

for $n \geq 2$. Since there is no first-order cyclical term, it follows that

$$I_C(\overline{r}) = I_C^{(2)} + I_C^{(3)} + I_C^{(4)} + \cdots = I_L^{(2)} + I_L^{(3)} + I_L^{(4)} + \cdots = I_L - I_L^{(1)} \quad (8.3.46)$$

At $\alpha = 0$,

$$I_L = I_L^{(1)} + I_C \quad (8.3.47)$$

Thus the equality in (8.3.40) is true. At $\alpha = 0$, we have

$$\gamma = 2\gamma_C + \gamma^{(1)} = 2\gamma_L - \gamma^{(1)} \quad (8.3.48)$$

The accuracy of the numerical solution of the cyclical transfer integral equation can be tested by checking that the condition (8.3.40) is satisfied at the backscattering direction of $\alpha = 0$. The numerical results satisfy (8.3.40) to within 1%.

In Fig. 8.3.3 we plot γ, γ_L, $\gamma^{(1)}$, and $\gamma^{(2)}$ as a function of scattering angle θ_s for $\tilde{\omega} = 0.95$ and $\tau_d = 2$. We note that single scattering and ladder solutions have a much weaker angular dependence than the second-order multiple-scattering solution and the total multiple-scattering intensity γ. We also note that the second-order theory differs from the full multiple-scattering solution γ by close to 4 dB. Hence the second-order scattering theory is not accurate for large albedo and appreciable optical thickness. We note that the full multiple-scattering cyclical term γ_C has a much narrower peak than the second-order cyclical term $\gamma_C^{(2)}$. This can be accounted for physically by the fact that higher-order multiple scatterings are more sensitive to the path length difference created by angular deviation from

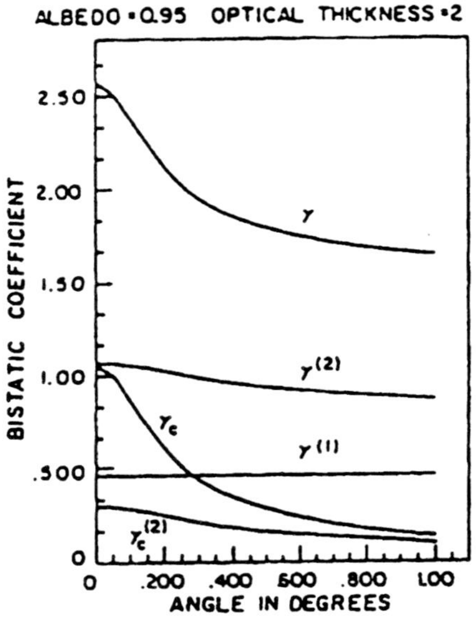

Figure 8.3.3 Bistatic scattering coefficients γ, $\gamma^{(1)}$, $\gamma^{(2)}$, γ_C, and $\gamma_C^{(2)}$ as a function of θ_s for $\tilde{\omega} = 0.95$ and $\tau_d = 2$.

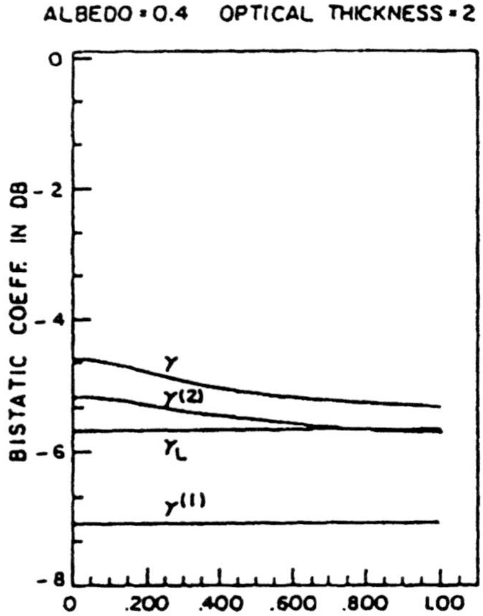

Figure 8.3.4 Bistatic scattering coefficients γ, γ_L, $\gamma^{(1)}$, and $\gamma^{(2)}$ as a function of θ_s for $\tilde{\omega} = 0.4$ and $\tau_d = 2$.

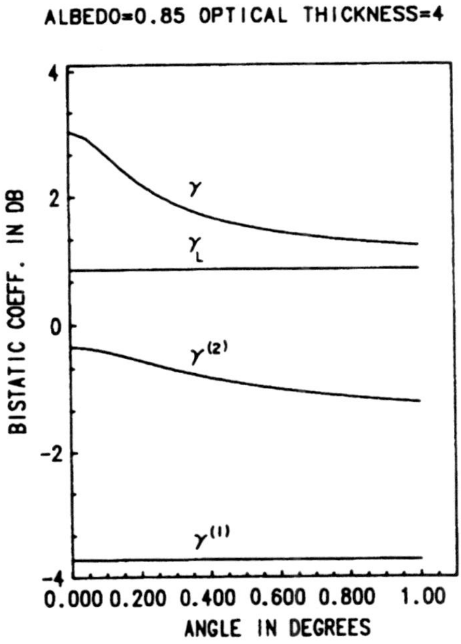

Figure 8.3.5 Same as Fig. 8.3.4 with $\tilde{\omega} = 0.85$ and $\tau_d = 4$.

the backscattering direction. The overall angular width of the peak in the backscattering direction is of the order of $\kappa_e/k = 3.18 \times 10^{-3} = 0.18°$.

In Fig. 8.3.4 the results are illustrated for a smaller $\tilde{\omega} = 0.4$. Multiple scattering plays a lesser role. Compared to the case $\tilde{\omega} = 0.95$, the backscattering peak here is wider. The second-order solution is also closer to the full multiple scattering solution than that of $\tilde{\omega} = 0.95$.

In Fig. 8.3.5 we plot the bistatic scattering coefficients with $\tau_d = 4$ at $\tilde{\omega} = 0.85$. The case of $\tau_d = 4$ in Fig. 8.3.5 approximates well the half-space solution. Comparison of the results in the two figures indicates that larger optical thickness gives a larger bistatic coefficient and a sharper backscattering peak.

4 Anisotropic Scatterers and Diffusion Approximation

In previous sections we studied the backscattering enhancement of isotropic scatterers. The approach was based on the second-order theory and also on a summation of all the ladder and cyclical terms. For isotropic scatterers the angular width of backscattering enhancement is of the order of the coherent wave-attenuation rate divided by the wavenumber. In this section we study the backscattering enhancement of waves by scatterers that do not scatterer isotropically. Nonisotropic scattering is characteristic of particles with sizes comparable with or larger than a wavelength. By restricting the observation angle to the vicinity of the backscattering direction, it is shown that both the ladder- and cyclical-term summations can be related to the unidirectional point-source Green's function of the transport equation. The Green's function is then solved by two methods. For the case of a small albedo or a small optical thickness, the second-order theory is applied. The angular width of enhancement is of the order of the coherent wave-attenuation rate divided by the wavenumber. For the case of a large albedo and a large optical thickness, diffusion approximation is used to calculate the unidirectional point-source Green's function. The angular width in this case is of the order of the transport rate divided by the wavenumber. The transport rate is equal to the product of the extinction rate and one minus the mean cosine of the scattering angle $\overline{\mu}$. For strong forward scattering, the mean cosine $\overline{\mu}$ is close to unity. Hence the angular width is substantially smaller for particles with strong forward scattering, a result that is shown to be in good agreement with experimental observations.

4.1 Summation of Ladder Terms and Cyclical Terms

Consider a plane wave incident upon a slab of discrete scatterers lying between $z = 0$ and $z = -d$ (Fig. 8.4.1). The incident wave is in the direction $\hat{k}_{id}$, and the scattered wave is measured in the direction $\hat{k}_s$. The subscript d is used to denote downward motion. We assume that the particles are uncorrelated in positions.

Let

$$\hat{k}_{id} = \sin\theta_i \cos\phi_i \hat{x} + \sin\theta_i \sin\phi_i \hat{y} - \cos\theta_i \hat{z} \tag{8.4.1}$$

$$\hat{k}_s = \sin\theta_s \cos\phi_s \hat{x} + \sin\theta_s \sin\phi_s \hat{y} + \cos\theta_s \hat{z} \tag{8.4.2}$$

In the backscattering direction, $\hat{k}_s = -\hat{k}_{id}$, so that $\theta_s = \theta_i$ and $\phi_s = \pi + \phi_i$. The scattering properties of the particles are described by the scattering amplitude $f(\hat{\alpha}, \hat{\beta})$ from direction $\hat{\beta}$ into direction $\hat{\alpha}$. Propagation in region 1 is described by the effective propagation constant K, which shall be approximated by $K = k + i\kappa_e/2$, where k is the background wavenumber and is the same as the wavenumber of regions 0 and 2, and κ_e is the coherent wave extinction rate. The thickness of the slab can be normalized by the optical

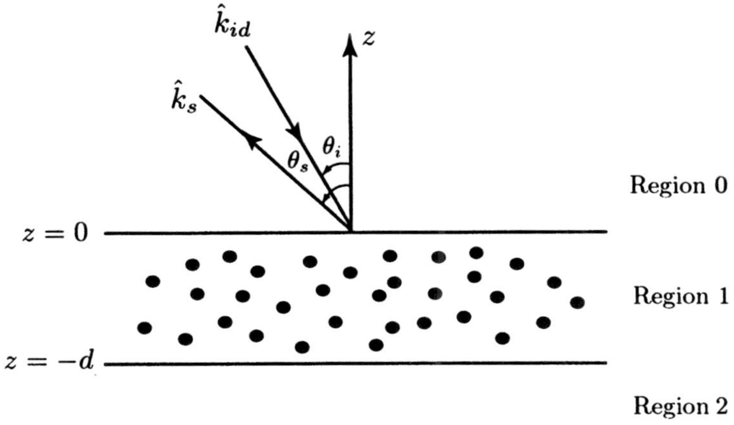

Figure 8.4.1 Geometric configuration of the problem: a plane wave incident upon a slab of particles, the scattering properties of which are anisotropic.

thickness $\tau_d = \kappa_e d$. The mean Green's function squared in medium 1 is

$$|\langle G(\overline{r}, \overline{r}')\rangle|^2 = \frac{\exp(-\kappa_e|\overline{r} - \overline{r}'|)}{16\pi^2|\overline{r} - \overline{r}'|^2} \tag{8.4.3}$$

The scattering phase function $p(\hat{\alpha}, \hat{\beta})$ from direction $\hat{\beta}$ into direction $\hat{\alpha}$ is

$$p(\hat{\alpha}, \hat{\beta}) = \frac{4\pi n_0}{\kappa_e}|f(\hat{\alpha}, \hat{\beta})|^2 \tag{8.4.4}$$

where n_0 is the number of particles per unit volume and the albedo $\tilde{\omega}$ is

$$\tilde{\omega} = \frac{1}{4\pi}\int d\hat{\alpha}\; p(\hat{\alpha}, \hat{\beta}) \tag{8.4.5}$$

where $\int d\hat{\alpha}$ denotes integration over a solid angle 4π of direction $\hat{\alpha}$. Let $I_L^{(n)}$ and $I_C^{(n)}$ denote, respectively, the nth-order ladder term and the nth-order cyclical term, and I_L and I_C denote, respectively, the sums of the ladder terms and the cyclical terms. The summation of ladder terms begins with $n = 1$, and the summation of cyclical terms starts with $n = 2$.

$$I_L = \sum_{n=1}^{\infty} I_L^{(n)}, \tag{8.4.6}$$

$$I_C = \sum_{n=2}^{\infty} I_C^{(n)}. \tag{8.4.7}$$

The first-order ladder term is, analogous to (8.2.16)

$$I_L^{(1)} = $$ 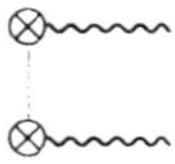 $$= \frac{n_o}{r^2} \int d\overline{r}_j \left| e^{-i\overline{K}_s \cdot \overline{r}_j} 4\pi f(\hat{k}_s, \hat{k}_{id}) e^{i\overline{K}_i \cdot \overline{r}_j} \right|^2 \tag{8.4.8}$$

where

$$\overline{K}_i = k\hat{k}_{id} - iK''_{iz}\hat{z} \tag{8.4.9}$$

$$\overline{K}_s = k\hat{k}_s + iK''_{sz}\hat{z} \tag{8.4.10}$$

$$K''_{iz} = \frac{\kappa_e}{2} \sec\theta_i \tag{8.4.11}$$

$$K''_{sz} = \frac{\kappa_e}{2} \sec\theta_s \tag{8.4.12}$$

and r is the distance from the slab to the receiver. Carrying out the integration as in Section 2 of (8.2.16)–(8.2.22), but for a slab of finite optical thickness of $\tau_d = \kappa_e d$, we have

$$I_L^{(1)} = \frac{n_0 A_o}{r^2} |f(\hat{k}_s, \hat{k}_{id})|^2 \frac{\{1 - \exp[-\tau_d(\sec\theta_s + \sec\theta_i)]\}}{\kappa_e(\sec\theta_s + \sec\theta_i)} \tag{8.4.13}$$

where A_o is the area of the slab.

The bistatic scattering coefficient is $\gamma(\hat{k}_s, \hat{k}_{id}) = 4\pi r^2 I/(A_o \cos\theta_i)$. Let γ_L and γ_C denote, respectively, the summation of ladder and cyclical bistatic coefficients, and $\gamma_L^{(n)}$ and $\gamma_C^{(n)}$ denote the respective nth-order solutions. The first-order ladder bistatic scattering coefficient is, from (8.4.13)

$$\gamma_L^{(1)}(\hat{k}_s, \hat{k}_{id}) = \frac{\sec\theta_i}{\sec\theta_s + \sec\theta_i} \{1 - \exp[-\tau_d(\sec\theta_s + \sec\theta_i)]\} p(\hat{k}_s, \hat{k}_{id}) \tag{8.4.14}$$

We next consider the rest of the terms in the series for I_L and I_C. Let

$$\overline{k}_D = k(\hat{k}_s + \hat{k}_{id}) \equiv \overline{k}_{Dt} + \hat{z}k_{Dz} \tag{8.4.15}$$

denote the deviation from the backscattering direction. The subscript t in (8.4.15) denotes the transverse direction. We first examine and compare two typical terms in the two series—for example, third-order terms $I_L^{(3)}$ and $I_C^{(3)}$. The third-order ladder term is

$$I_L^{(3)} = $$ 

$$= \frac{n_0^3}{r^2} \int d\overline{r}_j \int d\overline{r}_l \exp(2K''_{sz} z_j) \exp(2K''_{iz} z_l)$$
$$\times \left[\int d\overline{r}_m |f(\hat{k}_s, \hat{r}_{jm})|^2 \times \frac{\exp(-\kappa_e r_{jm})}{r_{jm}^2} |f(\hat{r}_{jm}, \hat{r}_{ml})|^2 \right.$$
$$\left. \cdot \frac{\exp(-\kappa_e r_{ml})}{r_{ml}^2} |f(\hat{r}_{ml}, \hat{k}_{id})|^2 \right] \tag{8.4.16}$$

where $r_{jm} = |\overline{r}_j - \overline{r}_m|$ is the distance between the jth and the mth scatterers, and $\hat{r}_{jm} = (\overline{r}_j - \overline{r}_m)/r_{jm} = -\hat{r}_{mj}$ is the unit vector pointing form particle m to particle j. The third-order cyclical term is

$I_C^{(3)} =$ 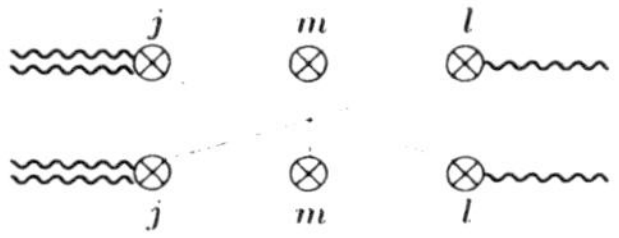

$$= \frac{n_0^3}{r^2} \int d\overline{r}_j \int d\overline{r}_l \exp[i(\overline{K}_s^* + \overline{K}_i) \cdot \overline{r}_l] \exp[-i(\overline{K}_s + \overline{K}_i^*) \cdot \overline{r}_j]$$
$$\times \left[\int d\overline{r}_m f^*(\hat{k}_s, \hat{r}_{lm}) f(\hat{r}_{ml}, \hat{k}_{id}) \frac{\exp(-\kappa_e r_{ml})}{r_{ml}^2} f^*(\hat{r}_{lm}, \hat{r}_{mj}) f(\hat{r}_{jm}, \hat{r}_{ml}) \right.$$
$$\left. \times \frac{\exp(-\kappa_e r_{jm})}{r_{jm}^2} f^*(\hat{r}_{mj}, \hat{k}_{id}) f(\hat{k}_s, \hat{r}_{jm}) \right] \tag{8.4.17}$$

To compare the two terms $I_L^{(3)}$ and $I_C^{(3)}$, we note that, by reciprocity, the scattering amplitude obeys the symmetry relation

$$f(\hat{\beta}, \hat{\alpha}) = f(-\hat{\alpha}, -\hat{\beta}). \tag{8.4.18}$$

By using (8.4.18) in (8.4.16) and (8.4.17), it readily follows that the middle parts of $I_L^{(3)}$ and $I_C^{(3)}$, as represented by the terms inside the brackets in (8.4.16) and (8.4.17), are almost equal. For backscattering enhancement, we are interested in $\hat{k}_s$ differing from $-\hat{k}_{id}$ by at most 2° to 3°. Hence we make the approximation

$$f^*(\hat{r}_{mj}, \hat{k}_{id}) \simeq f^*(\hat{r}_{mj}, -\hat{k}_s). \tag{8.4.19}$$

That is, in the argument of the scattering amplitude, we equate $\hat{k}_{id}$ to $-\hat{k}_s$. The approximation in relation (8.4.19) is valid (a) if $\hat{k}_s$ is in the vicinity of $-\hat{k}_{id}$, and (b) if the scattering amplitude $f(\hat{\beta}, \hat{\alpha})$ is not sharply peaked in the backward direction. This is a reasonable approximation because for a sharply peaked phase function, the sharp peak is in the forward direction and not in the backward direction. If (8.4.18) and the approximation (8.4.19) are inserted into (8.4.17), and we also use (8.4.9)–(8.4.12) and (8.4.15), it follows that

$$I_C^{(3)} = \frac{n_0^3}{r^2} \iint d\overline{r}_j d\overline{r}_l \exp[i\overline{k}_{Dt} \cdot (\overline{r}_{lt} - \overline{r}_{jt}) + ik_{Dz}(z_l - z_j)]$$
$$\cdot \exp[(K''_{sz} + K''_{iz})(z_l + z_j)] \times \left[\int d\overline{r}_m |f(\hat{k}_s, \hat{r}_{jm})|^2 \right.$$
$$\left. \cdot \frac{\exp(-\kappa_e r_{jm})}{r_{jm}^2} |f(\hat{r}_{jm}, \hat{r}_{ml})|^2 \frac{\exp(-\kappa_e r_{ml})}{r_{ml}^2} |f(\hat{r}_{ml}, \hat{k}_{id})|^2 \right]. \quad (8.4.20)$$

Comparing (8.4.16) and (8.4.20), it follows that by using the approximation of (8.4.19), the middle parts of $I_L^{(3)}$ and $I_C^{(3)}$, as represented by the terms inside the large square brackets, are equal. Applying (8.4.18) and relation (8.4.19) to all the terms in $I_L^{(n)}$ and $I_C^{(n)}$ for $n \geq 2$ yields the same conclusion, that the middle parts of $I_L^{(n)}$ and $I_C^{(n)}$, for $n \geq 2$, are equal.

Next we define $A(\overline{r}_j, \beta; \overline{r}_l, \alpha)$ as the sum of the middle parts of the series $\sum_{n=2}^{\infty} I_L^{(n)}$ and $\sum_{n=2}^{\infty} I_C^{(n)}$:

$$A(\overline{r}_j, \hat{\beta}; \overline{r}_l, \hat{\alpha}) \equiv n_0^2 |f(\hat{\beta}, \hat{r}_{jl})|^2 \frac{\exp(-\kappa_e r_{jl})}{r_{jl}^2} |f(\hat{r}_{jl}, \hat{\alpha})|^2$$
$$+ n_0^3 \int d\overline{r}_m |f(\hat{\beta}, \hat{r}_{jm})|^2 \frac{\exp(-\kappa_e r_{jm})}{r_{jm}^2} |f(\hat{r}_{jm}, \hat{r}_{ml})|^2$$
$$\cdot \frac{\exp(-\kappa_e r_{ml})}{r_{ml}^2} |f(\hat{r}_{ml}, \hat{\alpha})|^2 + n_0^4 \iint d\overline{r}_p d\overline{r}_m |f(\hat{\beta}, \hat{r}_{jp})|^2$$
$$\cdot \frac{\exp(-\kappa_e r_{jp})}{r_{jp}^2} |f(\hat{r}_{jp}, \hat{r}_{pm})|^2 \frac{\exp(-\kappa_e r_{pm})}{r_{pm}^2} |f(\hat{r}_{pm}, \hat{r}_{ml})|^2$$
$$\cdot \frac{\exp(-\kappa_e r_{ml})}{r_{ml}^2} |f(\hat{r}_{ml}, \hat{\alpha})|^2 + \cdots \quad (8.4.21)$$

Using the series in (8.4.21), the ladder intensity and the cyclical intensity can be represented as follows:

$$I_L = \sum_{n=1}^{\infty} I_L^{(n)} = I_L^{(1)} + \frac{1}{r^2} \int d\overline{r}_j \int d\overline{r}_l \exp(2K''_{sz}\, z_j + 2K''_{iz}\, z_l)$$
$$\cdot A(\overline{r}_j, \hat{k}_s; \overline{r}_l, \hat{k}_{id}), \quad (8.4.22)$$

$$I_C = \sum_{n=2}^{\infty} I_C^{(n)} = \frac{1}{r^2} \int d\overline{r}_j \int d\overline{r}_l \exp[i\overline{k}_{Dt} \cdot (\overline{r}_{lt} - \overline{r}_{jt}) + ik_{Dz}(z_l - z_j)]$$
$$\times \exp[(K''_{sz} + K''_{iz})(z_l + z_j)] A(\overline{r}_j, \hat{k}_s; \overline{r}_l, \hat{k}_{id}). \quad (8.4.23)$$

Hence the ladder intensity is equal to the integral of A weighted by attenuation factors, whereas the cyclical intensity is equal to the Fourier transform of A weighted by the same factor. In the backscattering direction, $\overline{k}_D = 0$; that is, the Fourier transform variable is zero. From (8.4.22) and (8.4.23), it then follows that , $I_L - I_L^{(1)} = I_C$.

4.2 Unidirectional Point Source Green's Function

The function $A(\overline{r}_j, \hat{\beta}; \overline{r}_l, \hat{\alpha})$ can be related to the unidirectional point-source Green's function for the transport equation. The unidirectional point-source Green's function $G(\overline{r}, \hat{s}; \overline{r}_0, \hat{s}_0)$ is the specific intensity at $\overline{r}$ in the direction $\hat{s}$ that is due to a point source at $\overline{r}_0$ in the direction $\hat{s}_0$. The Green's function in transport equations has both positions and directions. Consider a scattering volume V that is enclosed by the surface S. The definition of the unidirectional point source Green's function is that it obeys the equation

$$\frac{dG(\overline{r}, \hat{s}; \overline{r}_0, \hat{s}_0)}{ds} = -\kappa_e G(\overline{r}, \hat{s}; \overline{r}_0, \hat{s}_0) + n_0 \int d\hat{s}' |f(\hat{s}, \hat{s}')|^2 G(\overline{r}, \hat{s}'; \overline{r}_0, \hat{s}_0) + \delta(\overline{r} - \overline{r}_0)\delta_2(\hat{s} - \hat{s}_0) \qquad (8.4.24a)$$

subject to the boundary conditions that

$$G(\overline{r}, \hat{s}; \overline{r}_0, \hat{s}_0) = 0 \qquad (8.4.24b)$$

for $\overline{r}$ on S and $\hat{n} \cdot \hat{s} < 0$, where $\hat{n}$ is the outward normal to the surface S. In (8.4.24a), $\delta_2(\hat{s} - \hat{s}_0)$ is the Dirac delta function for solid angle integration, so that $\int d\hat{s}\delta_2(\hat{s} - \hat{s}_0) = 1$.

The differential equation (8.4.24a) and boundary condition (8.4.24b) are next cast into an integral equation

$$G(\overline{r}_a, \hat{s}; \overline{r}_b, \hat{s}_0) = \int_{\overline{r}_s}^{\overline{r}_a} ds_1 \exp(-\kappa_e|\overline{r}_a - \overline{r}_1|) \int d\hat{s}' n_0 |f(\hat{s}, \hat{s}')|^2 G(\overline{r}_1, \hat{s}'; \overline{r}_b, \hat{s}_0) + \int_{\overline{r}_s}^{\overline{r}_a} ds_1 \delta(\overline{r}_1 - \overline{r}_b)\delta_2(\hat{s} - \hat{s}_0) \exp(-\kappa_e|\overline{r}_a - \overline{r}_1|) \qquad (8.4.25)$$

where $\overline{r}_1 = \overline{r}_s + \hat{s}s_1$, with $\overline{r}_s$ being a point on the boundary S. The line integral $\int_{\overline{r}_s}^{\overline{r}_a} ds_1$ in (8.4.25) is integrating along a straight line in the direction $\hat{s}$ from $\overline{r}_s$ to $\overline{r}_a$.

The line integral can be converted to a volume integral,

$$\int_{\overline{r}_s}^{\overline{r}_a} ds_1 = \int_V d\overline{r}_1 \frac{\delta_2(\hat{s} - \hat{r}_{a1})}{r_{a1}^2} \qquad (8.4.26)$$

where V is the volume containing the random media. Substituting (8.4.26) into (8.4.25) gives the integral equation

$$G(\overline{r}_a, \hat{s}; \overline{r}_b, \hat{s}_0) = \frac{\exp(-\kappa_e r_{ab})}{r_{ab}^2}\delta_2(\hat{s} - \hat{r}_{ab})\delta_2(\hat{s} - \hat{s}_0) + \int_V d\overline{r}_1 \frac{\delta_2(\hat{s} - \hat{r}_{a1})\exp(-\kappa_e r_{a1})}{r_{a1}^2} \int d\hat{s}' n_0 |f(\hat{r}_{a1}, \hat{s}')|^2 G(\overline{r}_1, \hat{s}'; \overline{r}_b, \hat{s}_0). \tag{8.4.27}$$

Next, we iterate the integral equation of (8.4.27):

$$\begin{aligned} G(\overline{r}_a, \hat{s}; \overline{r}_b, \hat{s}_0) &= \frac{\exp(-\kappa_e r_{ab})}{r_{ab}^2}\delta_2(\hat{s} - \hat{r}_{ab})\delta_2(\hat{s} - \hat{s}_0) \\ &+ n_0 \int_V d\overline{r}_1 \frac{\delta_2(\hat{s} - \hat{r}_{a1})}{r_{a1}^2} \exp(-\kappa_e r_{a1})|f(\hat{r}_{a1}, \hat{r}_{1b})|^2 \frac{\exp(-\kappa_e r_{1b})}{r_{1b}^2}\delta_2(\hat{r}_{1b} - \hat{s}_0) \\ &+ n_0^2 \int_V d\overline{r}_1 \int_V d\overline{r}_2 \frac{\delta_2(\hat{s} - \hat{r}_{a1})}{r_{a1}^2} \exp(-\kappa_e r_{a1})|f(\hat{r}_{a1}, \hat{r}_{12})|^2 \\ &\times \frac{\exp(-\kappa_e r_{12})}{r_{12}^2}|f(\hat{r}_{12}, \hat{r}_{2b})|^2 \frac{\exp(-\kappa_e r_{2b})}{r_{2b}^2}\delta_2(\hat{r}_{2b} - \hat{s}_0) + \cdots \end{aligned} \tag{8.4.28}$$

Comparing (8.4.28) to (8.4.21) gives the relation between $A(\overline{r}, \hat{s}; \overline{r}_0, \hat{s}_0)$ and $G(\overline{r}, \hat{s}; \overline{r}_0, \hat{s}_0)$:

$$A(\overline{r}, \hat{\beta}; \overline{r}_0, \hat{\alpha}) = n_0^2 \int d\hat{s} \int d\hat{s}_0 |f(\hat{\beta}, \hat{s})|^2 G(\overline{r}, \hat{s}; \overline{r}_0, \hat{s}_0)|f(\hat{s}_0, \hat{\alpha})|^2. \tag{8.4.29}$$

Thus the problem of calculating I_L and I_C is reduced to finding the solution of the unidirectional point source Green's function for the transport equation.

A simplification can be made by integrating over $d\hat{s}_0$ in (8.4.29). Let

$$B(\overline{r}, \hat{s}; \overline{r}_1, \hat{\alpha}) = \frac{1}{4\pi} \int d\hat{s}_0 G(\overline{r}, \hat{s}; \overline{r}_1, \hat{s}_0) p(\hat{s}_0, \hat{\alpha}); \tag{8.4.30}$$

Then

$$A(\overline{r}, \hat{\beta}; \overline{r}_1, \hat{\alpha}) = \frac{\kappa_e^2}{4\pi} \int d\hat{s}\; p(\hat{\beta}, \hat{s}) B(\overline{r}, \hat{s}; \overline{r}_1, \hat{\alpha}). \tag{8.4.31}$$

Multiplying (8.4.24*a*) by $p(\hat{s}_o, \hat{\alpha})/4\pi$ and integrating over $d\hat{s}_o$ give the transport equation for $B(\overline{r}, \hat{s}; \overline{r}_1, \hat{\alpha})$.

$$\frac{dB(\overline{r}, \hat{s}; \overline{r}_1, \hat{\alpha})}{ds} = -\kappa_e B(\overline{r}, \hat{s}; \overline{r}_1, \hat{\alpha}) + \frac{\kappa_e}{4\pi} \int d\hat{s}' p(\hat{s}, \hat{s}') B(\overline{r}, \hat{s}'; \overline{r}_1, \hat{\alpha}) + \frac{\delta(\overline{r} - \overline{r}_1) p(\hat{s}, \hat{\alpha})}{4\pi} \tag{8.4.32}$$

with the boundary conditions that at $z = 0$

$$B(\overline{r}, \hat{s}; \overline{r}_1, \hat{\alpha}) = 0 \tag{8.4.33}$$

for $\hat{s} \cdot \hat{z} < 0$ and that at $z = -d$

$$B(\overline{r}, \hat{s}; \overline{r}_1, \hat{\alpha}) = 0 \tag{8.4.34}$$

for $\hat{s} \cdot \hat{z} > 0$.

In the next two sections we shall calculate the solutions of (8.4.32)–(8.4.34) for two limiting cases, namely, the second-order theory and the diffusion approximation. Once $B(\overline{r}, \hat{s}; \overline{r}_1, \hat{\alpha})$ is calculated, $A(\overline{r}, \hat{\beta}; \overline{r}_1, \hat{\alpha})$ is evaluated by using (8.4.31), and then I_L and I_C are calculated by using (8.4.22) and (8.4.23), respectively.

4.3 Second-Order Multiple-Scattering Theory

In the second-order multiple-scattering theory, only the second-order terms are included in the multiple-scattering process. It is valid for a small albedo and/or a small optical thickness. The transport equation for B is solved by ignoring the second term on the right-hand side of (8.4.32). Solving (8.4.32)–(8.4.34) then gives

$$B^{(2)}(\overline{r}_a, \hat{s}; \overline{r}_b, \hat{s}_0) = \frac{\exp(-\kappa_e r_{ab})}{r_{ab}^2} \delta_2(\hat{s} - \hat{r}_{ab}) \frac{p(\hat{s}, \hat{s}_0)}{4\pi} \tag{8.4.35}$$

where the superscript (2) denotes the second order. Using (8.4.31) gives

$$A^{(2)}(\overline{r}_a, \hat{\beta}; \overline{r}_b, \hat{\alpha}) = \left(\frac{\kappa_e}{4\pi}\right)^2 p(\hat{\beta}, \hat{r}_{ab}) \frac{\exp(-\kappa_e r_{ab})}{r_{ab}^2} p(\hat{r}_{ab}, \hat{\alpha}) \tag{8.4.36}$$

On substituting (8.4.35) into (8.4.23), the integral can be simplified by setting $(\overline{r}_{jt} - \overline{r}_{lt})/2 = \overline{r}_t$ and $(\overline{r}_{jt} + \overline{r}_{lt})/2 = \overline{r}_{st}$. The integrand is independent of $\overline{r}_{st}$, the integration of which gives A_0, the area of the slab. Hence

$$I_C^{(2)} = \frac{A_0}{r^2} \int_{-d}^{0} dz_a \int_{-d}^{0} dz_b \left(\frac{\kappa_e}{4\pi}\right)^2 \exp[-ik_{Dz}(z_a - z_b)] \cdot \exp[(K''_{sz} + K''_{iz})(z_a + z_b)] J(z_a, z_b), \tag{8.4.37}$$

where

$$J(z_a, z_b) = \int_{-\infty}^{\infty} d\overline{r}_t \exp(-i\overline{k}_{Dt} \cdot \overline{r}_t) p(\hat{k}_s, \hat{r}) \cdot \frac{\exp\left\{-\kappa_e [r_t^2 + (z_a - z_b)^2]^{1/2}\right\}}{r_t^2 + (z_a - z_b)^2} p(\hat{r}, \hat{k}_{id}) \tag{8.4.38}$$

$$\hat{r} = \frac{\overline{r}_t + \hat{z}(z_a - z_b)}{[r_t^2 + (z_a - z_b)^2]^{1/2}}. \tag{8.4.39}$$

The integral in (8.4.38) can be evaluated over the angular variables θ and ϕ. For $z_a > z_b$, let $x_t = (z_a - z_b)\tan\theta\cos\phi$ and $y_t = (z_a - z_b)\tan\theta\sin\phi$; then $d\overline{r}_t = (z_a - z_b)^2 \sec^2\theta \tan\theta\, d\theta d\phi$, and $r_t^2 + (z_a - z_b)^2 = (z_a - z_b)^2\sec^2\theta$. Also, let the angular directions of $\hat{k}_s$ and $\hat{k}_{id}$ be (θ_s, ϕ_s) and $(\pi - \theta_i, \phi_i)$, respectively. Thus

$$J(z_a, z_b) = \int_0^{\pi/2} d\theta\ \tan\theta \int_0^{2\pi} d\phi\ p(\theta_s, \phi_s; \theta, \phi) \times \exp[-\kappa_e(z_a - z_b)\sec\theta]\ p(\theta, \phi; \pi - \theta_i, \phi_i) \times \exp[-i(k_{Dx}\cos\phi + k_{Dy}\sin\phi)(z_a - z_b)\tan\theta] \tag{8.4.40}$$

for $z_a > z_b$. Similarly, for $z_a < z_b$,

$$J(z_a, z_b) = \int_0^{\pi/2} d\theta\ \tan\theta \int_0^{2\pi} d\phi\ p(\theta_s, \phi_s; \theta, \phi) \times \exp[-\kappa_e(z_b - z_a)\sec\theta]\ p(\theta, \phi; \pi - \theta_i, \phi_i) \times \exp[i(k_{Dx}\cos\phi + k_{Dy}\sin\phi)(z_b - z_a)\tan\theta] \tag{8.4.41}$$

In deriving (8.4.41), we have made use of the reciprocity relation for phase functions as well as the approximations $p(\theta_s, \phi_s; \pi - \theta, \phi) \simeq p(\theta_i, \pi + \phi_i; \pi - \theta, \phi)$ and $p(\pi - \theta, \phi; \pi - \theta_i, \phi_i) \simeq p(\theta_s, \phi_s; \theta, \pi + \phi)$. By substituting (8.4.40) and (8.4.41) into (8.4.37), the dz_a and dz_b integrations can be carried out. This gives

$$\begin{aligned}\gamma_C^{(2)}(\hat{k}_s, \hat{k}_{id}) = {} & \frac{\kappa_e^2}{4\pi\cos\theta_i}\int_0^{\pi/2} d\theta\ \sin\theta \int_0^{2\pi} d\phi\ \frac{p(\theta_s, \phi_s; \theta, \phi) p(\theta, \phi; \pi - \theta_i, \phi_i)}{K_{sz}'' + K_{iz}''} \\ & \times \mathrm{Re}\left\{ \frac{\exp[i(k_{Dz} + t)d - \kappa_e d\sec\theta - (K_{sz}'' + K_{iz}'')d] - 1}{i(k_{Dz} + t)\cos\theta - \kappa_e - (K_{sz}'' + K_{iz}'')\cos\theta} \right. \\ & - \frac{\exp[i(k_{Dz} + t)d - \kappa_e d\sec\theta - (K_{sz}'' + K_{iz}'')d]}{i(k_{Dz} + t)\cos\theta - \kappa_e + (K_{sz}'' + K_{iz}'')\cos\theta} \\ & \left. + \frac{\exp[-2(K_{sz}'' + K_{iz}'')d]}{i(k_{Dz} + t)\cos\theta - \kappa_e + (K_{sz}'' + K_{iz}'')\cos\theta} \right\}\end{aligned} \tag{8.4.42}$$

where

$$t = k_{Dx}\tan\theta\cos\phi + k_{Dy}\tan\theta\sin\phi \tag{8.4.43}$$

For the ladder term, we assume that it has a much weaker angular dependence than the cyclical term. Then

$$\gamma_L^{(2)}(\hat{k}_s, \hat{k}_{id}) = \gamma_C^{(2)}(\hat{k}_s, \hat{k}_{id})\Big|_{k_{Dx} = k_{Dy} = k_{Dz} = 0} \tag{8.4.44}$$

In the second-order theory

$$\gamma = \gamma^{(1)} + \gamma_L^{(2)} + \gamma_C^{(2)} \tag{8.4.45}$$

4.4 Diffusion Approximation

For the case of a large albedo and a large optical thickness, the diffusion approximation can be applied. The diffusion approximation is discussed in Chapter 8, Section 4 of Volume I. From (8.4.32),

$$\frac{dB(\overline{r},\hat{s};\overline{r}_1,\hat{\alpha})}{ds} = -\kappa_e B(\overline{r},\hat{s};\overline{r}_1,\hat{\alpha}) + \frac{\kappa_e}{4\pi}\int d\hat{s}' p(\hat{s},\hat{s}')\overline{B}(\overline{r},\hat{s}';\overline{r}_1,\hat{\alpha}) + \epsilon(\overline{r},\hat{s};\overline{r}_1,\hat{\alpha}) \tag{8.4.46}$$

where

$$\epsilon(\overline{r},\hat{s};\overline{r}_1,\hat{\alpha}) = \delta(\overline{r}-\overline{r}_1)\frac{p(\hat{s},\hat{\alpha})}{4\pi} \tag{8.4.47}$$

With a slight change of notation, this is the form of (8.4.4) of Volume I. We apply the same derivation as in Chapter 8, Section 4 of Volume I.

In the diffusion approximation, it is assumed that B is almost uniform in all directions. Then

$$B(\overline{r},\hat{s};\overline{r}_1,\hat{\alpha}) \approx U_d(\overline{r};\overline{r}_1;\hat{\alpha}) + \frac{3}{4\pi}\overline{F}_d(\overline{r};\overline{r}_1,\hat{\alpha})\cdot\hat{s} \tag{8.4.48}$$

where the diffuse flux vector $\overline{F}_d$ is, from (8.4.14) of Volume I,

$$\overline{F}_d(\overline{r};\overline{r}_1,\hat{\alpha}) = \int d\hat{s}\hat{s}B(\overline{r},\hat{s};\hat{r}_1,\hat{\alpha}) \tag{8.4.49}$$

On the other hand, the average diffuse intensity U_d is, from (8.4.10) of Volume I,

$$U_d(\overline{r};\overline{r}_1,\hat{\alpha}) = \frac{1}{4\pi}\int d\hat{s}B(\overline{r},\hat{s};\overline{r}_1,\hat{\alpha}) \tag{8.4.50}$$

The diffuse flux vector $\overline{F}_d$ can also be expressed in terms of U_d. Using (8.4.33) of Volume I,

$$\frac{3}{4\pi}\overline{F}_d(\overline{r};\overline{r}_1,\hat{\alpha}) = \frac{1}{\kappa_{tr}}\left[-\nabla U_d(\overline{r};\overline{r}_1,\hat{\alpha}) + \frac{3}{4\pi}\int d\hat{s}\hat{s}\epsilon(\overline{r},\hat{s};\overline{r}_1,\hat{\alpha})\right] \tag{8.4.51}$$

where

$$\kappa_{tr} = \kappa_e(1-p_1) \tag{8.4.52}$$

is the transport rate, and

$$p_1\hat{s} = \frac{1}{4\pi}\int d\hat{s}'\hat{s}' p(\hat{s},\hat{s}') \tag{8.4.53}$$

In terms of the mean cosine of the scattering angle $\overline{\mu}$, we have

$$p_1 = \tilde{\omega}\overline{\mu} \tag{8.4.54}$$

Thus from (8.4.48) and (8.4.51), B can be expressed entirely in terms of U_d. Following the derivation in Chapter 8, Section 4 of Volume I, U_d obeys the diffusion equation of

$$\nabla^2 U_d(\overline{r};\overline{r}_1,\hat{\alpha}) - \kappa_d^2 U_d(\overline{r};\overline{r}_1,\hat{\alpha}) = -Q(\overline{r};\overline{r}_1,\hat{\alpha}) \tag{8.4.55}$$

where

$$\kappa_d = [3\kappa_e(1-\tilde{\omega})\kappa_{tr}]^{1/2} \tag{8.4.56}$$

and

$$Q(\overline{r};\overline{r}_1,\hat{\alpha}) = \frac{3}{4\pi}\left[\kappa_{tr}\int d\hat{s}\epsilon(\overline{r},\hat{s};\overline{r}_1,\hat{\alpha}) - \nabla\cdot\int d\hat{s}\hat{s}\epsilon(\overline{r},\hat{s};\overline{r}_1,\hat{\alpha})\right] \tag{8.4.57}$$

The boundary condition is [Eq. (8.4.44) of Volume I],

$$U_d + \frac{\overline{F}_d\cdot\hat{n}}{2\pi} = 0 \tag{8.4.58}$$

with $\hat{n}$ being the normal on the boundary that points inward into the scattering medium. Using (8.4.51), this means that

$$U_d - h\nabla U_d\cdot\hat{n} = 0 \tag{8.4.59}$$

where

$$h = \frac{2}{3\kappa_{tr}} \tag{8.4.60}$$

The solution of U_d can be expressed in terms of the diffusion Green's function $G_d(\overline{r},\overline{r}')$ which obeys the following equation

$$\nabla^2 G_d(\overline{r},\overline{r}') - \kappa_d^2 G_d(\overline{r},\overline{r}') = -\delta(\overline{r}-\overline{r}'). \tag{8.4.61}$$

For slab geometry, $\hat{n} = -\hat{z}$ at $z = 0$ and $\hat{n} = \hat{z}$ at $z = -d$. The boundary conditions are

$$G_d(\overline{r},\overline{r}') + h\frac{\partial}{\partial z}G_d(\overline{r},\overline{r}') = 0 \qquad \text{at } z = 0 \tag{8.4.62}$$

and

$$G_d(\overline{r},\overline{r}') - h\frac{\partial}{\partial z}G_d(\overline{r},\overline{r}') = 0 \qquad \text{at } z = -d \tag{8.4.63}$$

In terms of G_d, we have

$$U_d(\overline{r};\overline{r}_1,\hat{\alpha}) = \int d\overline{r}' G_d(\overline{r},\overline{r}')Q(\overline{r}';\overline{r}_1,\hat{\alpha}) \tag{8.4.64}$$

In what follows, we examine the solution of γ_C and γ_L for the case of a normal incident plane wave with $\hat{k}_{id} = -\hat{z}$. For this case, the diffusion Green's function G_d can be calculated by solving (8.4.61) in the Fourier transform domain. Let

$$G_d(\overline{r},\overline{r}') = \frac{1}{8\pi^2}\int_{-\infty}^{\infty} d\overline{\lambda}_t \frac{\exp[i\overline{\lambda}_t\cdot(\overline{r}_t-\overline{r}_t')]}{\lambda_z} g(\lambda_z;z,z') \tag{8.4.65}$$

where

$$\lambda_z = (\lambda_t^2+\kappa_d^2)^{1/2} \tag{8.4.66a}$$
$$\lambda_t = |\lambda_t| \tag{8.4.66b}$$

Then $g(\lambda_z,z,z')$ obeys the equation

$$\frac{\partial^2 g}{\partial z^2}(\lambda_z;z,z') - \lambda_z^2 g(\lambda_z;z,z') = \frac{\lambda_z}{2}\delta(z-z') \tag{8.4.67}$$

with boundary conditions

$$g(\lambda_z;z,z') + h\frac{\partial}{\partial z}g(\lambda_z;z,z') = 0 \tag{8.4.68a}$$

at $z=0$, and

$$g(\lambda_z;z,z') - h\frac{\partial}{\partial z}g(\lambda_z;z,z') = 0 \tag{8.4.68b}$$

at $z=-d$. Solving (8.4.67)–(8.4.68) gives

$$\begin{aligned} g(\lambda_z;z,z') = \exp(-\lambda_z|z-z'|) + \frac{R(\lambda_z)}{D(\lambda_z)}\Big\{ &\exp[\lambda_z(z+z')] \\ &+ R(\lambda_z)\exp[-\lambda_z(z'-z+2d)] + \exp[-\lambda_z(z+z'+2d)] \\ &+ R(\lambda_z)\exp[-\lambda_z(2d+z-z')]\Big\} \end{aligned} \tag{8.4.69a}$$

$$R(\lambda_z) = \frac{h\lambda_z-1}{h\lambda_z+1}, \tag{8.4.69b}$$

$$D(\lambda_z) = 1 - R^2(\lambda_z)\exp(-2\lambda_z d), \tag{8.4.69c}$$

By using (8.4.65) and (8.4.69) in (8.4.64), U_d can be calculated.

In the vicinity of the backscattering direction,

$$A(\overline{r},\hat{k}_s;\overline{r}_1,-\hat{z}) \simeq A(\overline{r},\hat{z};\overline{r}_1,-\hat{z}) \tag{8.4.70}$$

The idea is that A is slowly varying on an angular scale of 1° or 2°, while the important phase variation that is due to deviation from the backscattering direction is given by the phase terms in the integrand of (8.4.23). We calculate $\overline{F}_d$ from (8.4.51). Then $B(\overline{r},\hat{s};\overline{r}_1,\hat{\alpha})$ and $A(\overline{r},\hat{z};\overline{r}_1,-\hat{z})$ can be calculated by using relation (8.4.48) and (8.4.31). We get

$$A(\overline{r}, \hat{z}; \overline{r}_1, -\hat{z}) = \frac{3\kappa_e^2}{4\pi}\left[\kappa_{tr}\tilde{\omega}^2 G_d(\overline{r}, \overline{r}_1) - p_1\tilde{\omega}\frac{\partial G_d(\overline{r}, \overline{r}_1)}{\partial z_1} - p_1\tilde{\omega}\frac{\partial G_d(\overline{r}, \overline{r}_1)}{\partial z} - \frac{p_1^2}{\kappa_{tr}}\delta(\overline{r} - \overline{r}_1) + \frac{p_1^2}{\kappa_{tr}}\frac{\partial^2}{\partial z \partial z_1}G_d(\overline{r}, \overline{r}_1)\right] \quad (8.4.71)$$

Substituting (8.4.65) and (8.4.71) into (8.4.23) and carrying out the integration of $d(\overline{r}_{lt} - \overline{r}_{jt})$ give $\delta(\overline{k}_{Dt} - \lambda_t)$, so that λ_t corresponds to $\overline{k}_{Dt}$. In what follows, we shall retain only the first term in (8.4.71) as the lowest-order solution for the diffusion approximation. The rest of the terms in (8.4.71) are higher-order corrections and should be small for the diffusion approximation to be valid. The dz_j and dz_l integration can also be carried out readily. We have, after some algebra,

$$\begin{aligned}\gamma_C(\hat{k}_s, -\hat{z}) = \frac{3\kappa_e^2}{2\alpha}\Bigg(&\frac{1}{\kappa_e}(\kappa_{tr}\tilde{\omega}^2)\mathrm{Re}\left[\frac{\exp(ik_{Dz}d - \alpha d - \kappa_e d) - 1}{ik_{Dz} - \alpha - \kappa_e} - \frac{\exp(ik_{Dz}d - \alpha d - \kappa_e d) - \exp(-2\kappa_e d)}{ik_{Dz} - \alpha + \kappa_e}\right] \\ &+ \frac{R(\alpha)}{D(\alpha)}(\kappa_{tr}\tilde{\omega}^2)\left|\frac{1 - \exp(ik_{Dz}d - \kappa_e d - \alpha d)}{-ik_{Dz} + \kappa_e + \alpha}\right|^2 \\ &+ \frac{R(\alpha)}{D(\alpha)}(\kappa_{tr}\tilde{\omega}^2)\left|\frac{e^{-\alpha d} - \exp(ik_{Dz}d - \kappa_e d)}{-ik_{Dz} + \kappa_e - \alpha}\right|^2 \\ &+ \frac{R^2(\alpha)}{D(\alpha)}(\kappa_{tr}\tilde{\omega}^2)2\mathrm{Re}\left\{\frac{[e^{-\alpha d} - \exp(-ik_{Dz}d - \kappa_e d)]}{(ik_{Dz} + \kappa_e - \alpha)} \times \frac{[e^{-\alpha d} - \exp(ik_{Dz}d - \kappa_e d - 2\alpha d)]}{(-ik_{Dz} + \kappa_e + \alpha)}\right\}\Bigg)\end{aligned} \quad (8.4.72)$$

In (8.4.72) we have

$$\alpha = (k_{Dt}^2 + \kappa_d^2)^{1/2} \quad (8.4.73a)$$

$$k_{Dt} = |\overline{k}_{Dt}| \quad (8.4.73b)$$

For $\hat{k}_{id} = -\hat{z}$, then, $k_{Dt} = k \sin\theta_s$, and $k_{Dz} = k(\cos\theta_s - 1)$.

The ladder intensity is given approximately by

$$\gamma_L = \gamma_L^{(1)} + \sum_{n=2}^{\infty}\gamma_L^{(n)} = \gamma_L^{(1)} + \gamma_C(\hat{z}, -\hat{z}) \quad (8.4.74)$$

For the half-space case with $d \to \infty$. Then, from (8.4.72) we obtain

$$\gamma_C(\hat{k}_s, -\hat{z}) = \frac{3\kappa_e^2\kappa_{tr}\tilde{\omega}^2}{2[k_{Dz}^2 + (\kappa_e + \alpha)^2]}\left(\frac{2h}{h\alpha + 1} + \frac{1}{\kappa_e}\right) \quad (8.4.75)$$

If we further assume that, in (8.4.75), $\tilde{\omega} = 1$, then $\alpha = k_{Dt}$ and $p_1 = \overline{\mu}$. Furthermore, if $\overline{\mu}$ is larger than 0.85, then $\kappa_e \gg \kappa_{tr}$. Also, $k_{Dt} = \mathcal{O}(\kappa_{tr})$. Equation (8.4.75) then further simplifies to

$$\gamma_C(\hat{k}_s, \hat{k}_{id}) = 3\frac{h\kappa_{tr}}{1 + hk_{Dt}} \tag{8.4.76}$$

At $\theta_s = 0$ and $k_{Dt} = 0$, $\gamma_C = 3h\kappa_{tr}$. If we plot γ_C against angle, we see a sharp peak at $k_{Dt} = 0$. To get the half-power point, we find $\theta = \theta_{sh}$ such that $\gamma_C = 3h\kappa_{tr}/2$. We get $k \sin\theta_{sh} = 3\kappa_{tr}/2$, where the subscript sh denotes the half-power point-scattered angle. Thus

$$\theta_{sh} \cong \frac{3\kappa_{tr}}{2k} = \frac{3\kappa_e(1 - \overline{\mu})}{2k} \tag{8.4.77}$$

The results of (8.4.77) are based on the assumption that the observation angle is in the vicinity of the backscattering direction, and the higher-order terms of the diffusion approximation can be neglected. Note that for the case of isotropic scattering $\overline{\mu} = 0$, we have $\theta_{sh} \simeq 3\kappa_e/(2k)$, a result that is consistent with the results of Sections 2 and 3.

4.5 Numerical Results

In this section, the results for anisotropic scatterers obtained using the second-order theory and the diffusion approximation are illustrated numerically. In Fig. 8.4.2 the bistatic coefficients, based on the second-order theory, are plotted as functions of θ_s for the incident angles $\theta_i = 0.1°, \phi_i = 0°$, and $\phi_s = 180°$. The peak value of the sharp peak is in the backscattering direction at $\theta_s = 0.1°$. The calculations are based on the Henyey-Greenstein phase function,

$$p(\mu) = \frac{\tilde{\omega}(1 - \overline{\mu}^2)}{(1 + \overline{\mu}^2 - 2\mu\overline{\mu})^{3/2}} \tag{8.4.78}$$

where μ is $\hat{s} \cdot \hat{s}'$ and $\overline{\mu}$ is the mean cosine of the scattering angle. For strong forward scattering, $\overline{\mu}$ is close to 1. For isotropic scattering, $\overline{\mu}$ is zero. In the second-order theory, the bistatic scattering coefficient is approximated as $\gamma \simeq \gamma_L^{(1)} + \gamma_L^{(2)} + \gamma_C^{(2)}$.

The optical thicknesses are $\tau_d = 1$ and $\tau_d = 2$, which are within the regime of validity of the second-order theory. The angular width of the backscattering enhancement peak, as seen from the figure, is of the order of 0.1° and is of the same order of magnitude as κ_e/k. Since the extinction rate κ_e is a function of n_0, the number of particles per unit volume, theoretical results depend on n_0 through κ_e. For a small concentration of particles, κ_e can be very small giving a very small angular width.

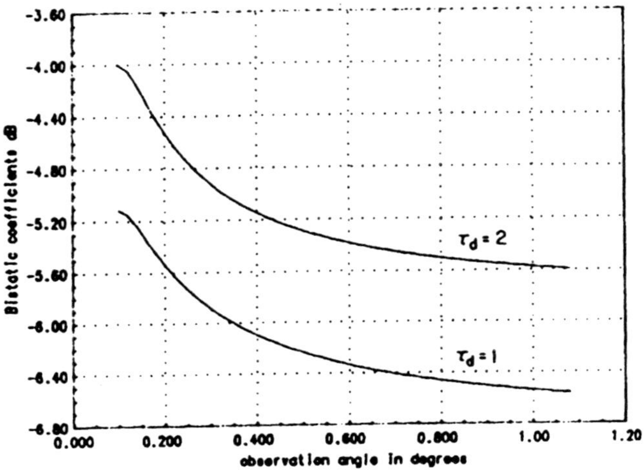

Figure 8.4.2 Bistatic scattering coefficients as a function of observation angles θ_s. The result is based on second-order theory, that is, $\gamma = \gamma_L^{(1)} + \gamma_L^{(2)} + \gamma_C^{(2)}$. The incident angle is $\theta_i = 0.1°, \phi_i = 0°$, the scattered angle is θ_s, and $\phi_s = 180°$. The parameter values are $\tau_d = 1$ and $\tau_d = 2, \kappa_e/k = 7.56 \times 10^{-4}$, $\overline{\mu} = 0.5$, and $\tilde{\omega} = 0.999$.

In Figs. 8.4.3 and 8.4.4 the bistatic coefficients, under the diffusion approximation (which is valid for a large optical thickness), are plotted as functions of θ_s for $\kappa_e/k = 0.02923$ and $\kappa_e/k = 0.01835$. The incident angle is $\theta_i = 0°$. The peak value is at $\theta_s = 0°$. The optical thicknesses are $\tau_d =$ 7.734, 38.67, and 7734 for Fig. 8.4.3 and $\tau_d = 9.708$, 29.124, and 4854 for Fig. 8.4.4. The angular widths of Figs. 8.4.3 and 8.4.4 are approximately 0.19° and 0.12°, respectively. The parameters in Figs. 8.4.3 and 8.4.4 are chosen such that the results of the angular width can be compared with the experimental data [Kuga and Ishimaru, 1984]. In the experimental measurements at $0.6328\,\mu$m ($\lambda = 0.475\,\mu$m in water), the wavenumber in water is $13.228\,\mu\text{m}^{-1}$. The experimentally measured coherent attenuation rates κ_e are 0.3867 and $0.2427\,\mu\text{m}^{-1}$ at particle diameters of 1.101 and $2.02\,\mu$m, respectively. These numbers correspond to $\kappa_e/k = 0.02923$ and 0.01835, respectively, both of which are used as input parameters in the results shown in Figs. 8.4.3 and 8.4.4. The mean cosine for both cases, based on the scattering, is $\overline{\mu} = 0.924$. The thickness of the slab is 2 cm, corresponding to τ_d = 7734 and $\tau_d = 4854$, respectively, for Figs. 8.4.3 and 8.4.4. The experimentally observed angular widths are approximately 0.1° in both cases and agree well with the theoretical results.

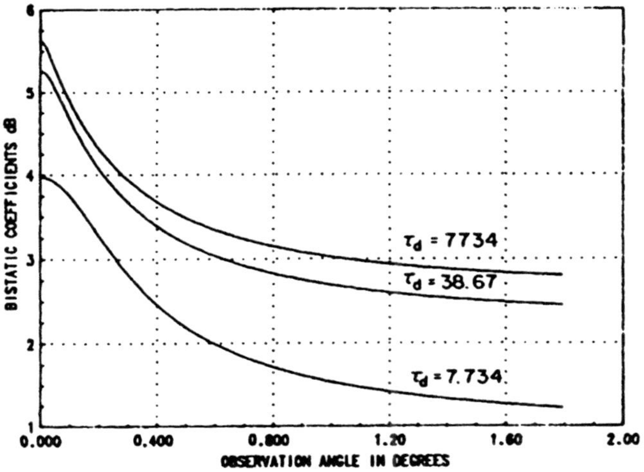

Figure 8.4.3 Bistatic scattering coefficients as functions of observation angle θ_s, based on the diffusion approximation, that is, $\gamma = \gamma_L^{(1)} + \sum_{n=2}^{\infty} \gamma_L^{(n)} + \sum_{n=2}^{\infty} \gamma_C^{(n)}$. The incident angle is $\theta_i = 0°$. The parameters are $\tau_d = 7.734$, 38.67, and 7734; $\kappa_e/k = 0.02923$; $\overline{\mu} = 0.924$; and $\tilde{\omega} = 0.999$.

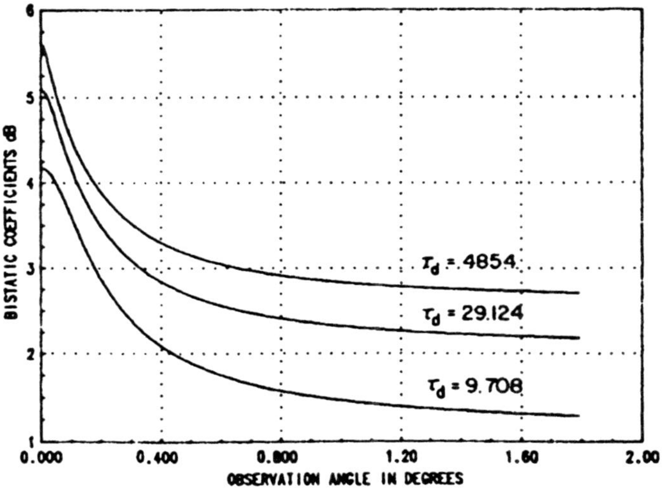

Figure 8.4.4 Bistatic scattering coefficients as functions of the observation angle θ_s, based on the diffusion approximation, that is , $\gamma = \gamma_L^{(1)} + \sum_{n=2}^{\infty} \gamma_L^{(n)} + \sum_{n=2}^{\infty} \gamma_C^{(n)}$. The incident angle is $\theta_i = 0°$. The parameters are $\tau_d = 9.708$, 29.124, and 4854; $\kappa_e/k = 0.01835$; $\overline{\mu} = 0.924$; and $\tilde{\omega} = 0.999$.

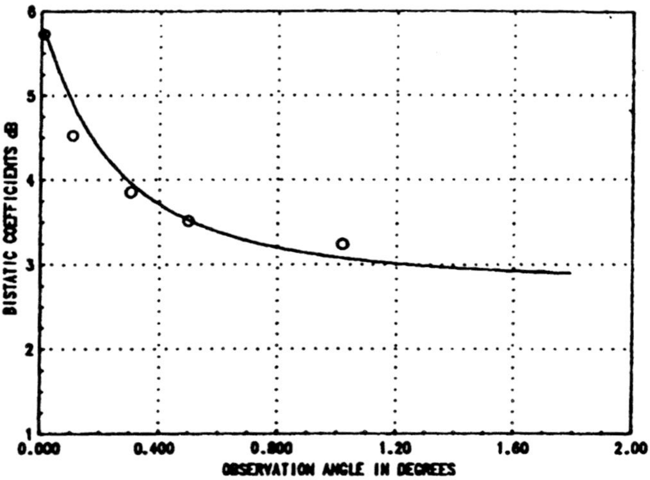

Figure 8.4.5 Bistatic scattering coefficients as functions of the observation angle θ_s based on the diffusion approximation, i.e., $\gamma = \gamma_L^{(1)} + \sum_{n=2}^{\infty} \gamma_L^{(n)} + \sum_{n=2}^{\infty} \gamma_C^{(n)}$. The incident angle is $\theta_i = 0°$, $\tau_d = 7143, \kappa_e/k = 0.022, \overline{\mu} = 0.9$, and $\tilde{\omega} = 0.999$. Comparison is made with experimental data from Wolf and Maret [1985] (circles).

In Fig. 8.4.5 we compare the results of the diffusion approximation with the experimental results in [Wolf and Maret, 1985], which have fine angular resolution. The parameters shown in the theoretical curve of diffusion theory are $\kappa_e = 0.357\ \mu\text{m}^{-1} = 1/$ (mean free path). The mean free path is $2.8\,\mu\text{m}$, the wavelength in water is $0.387\,\mu\text{m}, k = 16.23\,\mu\text{m}^{-1}$, $\overline{\mu} = 0.9$, and the optical thickness is $\tau_d = 7143$. The theory is in good agreement with experimental data.

REFERENCES AND ADDITIONAL READINGS

Akkermans, E., P. E. Wolf, and R. Maynard (1986), Coherent backscattering of light by disordered media, *Phys. Rev. Lett.*, **56**, 1471–1474.

Anderson, P. W. (1958), Absence of diffusion in certain random lattice, *Phys. Rev.*, **109**, 1492–1505.

Bahar, E. and M. A. Fitzwater (1989), Depolarization and backscatter enhancement in light scattering from random rough surfaces: Comparison of full-wave theory with experiment, *J. Opt. Soc. Am. A*, **6**, 33–43.

Barabanenkov, Yu. N. and M. Yu. Barabanenkov (1999), Effect of a pulse entrapping on weak localization of waves in a resonant random medium, *Waves in Random Media*, **9(1)**, 13–26.

Barabanenkov, Yu. N., Yu. A. Kravtsov, V. D. Ozrin, and A. I. Saichev (1991a), Enhanced backscattering: the universal wave phenomenon, *Proc. IEEE*, **79(10)**, 1367–1370.

Barabanenkov, Yu. N., Yu. A. Kravtsov, V. D. Ozrin, and A. I. Saichev (1991b), Enhanced backscattering in optics, *Progress in Optics*, **29**, 65–197.

Barabanenkov, Yu. N. and D. I. Kryukov (1992), Functional Fokker-Planck formalism for wave propagation in random media under delocalized and weak localized regions, *Waves in Random Media*, **2(1)**, 1–6.

Gu, Z. H., R. S. Dummer, A. A. Maradudin, and A. R. McGurn (1989), Experimental study of the opposition effect in the scattering of light from a randomly rough metal surface, *Appl. Optics*, **28**, 537–543.

Ishimaru, A. (1978), *Wave Propagation and Scattering in Random Media*, **1** and **2**, Academic Press, New York.

Ishimaru A. and L. Tsang (1988), Backscattering enhancement of random discrete scatterers of moderate sizes, *J. Opt. Soc. Am. A*, **5(2)**, 228–236.

Joannopoulos, J. D., R. D. Meade, and J. M. Winn (1995), *Photonic Crystals: Molding the Flow of Light*, Princeton University Press, Princeton.

John, S. (1991), Localization of light, *Physics Today*, **44(5)**, 32–40.

Johnson, J. T., L. Tsang, R. Shin, K. Pak, C. H. Chan, A. Ishimaru, and Y. Kuga (1996), Backscattering enhancement of electromagnetic waves from two-dimensional perfectly conducting random rough surfaces: A comparison of Monte Carlo simulations with experimental data, *IEEE Trans. Antennas Propagat.*, **44**, 748–756.

Jordan, D. L. and F. Moreno (1993), Enhanced backscattering and cross depolarization from multiscale surfaces, *J. Opt. Soc. Am. A*, **10(9)**, 1989–1995.

Kim, M.-J., J. C. Dainty, A. T. Friberg, and A. J. Sant (1990), Experimental study of enhanced backscattering from one- and two-dimensional random rough surfaces, *J. Opt. Soc. Am. A*, **7(4)**, 569–577.

Knotts, M. E. and K. A. O'Donnell (1993), Backscattering enhancement from a conducting surface with isotropic roughness, *Optics Comm.*, **99(1)**, 1–6.

Kuga, Y. and A. Ishimaru (1984), Retroreflectance from a dense distribution of spherical particles, *J. Opt. Soc. Am.*, **1**, 831–835.

Kuga, Y. and P. Phu (1995), Experimental techniques in random media and rough surface, *Progress in Electromagnetics Research*, **9**, Elsevier Science Publishers, Cambridge, Mass.

Kuga, Y., L. Tsang, and A. Ishimaru (1985), Depolarization effects of the enhanced retroreflectance from a dense distribution of spherical particles, *J. Opt. Soc. Am. A*, **2**, 616–618.

Lagendijk, A. and B. A. van Tiggelen (1996), Resonant multiple scattering of light, *Phys. Reports*, **270(3)**, 143–215.

Lawandy, N. M., A. M. Balachandran, A. S. L. Gomes, and E. Sauvan (1994), Laser action in strongly scattering media, *Nature*, **368**, 436–438.

Li, L., C. H. Chan, L. Tsang, K. Pak, P. Phu, and S. H. Lou (1994), Monte Carlo simulation and backscattering enhancement of random metallic rough surface at optical frequencies, *J. Electromag. Waves and Appl.*, **8(3)**, 277–293.

Mandt, C. and L. Tsang (1992), Backscattering enhancement from a random distribution of large discrete spherical scatterers with a size distribution, *J. Opt. Soc. Am. A*, **9(12)**, 2246–2251.

Mandt, C. E., L. Tsang, and A. Ishimaru (1990), Copolarized and depolarized backscattering enhancement of random discrete scatterers of large size based on second-order ladder and cyclical theory, *J. Opt. Soc. Am.*, **7(4)**, 585–592.

Maradudin, A. A. (1999), Multiple-scattering phenomena in the scattering of light from randomly rough surfaces, *Physica Status Solidi A*, **175(1)**, 241–252.

Maradudin, A. A., E. R. Mendez, and T. Michel (1990), Backscattering effects in the elastic scattering of p-polarization light from a large amplitude random grating, *Scattering in Volumes and Surfaces*, M. Nieto-Vesperians and J. C. Dainty, Eds., Elsevier Science Publishers, North-Holland.

Maradudin, A. A., T. Michel, A. R. McGurn, and E. R. Mendez (1990), Enhanced backscattering of light from a random grating, *Ann. Phys.*, **203(2)**, 255–307.

McGurn, A. R. and A. A. Maradudin (1993), Weak transverse localization of the light scattered incoherently from a one-dimensional random metal surfaces, *J. Optical Soc. Am. B*, **10(3)**, 539–545.

Mendez, E. R. and K. A. O'Donnell (1987), Observation of depolarization and backscattering enhancement in light scattering from Gaussian random surfaces, *Optics Comm.*, **61(2)**, 91–95.

Ogura, H. and N. Takahashi (1995), Wave scattering from a random rough surface: reciprocal theorem and backscattering enhancement, *Waves in Random Media*, **5(2)**, 223–242.

Pak, K., L. Tsang, C. H. Chan, and J. Johnson (1995), Backscattering enhancement of vector electromagnetic waves from two-dimensional perfectly conducting random rough surfaces based on Monte Carlo simulations, *J. Opt. Soc. Am. A*, **12(11)**, 2491–2499.

Pak, K., L. Tsang, and J. Johnson (1997), Numerical simulations and backscattering enhancement of electromagnetic waves from two-dimensional dielectric random rough surfaces with the sparse matrix canonical grid method, *J. Opt. Soc. Am. A*, **14(7)**, 1515–1528.

Phu, P., A. Ishimaru, and Y. Kuga (1993), Controlled millimeter wave experiments and numerical simulations on the enhanced backscattering from one dimensional very rough surfaces, *Radio Sci.*, **28(5)**, 533–548.

Phu, P., A. Ishimaru, and Y. Kuga (1994), Co-polarized and cross-polarized enhanced backscattering from two-dimensional very rough surfaces at millimeter wave frequencies, *Radio Sci.*, **29**, 1275–1291.

Sheng, P., Ed. (1990), *Scattering and Localization of Classical Waves in Random Media*, World Scientific, Singapore.

Sheng, P. (1995), *Introduction to Wave Scattering, Localization, and Mesoscopic Phenomena*, Academic, San Diego.

Tsang, L., C. H. Chan, and K. Pak (1994), Backscattering enhancement of a two-dimensional random rough surface (three-dimensional scattering) based on Monte Carlo simulations, *J. Opt. Soc. Am. A*, **11(2)**, 711–715.

Tsang, L. and A. Ishimaru (1984), Backscattering enhancement of random discrete scatterers, *J. Opt. Soc. Am. A*, **1**, 836–839.

Tsang, L. and A. Ishimaru (1985), Theory of backscattering enhancement of random discrete isotropic scatterers based on the summation of all ladder and cyclical terms, *J. Opt. Soc. Am. A*, **2**, 1331–1338.

Tsang, L. and A. Ishimaru (1985), Radiative wave and cyclical transfer equations for dense nontenuous media, *J. Opt. Soc. Am. A*, **2(12)**, 2187–2194.

van Albada, M. P. and A. Lagendijk (1985), Observation of weak localization of light in a random medium, *Phys. Rev. Lett.*, **55**, 2692–2695.

van Albada, M. P., M. B. van der Mark, and A. Lagendijk (1987), Observation of weak localization of light in a finite slab, *Phys. Rev. Lett.*, **58**, 361–364.

van Tiggelen, B. A., A. Lagendijk, A. Tip, Yu. N. Barabanenkov, and V. D. Ozrin (1993), Comment on problem of light diffusion in strongly scattering media (with reply), *Phys. Rev. Lett.*, **71(8)**, 1284–1285.

Wiersma, D. S. and A. Lagendijk (1996), Light diffusion with gain and random lasers, *Phys. Rev. E*, **54(4)**, part B, 4256–4265.

Wiersma, D. S. and A. Lagendijk (1997), Interference effects in multiple-light scattering with gain, *Physica A*, **241(1–2)**, 82–88.

Wiersma, D. S., M. P. van Albada, and A. Lagendijk (1995), Random laser?, *Nature*, **373**, 203–204.

Wolf, P. E. and G. Maret (1985), Weak localization and coherent backscattering of photons in disordered media, *Phys. Rev. Lett.*, **55(24)**, 2696–2699.

Yablonovitch, E. (1987), Inhibited spontaneous emission in solid-state physics and electronics, *Phys. Rev. Lett.*, **58**, 2059–2062.

Yablonovitch, E. (1993), Photonic band gap structures, *J. Optical Soc. Am. B*, **10**, 283–295.

INDEX

CPSIA information can be obtained at www.ICGtesting.com
Printed in the USA
BVOW03*0124130214

344782BV00007B/129/P

9 780471 388012